FUNDAMENTALS OF
VOICE AND DATA COMMUNICATION

By

K.K. Sharma

Assistant Professor
Deptt. of Electronics & Communication Engg.
LINGAYA's University
Faridabad

S.K. KATARIA & SONS®

Publishers of Engineering Books

4885/109, Prakash Mahal, Dr. Subhash Bhargav Lane,
Daryaganj, New Delhi-110002 (India)
Tel.: +91-11-43551243, Mobile: +91-9871775858, 7678425209
E-mail: katariabook@gmail.com, katariabooks@yahoo.com
website: www.skkatariaandsons.com

(e-books, subscriptions and bulk buying also available on our website)

Published by :

S.K. KATARIA & SONS®

4885/109, Prakash Mahal, Dr. Subhash Bhargav Lane,
Daryaganj, New Delhi-110002 (India)
Tel.: +91-11-43551243, Mobile: +91-9871775858, 7678425209
E-mail: katariabook@gmail.com, katariabooks@yahoo.com

First Edition: 2012, Reprint: 2026

Price: ₹ 595/-

Laser Typesetting by :
AISA ENTERPRISES, Delhi

Printed at :
Fine Offset Printers, Delhi

Preface

This text, '**Fundamentals of Voice and Data Communication Engineering**' is intended to serve as a one semester text for undergraduate course of Electronics and Communication Engineering. The purpose of this book is to extend and update the knowledge of the reader to the basic concepts of telecommunication systems and computer communication. In this book, the fundamental concepts of voice and data communication engineering are explained in a simple manner.

This book provides in depth knowledge on telecommunication switching and good background for advanced studies in communication networks. The entire subject is dealt with conceptual treatment and the analytical or mathematical approach is made only to some extent.

The telecommunication switching is the fast growing field and enormous research and development are undertaken by various organizations and firms. The communication networks have unlimited research potentials. Both telecommunication switching and communication networks develop new techniques and technologies everyday.

This book is an outgrowth of this need and hope it will be very helpful for the students.

Inspite of sincere efforts there may be some misprints, omission and errors. Any suggestion towards the improvement of this text will be highly appreciated.

FARIDABAD **K. K. SHARMA**
JULY, 2012

Acnowledgement

I take this opportunity to express my gratitude to the respected Late Prof G.V.K Sinha, founder Chairman of Lingaya's University, Dr. K. K. Aggarwal, Chancellor, Lingaya's University, Dr. P. Gadde, CEO and Pro Chancellor, Lingaya's University and Dr. G.S. Yadava, Pro Vice-Chancellor, Lingaya's University, Faridabad for providing innovative and encouraging environment to work on this project.

I would like to thank to the Management, teachers, students and supporting staff of Lingaya's University. I would like to express my gratitude to all the authors whose work has been freely consulted in the preparation of this text.

I am grateful to my wife Sunita, my daughters Aparna and Vaishali, and son Aman for supporting me to complete this project.

A special thanks to Mr Sanjeev Kataria, M/s S. K. Kataria & Sons, New Delhi and Mr. Sachin Arora for publishing this book in such a beautiful get-up and well in time.

Lastly, I am grateful to my father and mother without their blessing this book would have not come into existence.

K. K. Sharma **AUTHOR**

This Book is Dedicated

to

My Beloved Wife Sunita

Contents

UNIT-III .... 105—148

4. MULTIPLEXING OF TELEPHONE CHANNELS .... 107—148

UNIT-IV .... 149—318

5. INTRODUCTION TO DATA COMMUNICATION .... 151—223

I

UNIT

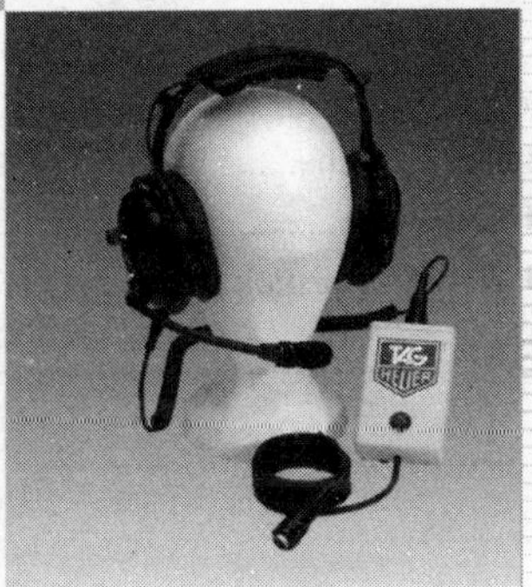

➤ Telephone Instruments and Signals

➤ Telephone Circuits

SYLLABUS

Telephone Instruments and Signals : Introduction, The Subscriber Loop, Standard Telephone Set, Basic Call Procedure, Call Progress Tones and Signals, Cordless Telephones, Caller Identification, Electronic Telephones.

Telephone Circuit : Introduction, The Local Subscriber Loop, Channel Noise and Units of Power Measurements, Transmission Parameters, Voice Frequency Circuit Arrangements, Crosstalk.

CHAPTER

1

TELEPHONE INSTRUMENTS AND SIGNALS

INSIDE THIS CHAPTER

1.1. INTRODUCTION

Telecommunication networks carry information signals among entities, which are geographically far apart. An entity may be a computer or human being, a facsimile machine, a teleprinter, a data terminal and so on. The entities are involved in the process of information transfer who may be in the form of a telephone conversation (telephony) or a file transfer between two computers or message transfer between two terminals etc. Today it is almost truism to state that telecommunication systems are the symbol of our information age.

With the rapidly growing traffic and untargeted growth of cyberspace, telecommunication becomes a fabric of our life. The future challenges are enormous as we anticipate rapid growth items of new services and number of users. What comes with the challenge is a genuine need for more advanced methodology supporting analysis and design of telecommunication architectures. Telecommunication has evaluated and growth at an explosive rate in recent years and will undoubtedly continue to do so.

Public and private telephone systems provide real-time information paths between two or more parties. The wire line public system is generally referred to as the PSTN (Public Switched Telephone Network) and private systems are created with PBX (Private Branch exchange) or KTS (Key Telephone System) switching technologies. Traditionally, these public and private information paths have taken the form of voice connections, originally through hardwired analog circuitry but later through an increasingly broad range of technologies such as radio transmission, digital signal encoding.

As telecommunications has moved from proprietary to open, standards-based systems, advanced voice solutions have grown richer and more cost effective. Several basic telephony concepts are critical to working with these solutions: call control, media processing, in-band and out-of-band signaling and local, dedicated first-party control versus shared, network-based third-party control.

1.2. HISTORICAL DEVELOPMENT

By the early 1800's scientists had developed ways to generate and transmit electricity. In 1819, oersted discovered the relation between magnetism and electricity. Ampere, Faraday and others continued this work in 1820. In 1834, Gauss and Weber wired over the roofs of Gottingen to make a telegraph system.

Samuel F.B. Morse's developed the first significant work in telecommunication. F.B. Morse developed code telegraphy in 1837. In 1844, a 40 mile telegraph line was setup between Baltimore and Washington by F.B. Morse. In 1845, Morse formed a telegraph company based on his technology. In 1849, the first slow telegraph printer link was setup. In 1874, Ban dot invented a "Multiplexes" system which enables up to six signals from telegraph machines to be transmitted together over the same line.

Alexander Graham Bell and Elisha Gray contributed significant works and filed paper related to telephony. The early stages of the development of telecommunication were due to A.G. Bell, G. Marconi and C.E. Shannon. In 1876, Bell invented a telephone system. In 1897 Marconi patented a wireless telephone system. Teletypewriter service was initiated in 1931.

In early days, very simple exchanges whose control is provided by a human operator and the elements of the switch assemblies are plugs and sacks. With increase in demand of service, human operator exchange was replaced by the invention of range of electromechanical switching devices. Of all the electromechanical switching devices that become available over the years, the step-by-step switching system invented by Almon B. Strowger in 1892 is still quite popular. The next automatic electromechanical switching system was crossbar switching.

First patent for crossbar device was granted in 1915 to J.N. Reynolds of Western Electric, USA. In 1919, two Swedish Engineers, Betulander and Palmgren got patent for crossbar switch. In 1938, AT&T laboratories in US introduced crossbar-switching system in the field. The electromechanical switching systems have been replaced by computer controlled switching systems referred to as stored program control (SPC). In SPC, switching is controlled by software program. The first computer controlled switch was introduced in 1960. Till 1965, computer controlled switching was used transistors and printed circuit technology. Since 1965 switching are based on microprocessors.

The use of computers to control the switching led to the designation "electronic" switching system (ESS) or Electronic automatic exchange (EAX). In 1970, first electronic switching system No. 1 ESS or No. 1 EAX was introduced. Digital electronic switching matrices were first introduced into the U.S. Public network in 1976 with AT&T's No. 4 ESS

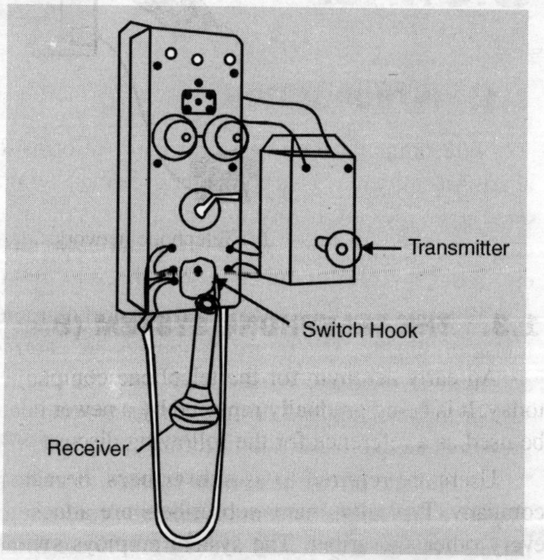

Fig. 1.1. Earlier Telephone With Separate Receiver Hanging on Switch Hook.

digital toll switch. By the mid 1980's the interoffice transmission environment has changed to almost exclusively digital. Figure 1.2 shows the various telephone networks.

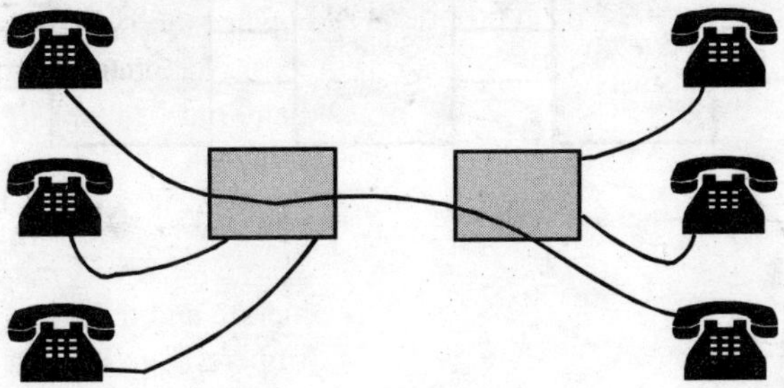

(*a*) Telephone networks in 1890

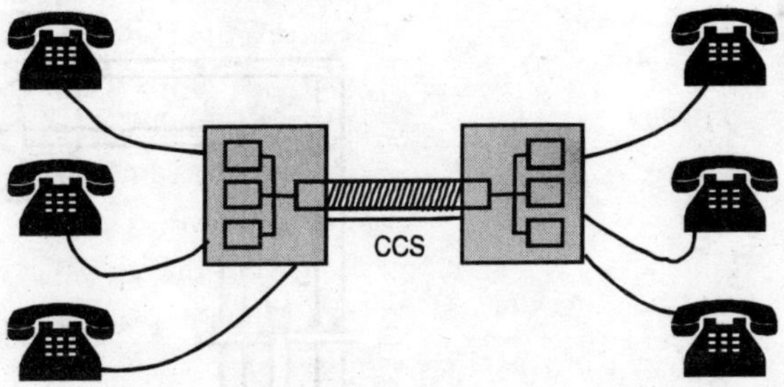

(*b*) Telephone networks in 1988

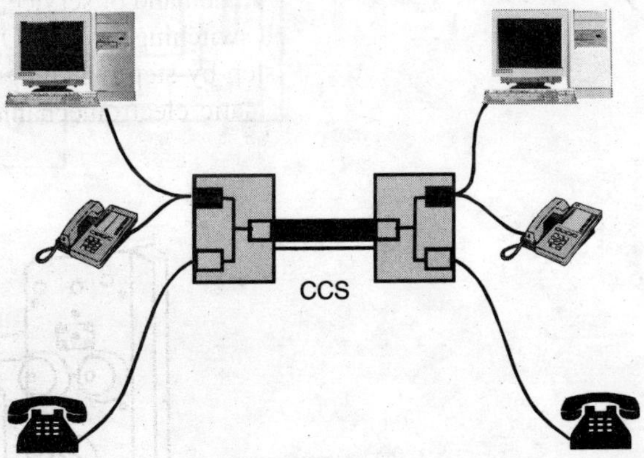

(*c*) Telephone networks after 1990 with the advent of ISDN

Fig. 1.2.

1.3. THE TELEPHONE SYSTEM (BASIC CONCEPTS)

An early acronym for the telephone company was **TELCO,** which is still being used to a fair extent today. It is being gradually replaced by a newer one— **POTS** (Plain Old Telephone System). Figure 1.3 will be used as a reference for the following discussion on the basic telephone system.

Users are referred to as **subscribers**, because they subscribe to a service provided by the telephone company. Typically, many subscribers are attached to the same network, and each subscriber can contact every other subscriber. The system employs **switches** to facilitate connections between subscribers. The arrangement of switches and subscribers and the interconnections between them determines the structure of the network.

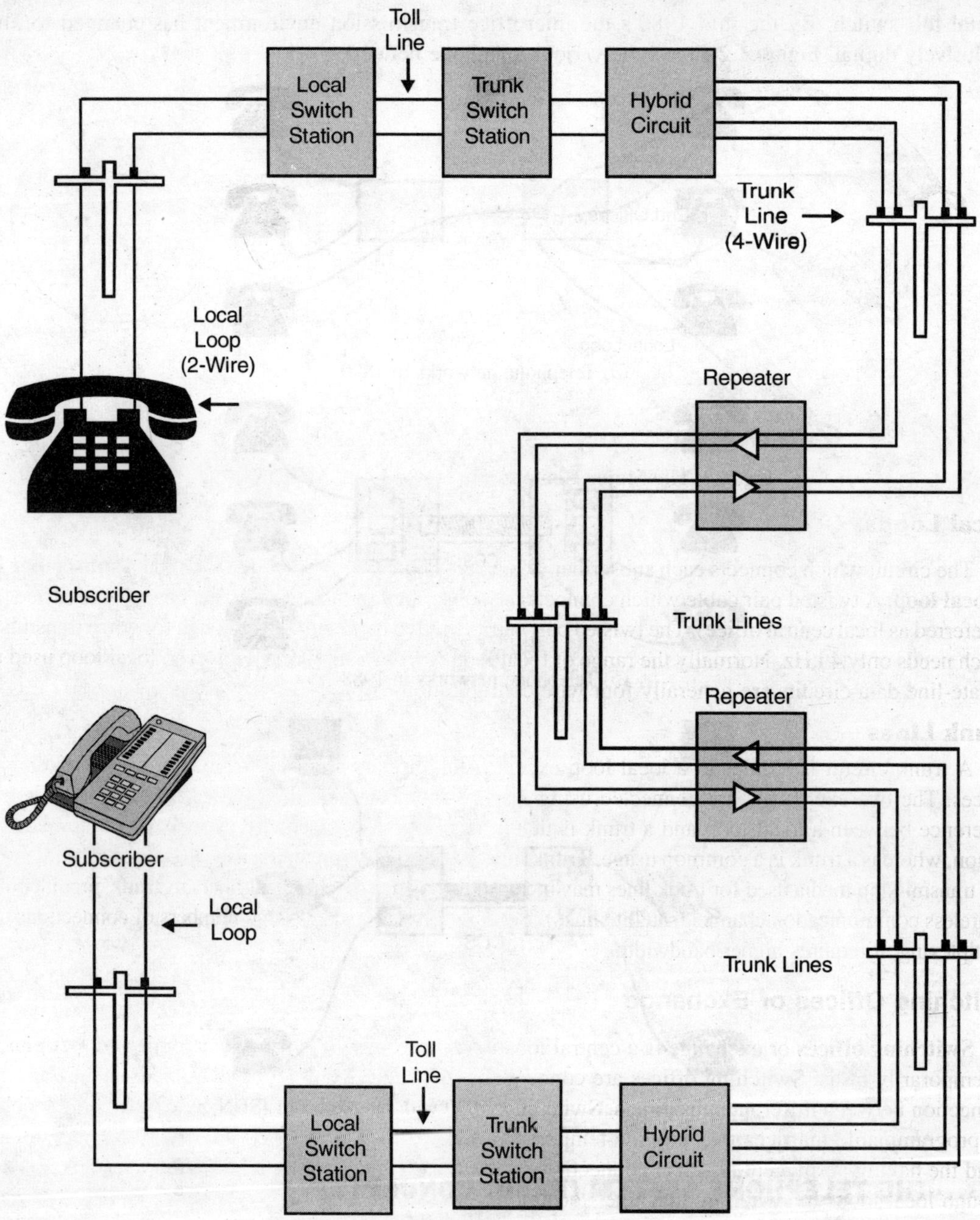

Fig. 1.3. Telephone System Elements.

1.4. MAIN COMPONENTS

Figure 1.4 shows the major components of the telephone network. They are as follows:

1. Local Loops
2. Trunks
3. Switching offices (end offices, tandem offices and regional offices)

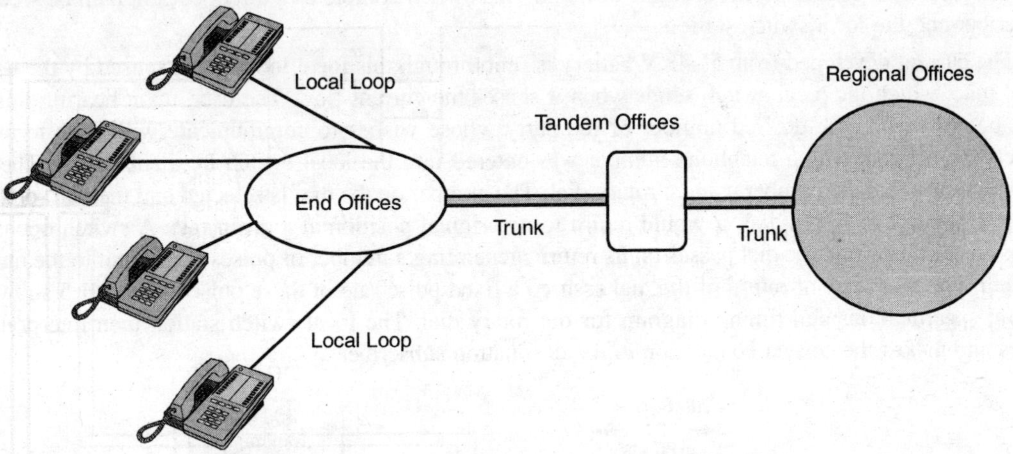

Fig. 1.4. The Major Components of the Telephone Network.

Local Loops

The circuit which connects each subscriber (via a pair of wires) to the switch is called a **subscriber loop** or **local loop**. A twisted pair cable which connects the subscriber telephone set to the nearest end office (may be referred as local central office). The twisted pair has its limited bandwidth but enough for voice transmission which needs only 4 kHz. Normally the range of local loop is from 1 to 10 km. However, local loop used with private-line data circuits are generally four wire configurations.

Trunk Lines

A trunk circuit is similar to a local loop except trunk circuits are used to interconnect two telephone offices. The two exchanges are connected using a set of interexchange lines called **trunk lines.** The main difference between a local loop and a trunk is that a local loop is permanently associated with a particular station, whereas a trunk is a common usage. Trunk lines are transmission media used to connects the interoffice. The transmission media used for trunk lines may be twisted copper wire or optical fibers. A trunk circuit could be a wireless communication channel (satellite links). Because it has to handles large numbers of connections using multiplexing, it requires higher bandwidth.

Switching Offices or Exchange

Switching offices or exchange is a central location where subscribers are interconnected, permanently or temporarily basis. Switching offices are connected with several local loops or trunk lines and provide a connection between different subscribers. Switching machines are located in exchange, switching machines are programmable matrices that provide temporary signal paths between two subscribers. This is done to avoid the having permanent physical connection between any two subscribers. Data modems are connected through local loops to switching machines located in exchange.

A **central exchange** is also known as central telephone exchange, central office or simply central. The purpose of a telephone exchange is to make a path for a call to be completed between two subscribers. While processing a call, a switch should provide three main functions:

(*i*) First function is to identify the subscribers

(*ii*) Establish a communications path

(*iii*) Monitor the calling process

The most common and familiar use of TELCO begins when a user, known as a subscriber, picks up the telephone handset and initiates a call by going off hook. In the telephone set, a multi-ganged switch activates with release of the handset from its cradle or hook. This switch completes a direct current path between the subscriber and the local switch station.

The current developed from a –48 V battery is sent through this **local loop** and is sensed by the user as a dial tone, which the local switch sends when it senses the current flow. The user, upon hearing this dial tone, begins to dial the desired number of the party whose wishes to communicate with. The term *dial* reflects the original way a telephone number was entered into the local switch by a subscriber. The user would literally dial the number using a rotary dial. The number on the dial is selected and the dial rotated to a stop. Upon releasing the dial, it would return to its original position at a given rate. A switch opens and closes for each position the dial passes on its return, generating a number of pulses equivalent to the number selected. The fixed rate of return of the dial assured a fixed pulse rate of these pulses. Figure 1.5 shows the general specifications and timing diagram for the rotary dial. The local switch station then interprets the pulses and makes the correct connection to the destination subscriber.

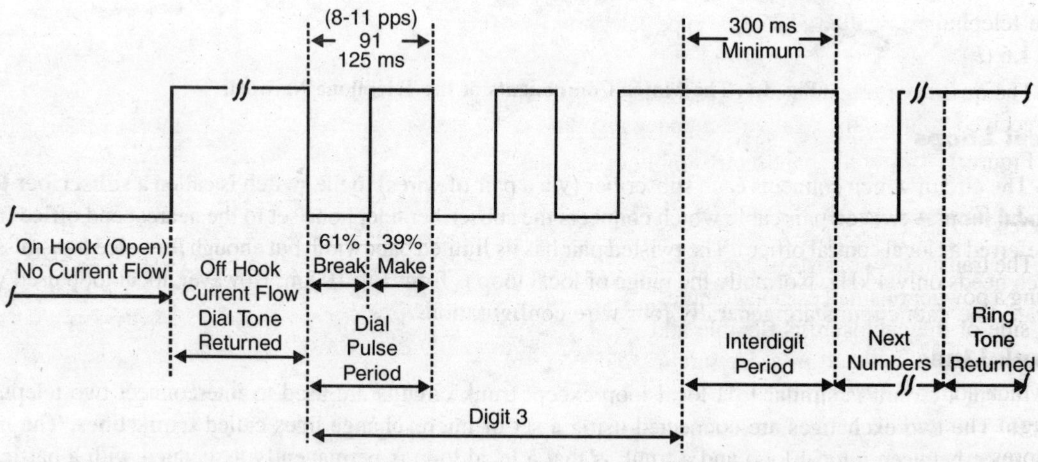

Fig. 1.5. Line Characteristics During Dialing.

The most familiar component of a telephone network is the telephone set provided for each user. It is a relatively simple device which can exchange control signals with the network to help establish and release calls, and to send and receive a user's speech signal.

1.5. STANDARD TELEPHONE SET

The telephone-set is a familiar end instrument in telecommuni-cation system. The telephone-set is basically a transducer. Transducer is a device that converts one form of energy into a different form. The transmitter of telephone converts sound energy into electrical energy. The receiver converts electrical energy into sound waves.

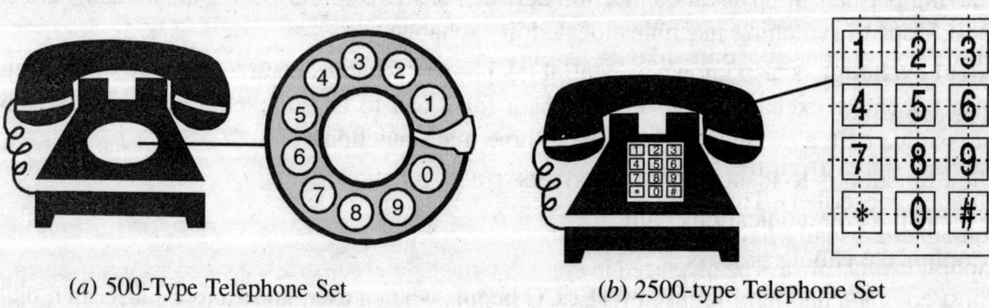

(*a*) 500-Type Telephone Set (*b*) 2500-type Telephone Set

Fig. 1.6. Telephone Standard.

The first telephone set that combined a transmitter and receiver into a single hand- held unit was introduced in 1878 and called the Butterstamp telephone. We talked into one end then turned the instrument around and listened with the other end. In 1951, western electric company introduced a telephone set that was the industry standard for nearly four decades. This telephone set is called the Bell system 500- type telephone and is shown in Fig. 1.6 (*a*). The 500 – type telephone set replaced the earlier 302 – type telephone set (the telephone with the hand-crank magneto, fixed microphone, hand-held earphone, and no dialing mechanism). Although there are very few 500 – type telephone set in use in the United State today, the basic function and operation of modern telephones are essentially the same. In modern-day telephone sets, the rotary dial mechanism is replaced with a Touch-Tone keypad. The modern Touch-Tone telephone is called a 2500 – type telephone set and is shown in Fig. 1.6 (*b*).

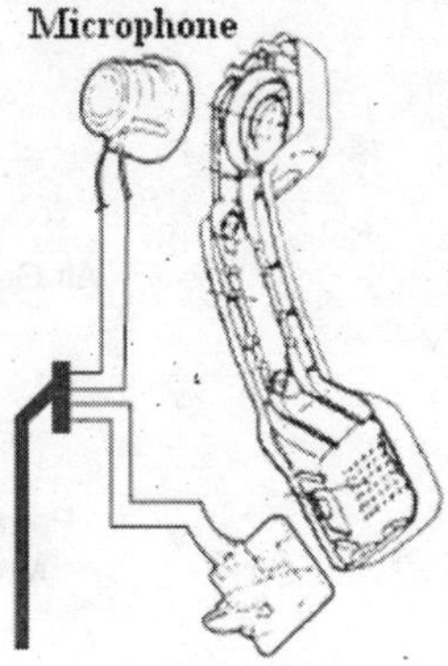

Fig. 1.7. Handset Assembly of Telephone.

The quality of transmission over a telephone connection depends on the received volume, the relative frequency response of the telephone circuit, and the degree of interference.

Figure 1.7 show the handset assembly of telephone. The handset consists of transmitter and receiver.

1.5.1. Transmitter

The transmitter consists of a box containing a powder of small carbon granules. One side of the enclosure is flexible and is mechanically attached to a diaphragm on which sound wave impinges. The diaphragm causes the carbon granules to compress or allow them to expand and consequently the resistance of the carbon granules decreases or increases in the box. The carbon granules conduct electricity and the resistance offered by them is dependent upon the density with which they are packed. If a voltage is applied to microphone, the circuit in the circuit varies according to the vibrations of the diaphragm. The varying electrical signal is similar to the varying sound signal. For this reason, it is called an analog signal. The theory of the carbon microphone indicates that the microphone functions like an amplitude modulator.

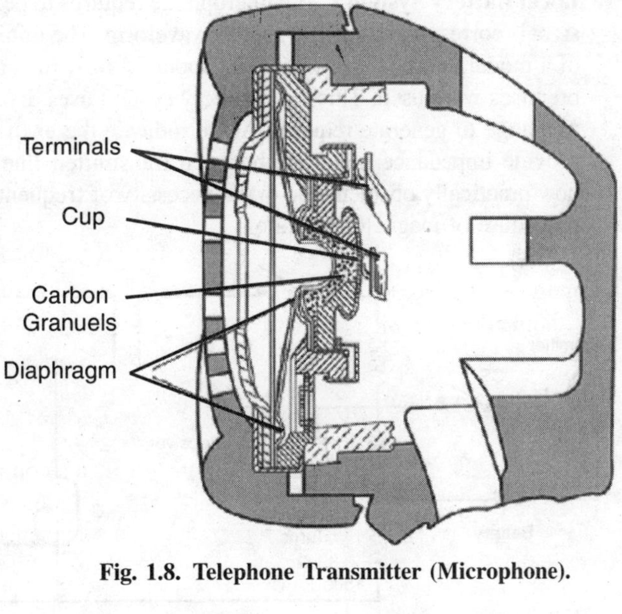

Fig. 1.8. Telephone Transmitter (Microphone).

1.5.2. Receiver

The varying signal from the handset ***A*** (calling subscriber) is coupled by wires to a receiver of the handset ***B*** (called subscriber). The receiver is an electromagnet with an accompanying magnetic diaphragm. The electromagnetic usually have two coils of about 100 turns with nominal resistance of 400 ohms. The receiver diaphragm must always be displaced in one direction from its unstressed position. It must be positioned with an air gap between it and the poles of the electromagnet. The diaphragm is made of cobalt iron and it is slightly conical shaped near the ear for uniform pressure distribution and hence the sound.

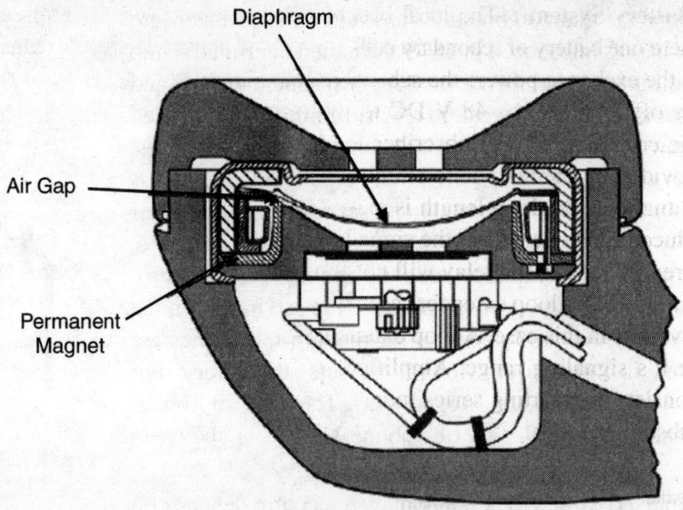

Fig. 1.9. Telephone Receiver (Loud-speaker).

1.5.3. Battery System

There are two types of battery systems used in telephone system which are as follows:

1. **Local Battery System :** The microphone requires to be energized in order to produce the electrical signals corresponding to the speech waveform. The minimum current required for proper operation of a modern carbon microphone is about 23 mA. In early system, local batteries at the subscriber premises were used. This local battery system uses dry cells to power the microphone, a magneto exchange to generate required AC to indicate the exchange for service and an autotransformer to provide impedance matching between transmitter, line and receiver. The local battery system is now practically obsolete due to the necessity of frequent replacement of dry cells and cumbersome procedure of magneto generator.

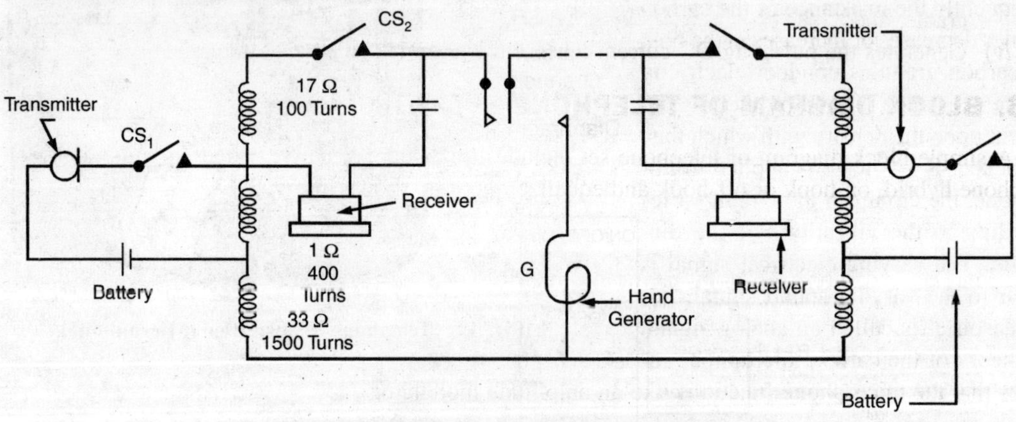

Fig. 1.10. Circuit for Local Battery Telephone Set.

A typical circuit for local battery telephone set is illustrated in Fig. 1.10. The autotransformer has three windings of the induction coils and has different resistance values because of different number of terms. The proper impedance matching is done with the receiver, transmitter and lines. Cradle switch (*S*1) and cradle switch (*S*2) are made when the telephone set is lifted out of the cradle. The magneto generator (*G*) is normally cut from the lines by its charge over contacts. Whenever, the magneto generator is operated, however, it is connected to the lines and the same time the telephone set are disconnected from them.

2. **Central Battery System :** The local battery system was completely replaced by the central battery system where one battery of secondary cells is provided at the telephone exchange. Thus, a battery of 48 volts from the exchange powers the subscriber loop. The central battery system is shown in Fig. 1.10. The center office supplies –48 V DC to the ring circuit. A negative voltage is used to minimize electrolytic corrosion of the subscriber loop wires. When a caller lifts the handset, a DC current path is provided by the telephone circuit.
In many situations, the loop length is extended while keeping the battery voltage is constant. This causes reduced effectiveness in the signaling. Also, if the current is insufficient (more attenuation) due to increased length, the relay will not function properly and the line seizure cannot be affected. The amplifiers and/or loop extenders as well as the inductive loading are the method used to increase the effectiveness in this case. A loop extender is a device that increases the battery voltage on a loop and hence it's signaling range. Amplifiers are used to counter balance the attenuation. Inductive loading consists of inserting series inductance in loop at fixed intervals to reduce the transmission loss on subscriber loops.

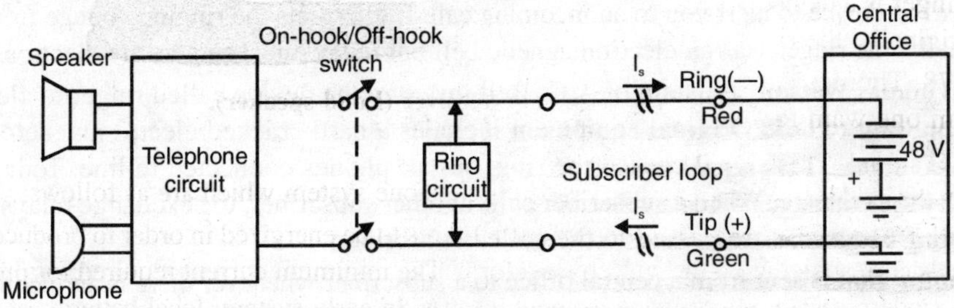

Fig. 1.11. Central Battery Subscriber Line Local Exchange Telephone.

The central battery system provides :

(*i*) An AC ringing voltage for the bell on the telephone,

(*ii*) Metallic path to provide power to the carbon granules,

(*iii*) A means to indicate the exchange for the need of service when the handset is removed from the cradle, and

(*iv*) Generates the pulsating DC current when the dialing is made.

1.6. BLOCK DIAGRAM OF TELEPHONE SET

A simple block diagram of telephone set includes transmitter, receiver, ringer circuits, dialer circuits, telephone hybrid, on-hook or off-hook and equalizer circuits, shown in Fig. 1.12.

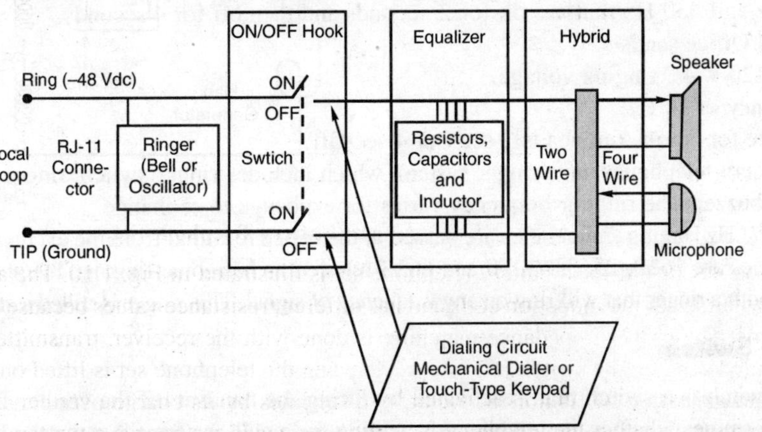

Fig. 1.12. A Simple Block Diagram of Telephone System.

Plain Old Telephone Set consists of five sections:

(*i*) Ringer Unit (*ii*) Hook Switch

(*iii*) Dialer Unit (*iv*) Hybrid/Speech Network

(*v*) Hand Set

The connection to the CO (Central Office) is with only 2 wires: Tip and Ring. This connec-tion is called the Local Loop. The Tip is +ve and coloured green. The Ring is –ve and coloured Red. If you look at a phone jack in your house, you will see that it is wired for 4 wires: Red, Green, Black and Yellow. Black and Yellow are not normally used.

The black and yellow wires can be used for a second telephone line or they can be used for running a Network Physical layer protocol called Phonenet by Farralon. Phonenet uses the Black and Yellow for Network communications. It is for use with Appletalk and is a replacement for Localtalk. It runs at the Localtalk speed of 230 Kbps which is reasonable for small networks.

1.6.1. Ringer Unit

The ringer is a device to alert you to an incoming call. It interprets the ringing voltage from the Central Office. Originally, the ringer was an electromagnetic bell but today, most ringers are electronic devices.

In 1878, Thomas Watson, assistant of A.G. Bell invented the device called ringer to alert the people with whom one want to talk. Original equipment includes a hard cracked electro magneto and a ringer to generate AC signal. This signal caused the ring, but all phones connected to line. Today the ringing signals sent by the exchange. When a subscriber calls another subscriber, the exchange relays are activated and the ringing frequencies are passed to the called subscriber.

The **ringing signal** is sent from a central office to a subscriber whenever there is an incoming call. The purpose of the ringing signal is to ring the bell in the telephone set to alert the subscriber that there is an incoming call. If there is no bell in the telephone set, the ringing signal is used to trigger another audible mechanism, which is usually a tone oscillator circuit. The ringing signal is normally a 20-Hz, 90-V_{rmsn} signal that is on for 2 seconds and then off for 4 seconds. The ringing signal should not be confused with the actual ringing sound the bell makes.

The audible ring produced by the bell was originally made as annoying as possible so that the called end would answer the telephone as soon as possible thus tying up common usage telephone equipment in the central office for the minimum length of time.

The **ring back** signal is sent back to the calling party at the same time the ringing signal is sent to the called party. However, the ring and ring back signals are two distinctively different signals. The purpose of the ring back signal is to give assurance to the calling party that the destination telephone number has been accepted, processed and is being ring. The ring back signal is an audible combination of two tones at 440 Hz and 480 Hz that are on for 2 seconds and then off for 4 seconds.

The Central Office sends :

- 90 to 120 V AC ringing voltage.
- Frequency of 20 Hz.
- Cadence for North America is 2 sec On/ 4 sec Off.

Present modern telephones have ringer system, which includes ringer switch, ringer IC, ringer volume control and the buzzer. The ringing frequency varies for exchange to exchange.

In India 162/3 Hz ringing frequencies are passed to the called terminal from the exchange. Other possible ringing frequencies are 162/3, 25, 332/3, 50 and 662/3 referred as harmonic ringing. Telephones manufactured today are made with a ringer that will ring on any frequency of signal. This ringer is called a **straight-line ringer.**

1.6.2. Hook Switch

The hook switch is a switch that is activated by lifting the handset off the cradle. The position of the hook switch determines whether the telephone is waiting for a call or actively using the line. The Off-hook position informs the network of a request for use. The On-hook position releases the use of the network.

On-Hook : Figure 1.11 is a block diagram showing the major functions of a telephone set. The ringer circuit is always connected across the line so it can signal an incoming call. The remainder of the telephone set is isolated from the line by the open contacts of the switchhook when the handset is on hook. No dc current flows since the ringer has a capacitor that blocks dc flow through it.

Off-Hook : When the handset is lifted off-hook to make call, the switch hook contacts close. Loop current flows from the central office battery through the telephone set and through a relay coil at the central office. When sufficient current flows in the relay coil, this relay is energized and its closed contacts signal to other central office equipment that a subscriber telephone is off-hook. A line finder searches until it finds the line with the off-hook signal. The line finder then sets up a connection for the switching equipment to begin receiving the telephone number. At this point, a dial tone generator is connected to the line to signal the caller to proceed with dialing. Dialing may be done by pulsing (interrupting) the loop current or by sending audio tones. When the first dialed digit is received at the central office, the dial tone is removed from the line.

Pulse dialing has been used from the time of the Strowger switch, but is rarely used now. Common control switches allow the use of tone dialing, which is much faster and allows the user to signal commands for additional services such as routing call through call center switch boards.

1.6.3. Dialer Unit

At the early stage, there was no numbering system. The calling person has to give all information about the called subscriber to the operator of manual exchange to make connection. During those times, a measles disease was spread over the Lovel town in United States. At that time, Moses Barker, a doctor raised a question that if the decease affects the telephone operators of the exchange, who will give the connection to the subscribers. By his idea, the numbering system arises. Now the number of digits extends to eight digits in India.

The mechanisms that transmit the identity (number) of the called subscriber are

1. Pulse dialing and
2. Multi-frequency dialing.

1. Pulse Dialing

A rotary dial telephone is used for implementing the pulse dialing. In the pulse dialing, a train of pulses is used to represent a digit of the subscriber number. The basic idea is to interrupt the D.C. path of the subscriber's loop for specific number of short periods to indicate the number dialed. This is called loop-disconnect (or rotary) signaling and in most countries the dial operates at about ten impulses per second with a bread of $66\frac{2}{3}$ msec and made of about $33\frac{1}{3}$ msec. Pulse dialing of digit 3 and 2 is shown in Fig. 1.13.

It is necessary to indicate the end of a pulse train so that a decoding circuit can know where one digit ends and the next begin. This is done by mechanical design, which ensures a minimum make period between any two consecutive digits. This period is called the inter-digit pause and is typically 200 msec minimums. The completion of dialing is identified if any break, last for few hundred milliseconds.

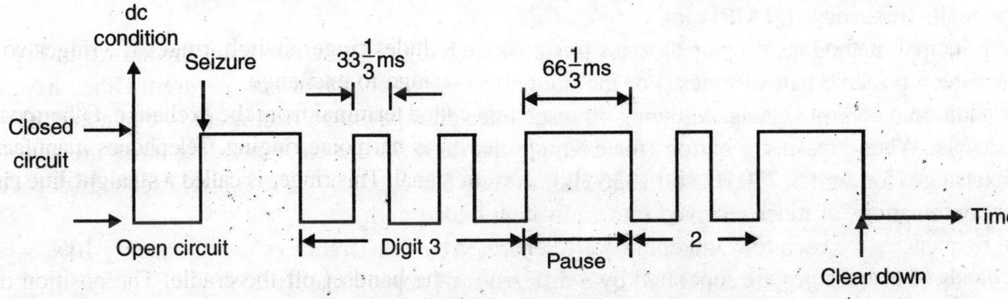

Fig. 1.13. Pulse Train of Rotary Dial.

The rotary dial telephone is classified either as cam type or trigger type depending on whether a cam mechanism or a trigger mechanisms is used for operating the impulsing contact.

This method of dialing is slow. For nationwide and international dialing, the routing signals to the switching center are fairly slow and inconvenient. The method, which is replacing the rotary dial telephone, is the push button telephone, which uses the multi-frequency dialing.

Pulse dialing telephone dial shown in Fig. 1.14.

The mechanism involve in a rotary telephone is shown in Fig. 1.15.

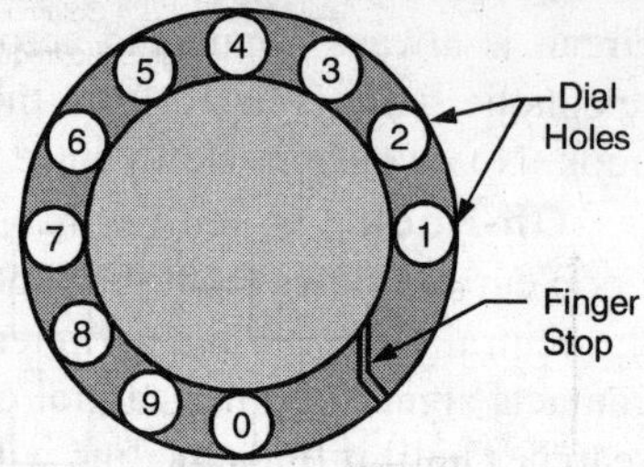

Fig. 1.14. Finger plate arrangement of rotary dial telephone.

The dial is being operated by placing a finger in the hole of appropriate digit to be dialed, drawing the finger plate round the clockwise directions to the finger stop position and allow the dial free by withdrawing the finger. The finger plate and the associated mechanism now return to the initial position under the tension of a spring. The dial pulses are produced during the return travel of the finger plate, then ignoring the human element in pulse timings.

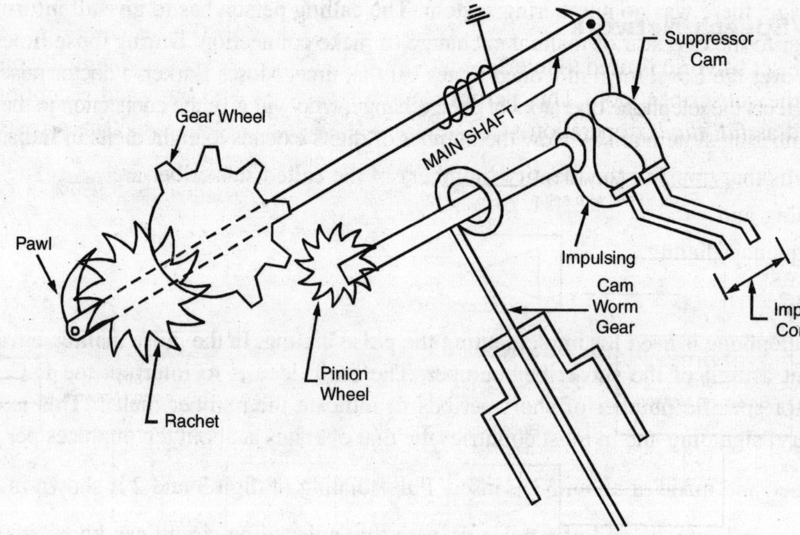

Fig. 1.15. Mechanism of Rotary Dial Telephone.

2. Multi-frequency Dialing

The touch-tone dialing scheme is shown in Fig. 1.16, the rotary dial is replaced by keypad. This is called a dual tone multi-frequency (DTMF) dial.

The principal method uses a pair of tones to signal each digit. These two tones are shown in Fig. 1.16. There is sixteen possible pair of tones. For the digits 0 to 9, ten combinations are used. Other six may be used for additional control signals. Allowing 40 msec inter-digit pauses, a maximum rate of about 12 digits/sec is feasible. When pressing a button (for example the digit 5), the corresponding tones are sent to the central exchange (for digit 5, 770 Hz and 1336 Hz), as shown in Fig. 1.17. The central exchange decodes the different combinations of tones received into equivalent dialed digit.

The frequencies for each row and column are separated by a difference of approximately 10%, whereas the two bands of frequencies are separated by a difference of approximately 25%.

Frequency tolerances are specified at ±1.5% for the telephone set's DTMF generator and ±2% at the center office. These prevent the telephone company from detecting erroneous digits.

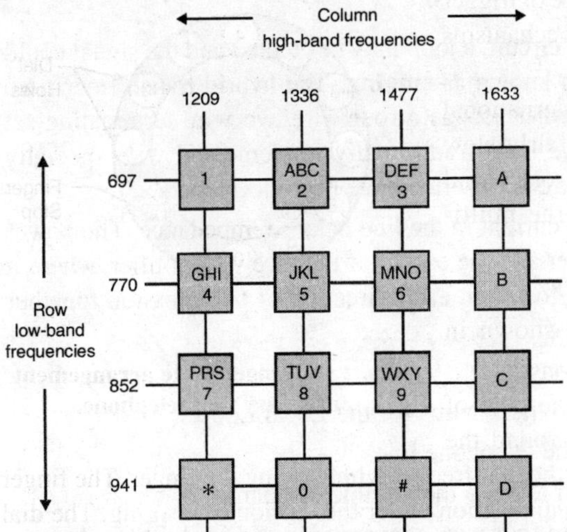

Fig. 1.16. Touch Tone Dialing Scheme (DTMF).

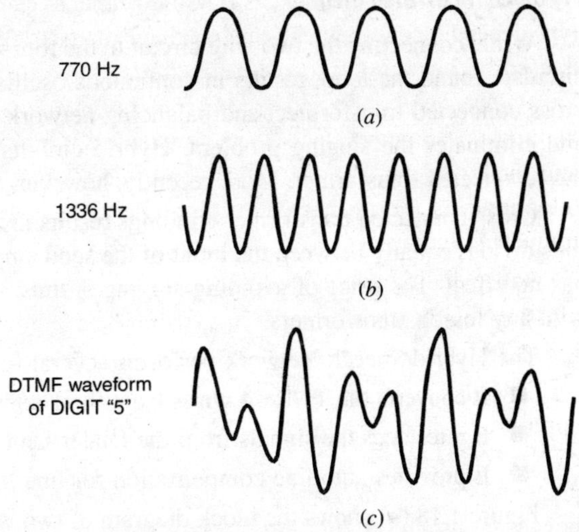

Fig. 1.17. DTMF Waveforms for Digit 5.

1.6.4. Hybrid/Speech Network

The telephone set's hybrid is used to interface the transmitter and receiver's individually paired wires to a single pair for the subscriber loop. This permits simultaneous transmission and reception of speech, commonly referred as *full-duplex* operation.

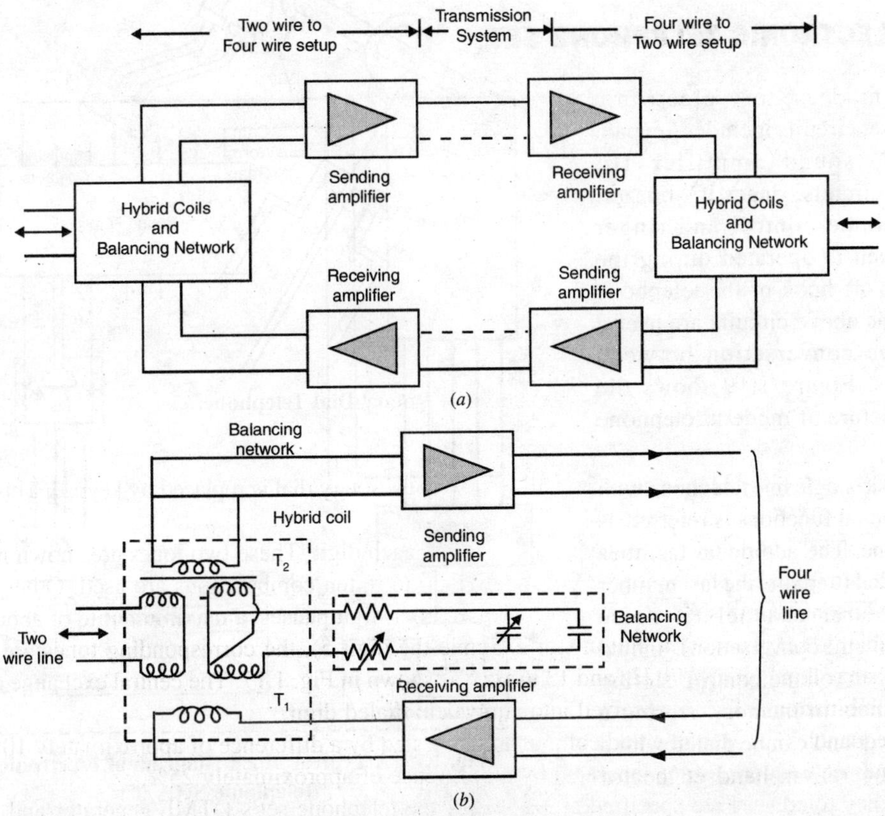

Fig. 1.18. (*a*) Block Diagram of Two Wire to Four Wire Conversion, (*b*) Hybrid Transformer Circuit.

Hybrid Transformer

While connecting the two wire circuit to the four wire circuit, a loop may be created and the signal could circulate round the loop, results in continuous oscillation known as **singing.** The hybrid transformer (two cross connected transformer) and balancing network together acts as a four wire/two wire terminating set and eliminates the singing problem. Hybrid circuits have been traditionally implemented with specially interconnected transformer. More recently, however, electronic hybrids have been developed.

Cross-connected transformer windings results in zero current in the line balance impedance. The power thus divides equally between the input of the send amplifier and the output of the receive amplifier, where it has no effect. The price of avoiding singing is thus 3 dB losses in each direction of transmission together with any loss in transformers.

The Hybrid/Speech Network performs several functions :

- It converts the Tx/Rx 4 wires from the Handset to the 2 wires for the Local Loop.
- It interfaces the signals from the Dialer Unit to the telephone line.
- It provides auto line compensation for line length to keep the volume constant.

Figure 1.18 (*a*) shows the block diagram of two wire to four wire conversion, and Fig. 1.18 (*b*) shows a hybrid transformer circuit.

1.6.5. Handset

The Handset contains transducers for converting mechanical energy into electrical energy. The microphone converts speech into electrical energy. The diaphragm or speaker converts electrical signals into audible signals (already discussed).

1.7. ELECTRONIC TELEPHONE SET

The modern or electronic telephone set circuits includes keypad, dialer IC, sound amplifier IC, associated circuits, ringer IC, buzzer; buzzer volume control and ringer switch which is operated during the on-hook or off-hook of the telephone handset. The above circuits are useful in effective conversation between subscribers. Figure 1.19 shows the typical structure of modern telephone circuits.

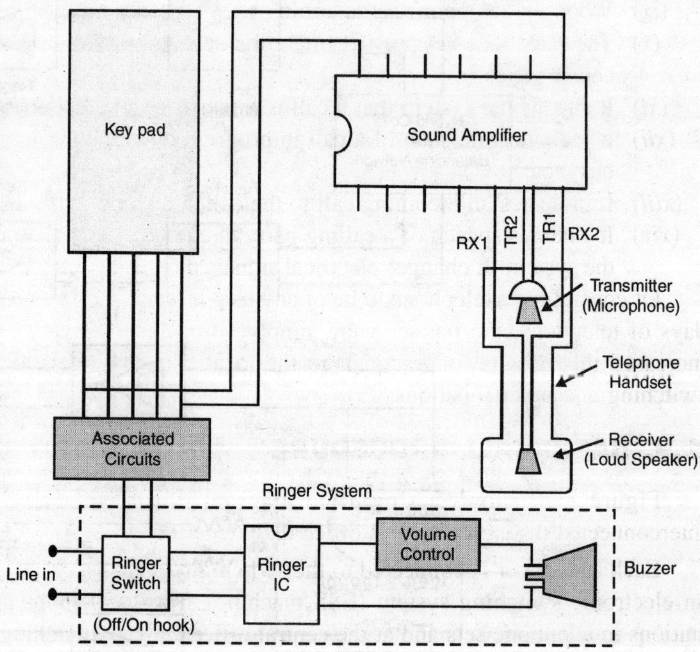

Fig. 1.19. A Typical Block Diagram of Electronic Telephone Set.

The basic single line telephone with many additional functions is referred as feature phone. The additional facilities includes redial (to redial the last number dialed), clearer channel (if noise interferes with the conversation), lighted keypad, ringer volume control, storing phone numbers in memory, storing a phone number and can be dialed with a one-touch operation, handset locator (cordless phone), call waiting service, automatic security code setting (used in cordless phone whenever the handset is placed on the base unit, the unit automatically selects one of many (in thousands) security codes. This codes help prevents the unauthorized use of your telephone line by another cordless telephone user) etc.

The modern telephone has the answering machine facility, message recording facility which uses PC multimedia card attached to microphone. Present day electronic feature phones are proprietary telephone set, hand free phone, integrated telephone, ISDN telephone, key system, and hybrid key system.

Leading telephone manufacture/assembler companies in India are ANCO, Beetal, BPL, Crompton, Godrej, GCEL, ITI, Panasonic, Punwire, Siemens, Shyam, Tata Fone, Telemate Telerisccon, Tushaco, Webcom etc.

The local exchange and the line interface requirements found in Local Switching Systems Generic Requirements (LSSGR) by AT&T and Bell Communications Research sets the standards for manufacturing telephone base units. This standards specify length of the local loop, type of ringer, minimum current for transmitter, DTMF pad etc.

1.8. FUNCTIONS OF TELEPHONE SET

(*i*) It requests the use of the telephone system when the handset is lifted.

(*ii*) It indicates that the system is ready for use by receiving a tone, called the Dial-Tone.

(*iii*) A method of dialing subscriber numbers must be incorporated.

(*iv*) The telephone set must provide a few amount of feedback from its microphone to its speaker.

(*v*) The telephone set must notify the user of an incoming call through a ring.

(*vi*) It sends the number of the telephone to be called to the system. This number is initiated by the caller when the number is dialed.

(*vii*) The telephone set must gain attention of the center office when a user request service by lifting the headset.

(*viii*) It automatically adjusts for changes in the power supplied to it.

(*ix*) When the telephone set is not in use, a DC path must be provided to the center office.

(*x*) The telephone set must regulate the speech amplitude for the varying distances to the telephone center office.

(*xi*) It signals the system that a call is finished when a caller 'hangs up' the handset.

(*xii*) It indicates the state of a call in progress by receiving tones indicating the status such as ringing, busy, etc.

(*xiii*) It indicates an incoming call to the called telephone by ringing bells or other audible tones.

(*xiv*) It changes speech of a calling party to electrical signals for transmission to a distant party through the system. It changes electrical signals received from a distant party to speech for the called party.

Of course, for a telephone to be of any use, it must be connected to another telephone. In the very early days of telephony, the phones were simply wired together with no switching. As the number of phones increased this became impractical, so the local exchange or central office was established to handle the switching and other functions.

1.9. BASIC CALL PROCEDURE

Figure 1.20 shows a simplified diagram illustrating how two telephone sets (subscribers) are interconnected through a central office dial switch.

Each subscriber is connected to the switch through a local loop. The switch is most likely some sort of an electronic switching system (ESS machine). The local loops are terminated at the calling and called stations in telephone sets and at the central office ends to switching machines.

When the calling party's telephone set goes off hook (*i.e.*, lifting the handset off the cradle), the switch hook in the telephone set is released, completing a D.C. path between the tip and the ring of the loop through the microphone. The ESS machine senses a D.C. current in the loop and recognizes this as an off-hook condition. This procedure is referred to as loop start operation since the loop is completed through the telephone set. The amount of D.C. current produced depends on the wire resistance, which varies with loop length, wire gauge, type of wires and the impedance of the subscriber's telephone. Typical loop resistance ranges from a few ohms up to approximately 1300 ohms, and typical telephone set impedances range from 500 ohms to 1000 ohms.

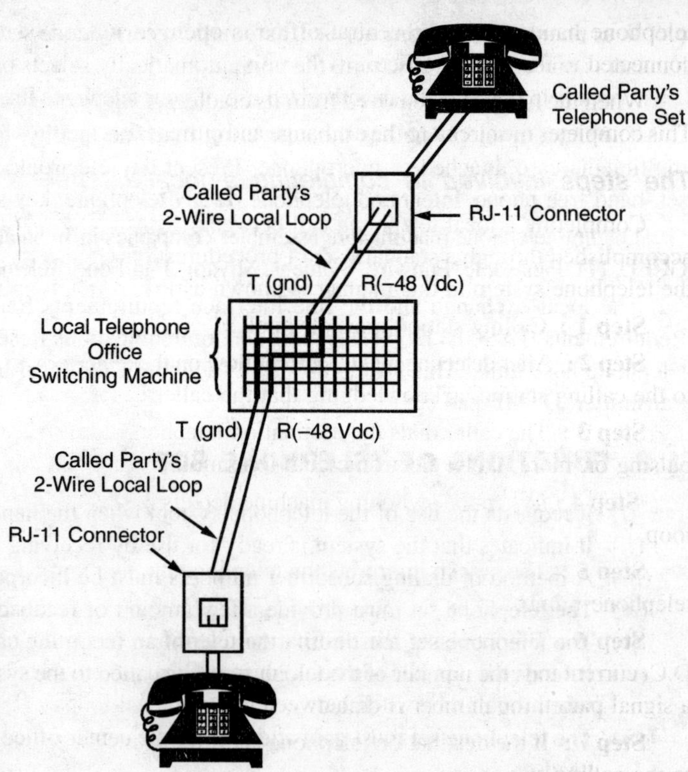

Fig. 1.20. Telephone Call Procedure.

Connecting the Phones

The central office has various switches and relays that automatically connects the calling and called phones. For now, assume that the connection has been made. The actual operation of switching systems will be covered in more detail a later. If the called phone handset is off-hook when the connection is attempted, a busy tone generated by the central office is returned to the calling phone. Otherwise, a ringing signal is sent to the called phone to alert the called party that a call is waiting. At the same time, a ring back tone is returned to the calling phone to indicate that the called phone is ringing.

Ringing the Called Phone

Early telephone circuits were point-to-point (not switched), and the caller gained the attention of the party at the other end by picking up the transmitter and saying "Hello". This was not very satisfactory, and schemes based on mechanical signaling arrangements were coon invented. The one in common use today, called the polarized ringer, or bell, was patented in 1878 by Thomas A. Watson. Electronic ringing circuits are quickly replacing polarized ringers in new telephone designs.

Answering the Call

When the called party removes the handset in response to a ring, the loop to the phone is completed by its closed switchhook and loop current flows through the called phone. The central office then removes the ringing signal and the ringback tone from the circuit.

Talking

The part of the telephone into which a person speaks is known as transmitter. It converts speech into variations of an electric signal accordingly.

Initiating a Call

When the handset of the telephone is resting in its cradle, the weight of the handset holds the switchhook buttons down and the switches are open. This is called the on-hook condition. The circuit between the telephone handset and the central office is open; however, the ringer circuit in the telephone is always connected to the central office.

When the handset is removed from its cradle, spring-loaded buttons come up and the switchhook closes. This completes the circuit to the exchange and current flows in the circuit. This is called the off-hook condition.

The steps involved in completing a local telephone call

Completing a local telephone call between two subscribers connected to the same telephone switch is accomplished through a standard set of procedures that includes the following steps listed below. Accessing the telephone system in this manner is known as POTS (plain old telephone service).

Step 1 : Calling station goes off hook.

Step 2 : After detecting a D.C. current flow on the loop, the switching machine returns an audible dial tone to the calling station, acknowledging that the caller has access to the switching machine.

Step 3 : The caller dials the destination telephone number using one of two methods – mechanical dial pulsing or, more likely, electronic dual-tone multifrequency (Touch-Tone) signals.

Step 4 : When the switching machine detects the first dialed number, it removes the dial tone from the loop.

Step 5 : The switch interprets the telephone number and then locates the local loop for the destination telephone number.

Step 6 : Before rising the destination telephone, the switching machine tests the destination loop for D.C. current to see if it is idle (on hook) or in use (off hook). At the same time, the switching machine locates a signal path through the switch between the two local loops.

Step 7 : If the destination telephone is off hook, the switching machine sends a station busy signal back to the calling station.

Step 8 : If the destination telephone is on hook, the switching machine sends a ringing signal to the destination telephone on the local loop and at the same time sends a ring back signal to the calling station to give the caller station to give the caller some assurance that something is happening.

Step 9 : When the destination answers the telephone, it completes the loop, causing D.C. current to flow.

Step 10 : The switch recognizes the D.C. current as the station answering the telephone. At this time, the switch removes the ringing and ring-back signals and completes the path through the switch, allowing the calling and called parties to begin their conversation.

Step 11 : When either end goes on hook, the switching machine detects an open circuit on that loop and then drops the connections through the switch.

The basic purpose of call progress tones and signals

Call progress tones and call progress signals are acknowledgment and status signals that ensure the processes necessary to set up and terminate a telephone call are completed in an orderly and timely manner. Call progress tones and signals can be sent from machines, machines to people, and people to machines. The people are the subscribers (*i.e.*, the calling and the called party), and the machines are the electronic switching systems in the telephone offices and the telephone sets themselves. When a switching machine output a call progress tone to a subscriber, it must be audible and clearly identifiable.

Signaling can be broadly divided into two major categories – station signaling and interoffice signaling. Station signaling is the exchange of signaling messages over local loops between stations (telephone) and Telephone Company switching machines. On the other hand, interoffice signaling is the exchange of signaling messages between switching machines. Signaling messages can be subdivided further into one of four categories: alerting, supervising, controlling and addressing. Alerting signals indicates a request for services, such as going off hook or ringing the destination telephone. Supervising signals provide information in the form of announcements, such as number changed to another number, a number no longer in service, and so on. Addressing signals provide the routing information, such as calling and called numbers.

Examples of essential call progress signals are dial tone, dual tone multifrequency tones, multifrequency tones, dial pulses, station busy, equipment busy, ringing, ring-back, receiver in hook, and receiver off hook. Table 1.1 and 1.2 summarize the most important call progress tones and their direction of propagation, respectively.

TABLE 1.1. Call Progress Tone Summary

Tone or Signal	Frequency	Duration/Range
Dial tone	350 Hz plus 440 Hz	Continuous
DTMF	697 Hz, 770 Hz, 852 Hz, 941 Hz, 1209 Hz, 1336 Hz, 1477 Hz, 1633 Hz	Two of eight tones On, 50-ms minimum Off, 45-ms minimum, 3-s maximum
MF	700 Hz, 900 Hz, 1100 Hz, 1300 Hz, 1500 Hz, 1700 Hz	Two of six tones On, 90-ms minimum, 120-ms maximum
Dial pulses	Open/closed switch	On, 39 ms Off, 61 ms
Station busy	480 Hz plus 620 Hz	On, 0.5 s Off, 0.5 s
Equipment busy	480 Hz plus 620 Hz	On, 0.2 s Off, 0.3 s
Ringing	20 Hz, 90 vrms (nominal)	On, 2 s Off, 4 s
Ring-back	440 Hz plus 480 Hz	On, 2 s Off, 4s
Receiver on hook	Open loop	Indefinite
Receiver off hook	D.C. Current	20-mA minimum 80-mA maximum
Receiver-left-off-hook alert	1440 Hz, 2060 Hz, 2450 Hz, 2600 Hz	On, 0.1 s Off, 0.1 s

TABLE 1.2. Call progress tone direction of propagation

Tone or Signal	Direction
Dial tone	Telephone office to calling station
DTMF	Calling station to telephone office
MF	Telephone office to telephone office
Dial pulses	Calling station to telephone office
Station busy	Telephone office to calling subscriber
Equipment busy	Telephone office to calling subscriber
Ringing	Telephone office to called subscriber
Ring-back	Telephone office to calling subscriber
Receiver on hoot	Calling subscriber to telephone office
Receiver off hook	Calling subscriber to telephone office
Receiver-left-off-hook alert	Telephone office to calling subscriber

1.10. CORDLESS TELEPHONES

A cordless telephone is basically a combination of telephone and radio transmitter/receiver. A cordless phone has two major parts:

1. Base Station
2. Handset.

The **base station** is attached to the phone jack through a standard phone wire connection, and as far as the phone system is concerned it looks just like a normal phone. The base receives the incoming call (as an electrical signal) through the phone line, converts it to an FM radio signal and then broadcasts that signal.

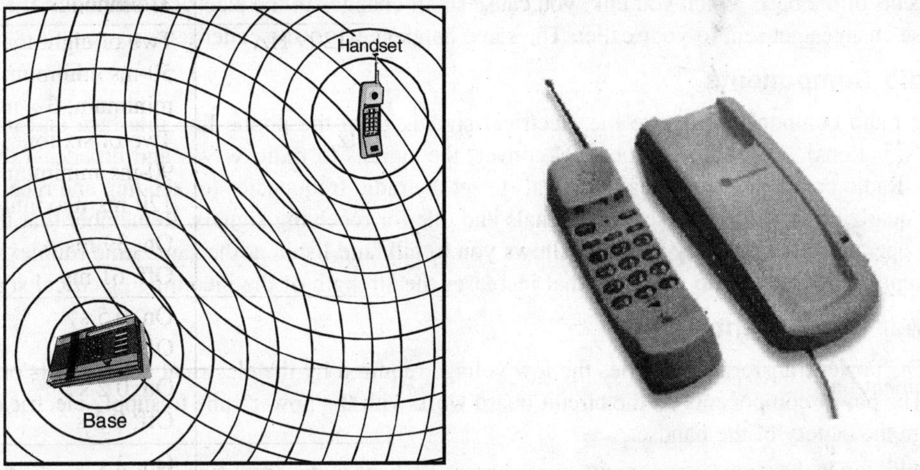

(*a*) Cordless Phone (*b*) Cordless Phone

Fig. 1.21.

The handset receives the radio signal from the base, converts it to an electrical signal and sends that signal to the speaker, where it is converted into the sound you hear. When you talk, the **handset** broadcasts your voice through a second FM radio signal back to the base. The base receives your voice signal, converts it to an electrical signal and sends that signal through the phone line to the other party.

The base and handset operate on a frequency pair that allows you to talk and listen at the same time, called duplex frequency.

1.10.1. Block Diagram of Base-Station

Figure 1.22 shows the simple block diagram of a cordless phone base-station. Following are the major elements :

1. **Phone line interface** : Receives and sends telephone signals through the phone line.

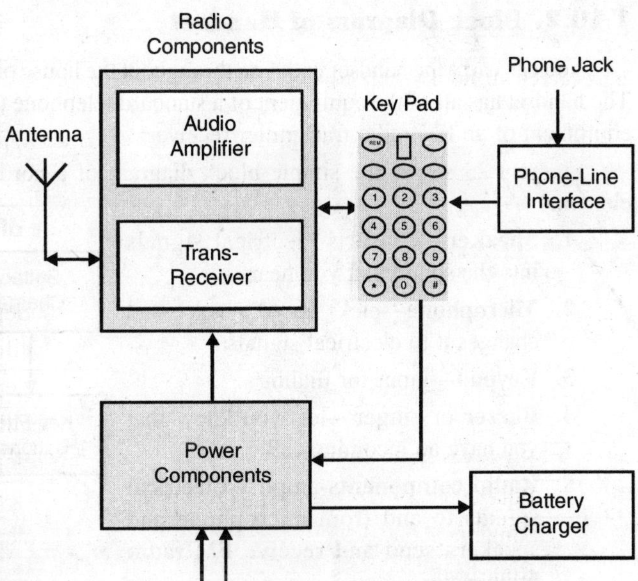

Fig. 1.22. Simple Block Diagram of Base-Station.

2. **Radio components :** It amplifies signals to and from phone-line interface, user controls and speaker phone (if present) broadcasts and receives radio signals to and from the handset.
3. **Power :** Supplies low voltage power to the circuits and recharges the battery of the handset.

1. Phone Line Interface

Phone line interface unit does two tasks.

First, they send the ringer signal to the bell (if it's on the base) or to the radio components for broadcast to the handset. This lets you know that you have an incoming call.

Second, they receive and send small changes in the phone line's electrical current to and from the radio components of the base. When you talk, you cause small changes in the electrical current of the phone line, and these changes get sent to your caller. The same happens when the caller talks to you.

2. Radio Components

The radio components receive the electrical signals from the phone line interface and user controls (keypads, buttons). The radio components convert the signals to radio waves and broadcast them via the antenna. Radio components use quartz crystals to set the radio frequencies for sending and receiving. There are two quartz crystals, one for sending signals and one for receiving signals. Remember that the base and handset operate on a frequency pair that allows you to talk and listen at the same time (duplex). The radio components include an audio amplifier that increases the strength of the incoming electrical signals.

3. Power Components

A DC power transformer supplies the low voltage required by the electrical components on the circuit board. The power components on the circuit board work with the power cube to supply electrical current to re-charge the battery of the handset.

In addition to the above components, some bases also have audio amplifiers to drive speakers for speaker phone features, keypads for dialing, liquid crystal displays (LCDs) for caller ID, light-emitting diodes (LEDs) for power/charging indicators, and solid state memory for answering machine or call-back features.

1.10.2. Block Diagram of Handset

You can carry the handset with you throughout the house or outside within the range of the base transmitter. The handset has all of the equipment of a standard telephone (speaker, microphone, dialing keypad), plus the equipment of an FM radio transmitter/receiver.

Figure 1.23 shows the simple block diagram of a cordless phone Headset. Following are the major elements :

1. **Speaker**—converts electrical signals into the sound that you hear
2. **Microphone**—picks up your voice and changes it to electrical signals
3. **Keypad**—input for dialing
4. **Buzzer** or **ringer**—lets you know that you have an incoming call
5. **Radio components-**amplify electrical signals to and from microphone and speakers send and receive FM radio frequencies
6. **LCD or LED displays**—indicator lights
7. **Re-chargeable battery**—supplies electrical power to handset

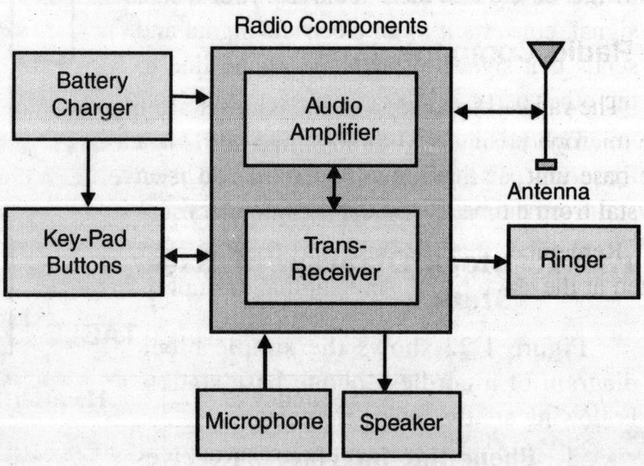

Fig. 1.23. Simple B/D of Handset.

1. Speaker

The speaker receives the electrical signals from the audio amplifier in the radio components and converts them into sound. When you remove the cover from the speaker, you see a large round permanent magnet with a hole in the middle and a deep groove surrounding the hole. Within this deep groove is a coil of fine copper wire that is attached to a thin plastic membrane. The plastic membrane covers the magnet and coil. To hear sounds, the following events happen:

(*i*) Electrical signals come from the radio components.

(*ii*) The electrical signals travel in the coil of copper wire.

(*iii*) The electrical signals induce magnetic currents in the coil of wire, thereby making it an electromagnet.

(*iv*) The electromagnetic coil moves in and out of the groove within the permanent magnet.

(*v*) The coil moves the attached plastic membrane in and out at the same frequencies as the changes in electric currents.

(*vi*) The movements of the membrane move air at the same frequencies, thereby creating sound waves that you can hear.

2. Microphone

The microphone changes the sound waves from your voice into electrical signals that are sent to the audio amplifier of the radio components. A microphone is essentially a speaker that works in reverse. When sound waves from your voice move the membrane, they make tiny electric currents either by moving a coil of wire within a magnet or by compressing the membrane against carbon dust.

3. Keypad

The keypad allows you to dial a number. It transfers the pressure from your fingertip on the appropriate key into an electrical signal that it sends to the radio components. Below the rubber keypad is a circuit board with black conductive material under each button. The keypad works like a remote control. When you press a button, it makes a contact with the black material and changes its electrical conductance. The conductance sends an electrical signal to the radio components indicating that you have selected that number.

4. Buzzer or Ringer

When the radio components of the handset receive the ringer signal from the base, they send electrical signals to the buzzer. The buzzer changes those electrical signals into sound much like the speaker does. You hear the buzzer sound and know that someone is calling you. In some phones, the speaker is used to make the ringer sound and there is no need for a separate ringer.

5. Radio Components

The radio components of the handset are like those of the base — they convert electrical signals from the microphone into FM radio signals and broadcast them at the same frequency as the receiving crystal of the base unit. The radio components also receive radio signals at the same frequency as the broadcasting crystal from the base, convert them to electrical signals and send them to the speaker and/or buzzer (ringer).

Remember that the base and handset operate on a duplex frequency pair that allows you to talk and listen at the same time. An example of duplex frequency pair is shown in Table 1.3.

TABLE 1.3.

Frequency	Handset	Base Station
Transmitter Frequency	49.28 MHz	44.32 MHz
Receiver Frequency	44.32 MHz	49.28 MHz

6. LCD or LED Displays

Most handsets have one or more light-emitting diodes (LED) that indicate various things, such as when the phone has an open line or when the battery is low. Some handsets have an LCD that can display numbers for caller ID features, similar to a cell phone. The LCD may be reflective or backlit so that you can see it when the room light is low.

7. Battery

The handset's battery supplies the power for all of the electrical components in the handset. All cordless phone handsets have a rechargeable battery (nickel-cadmium, nickel-metal hydride or lithium). When the battery runs low, an indicator light on the handset usually lights up or flashes. In some phones, a "beeping" sound may also indicate a low battery. You then recharge the battery on the base of the cordless phone.

Modern cordless phones have the same functions and much of the same hardware as we have discussed. However, many of the electronic circuits that were once achieved with transistors, resistors and capacitors have been replaced with integrated circuits. This advancement allows the handset to be either smaller with the same functions or the same size with more functions.

In summary, a cordless phone is basically a combination of a telephone and an FM radio transmitter/ receiver. Because it is a radio transmitter, it broadcasts signals over the open airways rather than specifically between the base and handset.

1.10.3. Major Features

Cordless phones have many of the same features as standard telephones, and there are many models, offering lots of different features.

Remember that a cordless telephone is a combination of a **telephone** and a **radio transmitter/receiver**. Because it is a radio transmitter/receiver, you have the following issues that you do not have on a standard cord phone:

- Range
- Sound quality
- Security

The **range** is the distance that the handset can be from the base. The **sound quality** can be affected by the distance, the way the information in the radio signal is transmitted and interfering structures such as walls and appliances. Security is an issue because the radio signals from both handset and receiver go over the open airways, where they can be picked up by other devices (other cordless phones, baby monitors, radio scanners).

The above issues relate to the following features of your cordless phone:

(*a*) Radio frequency
(*b*) Analog vs. digital technology
(*c*) Number of channels

(a) Radio Frequency

Because your cordless phone is a radio transmitter/receiver, it operates on various radio frequencies, which are set by the Federal Communications Commission (FCC) as with any other radio. Cordless phones operate over three major frequency bands (base and receiver use two closely related but separate frequencies within the band so that you can talk and listen at the same time):

(*i*) 43-50 MHz
(*ii*) 900 MHz
(*iii*) 2.4 GHz
(*iv*) 5.8 GHz

The **43-50 MHz band** was common to early cordless telephones and is still available in low-cost models. Because of the low frequency, these phones have short ranges (about 1,000 ft / 330 m) and poorer sound quality (due to interference from structures and appliances). The 43-50 MHz phone signals can also be picked up easily on radio scanners and nearby baby monitors.

The **900 MHz** band (actually 900-928 MHz) is the most common frequency for cordless phones today. The higher frequency gives it a greater range (5,000 to 7,000 ft/1,500 to 2,100 m) and better sound quality. However, 900 MHz signals can be picked up easily by most commercially available radio scanners.

In 1998, the FCC opened up the **2.4 GHz** range for cordless phone use. A 2.4 GHz or **5.8 GHz** cordless phone can operate over a greater distance and is above the frequencies that can be picked up by most commercially available radio scanners; therefore, it is more secure than lower frequency models.

(b) Analog vs. Digital

Analog technology is common in cordless telephones, especially in inexpensive models. Analog signals tend to be more noisy, or prone to interference with respect to sound quality. In addition, analog signals are easily picked up and interpreted by radio scanners.

In contrast, digital technology, like that found in a CD, allows the phone signals to sound clearer. Furthermore, digital signals are more secure. In 1995, digital spread spectrum (DSS) was introduced for cordless phones. DSS spread the digital information in pieces over several frequencies between the receiver and the base, thereby making it almost impossible to eavesdrop on cordless phone conversations.

(c) Channels

Each frequency band (43-50 MHz, 900 MHz, 2.4 GHz or 5.8 GHz) can be subdivided into different increments or channels. For example, on some models, when you're talking on your 900 MHz phone, the base searches for a pair of frequencies (channels) within that range, that is not already in use, in order to talk to the handset. So, if the base is capable of searching more increments, it can more easily find a frequency pair that is clear from interference, providing better sound quality. The number of cordless phone channels can vary as follows:

(*i*) **10 to 25 channels**—43-50 MHz phones, some inexpensive 900 MHz phones

(*ii*) **20 to 60 channels**—most 900 MHz phones

(*iii*) **50 to 100 channels**—high-end 900 MHz and 2.4/5.8 GHz phones

1.10.4. Brief History of Cordless Phones

Cordless phones first appeared around 1980. The earliest cordless phones operated at a frequency of 27 MHz. They had the following problems:

1. Limited range
2. Poor sound quality–noisy and ridden with static because walls and appliances interfered with the signals.
3. Poor security–people could easily intercept signals from another cordless phone because of the limited number of channels.

In 1986, the Federal Communications Commission (FCC) granted the frequency range of 47-49 MHz for cordless phones, which improved their interference problem and reduced the power needed to run them. However, the phones still had a limited range and poor sound quality.

Because the 43-50 MHz cordless phone frequency was becoming increasingly crowded, the FCC granted the frequency range of 900 MHz in 1990. This higher frequency allowed cordless phones to be clearer, broadcast a longer distance and choose from more channels.

In 1994, digital cordless phones in the 900 MHz frequency range were introduced. Digital signals allowed the phones to be more secure and decreased eavesdropping — it was pretty easy to eavesdrop on analog cordless phone conversations. In 1995, digital spread spectrum (DSS) was introduced for cordless phones. This technology enabled the digital information to spread in pieces over several frequencies between the receiver and the base, thereby making it almost impossible to eavesdrop on the cordless conversations.

In 1998, the FCC opened up the 2.4 GHz range for cordless phone use. This frequency has increased the distance over which a cordless phone can operate, and brought it out of the frequency range of most radio scanners, thereby further increasing security.

1.11. CALLER IDENTIFICATION

Caller ID enables a customer to see the caller's telephone number, name and time as the telephone is ringing. The service was originally conceived by AT & T Bell Lab in the late 1970s.

Caller ID, known also as **caller identification** (**CID**), or more properly **calling number identification** (**CNID**), is a telephone service, available on POTS (plain old telephone service) and most VOIP (Voice over Internet Protocol) lines, that transmits a caller's number to the called party's telephone equipment during the ringing signal, or when the call is being set up but before the call is answered.

Caller ID may be used to track down or limit the impact of prank calls, telemarketers, and other intrusions. However, it can also impede communication by enabling users to become evasive. The concept behind caller ID is the value of informed consent; however, it also poses problems for personal privacy. Another problem is that the possibility of caller ID spoofing makes it unreliable.

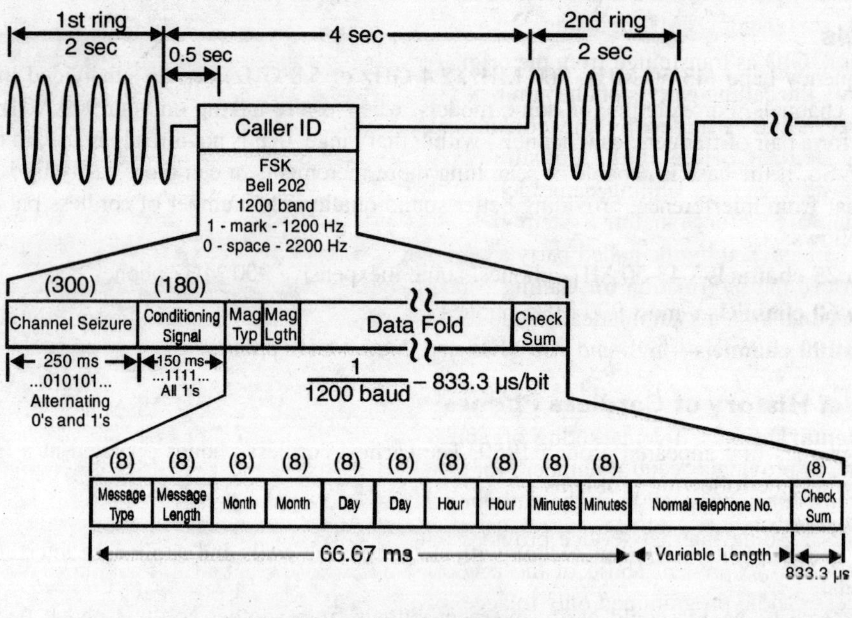

Fig. 1.24. Caller ID Signal Format.

If the call originates on a POTS line (a standard loop start line) caller ID is provided by the service provider's local switch. Since the network does not connect the caller to the caller until the phone is answered generally the caller ID signal cannot be altered by the caller. Most service providers however, allow the caller to block caller ID presentation through the vertical service code *67.

A call placed behind a private branch exchange (PBX) has more options. In the typical telephony environment a PBX connects to the local service provider through PRI trunks. Generally, although not absolutely, the service provider simply passes whatever calling line ID appears on those PRI access trunks transparently across the Public Switched Telephone Network (PSTN). This opens up the opportunity for the PBX administrator to program whatever number they choose in their external phone number fields.

The caller ID signal is sent to the customer as a 1200 baud FSK signal between the first and the second rings as shown in Fig. 1.24.

In the United States, caller ID information is sent to the called party by the telephone switch as an analog data stream (similar to data passed between two modems), using Bell 202 modulation between the first and second rings, while the telephone unit is still on-hook. If the telephone call is answered too quickly after the first ring, caller ID information will not be transmitted to the recipient.

There are two types of caller ID, number only and name alongwith number. Number only caller ID is called **Single Data Message Format** (SDMF), which provides the caller's telephone number, the date and time of the call. Name plus number caller ID is called **Multiple Data Message Format** (MDMF), which in addition to the information provided by SDMF format can also provide the directory listed name for the particular number. Caller ID readers which are compatible with MDMF can also read the simpler SDMF format, but an SDMF caller ID reader will not recognize an MDMF data stream, and will act as if there is no caller ID information present, *e.g.*, as if the line is not equipped for caller ID.

Instead of sending the caller ID in between the first and second ring, some systems use a "line reversal" to announce the caller ID, or caller ID signals are simply sent without any announcement. Instead of Bell 202, the European alternative V.23 is sometimes used, (without the 75 baud reverse channel) or the data is sent using DTMF signaling.

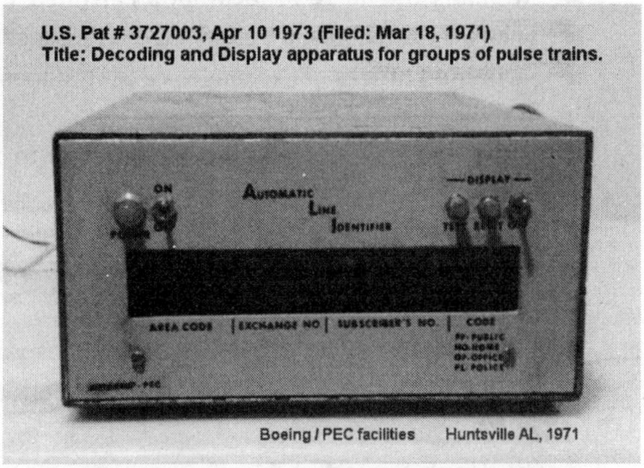

Fig. 1.25. The First Caller Identification Receiver.

In general, CID as transmitted from the origin of the call is only the calling party's phone number. The calling party name is added by the consumer's terminating central office if the consumer has subscribed to that service. Calling name delivery is not automatic. An SS7 (or Signaling System 7) TCAP query must be launched by the called party's central office to retrieve the information for Calling Name delivery to the caller ID equipment at the consumer's location. Canadian systems automatically send the calling name with the call set up and routing information at the time of the call.

Uses

1. **Telemarketing :** Telemarketing organizations often spoof caller ID. In some instances, this is done to provide a "central number" for consumers to call back, such as an 800 number, rather than having consumers call back the outbound call center where the call actually originated.
2. **ISPs :** Some Internet service providers (ISPs) providing dial-up access require the customer to use CNID to prevent abuse of the account by unauthorized callers. Some systems with dial-up access can be programmed only to accept calls with specific caller ID strings.
3. **Mobile providers :** Most mobile phone providers used the caller ID to automatically connect to voice mail when a call to the voice mail number was made from the associated mobile phone number, bypassing the need to enter a password. While this was convenient for many users, because of spoofing, this practice has been replaced by more secure authentication by many carriers.

Summary

- Telecommunication networks carry information signals among entities, which are geographically far apart. An entity may be a computer or human being, a facsimile machine, a teleprinter, a data terminal and so on.

- Alexander Graham Bell and Elisha Gray contributed significant works and filed paper related to telephony.
- In 1876, Bell invented a telephone system. In 1897 Marconi patented a wireless telephone system. Teletypewriter service was initiated in 1931.
- An early acronym for the telephone company was **TELCO,** which is still being used to a fair extent today. It is being gradually replaced by newer one— POTS (Plain Old Telephone System).
- The major components of the telephone network. They are as follows:
  1. Local Loops
  2. Trunks
  3. Switching offices (end offices, tandem offices and regional offices)
- The circuit which connects each subscriber (via a pair of wires) to the switch is called a **subscriber loop** or **local loop**.
- Normally the range of local loop is from 1 to 10 km.
- The two exchanges are connected using a set of interexchange lines called **trunk lines.**
- **Switching offices** or exchange is a central location where subscribers are interconnected, permanently or temporarily basis.
- The purpose of a telephone exchange is to make a path for a call to be completed between two subscribers.
- The telephone-set is basically a transducer.
- The transmitter of telephone converts sound energy into electrical energy. The receiver converts electrical energy into sound waves.
- There are two types of battery systems used in telephone system.
- Plain Old Telephone Set consists of 5 sections :
  (*i*) Ringer Unit
  (*ii*) Hook Switch
  (*iii*) Dialer Unit
  (*iv*) Hybrid/Speech Network
  (*v*) Hand Set
- The ringer is a device to alert you to an incoming call. It interprets the ringing voltage from the Central Office.
- The hook switch is a switch that is activated by lifting the handset off the cradle.
- The mechanisms that transmit the identity (number) of the called subscriber are **Pulse dialing** and **Multi-frequency dialing.**
- In the pulse dialing, a train of pulses is used to represent a digit of the subscriber number.
- The rotary dial is replaced by keypad. This is called a dual tone multi-frequency (DTMF) dial.
- The telephone set's hybrid is used to interface the transmitter and receiver's individually paired wires to a single pair for the subscriber loop.
- The Hybrid/Speech Network performs several functions :
  (*i*) It converts the Tx/Rx 4 wires from the Handset to the 2 wires for the local loop.
  (*ii*) It interfaces the signals from the Dialer Unit to the telephone line.
  (*iii*) It provides auto line compensation for line length to keep the volume constant.
- The modern or electronic telephone set circuits includes keypad, dialer IC, sound amplifier IC, associated circuits, ringer IC, buzzer; buzzer volume control and ringer switch which is operated during the on-hook or off-hook of the telephone handset.
- A cordless telephone is basically a combination of telephone and radio transmitter/receiver. A cordless phone has two major parts: Base Station and Handset.

- Caller ID enables a customer to see the caller's telephone number, name and time as the tele-phone is ringing. The service was originally conceived by AT&T Bell Lab in the late 1970s.
- There are two types of caller ID, *number only* and *name along with number*.

Review Questions

1. Explain the basic purpose of call progress tones and signals.
2. With the help of block diagram explain basic call procedure, call progress tones and signals.
3. Differentiate between the cordless and standard telephone set.
4. How is caller identification achieved in telephone sets.
5. Draw the block diagram of an electronic telephone and explain it.
6. What is meant by telecommunication network?
7. Explain the main components of PSTN in brief.
8. What is the range of voice frequencies?
9. Define bandwidth.
10. Draw the block diagram of a telephone set and explain the each block.
11. Briefly discuss the features of a standard telephone set.
12. Define decibel.
13. Discuss the dialer unit of a telephone set.
14. Explain the battery system of a telephone system.
15. What are the elements of telecommunication system?
16. List the end systems of instruments of telecommunication systems.
17. List the various methods of transmission systems.
18. Explain the necessity of hybrid network in a telephone set.
19. Explain the DTMF (Dual Tone Multiple Frequency) keypad layout and frequency allocation used in it.
20. Explain how caller ID operates and when it is used.
21. Draw the block diagram of a telephone system and briefly describe the essential components of a telephone set.
22. Enumerate the functions of a telephone system.
23. Discuss the basic call procedure in brief.
24. Explain the cordless telephone with the help of block diagram.
25. Write the features of cordless telephone set.
26. Explain how caller ID operates and when it is used.
27. Describe the difference between the operation of a cordless telephone and a standard telephone.
28. What is the purpose of dual tone multifrequency and multifrequency signaling?
29. Describe the dial pulsing.
30. What is the difference between ringing and ring back signal?
31. List and briefly describe the essential components of a standard telephone set.
32. Write short notes on:
    (*i*) Battery system
    (*ii*) Pulse dialing and multitone dialing
    (*iii*) Hybrid network
    (*iv*) Cordless telephone set
    (*v*) Caller ID

❑❑❑

CHAPTER

2

TELEPHONE CIRCUITS

INSIDE THIS CHAPTER

2.1. INTRODUCTION

A telephone circuits includes of two or more facilities such as interconnected in tandem, to provide a transmission path between a source and a destination. The facilities may be metallic cable pairs, optical fibers or wireless carrier systems. The interconnected facilities may be temporary or permanent.

The introduction of fiber cable needs a device at subscriber premises to convert electrical energy into light energy and this is the additional cost to the customer. But for high speed data transmission, switched cable TV, videophone, teleconferencing the fiber optic local loop has become essential.

The information transferred is called a message and the circuit or transmission path used is called the message channel. Nowadays telephone companies offer a wide range of message channels from a voice band circuit (a basic 4 kHz) to wideband microwave, satellite or optical fibre transmission systems which are capable of transferring wide band data or high resolution video.

In this unit we will be limiting our discussion to basic voice band circuits only.

2.2. LOCAL SUBSCRIBER LOOP

The circuit which connects each subscriber (via a pair of wires) to the switch is called a **subscriber loop** or **local loop**. A twisted pair cable which connects the subscriber telephone set to the nearest end office (may be referred as local central office).

Local Loop

A pair of wires connection a subscriber to the closest telephone office is called the local loop. One wire on the local loop is called the tip, and the other is called the ring (Fig. 2.1), the names tip and ring come from the 6 mm diameter two conductor phone plugs and patch cards used at telephone company switchboards to interconnect and test circuits. Tip and ring for a standard plug and jack are shown in Fig. 2.1, when a third wire is used, it is called the sleeve.

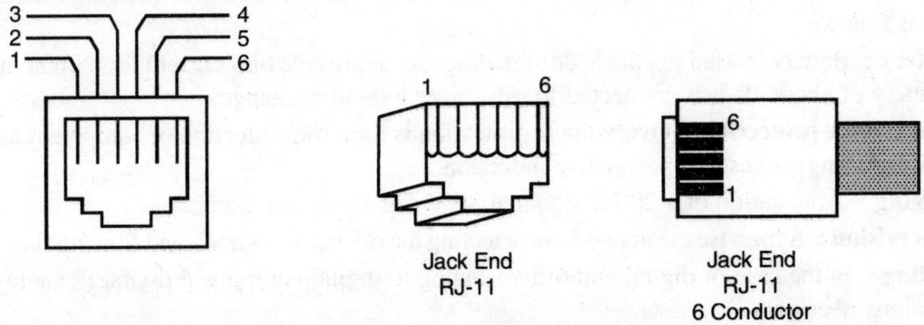

Fig. 2.1. RJ-11 Connector.

Since the 1960s, phone plugs and jacks have gradually been replaced in the home with a miniaturized plastic plugs knows as RJ-11 and a matching plastic receptacle. It is shown in Fig. 2.1. RJ stands for **Registered Jack**. It is sometimes described as **RJ-XX. RJ** is a series of telephone connection interfaces (receptacle and plug) that are registered with the U.S. Federal Communications Commission (FCC). The term jack sometimes describes both receptacle and the plug. It sometimes specifies only the receptacle. RJ-11 is the most common telephone jack in use. It can have up to six conductors. Although an RJ-11 plug is capable of holding six wires in a 5 mm × 5 mm body, only two wires (one pair) are necessary for a standard telephone circuit to operate. The other four wires can be used for a second telephone line and/ or for some other special function.

Every subscriber has his own pair of wires to the local switching office. The local loop is a metallic transmission line contains two insulated copper wires twisted together.

Twisted pair local loop is an excellent transmission medium for analog voice signals. The twisted pair has its limited bandwidth but enough for voice transmission which needs only 4 kHz.

The local loop is the primary cause of attenuation and phase distortion on a telephone circuit. Normally the range of local loop is from 1 to 10 km. However, local loop used with private-line data circuits are generally four wire configurations.

One end of each subscriber loop is terminated on a Main Distribution Frame (MDF) at the exchange. The drop wires (DW) from the telephones are connected to the distribution point (DP) which is located near the subscriber's premises. The distribution points at various locations are connected together by a distribution cables (DC) and terminated to the feeder points (FD).

The DC carries 10–500 pair of wires. Many feeder points related to a particular geographical area connected by a branch feeder (BF). From BF, through main feeder, all the subscriber loops are connected to MDF at the end office or local exchange. Typically, the MF carries 100– 2000 pair of wires to MDF. For the purpose of flexible interconnection such as transfer from location to other location or within the geographical area, the subscriber pair and exchange pairs are interconnected at the MDF by means of jumpers.

2.2.1. Fundamental Characteristics

The *transmission characteristics* of a cable pair depend on the wire diameter, conductance of the cable, conductor spacing and dielectric constant of the insulator separating the wires. These physical properties, in turn, determine the inductance, capacitance, resistance and conductance of the cable. The conductance and capacitance exist between the two wires, whereas the resistance and inductance are distributed along the length of wire.

The electrical characteristics of a cable such as capacitance, resistance and inductance are uniformly distributed along its length and are referred as *distributed parameters*. It is common practice to lump them in to discrete values per unit length such as microfarads / km, millihenrys / km, or ohms/1000 feet.

Figure 2.1 shows the electrical model for a copper wire transmission line. The amount of attenuation and phase delay experienced by a signal flowing through a metallic transmission line is a function of the frequency of the signal.

The subscriber loop is the most common interface in the network. The fundamental characteristic of this interface are as follows :

1. **Battery :** Battery is used to enable dc signaling and to provide bias current for carbon microphone, a battery of about 48 V is connected to subscriber loop at exchange.
2. **Overvoltage protection :** Overvoltage protection is used for protection of equipment and personal from lightning strikes and power line induction.
3. **Ringing :** Application of a 20 Hz signal at 86 V rms for ringer excitation.
4. **Supervision :** Supervise the network by detecting the off-hook/on-hook and flow/no flow dc current.
5. **Coding :** In the case of digital end office, analog to digital coding and digital to analog decoding functions necessary.
6. **Hybrid :** Hybrid is necessary for two wires to four wire conversion.
7. **Test :** Line test toward the subscriber disconnection of the switch.

The first letter of the above characteristics are combined together which is commonly known as **BORSCHT.**

2.2.2. Major Component Parts of a Traditional Local Loop

There are seven major component parts that make up a traditional local loop, which are as below :

1. Feeder Cable

The largest cable used in a local loop, normally 3600 pair of copper wire in conduit or underground.

2. Serving Area Interface

Serving area interface is a cross-connect point used to distribute the larger feeder cable into smaller distribution cables.

3. Distribution Cable

A type of feeder cable having less wire pairs.

4. Standard Network Interface

Standard network interface is a device that serves as the demarcation point between subscriber responsibility and local telephone company responsibility for telephone service.

5. Drop Wire

It is the final length of cable pair that terminate at standard or subscriber network interface.

6. Distribution Cable and Drop Wire Cross Connect Point

This is the location where individual cable pair are separated and extended to the subscriber's location on a drop wire.

7. Aerial

The portion of local loop which is strung between poles.

2.2.3. Limiting Factors of Subscriber Loop Design

While designing a subscriber loop, there are two limiting factors that we have to consider. Those factors are as follows :

1. Attenuation
2. Voltage drop

1. **Attenuation :** The attenuation refers to the energy loss in the line at a reference frequency, measured in decibels. If the length of the loop increases the attenuation also increases. The attenuation limit of the subscriber loop is normally 6 dB. The reference frequency is 800 Hz in Europe and 1000 Hz in America.
2. The second limiting factor is the voltage drop. The effectiveness of the signaling and conversation will be limited if the battery voltage is kept constant with increase in length. This is due to the IR drop of the line. The IR drop of the line varies with resistances of the battery used in the system, telephone set resistance and the allowable resistance of the subscriber loop.

The maximum allowable resistance in the subscriber loop and the loop resistance limit is calculated as follows. The loop resistance limit is calculated as follows. The loop resistance limit is used to determine the cable length and the cable gauge required.

$$R_{max} = \frac{V_B}{I_C} \quad \text{...(2.1)}$$

where, R_m = Maximum allowable resistance of subscriber loop.

V_B = Battery voltage

I_c = Minimum current required for proper operation of carbon microphone

The loop resistance limit is

$$R_L = R_m - (R_B + R_T) \quad \text{...(2.2)}$$

where, R_B = Approximate resistance maintained at the battery protect against short-circuit in the wire between subscriber and local office

R_T = Telephone set resistance.

2.2.4. Loading Coil

Attenuation limits arise from the ac response of the loop and refers to loop loss in decibels. The amplifiers/loop extenders or inductive loading coil methods are used to admit the loop length to increase beyond the limit. Inductive loading is the process of inserting series inductances (loading coils) into the loop at fixed intervals. These will reduce the transmission loss on subscriber loops. This method enables the designer to increase the loop length without using higher diameter copper wire.

The attenuation loss in twisted pair cable is because of the capacitance between wire pair. The wires of the local loop have a capacitance of approximately 0.1335 μ Farad per km regardless of the cable gauge. The longer the cable pairs the higher the capacitance. The capacitance can be reduced by separation of the conductors but practically difficult. Heavyside in 1887 proved that the distortion can be reduced by satisfying the equation $RC = LG$, where R is the resistance, C is the shunt capacitance, G is the conductance and L is the inductance. All are measured per unit length.

Figure 2.2 (*b*) shows the effect of frequency on attenuation for a 12000 feet length of 26-gauge copper cable. The figure shows, a 3 kHz signal experiences 6 dB more attenuation than a 500 Hz signal on the same cable. In fact, the cable acts as a low pass filter.

It has been experienced that by increasing the inductance of the cable, a substantial reduction in attenuation is achieved. Minimum attenuation requires a value of inductance 100 times the value obtained in ordinary twisted wire cable. It is impracticable to achieve such values on a uniformly distributed basis.

Instead, the desire effect can be obtained by adding inductors periodically in series with wire. This process is called **loading**, and the inductors are known as **loading coils.**

By inserting loading coils with proper inductance and at appropriate distances, the distortion can be minimized and high-speed transmission can be achieved.

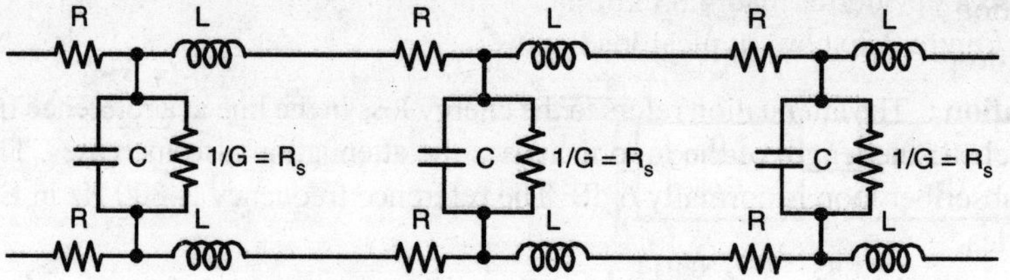

(*a*) Copper wire transmission line.

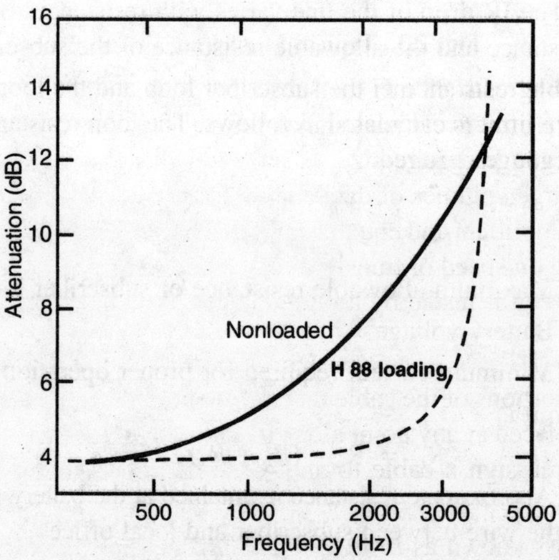

(*b*) Frequency–versus-attenuation characteristics for loaded and unloaded cables

Fig. 2.2.

Loading coils placed in a cable decrease the attenuation and increase the line impedance of the cable. It improves transmission levels for circuits longer than 18000 feet. By using local coils, local loop can be extended up to three to four times their previous length.

Table 2.1 gives the standard letters and its associated spacing. B, D and H are the most, commonly and spacing.

TABLE 2.1. Standard Letters and its Associated Spacing.

Letter Code	Spacing (ft)	Spacing (Km)
A	700	0.213
B	3000	0.914
C	929	0.283
D	4500	1.372
E	5575	1.7
F	2787	0.85
H	6000	1.83
X	680	0.2107
Y	2130	0.699

To offset the mutual capacitance (0.1335 μFarad), load coils having 88 mH inductances are placed at 1.8288 km (6000 ft) intervals on the cable. The first load coil placed at 0.9144 km (3000 ft) from the local exchange. Then for every 1.8288 km, one loading coil is placed till the subscriber premises. The loading is

not necessary, for the loop shorter than 4.83 km. For local loop length, above 18000 ft (5.48 km), loading coils are necessary. Figure 2.3 shows typical load coil spacing.

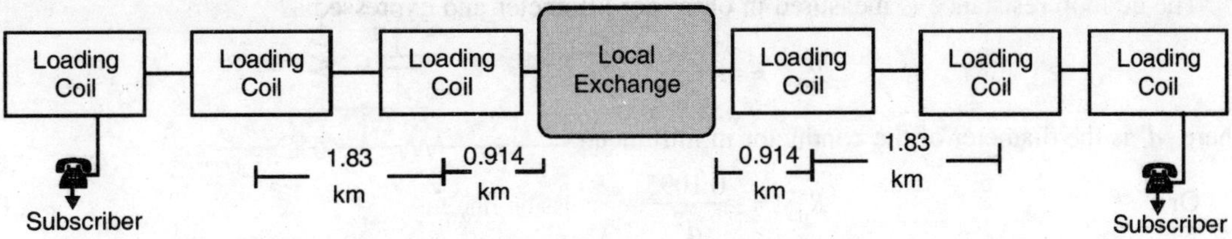

Fig. 2.3. Typical Load Coil Spacing.

Almost all loops are loading with 88 mH load coils. It results in a cutoff frequency of around 3800 Hz. Hence 88 mH coil loaded loops effectively pass signals above 3800 Hz. The loading coils are represented for example, as 19H88, 19 indicate the gauge size, H indicates the spacing of the coils and 88 indicate the inductance of the coils. For large volumes of data and high frequency transmission, special facilities and/or conditioned (introduction of amplifiers and equalizer) local loop will be leased. Also, it should be regenerated every 1.6 to 4.8 km depending on speed of transmission and volume of data is to be transmitted. The quality of reception at the subscriber end is tested by a rating system standardized by CCITT.

2.2.5. Bridge Taps

Bridge taps are unused portions of the cable that are connected in shunt to a working cable pair, such as a local loop. Bridge taps are placed at any point along a cable length. Bridge taps can increase the flexibility of cable by making it is to reassign a cable to another subscriber without working in the field to cross connects sections of cable.

Bridge taps produce a loss known as *bridging loss*. Bridge taps permit signals to split and propagate down more than one wire. Signals which propagate through the open circuited (un-terminated) cables reflect back from the open end of the cable, causing interference with the original signal. Bridge taps which are short and to the close vicinity of the originating ends often produces the most interference.

Although, the bridge taps and loading coils are not generally harmful to voice transmissions but in case of data transmission, if they are not properly used, they can completely destroy the integrity of a data signal. Therefore, bridge taps and loading coils are not used in the data transmission channel. It is sometimes difficult to locate a bridge tap even though it is estimated that the average local loop can have as many as 16 bridge taps.

2.2.6. Loop Length

The method of defining subscriber loop length by using the signal resistance limit as a base is called the *basic resistance design*. The maximum subscriber loop length, which is defined as the distance from the subscriber to the central office, is expressed as

$$L = \frac{R_L}{R_{dc}} \quad \text{...(2.3)}$$

where, R_L = loop resistance limit

R_{dc} = dc loop resistance

The dc loop resistance of a local loop mainly depends on the type of wire and wire size.

Generally local loops use 18 to 26-gauge, twisted-pair copper wire. The higher the wire gauge, the smaller the diameter, the more resistance, and the higher the attenuation. The lower the wire gauge, the larger the diameter, the less resistance, and the lower the attenuation. For example, 26-gauge unloaded copper wire has an attenuation of 2.67 dB/mile, whereas the same length of 19-gauge copper wire has only1.12 dB/mile. Therefore, the maximum local loop's length using 19-gauge wire is twice as long as a local loop using 26-gauge copper wire.

The dc loop resistance is measured in ohms per kilometer and expressed as

$$R_{dc} = \frac{21.96}{d^2} \quad ...(2.4)$$

where '*d*' is the diameter of the conductor in millimeters.

Or

$$R_{dc} = \frac{0.1095}{d^2} \quad ...(2.5)$$

where the dc loop resistance is measured in ohms per mile and '*d*' is the diameter of the conductor in inches.

2.2.7. Maximum Permissible Loop Length

The method of determining the maximum subscriber loop length using the attenuation or loop loss is called the basic transmission design. The total attenuation of a local loop is normally limited to a maximum value of 7.5 dB with a maximum dc resistance of 1300 Ω, which includes the resistance of the telephone set, *i.e.*, approximately 120 Ω.

The dc resistance of 26-gauge copper wire is approximately 41 Ω per 1000 feet, which restricts the round-trip loop length to approximately 5.6 mile.

The maximum loop length is calculated from the formula

$$L_{max} = \text{Attenuation limit/ loss per km} \quad ...(2.6)$$

2.3. TELEPHONE MESSAGE—CHANNEL NOISE

Noise is an unwanted electrical energy. In any real physical systems, the signal arrived at receiver may be accompanied by an unknown waveform which varies with time in an entirely unpredictable manner. This unpredictable waveform in a random process is called **noise**.

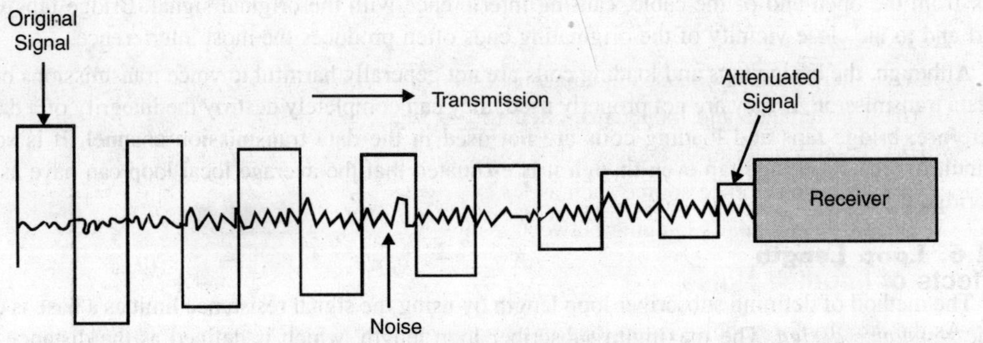

Fig. 2.4. Signal Plus Noise at Receiver.

At receiver whether the signal arrives over a communication channel or is received by an antenna, is accompanied by noise. At each stage of the switching system (local exchange, toll office, repeaters) additional noise added on the signal. Transmission lines have some amount of background noise generated by external sources. This noise combines with and distorts a transmitted signal. This noise generator may be fluorescent lights, motors, ovens, phones, copiers etc.

The common noises in telecommunication system are as follows:

1. White noise,
2. Impulse noise and
3. Inter modulation noise.

Figure 2.4 shows the signal plus noise at receiver.

2.3.1. White Noise

This is the most common noise in communication. This noise is easy to analyze and easy to find since it arises as thermal noise in all electrical components. The term white is used in analogy with white light, which is a superposition of all visible spectrum components. As the noise voltage and hence the noise is caused by the random movement of electrons in any of the active or passive device found in receivers, it is referred to variously as thermal, agitation, white or Johnson noise. White noise is truly random in the sense that a sample at any instant in time is completely uncorrelated to a sample taken at any other instant in time. Power spectral density of white noise is uniform over the entire frequency range of interest. Battery systems used to power subscriber loops are also a source of this type of noise.

The amplitude of the signal is kept sufficiently above the white noise to prevent an excess of hiss on radio or telephone circuits.

2.3.2. Impulse Noise and Intermodulation Noise

The most common form of noise in the telephone network is impulse noise and quantization noise. Impulse noise has peaks of amplitude that saturate channel and blot out data. Impulse noise is the main source of errors in data. This is so because, a 0.01 second of impulse may cause one bit lost for 75 bps speed transmission and 50 bits loss for 4800 bps speed of transmission.

Sources of Impulse Noise

The various sources of impulse noise generated internally and externally are been categorized in two categories which are as follows :

1. **Internal Sources :**
   (*i*) Poor relay contacts and jacks
   (*ii*) Poor quality soldering
   (*iii*) Non-soldered twisted joints
2. **External Sources :**
   (*i*) Sharp voltage changes in adjacent wires
   (*ii*) Capacitance and Inductance effects
   (*iii*) Electric interference causes by electric trains and electric machinery
   (*iv*) EM interference source such as RADAR
   (*v*) Open wire pairs hanging between telegraph poles can pickup atmospheric static

Effects of Impulse Noise

As the duration of impulse is too long in comparison with the speed of transmission, a sharp click like sound will be heard to the human listener. This noise removes two or more adjacent bits and thus the parity checking devices may not detect the error.

By quantization, we create a new signal, which is in approximation to the original signal.

If the actual signal is compared with quantized signal, some difference can be noticed. This error may be viewed as quantization noise or error and they are random in character. The quality of the approximation and reduction in quantization noise can be achieved by reducing the size of the steps, thereby increasing the number of allowable levels.

The signals from two independent channels intermodulate each other to form a product that falls in to a separate band of frequencies. This band may be reserved for another signal.

This distorts the signal receiver and referred as intermodulation noise. Also when a single frequency voice signal modulates voice signal in another channel, the voice might be clearly audible in a third channel. This problem can be reduced by proper design of modem.

2.3.3. Noise Weighting

The measurement of interference (noise), like the measurement of volume, is an effort to characterize a complex signal. Noise interference is comprised of two components which are the function of frequency, are as follows;

1. Annoyance

The noise that reaches a listener's ears affect the degree of annoyance to the receiver.

2. The Effect of Noise on Intelligibility

The power level of any disturbing signal, noise or interference is easily measured with a root-mean-square (rms) voltmeter. However, disturbances at some frequencies within the passband of a voice signal are subjectively more annoying than others. Thus, more useful measurements of noise or interference power in a speech network take into account the subjective effects of the noise as well as the power level.

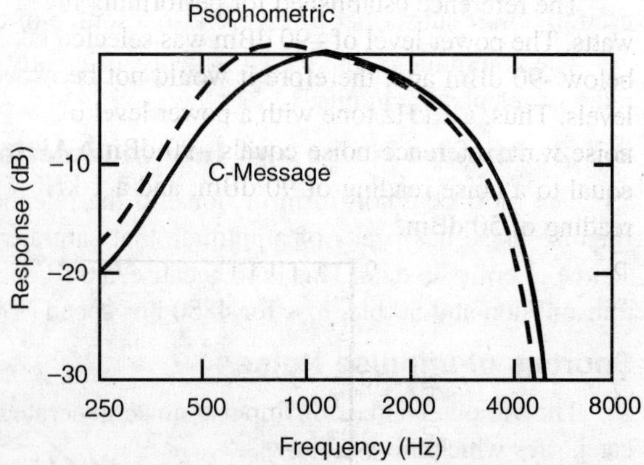

Fig. 2.5. C-Message and Psophometic Weighting,

The two most common such measurements in telephony use a *C-message weighting curve and a psophometric weighting curve*, as shown in Figure 2.5. These curves essentially represent filters that weight the frequency spectrum of noise according to its annoyance effect to a listener. C-message weighting represents the response of the 500-type telephone set. As far as perceived voice quality is concerned, only the noise that gets passed by the telephone set is important. Notice that disturbances between 1 and 2 kHz are most perceptible.

C-message weighting is used in North America while psophometric weighting is the European (ITU-T) standard.

Noise signals with equal interfering effect, the American Telephone and Telegraph Company (AT&T) introduced a weighting network referred as **C-message** weighting. The frequency response of neither a human ear nor a telephone receiver is flat and a telephone listener will respond differently to noise in different parts of the audio spectrum. Some of the noise that is present in bandwidth of voice channel will be unnoticed and the effective S/N ratio will be higher by a **weighting factor**. Its value depends on the frequency response of the telephone receiver and of the user ear. The Bell system uses C-message weighting.

When designing the C-message weighting network, team of observers were asked to adjust the loudness of 14 different frequencies between 180 Hz and 3500 Hz until the sound of each tone was judged to be equally annoying as a 1 kHz reference tone in the absence of speech. A 1 kHz is the best frequency response to human. The same team members were then asked to adjust the amplitude of the tones in the presence of speech until the effect of noise on articulation was equal to that of the 1 kHz reference tone. The result of the two observations were combined and plotted as shown in Fig. 2.6, and that is known as the **C-message weighting curve.**

The significance of the C-message weighting curve can be seen that a 200 Hz test tone of a given power is 25 dB less disturbing a 1 kHz test-tone of the same power, refer Fig. 2.5. Therefore, the C-message weighting network will produce 25 dB more loss for 200 Hz than it will for 1 kHz.

CCITT and common satellite practice use **psophometric weighting**. The symbol P is used for the psophemetric weighting factor and its value is 1.78, this corresponds to 2.5 dB.

In telephone engineering practice, signal power is expressed in terms of transmission level w.r.t. a reference point. The signal power at the reference point is indicated by the unit dBm0, where 0 represents zero transmission level point or the test point. Thus a –2 dBm0 signal means a signal producing an average power of –2 dBm at reference point.

The CCIR assumes that the long term average power carried by a single voice channel in a telephone system is taken to be –15 dBm0.

Telephone equipment is often adjusted by applying a 1 kHz tone at 0 dBm0 to the system to simulate peak power one channel; this is known as **Test tone.**

The reference established for performing message-channel noise measurements is –90 dBm *i.e.*, 10–12 watts. The power level of –90 dBm was selected because, at the time, power levels could not measure levels below -90 dBm and, therefore it would not be necessary to deal with negative values when reading noise levels. Thus, a 1 kHz tone with a power level of –90 dBm is equal to a noise reading of 0 dBrn (dB level of noise w.r.t. reference noise equals –90 dBm.) Alternatively, a 1 kHz tone with a power level of 0 dBm is equal to a noise reading of 90 dBm, and a 1 kHz tone with a power level of –40 dBm is equal to a noise reading of 50 dBm.

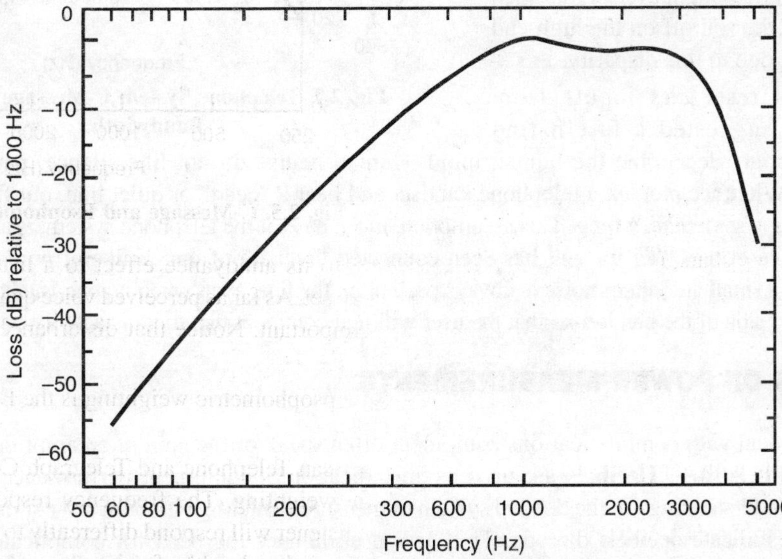

Fig. 2.6. C-Message Weighting Curve.

If the readings are made using C-message weighting, the power level is expressed by the abbreviation dBmC. Similarly, psophometrically weighted picowatts are expressed by the abbreviation pWp.

2.3.4. Bandwidth

Bandwidth limits the maximum frequency and switching speeds that can be applied to POTS lines. As we have already discussed the development of **C-message curve** in which numerous studies were conducted on a wide variety of people to determine what the POTS bandwidth should be. An example of these early studies was the one used to develop the **C-message** curve of Figure 2.7. People of varying ages, backgrounds, and sex were asked to listen to a set of tones and to place a mark on whether they:

(*a*) Just heard the sound.

(*b*) Heard it comfortably; or

(*c*) Heard it loudly.

The absence of a mark indicated that the tone was not heard at all. Each participant donned a set of headphones and was subjected to a number of tones at different frequencies. These frequencies were applied in random order and the results were compiled and graphed into the C-message curve that became the basis for the bandwidth specification that the telephone company adopted. Ideally, the voice bandwidth of the telephone system is 0 to 4 kHz, but a more practical range of 300 to 3 kHz is used in the design of telephone and data communications equipment. Another result of the test, which did not come as a great surprise to anyone, was that women hear higher tones that men do not. Most of the roll off on the high end of the band was due to this disparity.

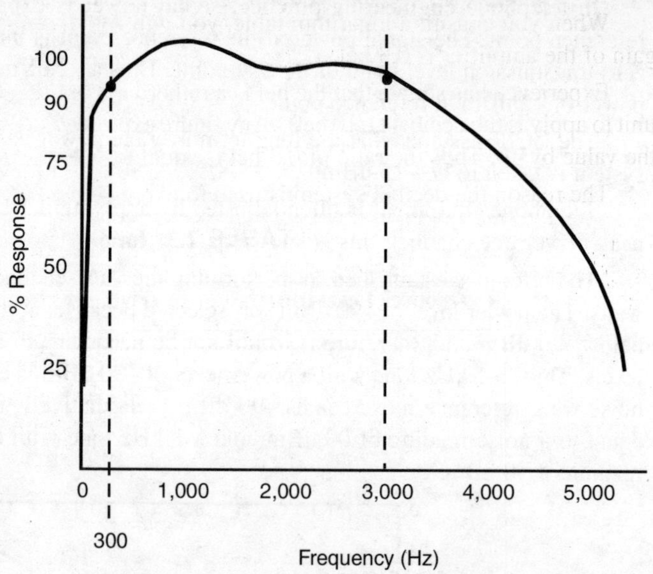

Fig. 2.7. Telephone System C-Message Response Bandwidth.

Additional tests and inputs from psychologists suggested a fascinating phenomenon about silence and the human mind. Human beings do not like silence, particularly on the telephone line. When we pick up a telephone handset and hear a "dead" or quiet line, our first impulse is to believe that there is something wrong. This assumption, in the case of the telephone system, is absolutely correct, even if the silence occurs after the call has been connected. Because of our dislike of pure silence, POTS is designed so that a small amount of noise is always present on the line, even when no one is talking. This noise is sensed in speaker side of the handset so that the user will feel confident that there is activity on the line.

2.4. UNITS OF POWER MEASUREMENTS

Users who deal with communications equipment often speak of the gain of an amplifier or a system in units called decibels (dB). **Decibels** are used as an indication of equipment performance; therefore, you need a basic understanding of the decibel system of measurement. Most modern test equipment is designed to measure and indicate decibels directly. This design eliminates the need for complicated mathematical calculations. Nevertheless, because many data link system alignment procedures center around dB readings and references, you need to understand the significance of an equipment gain rating as expressed in decibels. The equipment used in communications systems consists of several components, such as amplifiers, communications lines, antennas, couplers, and switches. Each component in the system will affect the signal by introducing a signal loss or gain.

These losses and gains can be described by a ratio of the power input and output by the equipment or cable.

The bel, named in honor of Alexander Graham Bell, expresses the logarithmic ratio between the input and output of any given component, circuit, or system. The **bel** is a unit that expresses the logarithmic ratio between the input and output of any given components, circuit or system and may be expressed in terms of voltage, current, sound levels, or power. The ratio can be calculated by using the following Power ratio formula :

$$N = \log_{10} \frac{P_1}{P_2} \text{ bel} \qquad \text{...(2.7)}$$

The gain of an amplifier can be expressed in bels (N) by dividing the output (P_1) by the input (P_2) and taking the base 10 logarithm ($\log_{10}$) of the resulting quotient. Thus, if an amplifier doubles the power, the quotient will be 2.

When you consult a logarithm table, you will find that the base 10 logarithm of 2 is 0.3; so the power gain of the amplifier is 0.3 bel.

Experience has shown that the bel is a rather large unit that is difficult to apply. A more practical, easier unit to apply is the decibel (1/10 bel). Any figure expressed in bels can be converted to decibels by multiplying the value by 10. Thus the ratio of 0.3 bel is equal to 3 decibels.

The reason the decibel system is used to express signal strength is shown in Table 2.2.

TABLE 2.2. Decibel Power Ratios.

Source Level (dB)		Power Ratio
1	=	1.3
3	=	2.0
5	=	3.2
6	=	4.0
7	=	5.0
10	=	$10 = 10^1$
20	=	$100 = 10^2$
30	=	$1.000 = 10^3$
40	=	$10,000 = 10^4$
50	=	$100,000 = 10^5$
60	=	$1,000,000 = 10^6$
70	=	$10,000,000 = 10^7$
100	=	10^{10}
110	=	10^{11}
140	=	10^{14}

For example, saying that a reference signal has increased 50 dB is much easier than saying that the output has increased 100,000 times. The basis of the decibel measuring system is the amount of increase or decrease from a reference level. Whether the input power is increased from 1 watt to 100 watts or from 1,000 watts to 100,000 watts, the amount of increase, or gain, is still 100 times or 20 dB.

Examine Table 2.2 again, taking particular note of the power ratios for source levels 3 dB and 6 dB. As the table illustrates, an increase of 3 dB represents a doubling of power. The reverse is also true. If a signal decreases by 3 dB, half of the power is lost. For example, a 1,000 watt signal decreased by 3 dB will equal 500 watts, while a 1,000 watt signal increased by 3 dB will equal 2,000 watts.

2.4.1. dBm

When we speak of the dB level of a signal, we are actually speaking of the logarithmic comparison between the input and output signals. The input signal is normally used as the reference signal. In some instances, a standard reference signal must be used in place of the input signal.

The most widely used reference level is a *1 milliwatt signal* (600-ohm load). When the 1-milliwatt reference is used, the standard decibel abbreviation of dB is changed to dBm; dBms are used as an indication of power, while dBs are used to indicate the ratio between the input and output.

A signal level of +3 dBm is 3 dB above 1 milliwatt, and a signal level of –3 dBm is 3 dB below 1 milliwatt. Whether you are using dB or dBm, a plus sign (+) or no sign indicates that the output level is greater than the reference (power gain), while a minus sign (–) indicates that the power level is less than the reference (power loss). The value 0 dBm indicates that the output power is equal to the 1-milliwatt reference. It is also used to express a definite amount of power (1 milliwatt). The value 0 dBm equates to 1 milliwatt.

Units common for signal and noise power measurements in the telephone industry

These terms are the units of measurements. These units common for signal and noise power measurements in the telephone industry.

1. **dBmO :** dBmO is dBm referenced to a zero transmission level point (0 TLP). dBmO is a power measurement adjusted to 0 dBm that indicates what the power would be if the signal were measured at a 0 TLP.
2. **rn (Reference Noise) :** rn is the dB value used as the reference for noise readings. Reference noise equals – 90 dBm or 1 pW (1×10^{-12} W).
3. **dBrn :** dBrn is the dB level of noise with respect to reference noise (–90dBm) dBrn is seldom used by itself since it does not specify a weighting.
4. **dBrnc :** dBrnc is similar to dBrn except dBrnc is the dB value of noise with respect to reference noise using C-message weighting. Noise measurements obtained with with a C-message filter are meaningful, as they relate the noise measured to the combined frequency response of a standard telephone and the human ear.
5. **dBrn-3-kHz Flat :** dBrn-3-kHz Flat noise measurements are noise readings taken with a filter that has a flat frequency response from 30 Hz to 3 kHz.
6. **dBrncO :** dBrncO is the amount of noise in dbrnc corrected to a 0 TLP. A noise reading of 34 dBrnc at a +7 dBm TLP equates to 27 dBrncO. dBrncO relates noise power readings (dBrnc) to a 0 TLP. This unit establishes a common reference point throughout the transmission system.

2.4.2. Line Impedance

Another important specification is the line impedance used as a standard for the telephone lines and equipment. POTS specify the impedance to be 600 ohms at both sending and terminating ends. Some telephone lines have been in existence since the telephone company first began. Others are recent additions and some are no longer twisted pair copper, but instead are now fiber optic lines. Most telephone switching equipment has been converted to solid-state devices and integrated circuits, replacing relays and older vacuum tubes. The end effect is that line and equipment impedances can fluctuate from lows of 100 ohms to highs above 5,000 ohms.

2.4.3. Types of Line Service

Complications concerning impedance arise as a result of the routing calls take to be completed. Physical routes are established at the time a call is initiated. This type of routing is known as connectionless service, which is performed through the telephone company's **direct distance dialing (DDD)** network. The advantage of this service is that lines are free when not in use. In contrast, a connection-oriented service is one in which a permanent pathway is established between two communicating stations. These connections are established through **leased** or **dedicated lines.** Anytime communication needs to occur, the users simply begin on this permanent connection. The downside of connection-oriented service is that resources are permanently allocated even when they are not used.

One problem with connectionless service is that each time a call is made; a different route is established to complete the call. This results in varying specifications on the connection each time it is established. You may have experienced this change by sensing a change in the quality of a voice connection. There have been times when you cannot hear a caller well and suggest that they hang up and redial the call. Immediately, the call comes through and is ten times better. Consider the implication of this when trying to send digital data over these same lines.

2.5. TRANSMISSION PARAMETERS AND PRIVATE-LINE CIRCUITS

Transmission parameters are refer to the dedicated or leased line private-line data circuits that's employed in the private sector of the public telephone network that do not utilize the public switched telephone network. Private-line data circuits have several advantages over using the switched public telephone network:

1. Transmission characteristics are more consistent because the same facilities are employed with every transmission.
2. The facilities are less prone to noise produced in telephone company switches.
3. Most private line data circuits use four wire, full duplex facilities. Therefore higher transmission bit rate and better performance is appreciated with private-line data circuit.
4. Line conditioning is available only to the private-line facilities.
5. Private-line data circuits are more economical for high volume of circuits.

Transmission parameters are divided into three broad categories :

1. **Bandwidth parameters**, which include envelope delay distortion and attenuation distortion.
2. **Interface parameters**, which include terminal impedance, in-band and out-of-band signal power and ground isolation.
3. **Facility parameters**, which include noise measurements, frequency distortion, amplitude distortion, phase distortion and non-linear distortion.

2.5.1. Line Conditioning

To reduce attenuation, envelope delay distortion, and improve the performance of data modem operating over standard message channel, the process is used to improve a basic telephone channel is referred as **line-conditioning**. Line conditioning is available only to private- line subscribers at an additional charge. Dedicated or leased lines, because of their permanent status, can be conditioned to tighten telephone company parameters so that they can transfer digital data more dependably.

Line conditioning improves the high frequency response of a message channel and reduces the power loss. The attenuation and delay characteristics of circuits are artificially altered to meet prescribed by the line conditioning requirements.

The parameters that can be affected by conditioning are :

1. Gain or attenuation fluctuations,
2. Propagation delay variables,
3. Signal-to-noise ratios and
4. Harmonic distortion.

Initial specifications for these parameters are listed by the telephone company as **basic conditioning.** These specifications are what the telephone company has designed their lines and equipment to satisfy. Changes in a signal's strength are dependent on many factors, not the least of which is the frequency of the signal. The reactance of the cables and equipments changes with respect to the signal's frequency, causing the amplitude of the signal to vary with respect to its frequency.

Gain or attenuation variations throughout the bandwidth of the telephone system are measured in reference to power of a standard signal frequency of 1004 Hz applied to the telephone lines. The actual power received is measured and given a reference value of 0 dB. **Decibel (dB)** is a measure of a power ratio between one power level and another used as a reference level. When the reference for the ratio is 1 mW, then the level of the power is in **dBm.** When the reference is any other power level then the difference between that level and a newer one is in dBs. A change of +3 dB is equivalent to doubling the power while a change of –3 dB represents a drop to half the power of the reference level. The reference level is usually designated 0 dB since there is no change between the reference and another signal measured having the same power level.

When the telephone company set up their specification for gain fluctuations across the bandwidth, they started with a 1,004 Hz signal generated by a sending station. The power at the receiving station was measured and that level designated as 0 dB. The frequency at the transmitter was altered in steps, holding the sending output level constant.

Received power was measured and compared to the 1,004-Hz power reference. From these measurements, a set of standards for basic line conditioning were formulated.

An example of results that could appear is shown in Figure 2.7. Here a reference of 0 dB is established at a power level of 12 dBm, or 12 dB above the 1 mW power level. As the frequency of the signal is altered, the power fluctuates. In the graph shown, it rises to a maximum level of 15 dBm, which is 3 dB above the 0-dB reference, and attenuates to –12 dB below the reference. These measurements can show how well the system performs over the range of the 300–3 kHz telephone bandwidth.

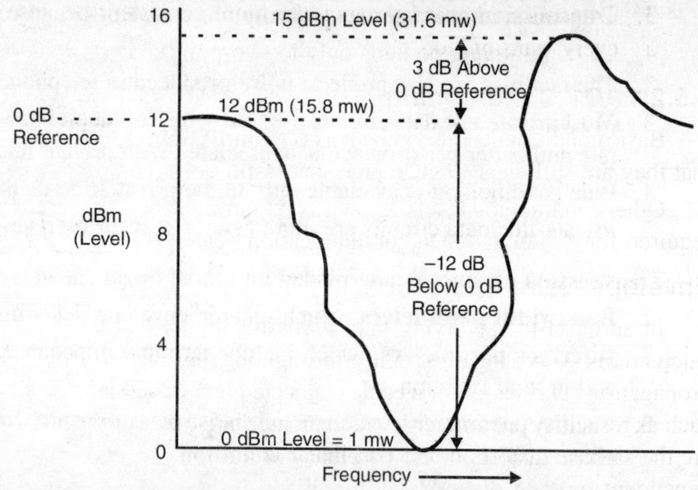

Fig. 2.8. Relationship of Level (dBm) and Gain/ Attenuation (dB).

A graph for the specifications for gain/attenuation variation for basic conditioning is shown in Fig. 2.7. A typical response curve for a line is included for purposes of illustration.

On the graph the 0 dB reference point is at 1,004 Hz. Note that the sample response curve always goes through this point. The straight lines on the graph show the minimum and maximum changes in power allowed for basic conditioning. For the entire bandwidth, the signal's level must remain within +3 dB and –12 dB of the reference level at 1004 Hz. Additionally, within the frequency range of 500 to 2,500 Hz, the variation limits are reduced to +2 dB and –8 dB. What this translates to is that for equipment and lines within the telephone system to operate satisfactorily, signals traveling on that system will not experience changes in their amplitudes beyond the limits specified with reference to levels at 1004 Hz.

2.5.2. C-type Line Conditioning

C-type line conditioning relates to line impairments for which compensation can be made with equalizers and filters. When a circuit is initially turned for service with a specific C-type conditioning that must meet the requirements for the type of conditioning.

There are five types or levels of C-type conditioning available. The levels of conditioning a subscriber select depends on the bit-rate, modulation technique and desired performance of the data modems used on the line. Types of C-type conditioning are as follows:

1. C1 and C2 conditioning pertain to two-point and multipoint circuits.
2. C3 conditioning is for access lines and trunk circuits associated private switched network.

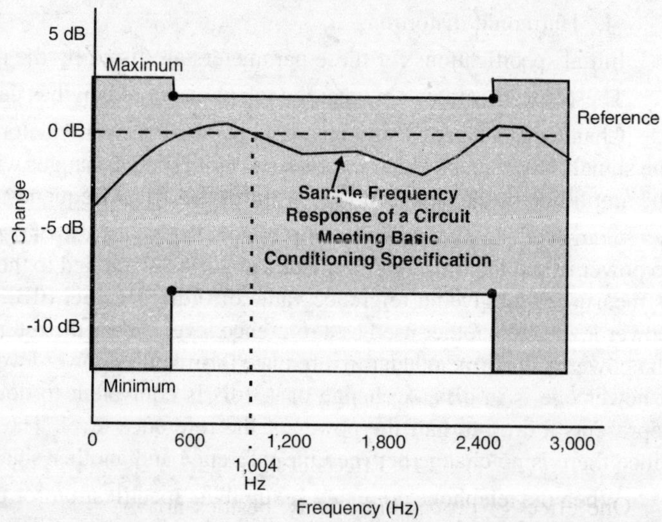

Fig. 2.9. Basic Conditioning Limits Graph.

3. C4 conditioning is belongs to two-point and multipoint circuits with a maximum of four stations.
4. C5 conditioning pertains only to two-point circuits.

2.5.2.1. C1 Conditioning

Basic conditioning, also known as **C1 conditioning**, is for voice-grade quality communications meaning that they are sufficient to guarantee successful voice communications.

Other conditioning levels, known as C1 through C5, offer more stringent guidelines for the gain variations required for digital and data communication networks.

Envelope Delay Distortion

In addition to specifying gain variation limits, C conditioning also details requirements concerning fluctuations in propagation delay throughout the 300 to 3 kHz bandwidth. To grasp the implication of propagation delay, consider a cable connected between two points. The cable has specific DC characteristics, such as resistance, and specific AC characteristics, such as impedance. The DC characteristics are those that are the same regardless of the frequency of the signal applied to the line. In contrast, AC characteristics are dependent on, and distinctly different for each signal frequency value. Line impedance is the effective resistance and capacitance or inductive reactance load presented to the transmitter by the cable in response to the signal frequency. These effects are created by the inherent reactive quality of the cable created by its physical makeup. One result of this reactive quality is the changes in gain or attenuation just described. A second effect involves the time it takes for a signal to propagate down a line.

Figure 2.10 shows a graph comparing the time of arrival of three signals denoted f_1, f_2, and f_3. The three signals, each at a different frequency, leave point A at the same time, t_0. All three propagate through the same medium, a pair of telephone wires for instance.

Because of the changes in reactance each signal feels, they arrive at point B at different times. The graph shows that when signal f_2 arrives at point B, signals f_1 and f_3 still have a way to go. It is interesting to note that the longer propagation delays may occur at both higher and lower frequencies from the signal with the shortest propagation time.

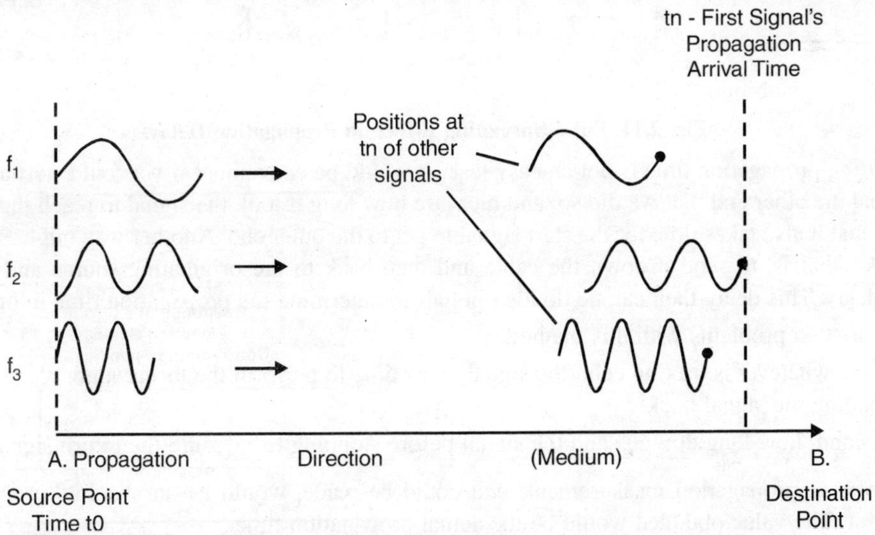

Fig. 2.10. Effects of Propagation Delay on Three Different Delay Signals.

One effect of propagation delay, besides arrival time, is pulse spreading. Consider the square wave signal shown in Figure 2.10 possibly created by sending a number of alternating 1s and 0s using non-return-to-zero (NRZ) format. A square wave is composed of a fundamental sine wave whose frequency is one-half that of the square wave, and numerous higher odd harmonics with frequencies that are an odd multiple of the fundamental.

As this square wave propagates down the line, those sine waves are delayed at different rates. The ones that experience shorter delays arrive at the receiving end first, while the ones having longer delays arrive measurably later. The effect is to spread the shape of the square wave so that it becomes difficult to decipher the logic 1 and 0 levels. Another effect of spreading is to reduce the additive effects of the sine waves, causing the amplitude to reduce. If the spreading is severe enough, part of the wave shape can become totally lost, which, for digital data, can result in the loss of one or more bits.

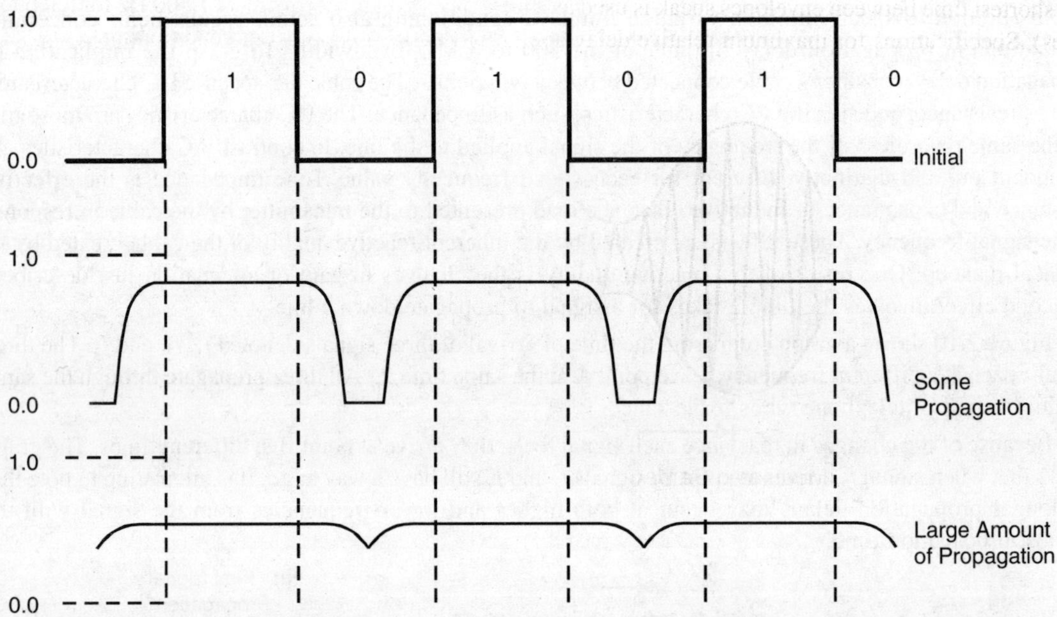

Fig. 2.11. Pulse Spreading Effect on Propagation Delay.

Measuring propagation time is not an easy task. It would be convenient if we could start a signal down a line, signal the other end that we did so, and measure how long it took the signal to reach the far end. The problem is that it also takes time for the start signal to get to the other end. Another way might be to echo the signal back—that is, to send it down the cable and then back to the originating source and measure the roundtrip delay. This delay then can be divided in half to determine the propagation time in one direction.

There are two problems with this method:

1. First, whatever is used to echo the signal takes time to perform the turnaround, receiving and then sending the signal back.
2. Second, how long do you send the signal before stopping to measure the return signal?

This type of propagation measurement, if it could be made, would be an absolute propagation time measurement. The value obtained would be the actual propagation time.

Fortunately, absolute propagation time is not a crucial quantity. If every signal arrived with the same propagation time, the overall effect would be a uniform delay in the communication.

The true concern is the relative delay between signals that could cause pulse spreading to occur. We need to measure and quantify this relative delay.

The method used to measure propagation time is called **envelope delay distortion,** which is a clever process used to make this measurement. The signals that are to be measured are used as carrier signals that are modulated with a low sine wave signal as shown in Figure 2.12 (*a*). The frequency and amplitude of the modulating sine wave is held constant for each carrier signal sent. This modulating signal is known as the *envelope* that surrounds the carrier frequency. As each carrier signal that is modulated and sent is received, the time between the peaks (T_{pk}) of the envelope is measured (Figure 2.12 (*a*)).

After all the carrier signals have been sent and measured, the T_{pk} times of each are compared to each other with the resulting difference (T_d) becoming a measure of the relative difference in the propagation time between any two signals (Figure 2.12 (*b*)). For basic and C conditioning, the carrier frequency that results in the shortest time between envelopes speak is used as a reference (usually occurring at 1,800 Hz for telephone lines). Specifications for maximum relative delay are then listed for each level of conditioning.

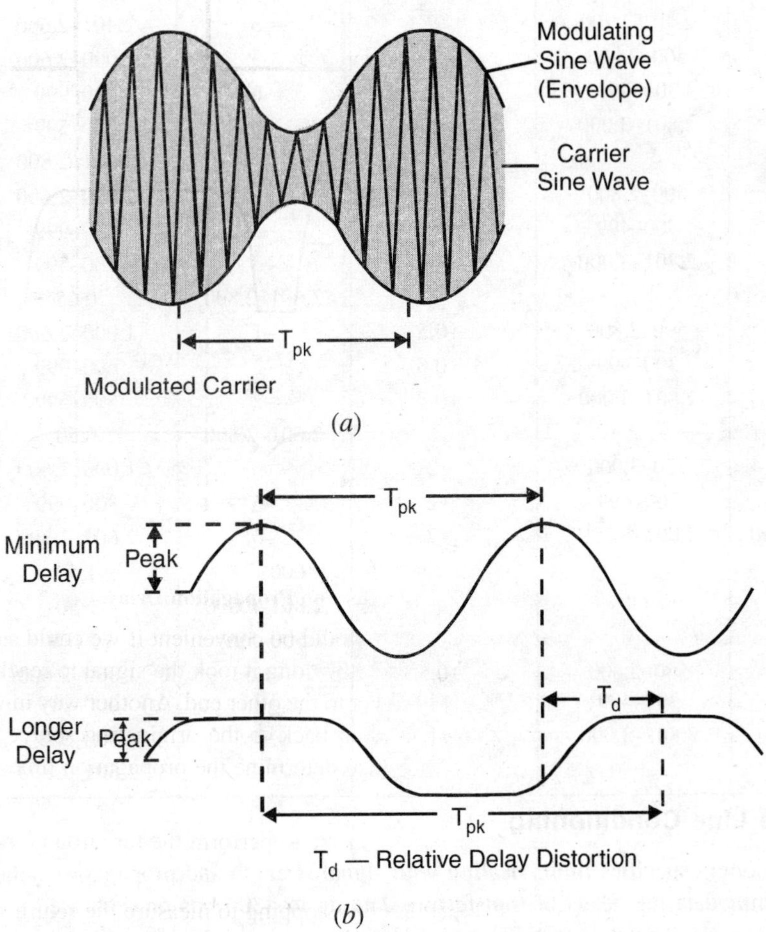

Comparison of Envelope Signals Affected By Delay of Carrier Signals

Fig. 2.12 (*a*) and (*b*). Propagation Delay Distortion.

Table 2.3 summarizes the specifications for basic and C-type line conditioning for gains and limits of envelope delay distortion.

TABLE 2.3. Telephone Line Conditioning Parameters

Condition Level	Gain Limitations			Propagation Delay	
	Frequency (Hz)	Limits (dB) Max.	Min.	Frequency (Hz)	Limit (ms) Max
Basic	500–2,500	+2	–8	800–2,600	1.75
Voice	300–499	+3	–12		
Grade	2,501-3,000	+3	–12		
C1	1,000–2,400	+1	–1	1,000–2,400	1.00
Station to One	300–999	+2	–6	800–999	1.75
or More Stations	2,401-2,700	+2	–6	2,401–2,600	1.75
C2	500–2,800	+3	–3	1,000–2,600	+0.50
Same as C1	300–499	+2	–6	600–999	1.50
	2,801–3,000	+2	–6	500–599	3.00
				2,601–2,800	3.00
C3	500–2,800	+0.5	–1.5	1,000–2,600	0.11
Local	300–499	+0.8	–3	600–999	0.30
Private Line	2,801–3,000	+0.8	–3	500–599	0.65
			2,601–2,800	0.65	
C3	500–2,800	+0.5	–1	1,000–2,600	0.08
Trunk	300–499	+0.8	–2	600–999	0.26
Private Line	2,801–3,000	+0.8	–2	500–599	0.50
			2,601–2,800	0.50	
C4	500–3,000	+2	–3	1,000–2,600	0.30
Maximum of	300–499	+2	–6	800–999	0.50
Four interconnected	3,001–3,200	+2	–6	2,601–2,800	0.50
Stations			600–799	1.50	
			2,801–3,000	1.50	
			500–599	3.00	
C5	500–2,800	+0.5	–1.5	1,000–2,600	0.10
Station-to-	300–499	+1	–3	600–999	0.30
Station Only	2,801–3,000	+1	–3	500–599	0.60
				2,601–2,800	0.60

2.5.2.2. D-type Line Conditioning

C-type conditioning specifies limits dealing with signal strength and propagation delay, which can be tightened to accommodate the needs of transferring data up to 4800 bps on dedicated telephone circuits. Additional factors concerning signal-to-noise ratio and harmonic distortion affect the manner in which higher data rates are transmitted using the telephone system. The symbols used to transport data at these rates are produced by modulating the phase and amplitude of a constant frequency analog signal. Since amplitude modulation is part of the process, the resulting signals are affected by a system's signal-to-noise ratio.

Detecting changes in the phase of these signals can be corrupted by the existence of harmonics of the original signal since these harmonics would appear to the receiver as changes in the signal's phase as well as in its frequency.

D-type conditioning neither reduces the noise on a circuit nor improves the SNR and nonlinear distortion. This conditioning is simply a requirement and does not add anything to the circuits, and it cannot improve a circuits. D-type conditioning simply places higher requirements on circuits used for high speed data transmission. This conditioning sometimes referred to as high performance conditioning and can be applied to private line data circuits in addition to either basic or C-conditioned requirements. D-type conditioning is mandatory when the data transmission rate is 9600 bps because without D-conditioning, it is highly unlikely that the circuit meet the minimum performance requirements specified by the telephone company.

There are of two types of D-conditioning which are as follows :

1. D1-type conditioning: It specifies requirement for two point circuits.
2. D2-type conditioning: It specifies requirement for multi point circuits

D conditioning addresses these concerns by tightening the telephone company's basic specifications for both signal-to-noise and harmonic distortion. TELCO specifies the acceptable signal-to-noise ratio (SNR) of their system to be 24 dB. That is the noise level is 24 dB below the desired signal level. D conditioning improves this specification by 4 dB to 28 dB.

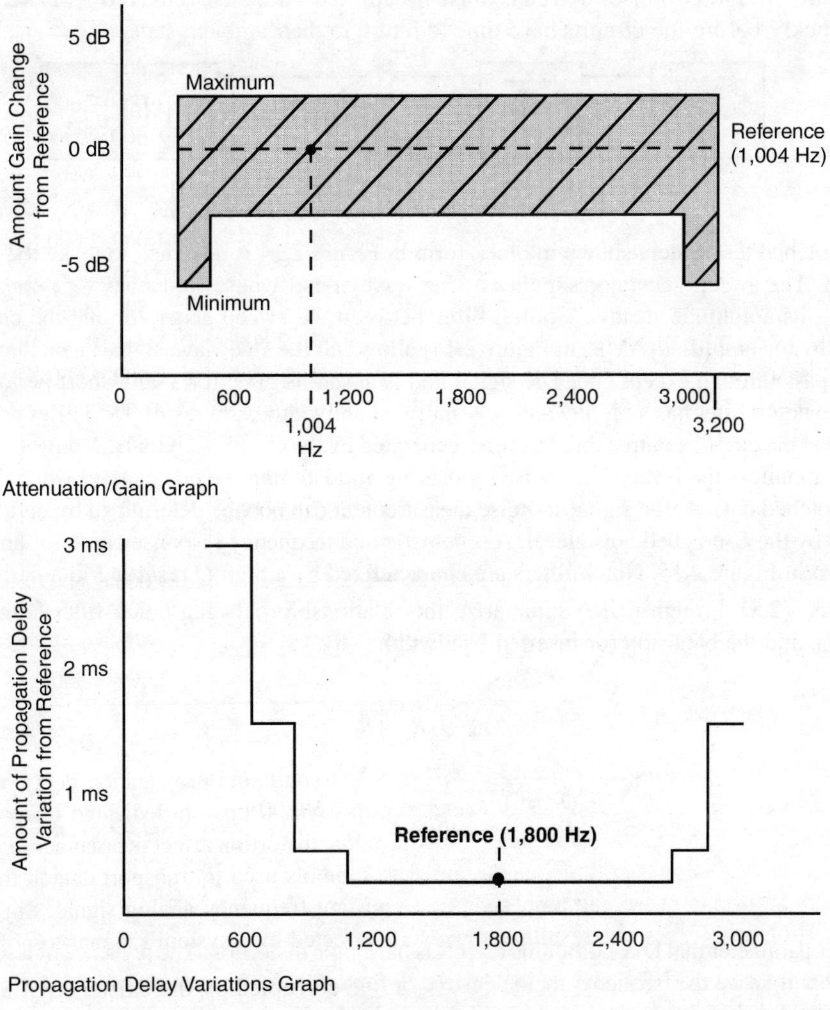

Fig. 2.13. C4 Conditioning Graphs.

SNR measurements can be done in one of two ways :

1. **Unweighted**, this is a measure of the existing noise on a line without any active signal on it, or weighted, which measures noise with a signal present. Unweighted noise is known as **inherent** or **white noise** and is the noise generated by circuits "at rest". A signal noise ratio figure can be determined by dividing the intended signal level by the unweighted noise level. It becomes apparent that this measure is not entirely accurate. The SNR value can be artificially inflated by applying enough signal power to present a high number. However, what is not taken into account is that when the signal is applied and the circuits become active, they generate additional noise because of the increased activity. When a circuit is fed a signal, the semiconductor and resistive devices, which are part of the circuit, become agitated.
2. **Weighted noise**: There is more electron activity within these devices because of the energy brought by the applied signal. As a result, these semiconductors and resisters generate additional noise above the inherent or white noise. Remove the signal and after a short period, these devices settle back to their ambient, or at-rest, state and the noise is reduced. The additional noise generated by applying a signal to a circuit or system is called **weighted noise.**

To measure weighted noise, a signal must be applied and then removed and the existing noise measured quickly before the circuits have time to return to their ambient state.

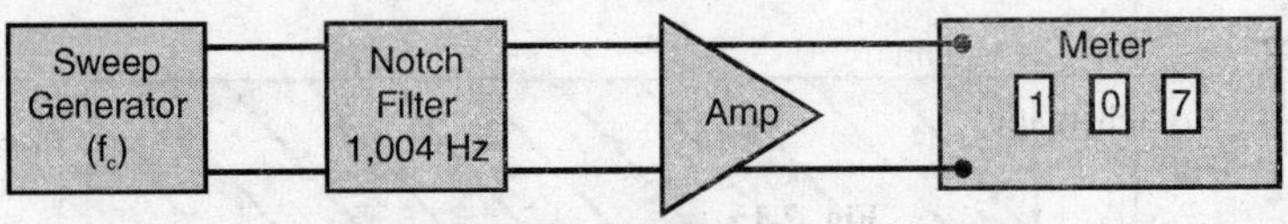

Fig. 2.14. Weighted Noise Measurement.

The C-notched test system shown in block form in Figure 2.14 is designed to make the weighted noise measurement. The sweep generator supplies a sine wave signal whose frequency (f_c) constantly changes while holding its amplitude steady. A **notch filter** between the sweep generator and the circuit under test (represented by the amplifier (AMP) in Figure 2.14) allows all the sine wave signals generated by the sweep generator to pass through except one. The signal that is absent is gone for a very short period of time. This is because the notch filter has a high Q and very narrow bandwidth. As a result, for a brief period, the signal is not present at the circuit's output, but the noise generated by the circuit activity is. A power meter connected to the output monitors the level of the noise by making and latching the power level when the signal is not present (or notched out). A true signal-to-noise measurement can now be determined by dividing the applied signal power by the C-notched noise level. A schematic and frequency response curve for an active notched filter is shown in Figure 2.15. Notch filters are characterized by a high Q resulting in a narrow bandstop.

Equations (2.8) through (2.10) summarize the relationship between notch filter components, notch frequency (f_n), and the bandstop (or inverted bandwidth, $-BW$).

$$f_n - \frac{1}{2\pi C\sqrt{(R_1)(R_2)}} \qquad ...(2.8)$$

$$Q = 0.5\sqrt{\frac{R_2}{R_1}} \qquad ...(2.9)$$

$$BW = \frac{fn}{Q} \qquad ...(2.10)$$

The other parameter that D conditioning affects is harmonic distortion. The presence of a second harmonic signal (one that is twice the frequency of the desired or fundamental frequency) must not exceed a level of –35 dB below the level of the fundamental signal to meet D-conditioning. Third harmonic signals (those three times the frequency of the fundamental signal) are required to be an additional –5 dB lower, for a total of –40 dB below the desired signal level.

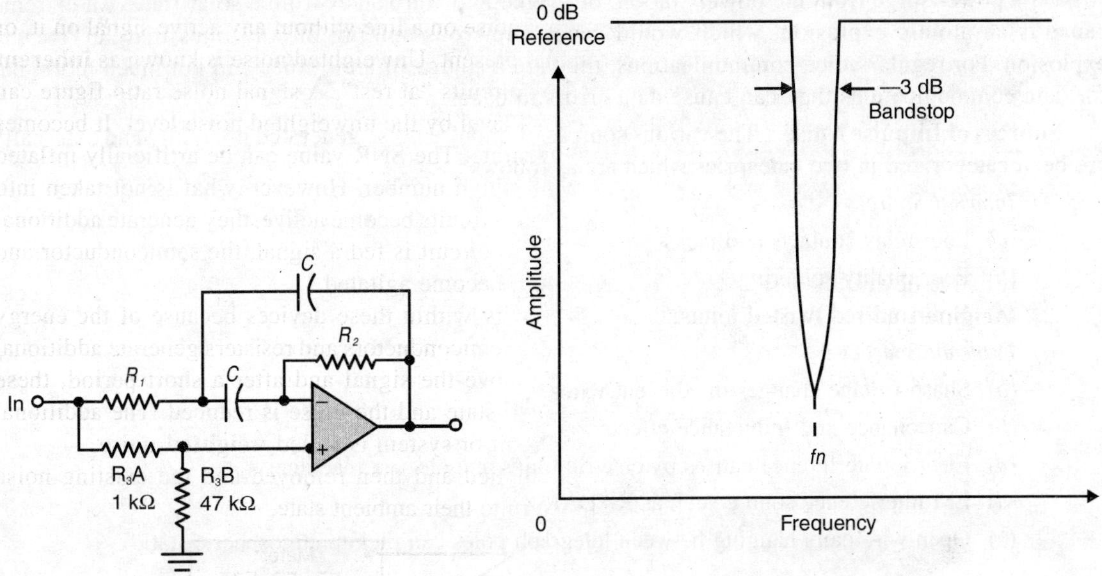

(*a*) Schematic (*b*) Frequency Response

Fig. 2.15. Active Notch Filter.

2.5.3. Facility Parameters

There are some potential impairment caused by telephone company equipment and the limits specified relate to all private-line data circuits using voice-band facilities regardless of line conditioning. Facility parameters represent potential impairments to a data signal and it includes:

1. Impulse noise,
2. Gain hits and dropouts,
3. Phase hits,
4. Jitter,
5. 1004-Hz variation,
6. C-message noise,
7. Frequency shift,
8. Phase intercept distortion and
9. Peak-to-average ratio.

1. Impulse Noise and Intermodulation Noise

The most common form of noise in the telephone network is impulse noise and quantization noise. Impulse noise has peaks of amplitude that saturate channel and blot out data. Impulse noise is the main source of errors in data. This is so because, a 0.01 second of impulse may cause one bit lost for 75 bps speed transmission and 50 bits loss for 4800 bps speed of transmission.

Once a line is conditioned, normal signal variations due to telephone equipment is minimized. However, sudden and unexpected changes in signal strength, caused by external factors, can result in creating numerous data errors. These changes, which are random and usually last for short periods of time, are termed **impulse impairments** because of their sudden and short (impulse) disturbance of line activity (impairment). Examples of impairment causes are induced currents into the lines from lightening strikes, a power surge from the power station, or magnetic disturbances from a solar flare. An extreme cause is an atomic explosion, which would cause numerous spiking on any lines that survive the explosion. For regular voice communications, they are a source of annoyance and not much more, but for data communications they can cause data errors to occur.

Sources of Impulse Noise : The various sources of impulse noise generated internally and externally are been categorized in two categories which are as follows :

(i) Internal Sources :

(*a*) Poor relay contacts and jacks

(*b*) Poor quality soldering

(*c*) Non-soldered twisted joints

(ii) External Sources :

(*a*) Sharp voltage changes in adjacent wires

(*b*) Capacitance and Inductance effects

(*c*) Electric interference causes by electric trains and electric machinery

(*d*) EM interference source such as RADAR

(*e*) Open wire pairs hanging between telegraph poles can pickup atmospheric static

Effects of impulse noise : As the duration of impulse is too long in comparison with the speed of transmission, a sharp click like sound will be heard to the human listener. This noise removes two or more adjacent bits and thus the parity checking devices may not detect the error.

By quantization, we create a new signal, which is in approximation to the original signal.

If the actual signal is compared with quantized signal, some difference can be noticed. This error may be viewed as quantization noise or error and they are random in character. The quality of the approximation and reduction in quantization noise can be achieved by reducing the size of the steps, thereby increasing the number of allowable levels.

The signals from two independent channels intermodulate each other to form a product that falls in to a separate band of frequencies. This band may be reserved for another signal.

This distorts the signal receiver and referred as intermodulation noise. Also when a single frequency voice signal modulates voice signal in another channel, the voice might be clearly audible in a third channel. This problem can be reduced by proper design of modem.

2. Impulse Gain Hit

One type of impulse impairment is an impulse gain variation known as an impulse gain hit, which the telephone company specifies as a change in signal gain of no more than ±3 dB, lasting no more than 4 ms as illustrated in Figure 2.16. These variations are in addition to the gain variations specified in the conditioning tables. The telephone company considers changes within the 4-ms time slot to be random and not in violation of the conditioning limits.

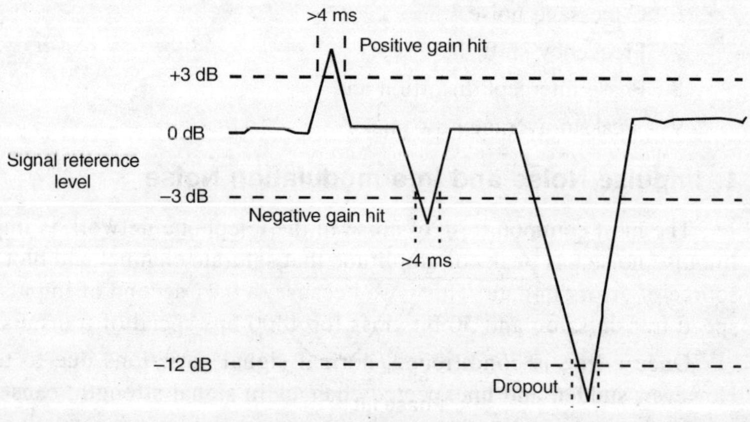

Fig. 2.16. Impulse Gain Hits

A dropout is a decrease in circuit gain or signal level of more than 12 dB lasting more than 4 ms. Dropouts are generally caused by deep fades on radio facilities or by switching delays, and it is characteristics of temporary open-circuit conditions.

3. Impulse Phase Hit

Electrical disturbances, such as those from electrical storms, also can cause changes in the phase of a signal for a short duration. Recall that a symbol is a single electrical parameter that represents a group of binary bits and that variation in symbols can be based on amplitude, frequency, or the phase of a signal. On a system that bases the formation of symbols on the signal's phase, a sudden change in the phase, known as an impulse phase hit, which lasts for a short period, can result in data errors. To minimize this effect, system designers must adhere to the telephone company specification for impulse phase changes.

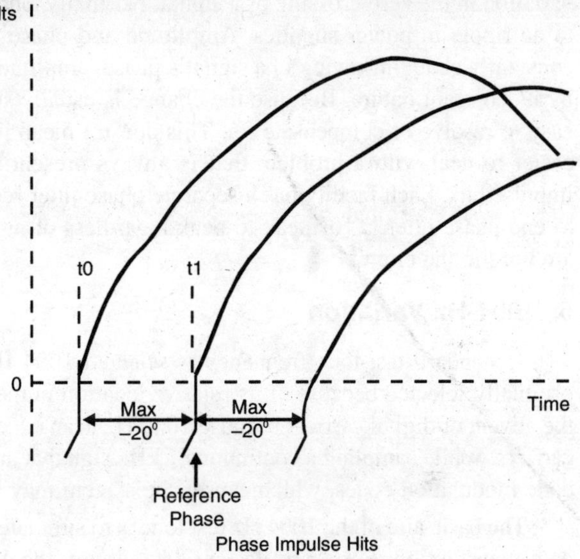

Fig. 2.17. Impulse Phase Hits.

These specifications limit impulse phase shifts to 20° in phase (positive or negative, as shown in Figure 2.17) as permissible on telephone lines. Again, this is for a short period of time, but it is a concern for reliable communications that depend on phase differences between signals to supply digital information.

4. Jitter

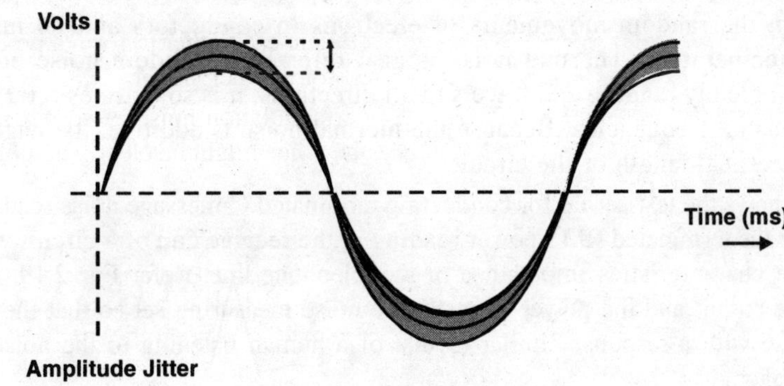

(*a*) Amplitude Jitter

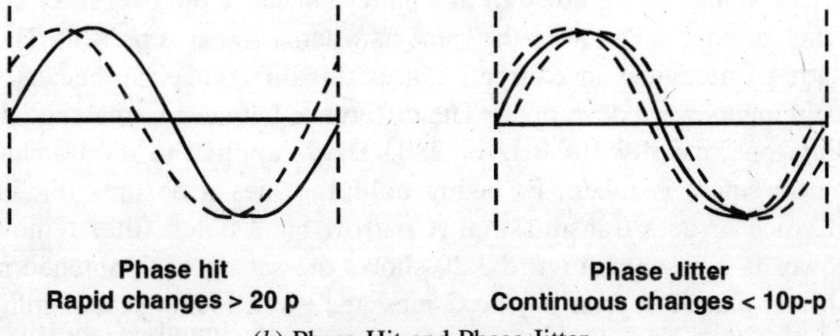

(*b*) Phase Hit and Phase Jitter

Fig. 2.18.

Another impairment to consider has direct impact on systems that use the phase or amplitude of signals to carry data symbols. Phase jitter is a form of incidental phase modulation-a continuous, uncontrolled variation in the zero crossing of a signal. Normally, phase jitter occurs at a 300 Hz or lower and mainly due to ac ripple in power supplies. Amplitude and phase **jitter** shown in Figure 2.18 (*a*) and (*b*) is a small, constantly changing swing in a signal's phase, amplitude, or both. It differs from impulse-type impairments by its constant nature. Because the change is usually small in size and predictable, this type of problem is easy to resolve or compensate for. This doesn't mean jitter can be ignored—it must be dealt with. It is just easier to deal with a problem that is always present rather than one that only occurs occasionally, like impulse hits. Each facility has a separate phase jitter requirements; however, the maximum acceptable end-to-end phase jitter is 10° peak-to-peak regardless of numbers of transmission facilities or telephone offices are used in the circuit.

5. 1004-Hz Variation

A standard test-tone frequency is selected 1004 Hz by the telephone company. Although 1 kHz was originally selected because of its relative location in the pass band of a standard voice band circuit but with the advent of digital carriers because 1000 Hz is an exact submultiples of the 8 kHz sample rate used with T-carriers. While sampling a continuous 1 kHz signal at an 8 kHz rate produced repetitive patterns in the Pulse code modulation codes, which causes the system may lose frame synchronization.

The basic aim of the 1004 Hz test tone is to simulate the combined signals power of a standard voiceband data transmission. For a private-line data circuit, the 1004 Hz channel loss is typically 16 dB. Long term variations in the gain of the transmission facility are referred as 1004 Hz variation and it should not exceed ± 4 dB. So, the received signal power should be within the specified limits of –12 dBm to –20 dBm (–16 dB + 4 dB = –12 dBm and –16 dB – 4 dB = –20 dB).

6. C-Message Noise

C-Message noise measurements estimate the average weighted RMS noise power. Thermal noise is produced from the random movements of electrons in conductors and its magnitude is directly proportional to temperature. Thermal noise is also referred as random noise because the electrons movements is completely random and travels in all directions, it is sometimes referred to as white noise because it contains all frequencies. Because the thermal noise is additive, its magnitude is dependent, in part, on the electrical length of the circuit.

Figure 2.19 shows the test-set-up for conducting terminated C-message noise reading. C-message noise measurements are the terminated RMS power reading at the receive end of a circuit with the transmit end terminated in the characteristics impedance of the telephone line. Refer Fig. 2.19, a C-message filter is place between the circuit and the power meter in the noise measuring set so that the noise measurements evaluator the noise with a response similar to that of a human listening to the noise through a standard telephone set speaker.

There is a problem to measuring noise in this pattern because the overall circuit characteristics, in the absence of a signal, are not necessarily the same as when a signal is present. The use of companding process and automatic gain devices in a circuit causes this difference. So because of this problem, C-notched noise measurements were developed. The difference between C-message noise measuring only in fact that a holding tone, usually 1004 Hz or 2804 Hz, is applied to the transmit end of the circuit while the noise measurement is taken. By using holding tones it ensures that the circuit operation simulates a loaded voice or data transmission. A narrow band notch filter removes the holding tone before the noise power is measured. Figure 2.20 shows the set up of C-notched noise measurements, where the notch filter is placed in front of the C-message filter, locking the holding tone from reaching to the power meter.

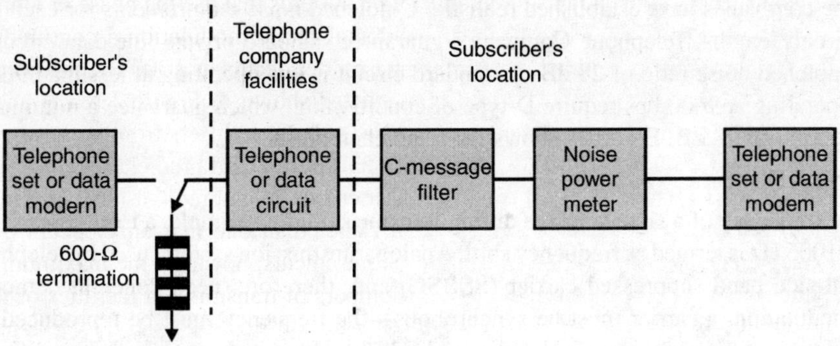

Fig. 2.19.

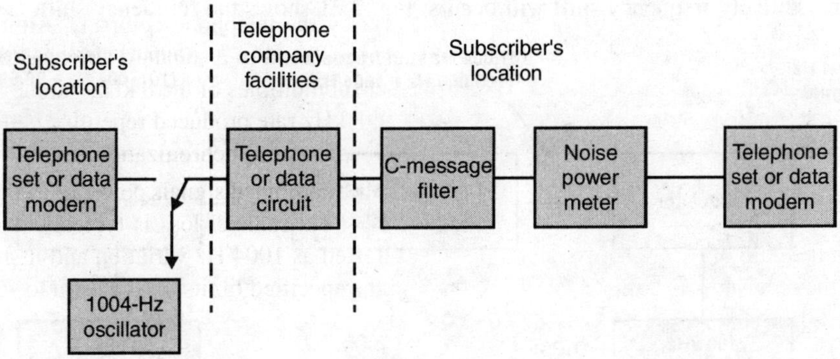

Fig. 2.20.

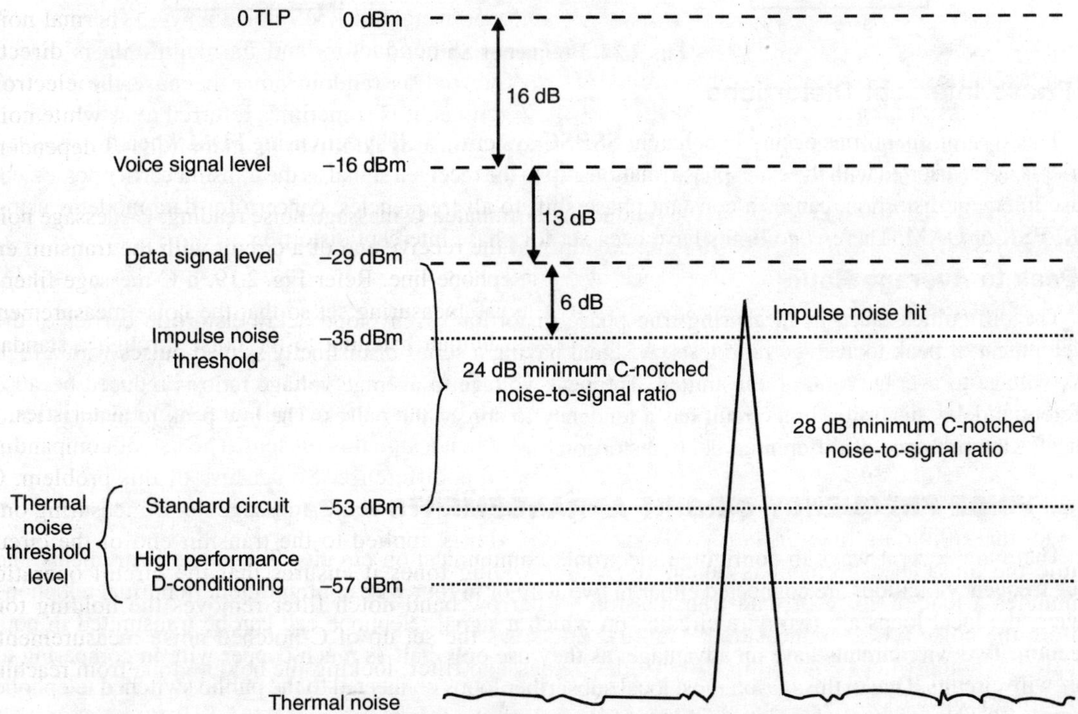

Fig. 2.21. C-notched Noise and Impulse Noise.

Telephone companies have established realistic, C-notched noise requirements for each type of facility for various circuit lengths. Telephone Company's guarantee standard private-line data circuits a minimum signal to –C-notched noise ratio of 28 dB. A standard circuit is one operating at less than 9600 bps. All the data circuit operating at 9600 bps require D-type of conditioning, which guarantee a minimum signal to C-notched noise ratio of 28 dB, Fig. 2.21 shows the C-notched noise.

7. Frequency Shift

When the frequency of a signal change during transmission, for example, a tone transmitted at 1004 Hz is received at 1006 Hz is termed as frequency shift. Analog transmission system, used by telephone companies operate signal-side band suppressed carrier (SSBSC) and, therefore; need coherent demodulation. With coherent demodulation, a carrier must be synchronous—the frequency must be reproduced exactly in the receiver. If this is not done, the demodulated signal will be offset in frequency by the difference between transmitted transmit and receive carrier frequencies. The longer a circuit, the more analog transmission systems and noise likely frequency shift will occurs. Fig. 2.21 shows the frequency shift.

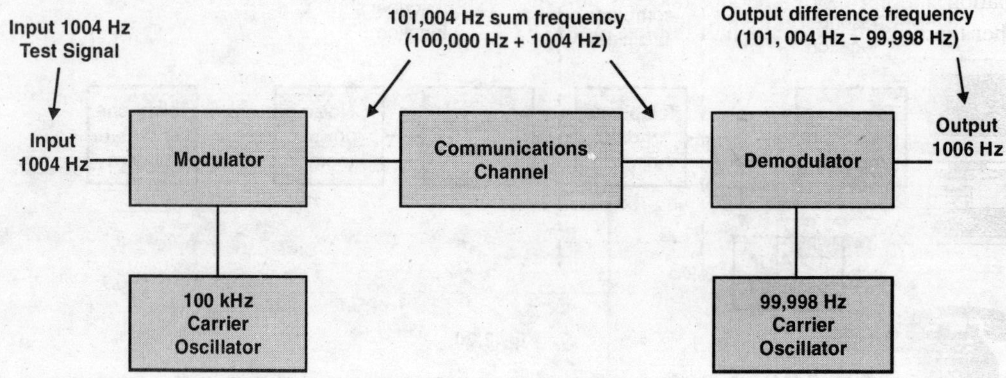

Fig. 2.22. Frequency Shift.

8. Phase Intercept Distortions

This type of distortions occurs in coherent SSBSC systems, such system using FDM when the received carrier is not reinserted with the exact phase relationship to the received signal as the transmit carrier possessed. Phase intercept distortion, causes a constant phase shift to all frequencies, concern for data modems using PSK, FSK or QAM. There is no limits have been set for phase intercept distortion.

9. Peak to Average Ratio

The difficulties faced in measuring true phase distortion or envelope delay distortion carrier to the development of peak to average ratio tests. A signal having a series of distinctly shaped pulses with a high peak voltage to average ratio is transmitted. The peak voltage to average voltage ratio is reduced because differential delay distortion in a circuit has a tendency to spread the pulses. The low peak to average ratio indicates the presence of differential delay distortion.

2.6. VOICE FREQUENCY CIRCUIT ARRANGEMENTS

There are several ways to configured electronic communication circuits. Telephone instruments and voice frequency facilities are connected either in two wire or in four wire configuration. In normal telephone service, the local loops are two wire circuits, on which a signal telephone call can be transmitted in both direction. Two wire circuits have an advantage, as they use only half as much copper wire in comparing to four- wire circuits. Due to this reason most local subscriber loops connected to the public switched telephone network (PSTN) are two wire circuits. Most of the private line data circuits are connected in a four wire configuration.

2.6.1. Two-Wire Voice Frequency Circuits

In two-wire transmission arrangements only two-wire are used (one for the signal and one for a ground or reference). In normal telephone service, the local loops are two-wire circuits, on which a single telephone call can be transmitted in both directions. If the distance between the subscribers is substantial, the amplifiers (repeaters) are necessary to compensate the attenuation. As the amplifiers are unidirectional for two way communications, four-wire configuration is necessary.

Figure 2.23 shows the block diagrams for two-wire circuit configurations. There are two ways of two-wire configurations namely;

1. Passive circuit
2. Active circuit

Figure 2.23 (*a*) shows simplest two wire configurations, which is a passive circuit consisting of two copper wires connecting a telephone at station through a telephone company interface to a telephone at the destination stations. The telephone, modems, and circuit configurations are capable of two way of transmission in either the full duplex or the half duplex mode.

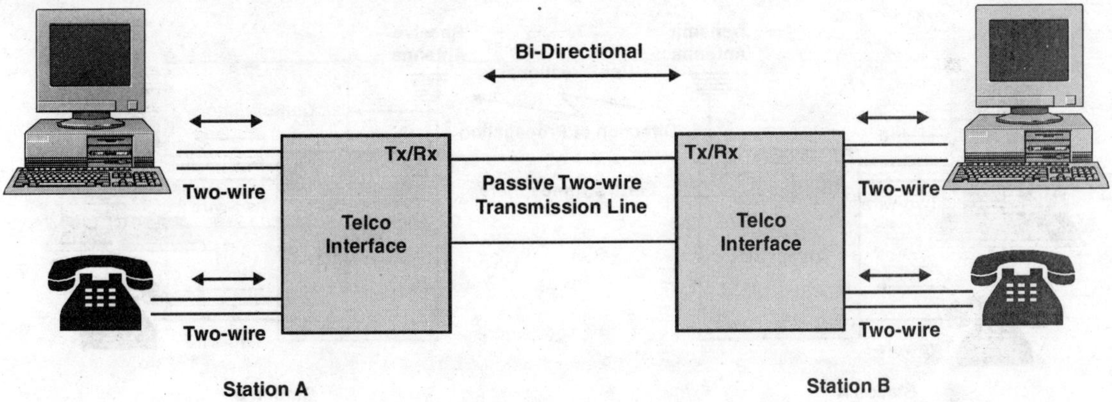

Fig. 2.23 (*a*). Simplest Two Wire Configuration.

Figure 2.23 (*b*) shows two wire active circuit configuration in which only difference is the addition of an amplifier to compensate the attenuations. The amplifiers are unidirectional and thus limit transmission to one direction only, *i.e.*, simplex mode.

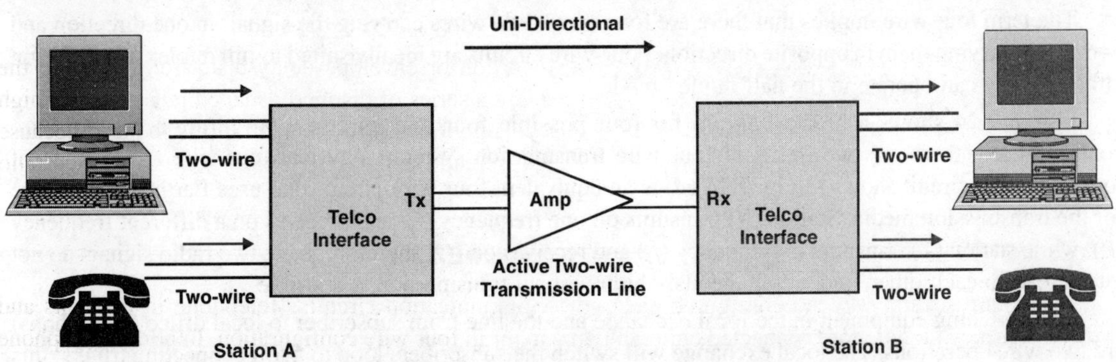

Fig. 2.23 (*b*). Two Wire Active Circuit Configuration of an Amplifier.

Figure 2.23 (*c*) shows a two-wire circuit using a digital T-carrier for the transmission media. In this, we have a T-carrier transmitter at one end and a T-carrier receiver at the other end.

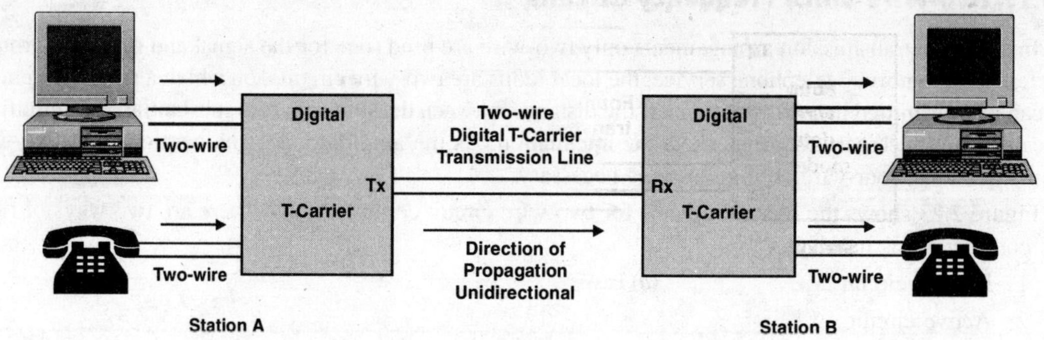

Fig. 2.23 (*c*). Two Wire Circuit Using a Digital T-Carrier for the Transmission Media.

Figure 2.23 (*d*) shows as equipment two wire circuits. In this unguided transmission media is used be Earth's atmosphere and there are no copper wires between the two stations.

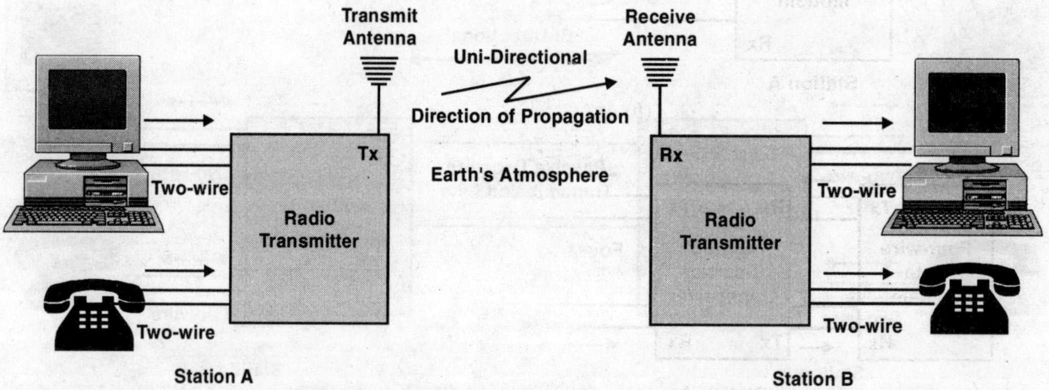

Fig. 2.23 (*d*). Two Wire Circuit Using Unguided Transmission Media.

2.6.2. Four Wire Voice Frequency Circuits

The term four wire implies that there are four wires, two wires carrying the signal, in one direction and two wires carrying them in opposite direction. Four-wire circuits are ideally suited to full duplex transmission although they can operate in the half duplex mode.

Figure 2.24 shows a block diagram for four possible four wire circuit configuration, as two wire configurations, there are two forms of four wire transmission systems: physical four wire and equivalent four wire. The circuit shows in Fig. 2.24 (*d*) is an equivalent four wire circuit that uses Earth's atmosphere for the transmission media. Station (*X*) transmits on one frequency (f_x) and receives on a different frequency (f_y), while station (*Y*) transmits in frequency (f_y) and receives on (f_x), therefore, have two radio signals do not interfere with each other, and simultaneously bidirectional transmission is possible.

The switching equipment in the local exchange and the line from subscriber to local office (local loops) are two wire operations. The local exchange will switch the subscribers loop to a toll connecting trunks; this is also a two wire transmission. The tolls offices are interconnected with inter toll trunks (which connect towns and cities). These trunks are of four wire transmissions. Figure 2.25 shows the simple arrangement of the two wires and four wire transmissions.

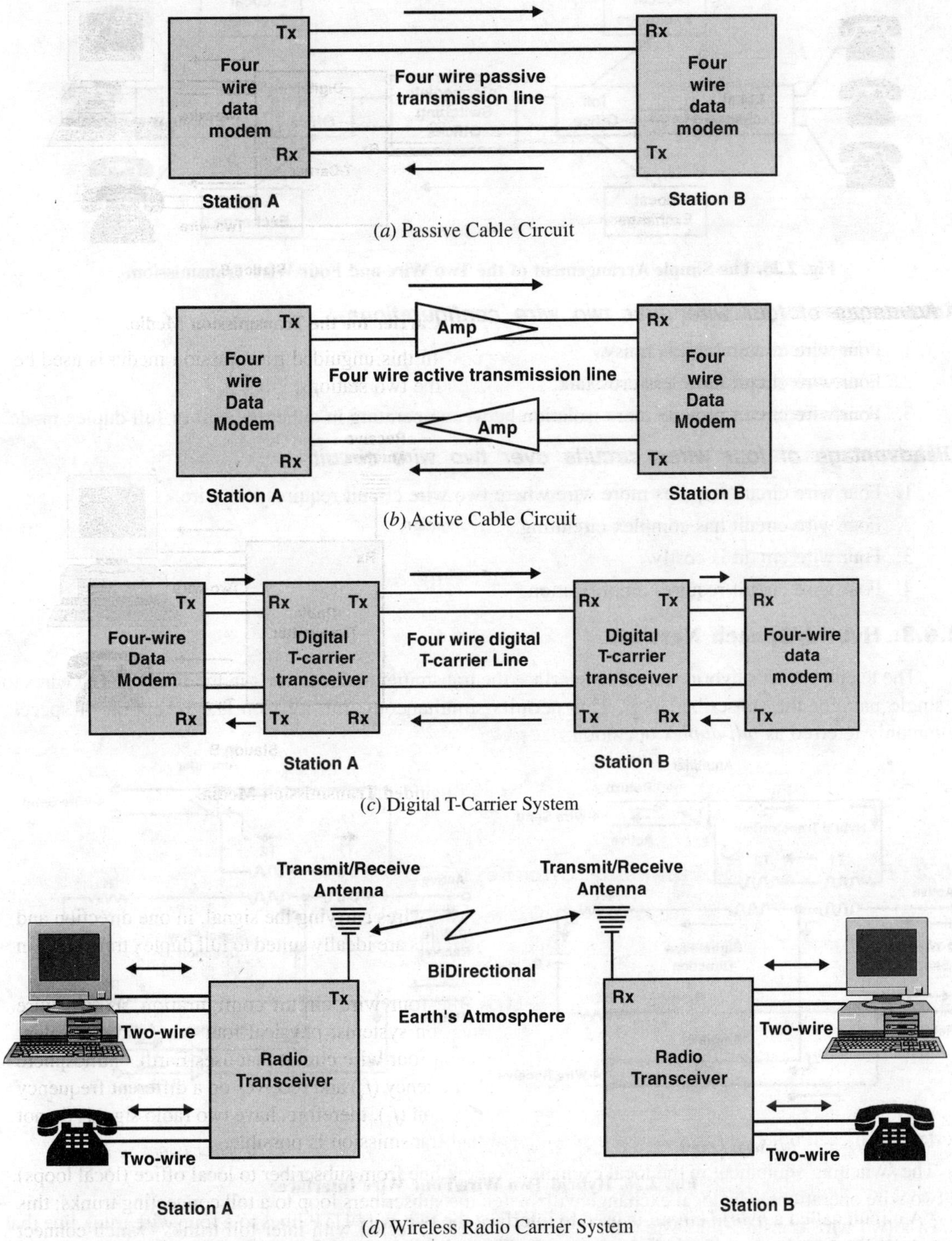

(*a*) Passive Cable Circuit

(*b*) Active Cable Circuit

(*c*) Digital T-Carrier System

(*d*) Wireless Radio Carrier System

Fig. 2.24. Four wire configurations: (*a*) Passive cable circuit (*b*) Active cable circuit (*c*) Digital T-carrier system (*d*) Wireless radio carrier system

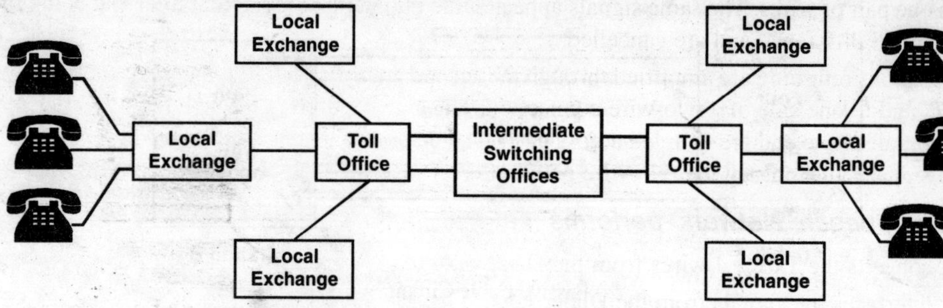

Fig. 2.25. The Simple Arrangement of the Two Wire and Four Wire Transmission.

Advantages of four wire over two wire configurations

1. Four wire circuit are less noisy.
2. Four wire circuit have less crosstalk.
3. Four wire circuit provide more isolation between operating in either the half or full duplex mode.

Disadvantage of four wires circuits over two wire circuit

1. Four wire circuit requires more wire where two wire circuit requires less wire.
2. Four wire circuit has complex circuiting.
3. Four wire circuit is costly.
4. Four wire circuit requires amplification.

2.6.3. Hybrid/Speech Network

The telephone set's hybrid is used to interface the transmitter and receiver's individually paired wires to a single pair for the subscriber loop. This permits simultaneous transmission and reception of speech, commonly referred as *full-duplex* operation.

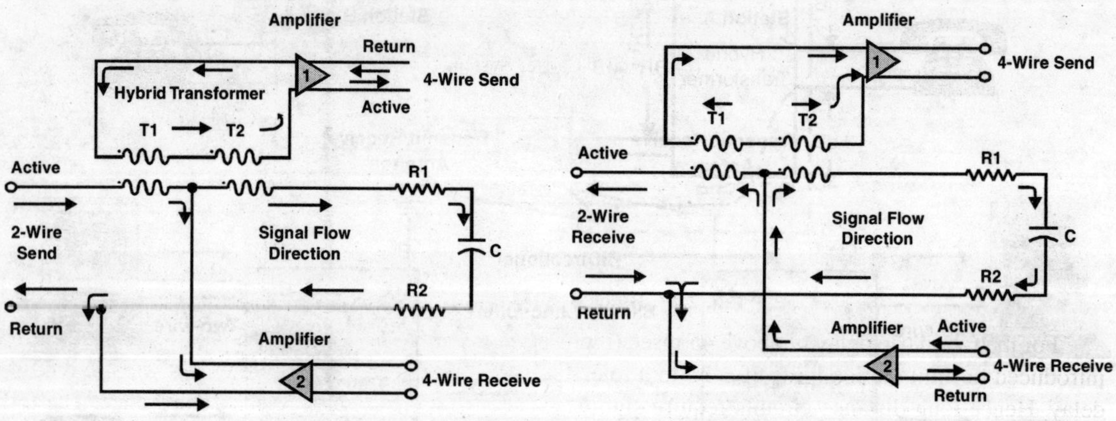

(a) Send Loops *(b) Receiver Loops*

Fig. 2.26. Hybrid Two Wire/Four Wire Interface.

A circuit called a *hybrid circuit* is used to interface the two-wire UTP lines to a four-wire trunk line that transports the conversation in a half duplex mode. The hybrid circuit shown in Figure 2.26 receives the call signals on a pair of wires, sending them through the primary windings of transformers T_1 and T_2 and impedance matching components, R_1, R_2 and C. The calls are coupled to the transformer secondary's, amplified, and sent out on one pair of wires. The same signals appear at the output of a second amplifier, but being in phase, have no potential difference and are cancelled.

Conversations returning are amplified through the second amplifier and fed, out of phase, to the junction of T_1 and T_2 and to one side of a two-wire connection. The signal traveling through the primaries of T_1 and T_2 are now out of phase and are coupled across to the secondary's and cancelled at the input of amplifier 1. Hence, the signal returns along the two-wire path but is not echoed back along the four-wire line.

The Hybrid/Speech Network performs several functions

- It converts the Tx/Rx 4 wires from the Handset to the 2 wires for the Local Loop.
- It interfaces the signals from the Dialer Unit to the telephone line.
- It provides auto line compensation for line length to keep the volume constant.

2.6.4. Echoes and Singing

Echoes and singing both occurs as a result of transmitted signals being coupled into a return path and fed back to the respective sources. Coupling will be zero only when perfect impedance matching occurs. Impedance matching between trunks and subscriber loop (two wires to four wires at hybrid) is difficult due to various subscriber loop lengths. A signal reflected to the speaker's end of the circuit is called **talker echo** and at the listener's end is called **listeners.**

Echo : The talker echo is more troublesome. When the returning signal is repeatedly coupled back into the forward path to produce oscillations, **singing** occurs. Basically singing results if the loop gain at some frequency is greater than unity. An echo coming 0.5 msec after the speech is not much effect. The echoes with a round trip delay of more than 45 msec cannot be tolerated. Figure 2.27 explains the path of the echo and the losses and gain of the signals at various parts of the system.

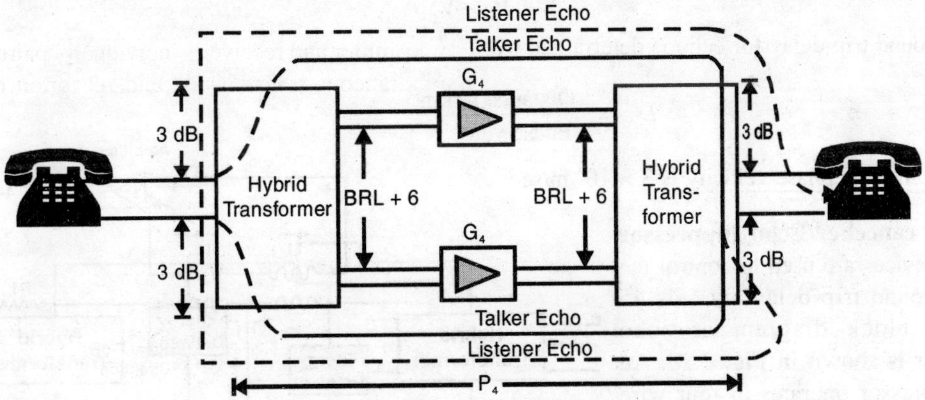

Fig. 2.27. Block Diagram of Echo Path.

For the round trip delay of above 45 msec (representing approximately 2900 km of wire) attenuators are introduced to limit the loudness of echo to a tolerable level. The attenuation required is related to the time delay. Hence long distance circuits require significant attenuation to minimize echo annoyance. The total attenuation from one two wire circuit to the other (α_2) is

$$\alpha_2 = \alpha_{24} + \alpha_{42} - G_4 \quad \text{...(2.11)}$$

where, α_{24} = attenuation between 2 to 4 wire line

α_{42} = attenuation on between 4 to 2 wire line

G_4 = total gain of one side of four wire circuit (in dB)

The α_{24} and α_{42} are normally 3 dB. Thus

$$\alpha_2 = 6 - G_4 \quad \text{...(2.12)}$$

The Trans hybrid loss (TL) is the attenuation through the hybrid coil from one side of the four wire circuit to the other and given as

$$TL = 6 + BRL \quad ...(2.13)$$

The loss of reflected signal due to the mismatch of the network is given by balance return loss (BRL). If impedance of the four wire circuit is Z_B and the impedance of two wire circuit is Z_2, the BRL is given as

$$BRL = 20 \log_{10} \left| \frac{Z_B + Z_2}{Z_B - Z_2} \right| \text{ dB} \quad ...(2.14)$$

If Z_B and Z_2 matches BRL = ∞. It indicates the attenuation on echo is infinite and no necessity of any attenuators.

The total attenuation on echo is given by net attenuation offered by the four-line circuit minus the gain of the amplifiers. Hence by clockwise movement from talker to talker circuit,

$$\alpha_t = [3 - G_4 + (BRL + 6) - G_4 + 3] \text{ dB}$$

$$\alpha_t = 2\alpha_2 + BRL$$

The echo delayed is $D_t = 2D_4$ ...(2.15)

where, D_4 is the delay of four-wire circuit.

If P_4 is incoming power on the 4 wire circuits, P_2 power reaching the 2 wire circuit and $P_4 - P_2$ is the power reflected onto the return path, the BRL in terms of power is

$$BRL = 10 \log_{10} \frac{P_4}{P_4 - P_2} \text{ dB}$$

BRL in terms of voltage is, $BRL = 20 \log_{10} \frac{V_4}{V_4 - V_2}$ dB

The round trip delay for echo is determined by

$$D_c = \frac{\text{Distance in km}}{\text{Phase velocity}}$$

Normally, the phase velocity is 3×10^7 m/sec.

Echo canceller/Echo suppressor : These devices are used to control the echo. If round trip delay exceeds 45 msec. A block diagram of echo suppressor is shown in Fig. 2.28. An echo suppressor operates in four wire circuits by measuring the speech power in each leg and inserts a large amount of loss (35 dB typically) in the opposite leg when the power level exceeds a threshold. Thus a returning an echo is essentially blocked by the high level of attenuation.

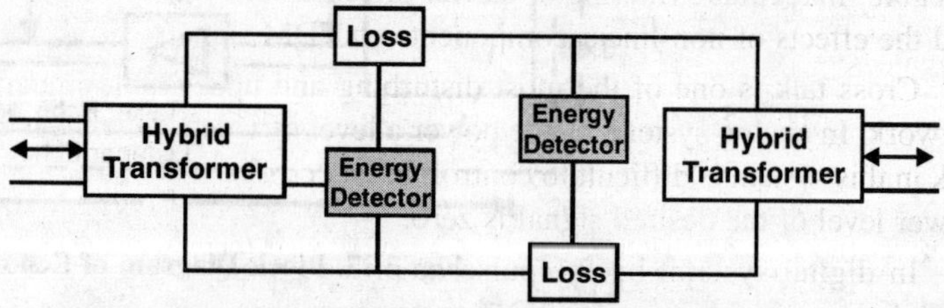

Fig. 2.28. Echo Suppressor Block Diagram.

A device that suppresses echoes would also suppress data. Hence when full duplex transmission is used, the echo suppressors must be disabled. An echo suppressor converts a full duplex circuit into a half duplex circuit with energy sensing being the means of turning the line around. The echo suppressors must be disabled when full duplex transmission is used.

The user's modem can disable the echo suppressors by transmitting a single frequency tone in the band 2010–2040 Hz for at least 400 m sec. During this time, no other signal or tone should be transmitted. The echo suppressor will remain disabled if a signal transmitted within 100 msec of disabling tone being removed. Thus there may be a clip at the beginning portions of speech segments. This is another drawback of echo suppressors for voice circuits.

Echo control or echo cancellation is the recent electronics technology. An echo canceller operates by simulating the echo path to subtract a properly delayed and attenuated copy of a transmitted signal. Echo cancellers do not physically insert attenuators. A typical echo canceller block diagram is shown in Fig. 2.29.

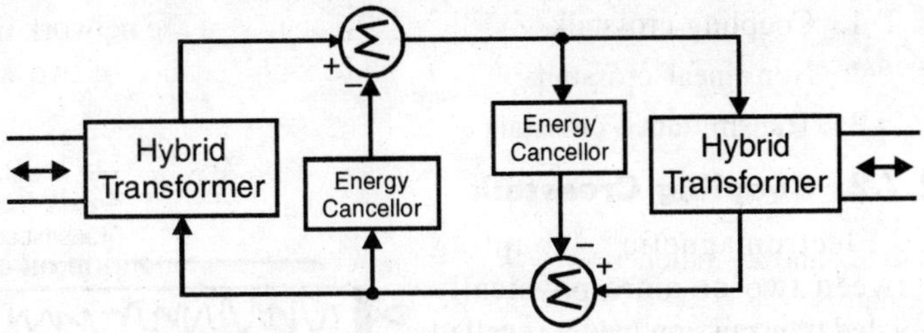

Fig. 2.29. A Block Diagram of Echo Cancellor.

The transmitted speech is stored for a period of time equal to the round trip delay of the circuit. The stored signal is attenuated and then subtracted from the incoming signal. This kind of circuits is available in satellite circuits.

2.7. CROSS TALK

An additional difficulty that can cause problems in using telephone lines is **cross talk** between neighbouring lines. Most of us have experienced this problem on long distance calls at one time or another. While talking with your party, you can distinctly hear, albeit at a very low volume, another phone conversation going on. What you are experiencing is cross talk, in which one line is inducing current into the other in the same manner as a transformer inducing current from its primary to secondary winding. Once again, for voice conversations, this is a small annoyance, but for data communications, it is a possible source of data errors. There are no telephone company specifications addressing lines that cross talk.

The current from the battery in the subscriber loop (when telephone handset is off hook) is limited to the range of minimum 20 mA to maximum of 60 mA. The current variation depends on the length of the subscriber loop. In long loops the current is less and in short loops the current may exceed 60 mA (an electronic component varistor in telephone set is used to limit the current within 60 mA). The large current flow causes electromagnetic fields and thus creates signal distortions in adjoining wires. This distortion is called **crosstalk**. Some of the major sources of cross talk are coupling between wire pairs in cable, inadequate filtering or carrier offsets in older frequency division multiplexing (FDM) equipments and the effects of non-linear components on FDM signals.

Cross talk is one of the most disturbing and undesirable imperfections that can occur in a telephone network. In analog system, as the power a level of voice signal considerably varies (40 dB ranges) the cross talk in this system is difficult to control. In fact cross talk is more noticeable during speech pauses, where the power level of the desired signal is zero.

In digital system, by the pulse amplitude modulation, pulse length modulation and pulse position modulation, which are used in TDM system results in attenuation and delay distortion. This causes dispersion of the transmitted pulses. They spread in time and interfere with the pulses of adjacent channels and causes inter-channel cross talk. Pulse code modulation is used to overcome the interchannel cross talk problem.

2.7.1. Unit of Crosstalk

Crosstalk interference is measured in a special decibel unit of measurements, dBx, where dBx is referred to the level on the cable that is being interfered with.

In telephone communication, 90 dB is considered the ideal isolation between adjacent lines. Mathematically, dBx is given as

$$\text{dB}x = 90 - (\text{crosstalk loss in dB})$$

For example, the magnitude of the cross talk on a circuit is 60 dB lower than the power of the signal on the same circuit. The crosstalk is then 90dB – 60 dBx = 30 dBx.

There are three primary types of crosstalk in telephone systems which are as following :

1. Coupling crosstalk
2. Non-linear crosstalk
3. Transmittance crosstalk

2.7.2. Coupling Crosstalk

Electromagnetic coupling between two or more physically isolated transmission media is called coupling crosstalk.

The basic forms of crosstalk concern to telecommunication engineers are near end crosstalk (NEXT) and far end crosstalk (FEXT).

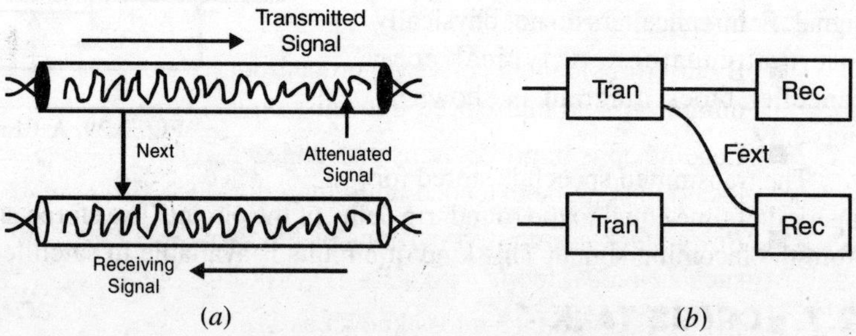

Fig. 2.30. (*a*) NEXT and (*b*) FEXT.

NEXT occurs near the transmitter and creates distortions that affect the signal on adjacent receive pairs. This type of noise can be generated when a transmission line carrying a strong signal is coupled with a transmission line carrying a weak signal. NEXT is measured in dB, with higher values being better. NEXT is measured for all frequencies between 0 and 100 MHz. NEXT is measured by injecting a signal on a wire pair and measuring its crosstalk on another wire pair. NEXT should be measured at both ends. NEXT is more troublesome because of a large difference in power levels between the transmitted and receive signals. Twisted wire pairs reduce this type of cross talk.

FEXT is a measure of the crosstalk that exists at the receiver end of the cable. FEXT refers to unwanted coupling into a received signal from a transmitter at a distant location.

This noise is relevant on networks that transmit on multiple pairs in the same direction in the same time. For example, Gigabit Ethernet, which enable the organizations to upgrade to 1000 Mbps while using the same operation system and software. One example of transmission media of Gigabit Ethernet is 1000 Base T and it uses four pairs of CAT–5 GB cable with RJ–45 connector. The ELFEXT (Equal Level Far End Cross Talk) test provides standard way to measure far end cross-talk (irrespective of cable length) and is required for Gigabit Ethernet cable certification.

Nonlinear Crosstalk

Nonlinear crosstalk is due to the nonlinear amplification in analog communication systems. Nonlinear amplification results harmonics and cross product. If the non linear frequency components fall into the passband of another channel, they are considered crosstalk.

Transmittance Crosstalk

Crosstalk can also be result by inadequate control of the frequency response of a transmission system such as poor filter performance. This type of interference is caused by inadequate control of the transfer characteristics or transmittance of network, it is called *transmittance crosstalk.*

Summary

- The circuit which connects each subscriber (via a pair of wires) to the switch is called a **subscriber loop** or **local loop**.
- The local loop is the primary cause of attenuation and phase distortion on a telephone circuit. Normally the range of local loop is from 1 to 10 km.

- The *transmission characteristics* of a cable pair depend on the wire diameter, conductance of the cable, conductor spacing and dielectric constant of the insulator separating the wires.
- While designing a subscriber loop, there are two limiting factors that we have to consider. Those factors are as follows: *Attenuation and voltage drop.*
- The loop resistance is $R_{max} = V_B/I_C$
- The loop resistance limit is $R_L = R_m - (R_B + R_T)$
- Inductive loading is the process of inserting series inductances (loading coils) into the loop at fixed intervals.
- By inserting loading coils with proper inductance and at appropriate distances, the distortion can be minimized and high-speed transmission can be achieved.
- Loading coils improves transmission levels for circuits longer than 18000 feet. By using local coils, local loop can be extended up to three to four times their previous length.
- *Bridge taps* are unused portions of the cable that are connected in shunt to a working cable pair, such as a local loop.
- Bridge taps produce a loss known as *bridging loss.*
- Bridge taps permit signals to split and propagate down more than one wire.
- The maximum subscriber loop length, which is defined as the distance from the subscriber to the central office, is expressed as $L = R_L / R_{dc}$
- The maximum loop length is calculated from the formula,

$$L_{max} = \text{Attenuation limit/ loss per km}$$

- The common noises in telecommunication system are as follows:
  1. White noise,
  2. Impulse noise and
  3. Inter modulation noise.
- The two most common measurements in telephony use a C-message weighting curve and a psophometric weighting curve.
- C-message weighting is used in North America while psophometric weighting is the European (ITU-T) standard.
- The significance of the C-message weighting curve can be seen that a 200 Hz test tone of a given power is 25 dB less disturbing a 1 kHz test-tone of the same power.
- CCITT and common satellite practice use **psophometric weighting**. The symbol P is used for the psophemetric weighting factor and its value is 1.78, this corresponds to 2.5 dB.
- Telephone equipment is often adjusted by applying a 1 KHz tone at 0dBmo to the system to simulate peak power one channel; this is known as *Test tone.*
- Bandwidth limits the maximum frequency and switching speeds that can be applied to POTS lines.
- The most widely used reference level is a *1-milliwatt signal* (600-ohm load).
- When the 1 milliwatt reference is used, the standard decibel abbreviation of dB is changed to dBm; dBms are used as an indication of power, while dBs are used to indicate the ratio between the input and output.
- Physical routes are established at the time a call is initiated. This type of routing is known as connectionless service, which is performed through the telephone company's **direct distance dialing (DDD)** network.

- The process is used to improve a basic telephone channel is referred as **line-conditioning**.
- Line-conditioning improves the high frequency response of a message channel and reduces the power loss.
- C-type line-conditioning relates to line impairments for which compensation can be made with equalizers and filters.
- Basic conditioning, also known as **C1 conditioning**, is for voice-grade quality communications meaning that they are sufficient to guarantee successful voice communications.
- The method used to measure propagation time is called **envelope delay distortion,** which is a clever process used to make this measurement.
- Unweighted, this is a measure of the existing noise on a line without any active signal on it, or weighted, which measures noise with a signal present.
- The most common form of noise in the telephone network is impulse noise and quantization noise.
- Telephone instruments and voice frequency facilities are connected either in two wires or in four wire configuration.
- In normal telephone service, the local loops are two-wire circuits, on which a single telephone call can be transmitted in both directions.
- The term four wire implies that there are four wires, two wires carrying the signal, in one direction and two wires carrying them in opposite direction.
- A circuit called a *hybrid circuit* is used to interface the two-wire UTP lines to a four-wire trunk line that transports the conversation in a half duplex mode.
- The echo suppressors must be disabled when full duplex transmission is used.
- Cross talk is one of the most disturbing and undesirable imperfections that can occur in a telephone network.
- Crosstalk interference is measured in a special decibel unit of measurements, dBx, where dBx is referred to the level on the cable that is being interfered with.

Review Questions

1. What is channel noise ? Explain cross-talk and its sub-types.
2. Briefly describe C-message noise weighting and its significance.
3. Briefly explain transmission parameters and private line circuits.
4. Explain the working of voice frequency circuit arrangement.
5. Explain, what is cross talk and how it could be reduced.
6. Briefly discuss a local subscriber loop.
7. Write down the fundamental characteristics of local subscriber loop.
8. Discuss the limiting factors of subscriber local loop.
9. What components of the telephone system make up the local loop?
10. What are the two names for long distance switching stations?
11. What is the purpose of repeaters in long distance trunk lines?
12. What function does a hybrid circuit perform?
13. Explain loading coils and bridge taps and discuss effect on performance of a telephone circuit.
14. Briefly describe the difference between c-message weighting and psophometric noise weighting.
15. List the end instruments and its use.

16. What are the primary functions of transmission systems?
17. Define local loop and trunks.
18. What is known as singing? How singing can be reduced?
19. List various impairments which affect the signals.
20. Draw echo canceller circuit and explain.
21. Define noise related to telecommunication.
22. What are the various sources of impulse noise?
23. Discuss C-message weighting and describe its significance.
24. What is the difference between dB and dBm?
25. Enumerate the three categories of transmission parameters.
26. Discuss attenuation distortion; envelope distortion.
27. What is meant by line-conditioning?
28. What kind of circuits can have C-type line conditioning; D-type line-conditioning?
29. When is D-type conditioning is mandatory? Discuss the limitations imposed with D-type conditioning.
30. What considerations are addressed by the interface parameters?
31. What considerations are addressed by the facility parameters?
32. What is the purpose of echo canceller and echo suppressor?
33. Describe what is meant by a two-wire circuit and four-wire circuit.
34. Discuss the units of power measurements.
35. How do you decide the maximum subscriber loop length?
36. What is the benefit to digital data links in using dedicated lines?
37. What is the TELCO bandwidth and specified line impedance?
38. What type of telephone lines can be conditioned?
39. What are the reference frequencies for gain variation and envelope delay measurements?
40. What are the tightest limits for conditioning that can be applied to telephone lines? What frequencies do these limits apply to?
41. Explain the D-type line-conditioning and describe its main features.
42. What two electrical parameters are specified using D-conditioning?
43. What are the characteristics of a notch filter?
44. What is the difference between weighted and unweighted noise?
45. Enumerate the facility parameters.
46. What do you understand by voice frequency circuit arrangements and discuss various types of circuits?
47. Compare the four-wire voice frequency circuit with the two-wire frequency circuits.
48. Briefly discuss the function of a two-wire to four-wire hybrid set.
49. What do you understand by crosstalk in telephone system?
50. List and describe three types of crosstalk.
51. What are called crosstalk?
52. Explain briefly NEXT and FEXT.
53. Explain the various types of distortion.
54. What is BORSCHT?
55. Consider a subscriber loop of 12 km long. The loop resistance is 1600 ohm. Calculate the d.c. loop resistance and determine the cable gauge for the loop.
56. Explain the concept of inductive loading.

57. Write short notes on following :

(*i*) Transmission characteristics
(*ii*) BORSCHT
(*iii*) Loop resistance
(*iv*) Maximum permissible loop length
(*v*) Loading coils
(*vi*) Bridge taps
(*vii*) Line impedance
(*viii*) Line-conditioning
(*ix*) C-message weighting and Psophometric noise weighting
(*x*) Crosstalk
(*xi*) NEXT and FEXT
(*xii*) Echo and singing
(*xiii*) Echo suppressor
(*xiv*) Envelope delay distortion
(*xv*) Impulse noise
(*xvi*) Nonlinear distortions
(*xvii*) Hybrid circuit
(*xviii*) D-type conditioning
(*xix*) 1004 Hz variation
(*xx*) Impulse gain hits and dropout
(*xxi*) Phase jitter

❑❑❑

II

UNIT

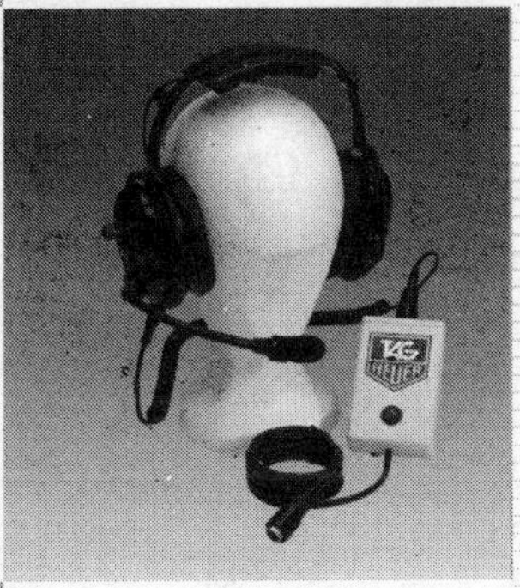

Public Telephone Networks

SYLLABUS

Public Telephone Network : Introduction, Transmission System Environment, Public Telephone Network, Instruments, Local Loops, Trunk Circuits, Local Central and Operator-Assisted Exchanges, Automated Central Office Switches and Exchanges, Telephone Numbering Plan, Telephone Services, Telephone Switching Hierarchy, Common Channel Signaling System.

C
H
A
P
T
E
R

3

PUBLIC TELEPHONE NETWORKS

INSIDE THIS CHAPTER

3.1. INTRODUCTION

The largest communication network on this earth is the global Public Telephone Network (PTN).The telephone network had its beginnings in the late 1876s. The entire network, which is referred to as plain old

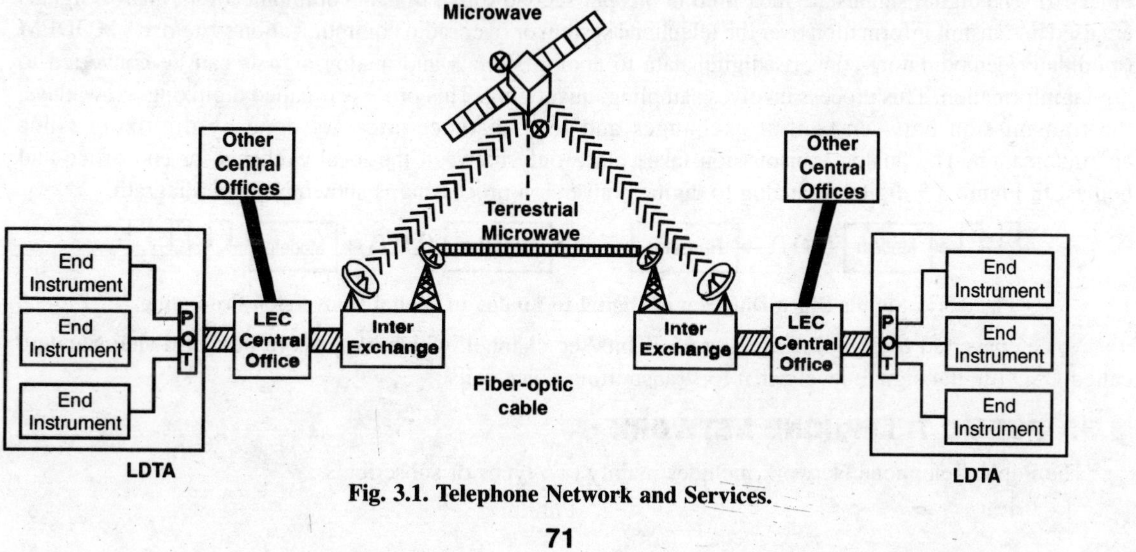

Fig. 3.1. Telephone Network and Services.

telephone set (POTS) was originally designed to carry analog voice signals, the advent of computers and data communication in the 1980s, has lead to it being increasingly used to also carry computer data traffic. This network brings together about a billion telephone sets in a highly connected fashion which allows almost every set to be reached by every other set. The global telephone network is the center-piece of the field of telecommunication.

One characteristics of the PTN is that every piece of equipment, technique, or procedure, old or new, is capable of working with the rest of the system.

Figure 3.1 shows the telephone network and services.

3.2. TRANSMISSION SYSTEMS

The transmission system provides circuits and path between two subscribers. The transmission circuits are capable of receiving the electrical signals at one point and delivering them to a destination with good quality. If necessary, more transmission system can be setup for long distant transmission for good quality. The transmission path, which is also referred as telephone channel or transmission media, is designed to provide voice-grade communication. It has evolved into a worldwide network that encompasses a variety of transmission media and switching systems. Thus the transmission system behaves as an excellent candidate for a data communication over long distances.

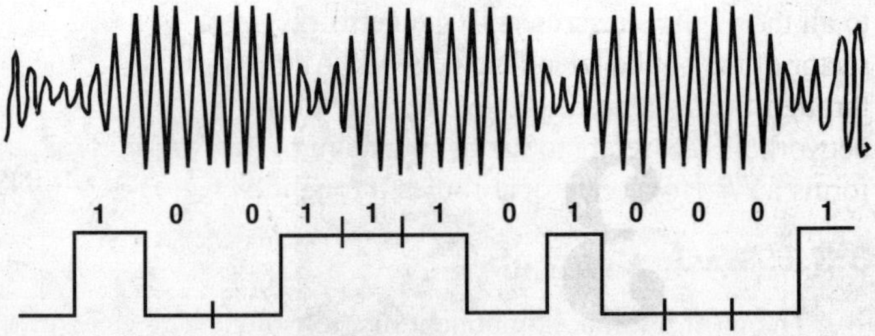

Fig. 3.2. Shows the Voice Signal and Digital Signals.

The transmission system involves analog transmission (voice communication) and digital transmission. The analog signals are characterized by frequency, amplitude and phase. In analog transmission system, signals propagate through the medium as continuously varying electromagnetic waves. The medium for an analog transmission may be twisted pair cable, coaxial cable, optical-fiber cable, microwave radio and satellites. The analog signals are subjected to deterioration due to attenuation and noise addition in the channel. Hence amplifiers, filters and necessary circuits are added in transmission system to upgrade the analog signal. Figure 3.2 shows the voice signal and digital signals. In digital transmission system, signals propagate as a discrete voltage pulses (positive voltage represents binary 1, and negative voltage represents binary 0). The digital signals are measured in bits per second (bps). In data communications, analog signals are used to transmit information over the telephone system or over radio communication systems. A MODEM (modulator/demodulator) converts digital data to analog signals and analog signals can be converted to digital information. This process involves sampling, quantizing. This process is called digitizing. Nowadays, the transmission between central exchanges and long distance sites are done by digitizing voice communication. The analog transmission takes place only between the local exchange or end office and homes. In Figure 3.3 digital to analog to digital conversion processing is shown as block diagram.

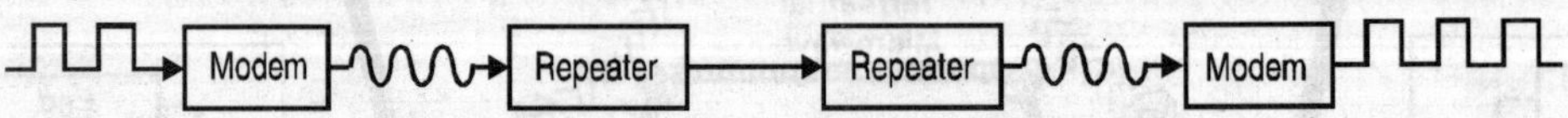

Fig. 3.3. A Simple Block Diagram of Digital to Analog to Digital Conversion Processing.

Voice converted to digital requires a 64 K bits/sec channel, which happens to be worldwide standard called DSO (digital signal, level zero) for transmitting voice calls.

3.3. PUBLIC TELEPHONE NETWORK

The Public Telephone Network includes mainly two types of subscribers :

1. Private
2. Public

Private

All those customers who lease transmission media, equipments and services from the telephone companies on permanent basis have known as **private subscribers** or private line circuit or dedicated circuits. The leased circuits are specifically designed and configured for their use, for example, banks do not like to share their communication network with other users. Therefore, they lease facilities and equipments from public telephone companies and operate a private telephone or network within the public telephone network. Public telephone companies also referred as service providers. They provide services and lease equipments to other private companies, government agencies and organizations.

Public

Subscribers to the public sector of the telephone network share facilities and equipments that are available to all the public subscribers to the network. Anyone with the telephone number is a subscriber to the public sector of the public telephone network. All the subscribers to public network are interconnected temporarily through the switches. So they are also known as **public switched telephone network (PSTN)** or dial-up network. PSTN refers to the combination of transmission lines and switches (may be central offices) which forms a system of electrical routes through the network.

3.3.1. Basic Concepts

The most familiar component of a telephone network is the *telephone set/ instrument* provided for each user. It is a relatively simple device which can exchange control signals with the network to help establish and release calls, and to send and receive a user's speech signal.

Users are referred to as ***subscribers***, because they subscribe to a service provided by the telephone company. Typically, many subscribers are attached to the same network, and each subscriber can contact every other subscriber. The system employs ***switches*** to facilitate connections between subscribers. The arrangement of switches and subscribers and the interconnections between them determines the structure of the network. We have already discussed the most familiar components of a telephone network in Chapter 1.

Main Components

1. Telephone Instrument
2. Local Loops
3. Trunk Circuits
4. Switching offices or exchanges (end offices, tandem offices and regional offices)

Local loops

The circuit which connects each subscriber (via a pair of wires) to the switch is called a **subscriber loop** or **local loop**. A twisted pair cable which connects the subscriber telephone set to the nearest end office (may be referred as local central office). The twisted pair has its limited bandwidth but enough for voice transmission which needs only 4 kHz. Normally the range of local loop is from 1 to 10 km. However, local loop used with private-line data circuits are generally four wire configurations.

Trunk lines

A trunk circuit is similar to a local loop except trunk circuits are used to interconnect two telephone offices. The two exchanges are connected using a set of interexchange lines called **trunk lines.** The main difference between a local loop and a trunk is that a local loop is permanently associated with a particular station, whereas a trunk is a common usage. Trunk lines are transmission media used to connects the interoffice**.** The transmission media used for trunk lines may be twisted copper wire or optical fibers. A trunk circuit could be a wireless communication channel (satellite links). Because it has to handles large numbers of connections using multiplexing, it requires higher bandwidth.

Switching Offices or Exchange

Switching offices or exchange is a central location where subscribers are interconnected, permanently or temporarily basis. Switching offices are connected with several local loops or trunk lines and provide a connection between different subscribers. Switching machine are located in exchange, switching machines are programmable matrices that provide temporary signal paths between two subscribers. This is done to avoid the having permanent physical connection between any two subscribers. Data modems are connected through local loops to switching machines located in exchange.

A **central exchange** is also known as central telephone exchange, central office or simply central. The purpose of a telephone exchange is to make a path for a call to be completed between two subscribers. While processing a call, a switch should provide three main functions:

(*i*) First function is to identify the subscribers.
(*ii*) Establish a communications path
(*iii*) Monitor the calling process

Telephone Services

Figure 3.4 shows a typical telephone circuit for the medium distances. A telephone connection may be as simple as two telephones and a single local switching office, or it may involves several switching offices, transmission facilities and telephone companies. Telephone set convert audio signals in to electrical signals and vice versa. It also generates supervisory signals and address information. The local loop provides a two way path for conveying speech and for exchanging ringing, switching and supervisory signals. The higher efficiencies of new telephone sets and modems compensate for increased loop loss, allowing longer loop lengths.

The small period of time approximately 10% during busy hours, that a subscriber loop is utilized led to the development of line concentrators between central offices and subscribers. A concentrator allows many subscribers to share a limited number of lines to a central office switch. The line from a concentrator to the central office is essentially a trunk circuit which is being shared among many subscribers on an as needed basis. Trunk circuits of various types are used to interconnect local offices, local offices to toll offices and toll offices to other toll offices. After appropriate switching and routing functions are performed at toll offices, messages are generally connected to intertoll trunks by means of a two wire to four wire interfacing set or hybrid, which splits the two direction of signal propagation so that the actual long distance segment of the route can be accomplished on four wire basis. Signal are connected through intertoll trunks to remote toll switching centers and finally reach the recipient of the call through a toll connecting trunk, a local office, another four wire to two wire term set, a local switching office and a final subscriber loop.

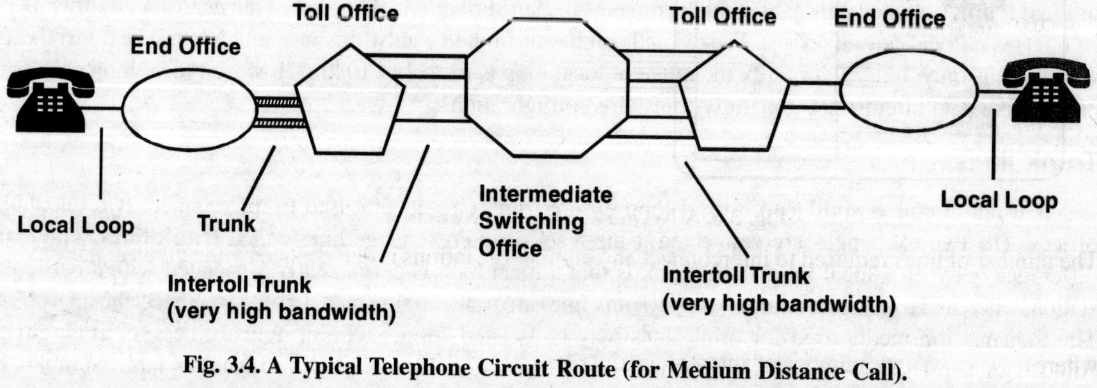

Fig. 3.4. A Typical Telephone Circuit Route (for Medium Distance Call).

3.3.2. A Simple Network

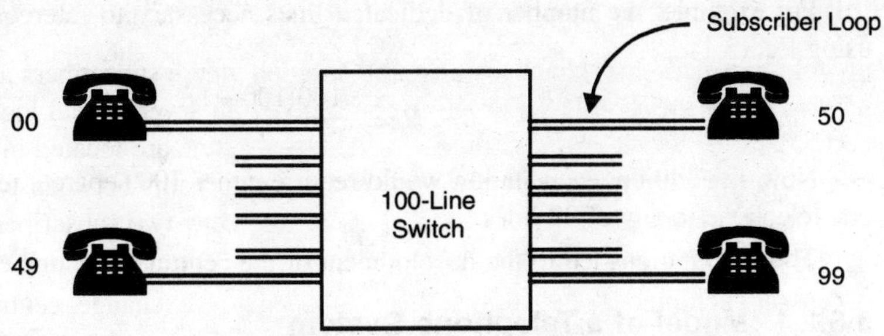

Fig. 3.5. Single Switch Network.

To better understand how the system works, we will begin by looking at a very simple network.

Consider a small community of 100 families. The communication needs of this community can be served by one switch, to which every family (or subscriber) is directly connected (see Fig. 3.5). The circuit which connects each subscriber (via a pair of wires) to the switch is called a **subscriber loop**. The switch itself is called a **local exchange** (local because the connection is direct).

A numbering scheme is used to uniquely identify each subscriber and to gather call-related information for billing purposes. In this case, a two-digit number would suffice (*i.e.*, 00-99). When a subscriber dials the number of another subscriber, the switch follows a sequence of steps for establishing a communication path between the two subscribers. Provided the destination number is not engaged, a connection is guaranteed.

Now suppose that there is a similar community with its own local exchange, and that we wish to provide phone access between the two communities. To do this, the two exchanges are connected using a set of interexchange lines called **trunk lines** (see Figure 3.6). In practice, it is unlikely that all of the subscribers in one community would want to simultaneously contact subscribers in the other community.

If we establish that at most say six subscribers are likely to call the other community, then six trunk lines will be sufficient to connect them.

The numbering scheme is expanded to take into account the above changes.

Each subscriber is now allocated a three digit number. The first digit identifies the exchange (say 1 for the first community and 2 for the second community) and the next two digits identify a subscriber in that community as before. When a subscriber dials a number, the local exchange looks at the first digit of the number. If it matches its own number then it treats the remaining two digits as for a subscriber on the same exchange, and attempts to establish a direct connection between the two as before.

If the first digit identifies the other exchange, then it uses one of the trunk lines to inform the other switch of the call and passes to it the two remaining digits. The two switches then cooperate to establish a call between the two subscribers. If more than six subscribers attempt to simultaneously call subscribers on the other exchange, the seventh call and beyond will be blocked due to insufficient trunk lines.

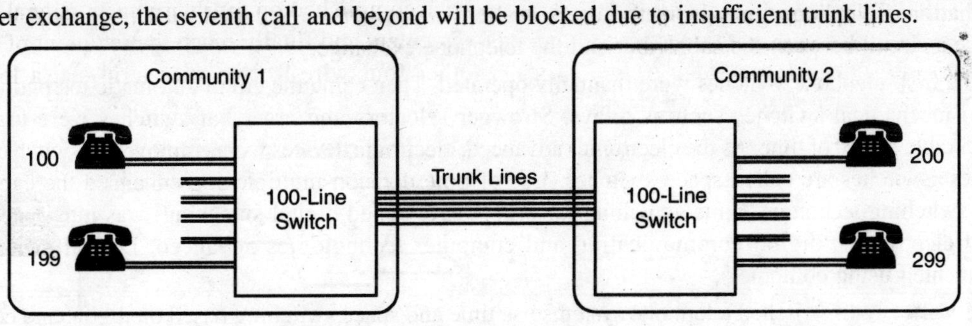

Fig. 3.6. A Network With Two Switches.

The number of lines required to interconnect any number of stations is determined by the given equation:

$$N = \frac{n(n-1)}{2} \qquad \text{...(3.1)}$$

where N = number of interconnecting lines

n = number of parties

For example, the number of dedicated lines necessary to interconnect 100 parties is determined by using Eq. (3.1)

$$N = \frac{100(100-1)}{2} = 4950$$

Now, in addition, each station would require either 100 separate telephones or capability of switching one telephone to any of 99 lines.

These limitations led to the development of the central telephone exchange.

3.3.2.1. Model of a Telephone System

A basic model of a telephone system is shown in Fig. 3.7. A telephone switch has some input and some output lines (4 are shown in the diagram). The switch is connected to the control block which identifies the particular cross point contact to be closed /opened. The control block is connected to all the input as well as output lines and keeps monitoring them. The control block closes/opens a contact by issuing appropriate control signals to the selected contact through the control lines. The user lines are numbered (telephone numbers). The control block identifies the calling line and after enquiring the called number and checking that the called party is free, issues control command to the switch to close the relevant contact. Similarly the control block identifies the end of conversation and disconnects the particular contact. The control is usually stored program control (SPC) using a computer. Memory is used to store the called and calling numbers, users' data and much other information. Telephone lines from the users (called subscribers) are called subscriber local loops and provide a circuit between the subscriber and the telephone exchange.

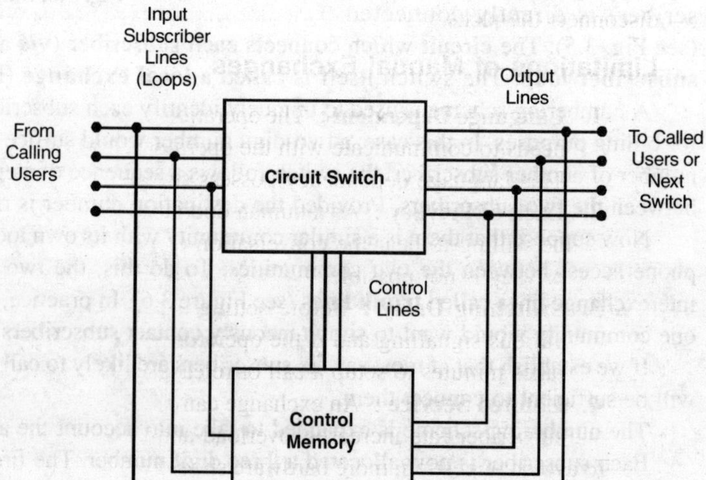

Fig. 3.7. Telehone System Basic Model

The first telephone switches were manually operated. Then came the era of automatic telephony where electro mechanical switches such as relays, Strowger selectors and cross bar switches were used for a considerable period of time. As the electronics advanced, electronic devices were employed as switch contacts. All these switches are called space switches. Use of time division multiplexing enhanced the capacity of space switching technique. Time switching (also popularly called digital switching) was introduced as the digital electronics, digital communication and computer technologies advanced. Digital switches are implemented using computers.

Modern circuit switched telephone systems use time and space switching based on digital and computer technologies. The control is in any case stored program computer based. Digital switching will be covered in more detail in later lessons.

3.3.3. Manual Exchanges/Operator Assisted Local Exchanges

In the manual exchange (until 1892), the control was provided by a human operator and the elements of the switch assemblies are plugs and jacks. All the local exchange hardware's are duplicated for each subscriber except the ringing generator, operator's head set and the battery.

If a subscriber *A* initiates a call to the subscriber *B*, *A* lifts the telephone handset from the cradle. This action closes the subscriber's loop which includes transmitter and receiver of the handset. The closing circuit causes a dc current (from battery) to flow through line relay and illuminates the lamp of subscriber *A*. By seeing the light, the human operator, closes the speak key and ask the subscriber *A* "number please". By knowing the called subscriber is *B*,

The operator throws ring key *B* to the ringing generator.

The ringing generator provides a dc current to alert the subscriber *B*. If *B* does not pick up the phone after reasonable time, the operator reports to *A* that the call cannot be connected. If *B* lefts his phone from its cradle, the lamp of *B* glows, The operator then connect jack *A* to jack *B* and then say to *A* go ahead please. Both lights of *A* and *B* are 'on' till their conversation. If any one or both lamps go off, the operator will disconnect the jacks.

Limitations of Manual Exchanges

1. **Language Dependent :** The operation of a human exchange is language dependent as the subscriber needs to communicate with the operator. In multilingual areas (big towns, cities and tourist spots). This language dependency possess severe problems.
2. **Lack of Privacy :** As a human operator is involving in connecting two subscribers, he or she may be willing to hear the conversation of two VIP's or record the message. So in human exchanges, privacy is not possible.
3. **Switching Delay :** Before setting a path between two subscribers, the operator has to monitor various signalling and if the operator is not active, the delay in switching will be high normally it takes minutes to setup a call or release a call.
4. **Limited Service :** An exchange can provide service only to minimum number of subscriber. If the subscriber rate increases, overload and thus congestion are not unexpected.

To avoid congestion, more hardware should be duplicated and more human operator is necessary. These all will results in large overhead for the exchange.

3.3.4. Automated Central Office Switches

A manual exchange can provide service only to minimum number of subscriber. If the subscriber rate increases, overload and thus congestion are not unexpected. Thus, automated switching systems were developed. Early telephone switches were **electromechanical**, many of which are still in use.

They are so named because they use electromechanical relays to perform switching functions. The relays respond to the dial pulses from subscriber phones, and hence activate connections.

An **automated switching system** is a system of switches, sensors and devices that allows subscribers to give instructions directly to the switch without having to go through an operator.

The required features of any automatic switching system are as follows :

TABLE 3.1. Facilities of an Automatic Exchange.

Function	Performed by
To detect that a caller has lifted his handset	Subscriber Line Circuit
To 'busy' his line so that he is not interrupted	Subscriber Line Circuit
To allocate equipment to the caller, if available	Linefinder & Allotter
To indicate to that the caller that he may proceed with dialling	Dial Tone
To accept digits from the caller and route accordingly	Group Selectors
To connect the call through to the appropriate subscriber	Final Selector
To either return Busy Tone if busy or apply ring signal to the called party's phone and ring tone to caller and then cease ringing when the called party answers	Final Selector & Ring Generator
To detect the answering of the call and register it against the caller's account	Metering Circuits
To alert engineers in case of fault	PG/CSH Alarms etc.

3.3.5. Subscriber's Line Circuit (SLC)

Every subscriber is connected to his local exchange by one pair of wires; this single pair carries the voice in both directions and the ring current to ring the bell when a call is received. Within the subscriber's premises, in the UK, the line is actually then split into three wires to allow for an anti-tinkling circuit but it's important to bear in mind that only two wires run from exchange to subscriber (known as *a* and *b*). Once at the exchange, an additional one or two wires are added to the line. These are used for internal signalling are known as *P* (or *Private*) and *M* (or *Meter*).

At the exchange, every subscriber's line terminates into its own **Subscribers Line Circuit (SLC)**. This consists of a pair of relays dedicated to that subscriber; if there are 1000 subscribers on that exchange, then there are 1000 SLCs. All other equipment onwards in the chain is shared between all subscribers - otherwise if there were 1000 of everything the cost and size of an exchange would be astronomical and wasteful. It would only be necessary if it was expected that every subscriber was going to place a call at exactly the same time, and that would never happen. As such, when an exchange is designed, consideration is given to the maximum amount of traffic that would ever need to be carried at one time, and equipment is therefore provided to allow for that. Overflow meters are also installed so that if it can be seen that all available equipment is regularly in use, additional capacity can be added. Although SLCs normally consist of two relays (called 'LR' and 'K' or 'L' and 'K'), on unattended exchanges (ones without an engineer on-site 24 hours) a third relay (Called 'P') was provided to 'park' the line should a fault occur, to avoid tying up common equipment for a long period.

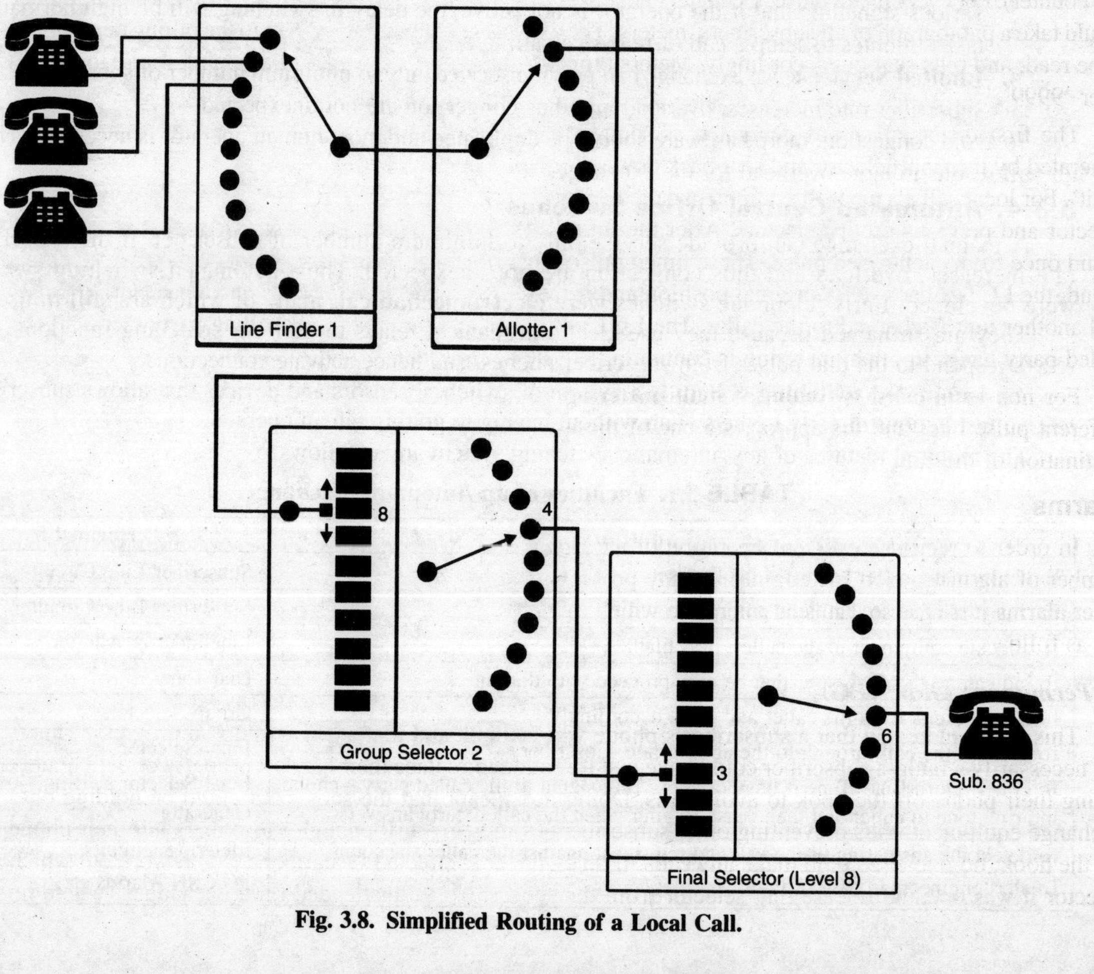

Fig. 3.8. Simplified Routing of a Local Call.

3.3.6. Line Finder and Allotter

When the subscriber lifts his handset, current start to flow on the line; this is detected by the SLC. As you will recall, dialing of the digits causes selectors to step up or round the corresponding number of pulses. As there are many subscribers, but only a few selectors, there has to be a method for (1) Finding a free (available) selector and (2) Connecting the calling subscriber to that selector.

Step (1) is done by the **Allotter**.

Step (2) is done by the **Line finder**.

Although the linefinder is shown looking like a Uniselector in the diagram it is in fact normally a two-motion selector which means that it can serve up to 100 or 200 subscribers. The Allotter, on the other hand is ususally a uniselector, with 25 or 50 outlets, thus allowing access to 25 or 50 first group selectors. The Subscriber's Line Circuits are not shown in the diagram, but would be in the line between the subscribers' telephones and the linefinders.

Another function of the SLC is to 'mark' the caller's line as 'busy' so that incoming calls will detect that the line is in use.

Charging

With the introduction of automatic exchanges, the need arose for automatic charging. Every subscriber was allocated a digit counter (meter) in the exchange. This consisted of an electromagnet which closed with every metering 'pulse'. The electromagnet's armature drove a set of numerical decimal cams (just like a car trip counter). For every meter pulse, the meter clicked one unit. At the end of each billing period, a photographer would take a photograph of all subscribers' meters. These photographs were then sent to the billing department to be read, and bills sent out accordingly. Meters typically had four or five digits, wrapping round to '0000' after '9999'.

The first meter pulse was generated as soon at the called subscriber answers the call. This pulse is generated by the final selector and sent back down the chain of selectors to the caller's meter to register one 'unit'. For local calls, a piece of equipment called a **Local Call Timer** (LCT) was in circuit between the final selector and previous group selectors. After the initial 'answering' pulse, the LCT starts rotating; clicking round once from each timed pulse. These timed pulses are generated by the Ring Generator. After 10 clicks round, the LCT is back to its original position and if the call is still active, it sends another metering pulse, and another unit is charged to the caller. The LCT only stops rotating when the calling party clears. If the called party hangs up, then metering is continuing.

For non-local calls, a system of multi-metering was introduced. A meter pulse generator produced different pulse rates and the appropriate one of these would be applied to the call timer depending on the destination of the call.

Alarms

In order to facilitate efficient operation of an exchange, all Strowger type exchanges are fitted with a number of alarms to alert the engineer to any problems. Some alarms indicate equipment failure, whereas other alarms just indicate unusual operation which might be cause for concern. The most important alarms are as follows:

1. Permanent Glow (PG)

This alarm indicated that a subscriber's phone was offhook and that a call was not in progress. This is not necessarily a fault; a subscriber could have just forgotten to replace their handset properly or deliberately taking their phone off the hook to avoid calls. Doing this meant that the subscriber could be holding onto exchange equipment, thus preventing other subscribers from using it. If enough subscribers left their phone off the hook, no-one else could make any calls. If an engineer spotted a PG alarm, he would work out which selector it was one and release that selector from the subscriber. In unattended rural exchanges (where no engineer was permanently on-site) the subscriber's line circuit differed slightly in that a 'parking' relay was provided so that a PG condition was automatically cleared. The term 'Permanent Glow' originates from the days of manual switchboards when a subscriber being off hook would be shown by his light 'glowing' on the board.

2. Called Subscriber Held (CSH)

This condition occurs when the called party hangs up his phone but the caller still remains active. A CSH alarm is not harmful, and is common—for instance, if someone puts the phone down to go and take the call in another room, a CSH condition will occur whilst their phone is back on the hook. As soon as a CSH condition is detected, a timer starts. If a period of (say) 3 minutes elapses without the called subscriber picking their handset up again then the call is cleared down and charging ceases.

3. Release Alarm (RA)

The release alarm is the important one. The earth supply to most selector electromagnets is connected via the release alarm. As these electromagnets are intended to operate for brief moments only (to drive the selectors), then if current is drawn for more than a few seconds, then it is assumed that the selector has jammed. In order to avoid burning out the selector, the PG alarm is raised, drawing the engineer's attention to the particular rack at fault and he can rush round and clear it. In an unattended exchange, circuitry is provided to automatically lower the current to the faulty selector in order to stop it burning out until the engineer arrives.

3.4. TELEPHONE NUMBERING PLAN

The numbering plan is used to identify the subscribers connected in a telecommunication network. The main objective of numbering plan by any nation is to standardize the number length wherever practical according to CCITT recommendations.

Other objectives includes

(*a*) to meet the challenges of the changing telecom environment

(*b*) to meet subscriber needs for a meaningful and user friendly scheme

(*c*) to reserve numbering capacity to meet the undefined future needs.

A **telephone numbering plan** is a type of numbering scheme used in telecommunications to allocate and route telephone numbers in a telephone network. A closed numbering plan, such as found in North America, features fixed length area codes and local numbers. An open numbering plan features variance in length of area code or local number, or both.

A **dial plan** establishes the expected number and pattern of digits for a telephone number. This includes country codes, access codes, area codes and all combinations of digits dialed. For instance, the North American public switched telephone network (PSTN) uses a 10-digit dial plan that includes a 3-digit area code and a 7-digit telephone number. Most PBXs support variable-length dial plans that use 3 to 11 digits. Dial plans must comply with the telephone networks to which they connect.

In early telephone systems, connections were made in the central office by telephone operators using patch cords to connect one party to another. If a person wanted to make a phone call, he or she would pick up a phone and wind a crank on the side. The crank was a small generator that would light a lamp at the central office. An operator would see the light and insert their patch cord into a socket and assist the customer with the call connection. The operator would use patch cords to connect the caller to the person being called. If the party being called was in another exchange, the operator would use a patch cord to connect to another exchange where an operator elsewhere would finish the connection. As technology advanced, electro-mechanical switches were introduced and calls were made using "rotary dials".

Initial use of area codes in the United States began in the 1950s with large cities. By 1966, the system was nationwide.

Area codes were assigned based on the length of time a rotary dial phone took to dial the area code. Densely populated areas like New York City, Chicago, Los Angeles, and Detroit had huge incoming call volume and were assigned numbers (212, 312, 213, 313) that could be quickly dialed from a rotary dial phone. On a rotary dial phone low digits (1, 2, 3, 4) could dial quickly as the time the rotary dial took to return to the home position was minimal. High digit numbers (7, 8, 9, 0) on rotary dial phones took much longer to return to the home position and were usually used in less densely populated areas like Vermont (802), rural Texas (915), Tennessee (901), and the Canadian Maritimes (902). This system became unnecessary when touch-tone phones arrived, as the tone allowed instant entry of digits.

3.4.1. Standards

Although the International Telecommunication Union (ITU) has attempted to promote common standards among nation states, numbering plans take different formats in different parts of the world. For example, the ITU recommends that member states adopt **00** as their international access code. However, as these recommendations are not binding on member states, some have not, such as the United States, Canada, and other countries and territories participating in the North American Numbering Plan.

3.4.2. ITU Recommendations in Numbering

Some important recommendations of ITU are described below :

1. Recommendation E.164

It provides the number structure and functionality for three categories of numbers used for international public telecommunication. The international numbering plan establishes country codes, that is, area codes that denote nations or groups of nations. The E.164 standard regulates country codes at the international level and sets a maximum length limit on a full international phone number (15 digits). However, it is each country's responsibility to define the numbering within its own network. As a result, regional area codes may be:

- A fixed length, e.g., three digits in the United States and Canada; two digits in Brazil; one digit in Australia and New Zealand,
- A variable length, e.g., between 2 and 5 in Germany, Argentina and in Austria; between 1 and 5 in Japan; 1 or 2 in Syria and Peru; between 2 and 5 in the United Kingdom, or
- Incorporated into the subscriber's number, as is the case in many countries, such as Spain or Norway. This is known as a "closed" telephone numbering plan. In some cases a trunk code (usually **0**) must still be dialed, as in Belgium, Italy, Switzerland, South Africa and some locations within the NANP.

The three categories of numbers are :

(*i*) **National telephone services :** An international public telecommunication number (for geographic areas) is also referred to as the national significant number (NSN). NSN consists of the country code (CC), national destination code (NDC) and the subscriber number (SN).

(*ii*) **Global telephone services :** An international public telecommunication number for global telephone service consists of a three digit country code and global subscriber number.

The country code is always in the 8XX or 9XX range.

(*iii*) **International networks :** An international public telecommunication number for international networks consists of three digit country code, a network identification code and a subscriber number. The country code is always in the 8XX range. The identification code is one to four digits.

Most telephone numbers belong to the E.164 numbering plan, though some PABXs have internal extensions.

The E.164 numbering plan for telephone numbers includes :

- Country calling codes
- Regional numbering plans, such as:
  - the European Telephony Numbering Space
  - the North American Numbering Plan

- Various national numbering plans, such as:
  - Telephone numbers in the United Kingdom

2. Recommendation E.123

This defines a standard way to write telephone numbers, email addresses and web addresses. It recommends how to use hyphen (-), space ( ), or period (•) . ( ) are used to indicate digits that are sometimes not dialed, / is used to indicate alternate numbers and • is used in web addresses.

3. Recommendation E.162.

This recommendation describes that the originating country must analyze a maximum of seven digits of the E.164 international number. When a number is being analyzed, it will be done according to this recommendation.

Also, the international numbering plan or world numbering plan has been defined in recommendations E.160 ; E.161 and E.162.

3.4.3. International Numbering Plan

This plan has to be implemented irrespective of a country's national numbering plan and implemented in accordance to the recommendations of ITU. With some standard international framework, subscribers from different countries can call each other. This plan makes it possible to access all countries with the same country code anywhere in the world.

For the international numbering plan, the world has been divided into nine geographical areas as given below. The general rule is that within each global region each country code starts with the same digit. Figure 3.9 shows the geographical map of world numbering zones.

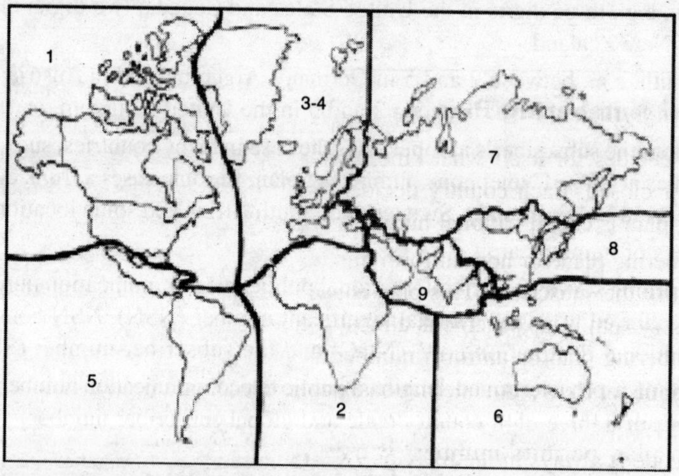

Fig. 3.9. The Geographical Map of World Numbering Zones.

The numbering format for international telephone number is shown in Fig. 3.10. An international telephone number starts with one to three digit country code followed by 9 to 12 subscriber number. The dialing procedure is that the international prefix '00' should be dialed first followed by the telephone number.

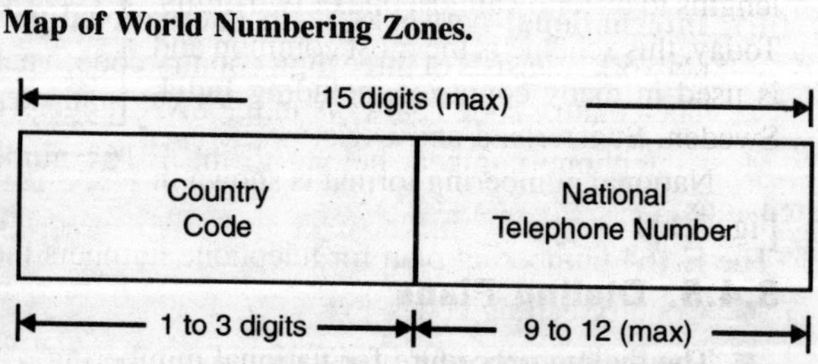

Fig. 3.10. International Telephone Number

Table 3.2 shows the zone code, Zone and two examples in each zone. Digit '0' is not used to indicate zone. Generally '0' is used as Trunk prefix and '00' is used for international prefix.

TABLE 3.2. World Numbering Zones.

Zone code	Zone	Example	
		Country	Country code
1.	North America, cribbean (33 countries)	Canada	1
		United States	1
2.	African continent (61 countries)	Egypt	20
		South Africa	27
3.	Europe (34 countries)	Italy	39
		France	33
4	Europe (15 countries)	Germany	49
		United Kingdom	44
		(Czech Republic)	420
5.	Central and South America (28 countries)	Costarica	506
		Brazil	55
6.	Oceania, South Pacific (32 countries)	Malaysia	60
		Signapore	65
7.	Former USSR (2 countries)	Khzakhsten	7
		Russian Federation	7
8.	East Asia (21 countries)	Japan	81
		International free phone service	800
		Bangladash	880
9.	Middle East, South-West Asia	India	91
		Pakistan	92
		Iraq	964
		Nepal	977

3.4.4. National Numbering Plan

Each country decides for itself what kind of numbering plan it can have. A numbering plan may be open, semi-open or closed. Each country decides what rules to follow when issuing telephone numbers. Such a numbering plan is called national numbering plan.

An open numbering plane or non-uniform numbering scheme allows variations in the number of digits to be used to identify the subscriber. This plan is used in countries equipped extensively with non-director strowger switching system. This scheme is almost extinct.

A closed numbering plan or uniform numbering plan refers to a numbering plan which only allows telephone numbers of a predetermined length. Special services (toll free, premium rate, etc.) are usually excluded from this rule.

A semi-open plan permits number lengths to differ by almost one or two digits. Today, this scheme is the most common and is used in many countries including India, Sweden, Switzerland and U.K.

National numbering format is shown in Fig. 3.11.

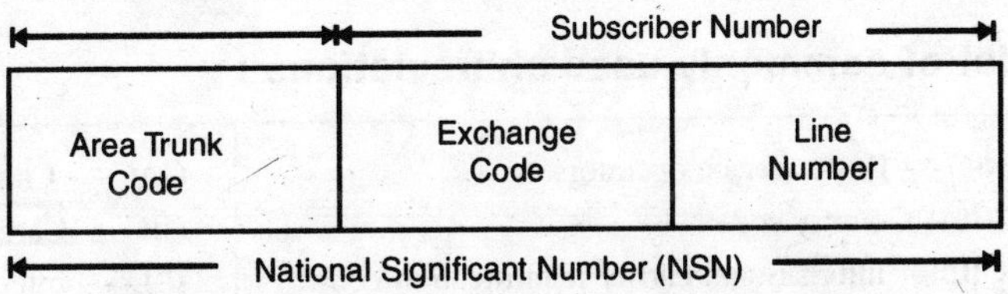

Fig. 3.11. National Telephone Number Format.

3.4.5. Dialing Plans

The dialing procedure for national numbering plan are also comes in two categories.

1. Open dialing plan
2. Closed dialing plans

1. Open Dialing Plan

In open dialing plan local calls can be placed without the trunk prefix and area code. An *open* dialing plan is one in which there different dialing arrangements for local are and long distance telephone calls. This means that to call another number within the same city or area, callers need dial only the number, but for calls outside the area, an *area code* is required. In this situation it is customary to show the area code in parentheses, signifying that in some cases the area code is optional or is not required, as suggested by ITU-T Recommendation E.123. The area code is prefixed by a trunk code (usually "0"), which is omitted when calling from outside the country.

2. Closed Dialing Plans

A closed numbering plan refers to a numbering plan which requires users to dial all numbers at all times. This means that local-local calling also requires the area code to be dialed, as well as the trunk prefix. A **closed numbering** plan is one in which the subscriber's number is a standard length, and a *closed dialing* plan is one in which the subscriber's number is used for all calls, even in the same area. This has traditionally been the case in small countries and territories where area codes have not been required. However, there has been a trend in many countries towards making all numbers a standard length, and incorporating the area code into the subscriber's number. This usually makes the use of a trunk code obsolete.

Thus, the National significant number (NSN) is the combination of trunk code, exchange code and the number. The exchange code and line number together called as subscriber number (SN). NSN length varies from country to country.

3.4.6. Numbering Plan in India

Department of Telecommunication (DOT) India has released its national numbering plan dated April 2003. It was last reviewed during 1993. This existing numbering plan was formulated at a time when there was no competition in the basic telecom services were not available in the country. Further, the existing numbering plan was meant to address monopolistic environment in national and international long distance dialing.

The new numbering plan has been formulated for a projected forecast of 50% teledensity by the year 2030 and thus making numbering space available for 75 crore telephone connections in the country comprising of 30 crore basic and 45 crore cellular mobile connections. The new national numbering plan will be able to meet the challenges of multi operator, multi service environment and will be flexible enough to allow for scalability for next 30 years without any change in basic structure. This plan is aimed at PSTN services, cellular mobile services and paging services. This section focuses mainly on the PSTN services.

List of commonly used abbreviations :

BSO — Basic service operators	CAC — Carrier access code
CC — Country code	CIC — Carrier identification code
ICIC — International carrier identification code	ILD — International long distance
LDCA — Long distance number charging area	LDCC — Long distance charging centre
NDC — National destination code	NLD — National long distance
NLDO — National long distance operators	NSN — National significant number
POI — Point of interconnect	SDCA — Short distance number charging area
SDCC — Short distance charging centre	SN — Subscriber number
TAX — Trunk automatic exchange	TC — Trunk code

LDCA, SDCA, NDA, SN and NSN : Broadly, our country is divided into eight regions with each region being identified by a single digit code as shown in Fig. 3.12. The above said notations are generally used in telecommunications and thus discussed to some extent.

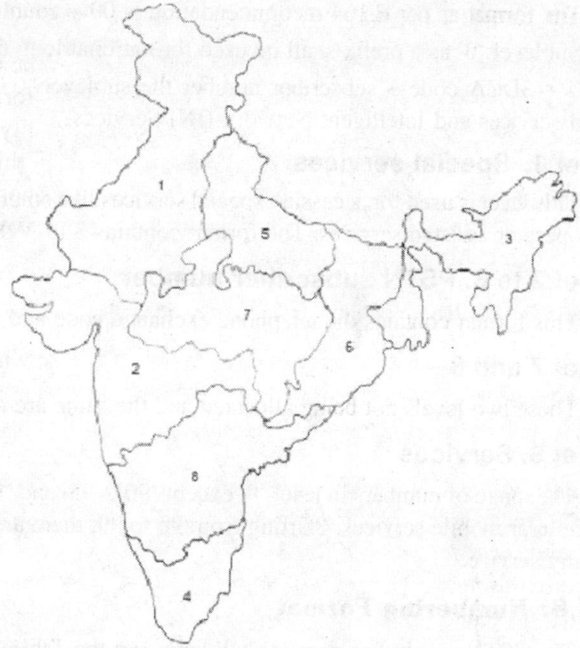

Fig. 3.12. Eight Regions of INDIA.

LDCA : Long distance charging centre comprises of one or several SDCA's. There are at present 322 LDCA's. Each LDCA has a long distance charging centre (LDCC) which is a Trunk Automatic Exchange (TAX).

NPA : Numbers in a Numbering Plan Area (NPA) are not duplicated and called 'subscriber numbers'. To make a call from one subscriber to another subscriber in the same NDA, only the subscriber number needs to be dialed. At present NDA is same as SDCA.

SN : Subscriber number (SN) is a 6 to 8 digit number which includes 2 to 4 digit Telephone exchange code.

SDCA : There are 2645 SDCA's. Each SDCA is allocated a unique trunk code. Each SDCA has one or more number of exchanges. Therefore, there are 2645 codes required to identity the complete country based on SDCA linked numbering scheme. The length of the trunk codes shall vary from 2 to 4 digits. SDCA has a short distance charging centre (SDCC).

SDCC is an integrated local cum tandem or a transit switch. The size of SDCA generally varies between 800 sq. kms. to 2000 sq. kms.

NSN : The numbering scheme envisages the use of SDCA based linked numbering with 10 digit National Significant Number (NSN).

SDCA code	Subscriber Number
2 digit	8 digit
3 digit	7 digit
4 digit	6 digit

3.4.7. National Numbering Scheme

There are ten levels of numbering schemes starting from 0 to 9. In each level, there are many sublevels as a classification of services. All the levels are defined briefly and some sublevels are explained in the following paragraph.

Level 0. Prefix codes

There are various sublevels. Some are described. Sublevel '000' as a prefix shall be used for home country direct service (Bilateral) and international toll free service (Bilateral). The format is 000 + country code + operator code '000800' is used for bilateral international toll free service. Sublevel '00' as a prefix shall be used for international dialing.

The format as per E.164 recommendation is 00 + country code + NSN.

Sublevel '0' as a prefix shall be used for national long distance calls. The format is

0 + SDCA code + subscriber number the sublevel '09' is used for cellular mobile services, satellite based services and Intelligent Network (IN) Services.

Level 1. Special services

This level is used for accessing special services like emergency services, supplementary services, inquiry and operator assisted services. The format contains 3 to *N* digit depending on service.

Level 2 to 6. PSTN subscriber number

This format contains the telephone exchange code and subscriber number.

Level 7 and 8

These two levels not being allocated and the same are reserved for new services.

Level 9. Services

The range of numbers in level '9' except '90', '95' and '96' ('96' is used in paging services) are reserved for cellular mobile services. Starting from 90 to 99, there are 9 sublevels. '90' is used as spare not allocated for any service.

3.4.8. Numbering Format

The PSTN numbering format shall be as per the Table 3.3 given below :

TABLE 3.3.

Trunk code (SDCA code)	+	Telephone exchange code	+	Last n digits of subscriber number
ABCD	+	EF	+	PQRS
ABCD	+	EFG	+	PQR
ABC	+	EF	+	PQRST
ABC	+	EFG	+	PQRS
AB	+	EFG	+	PQRST
AB	+	EFGH	+	PQRS

The trunk code is 2 to 4 digits. The telephone exchange code and last *n* digits of subscriber number together called subscriber number and is from 6 to 8 digits. Hence national number to call a subscriber is 8 to 12 digits. To call a subscriber in another SDCA, prefix '0' must be dialed first.

SDCA Code : Digit A can have any value from 1 to 8. Digit B, C and D can have any value between 0 to 9. AB codes can have any value between 11 to 89 (79 trunk codes). The code 10 is earmarked for carrier access code for NLD service and ILD service. Code 11, 20, 22, 33, 40, 44, 79 and 80 (8 codes) are used presently. 39, 50, 60, 69 and 70 (5 codes) are available for allotment to SDCA's with 3-digit code depending on the requirement. Certain three digit spare codes like 555, 666 and 888 are not to be used as SDCA codes. These are reserved for future services.

Telephone exchange code : Digit E can have any value between 2 to 6. The 0, 1, 7, 8, 9 are not allowed as '0' is used as trunk prefix, level '1' is used for special services, 7 and 8 are kept spare for future services and level '9' is used for cellular mobile services, paging services and access to adjacent areas. Digit F, G and H can have any value from 0 to 9. EFG can take value between 200 and 699 (500 exchange codes).

Last *n* digits of subscriber number : Digit P, Q, R, S and T can have any value from 0 to 9.

Dialing : For a call within a local area *i.e.* SDCA, subscriber number (Telephone exchange code + last *n* digits of subscriber code) only need to be dialed. Thus the number of digits needed to be dialing is 6 to 8.

For calls outside the SDCA, 0 + NSN is need to be dialed (NSN = SDCA code + Telephone exchange code + last n digit of subscriber number). However, access to adjacent areas can also be possible by dialing '95' followed by NSN.

It should be noted that the information given above are according to the materials from the websites, magazines and reference books only. This information's are subject to change according to the change in Government policies and the change in ITU recommendations.

The readers of this book are suggested to refer the related websites for updated changes in numbering plans.

3.5. TELEPHONE SWITCHING HIERARCHY

Figure 3.13 illustrates the North American hierarchical network configuration. It is made up of five classes of exchanges. A class 1 exchange represents a regional center and has the highest order in the hierarchy. A class 5 exchange represents an end office (local exchange) and has the lowest order in the hierarchy. Solid lines represent main trunk lines between exchanges. Dashed lines represent **high usage trunk** lines and can be established between any two exchanges, regardless of their levels in the hierarchy. Should the high usage trunks reach their maximum capacity, additional traffic is overflowed to the main trunks, from which further overflow is permitted. Because of this, the latter is also referred to as the **final route**.

Hierarchical networks have the advantage of requiring relatively simple switch designs. Their main disadvantage is that loss of connection at higher order trunks may severely disrupt the traffic. To reduce this risk, the highest order exchanges are usually interconnected in a mesh-like fashion.

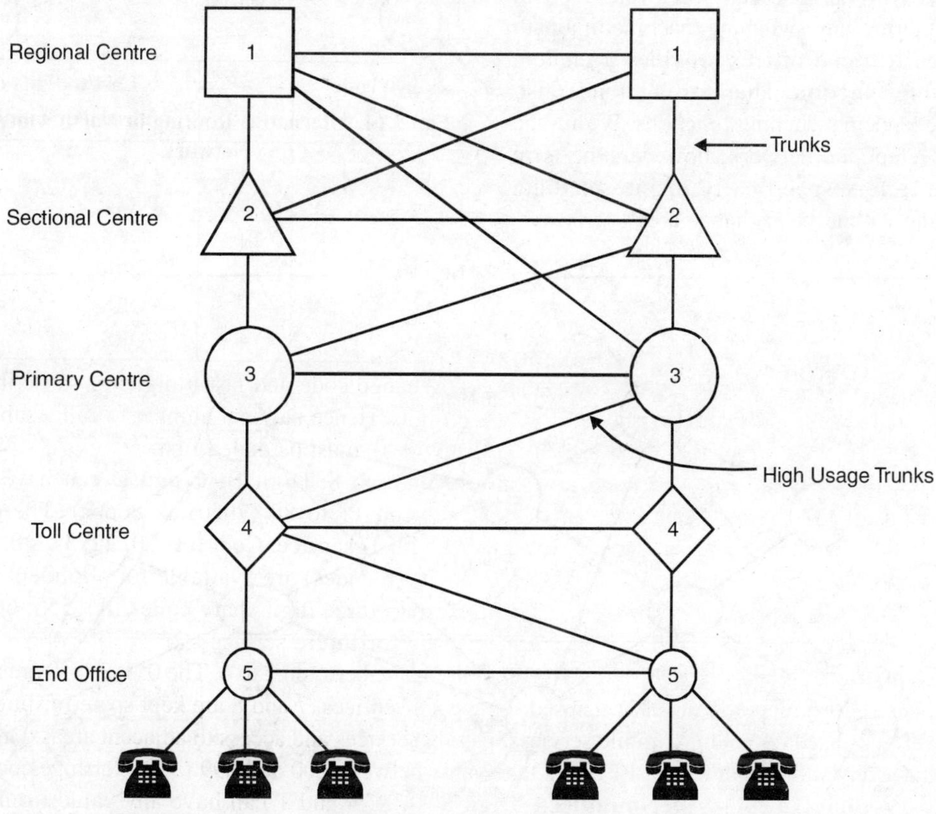

Fig. 3.13. North American Hierarchical Network Configuration.

Traffic was always routed through the lowest available level of the network. This procedure not only used fewer network facilities but also implied better circuit quality because of shorter paths and fewer switching points. Figure 3.14 shows the basic order of selection for alternate routes. The direct interoffice trunks are depicted as dashed lines, while the backbone, hierarchical network is shown with solid lines.

In addition to the high-usage trunks, the backbone network was also augmented with additional switching facilities called tandem switches. These switches were employed at the lowest levels of the network and provided switching between end offices.

Tandem switches were not part of the toll network, as indicated in Figure 3.15, but were (and are) part of what is referred to as an exchange area. Generally speaking, an *exchange area* is an area within which all calls are considered to be local calls (*i.e.*, toll free). In general terms, any switching machine in a path between two end offices provides a tandem switching function. Thus toll switches also provide tandem switching functions. Within the public telephone network, however, the term tandem refers specifically to intermediate switching within the exchange area'.

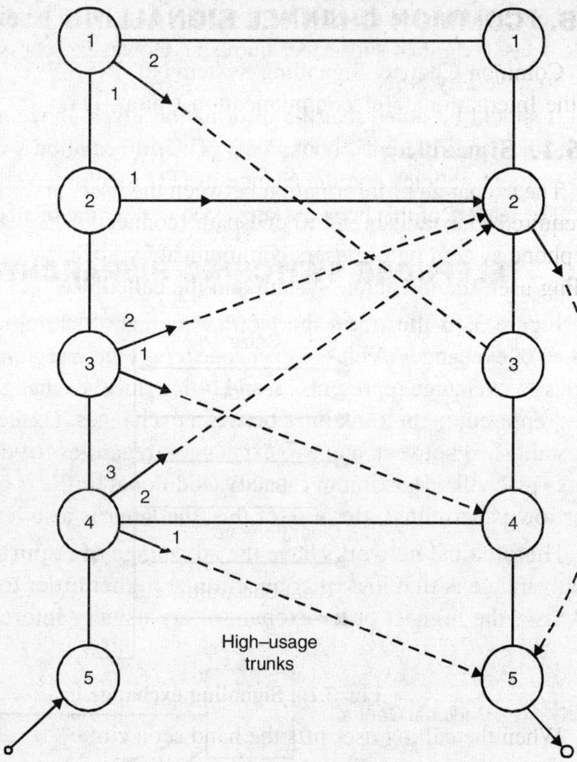

Fig. 3.14. Alternative Routing in North American Network.

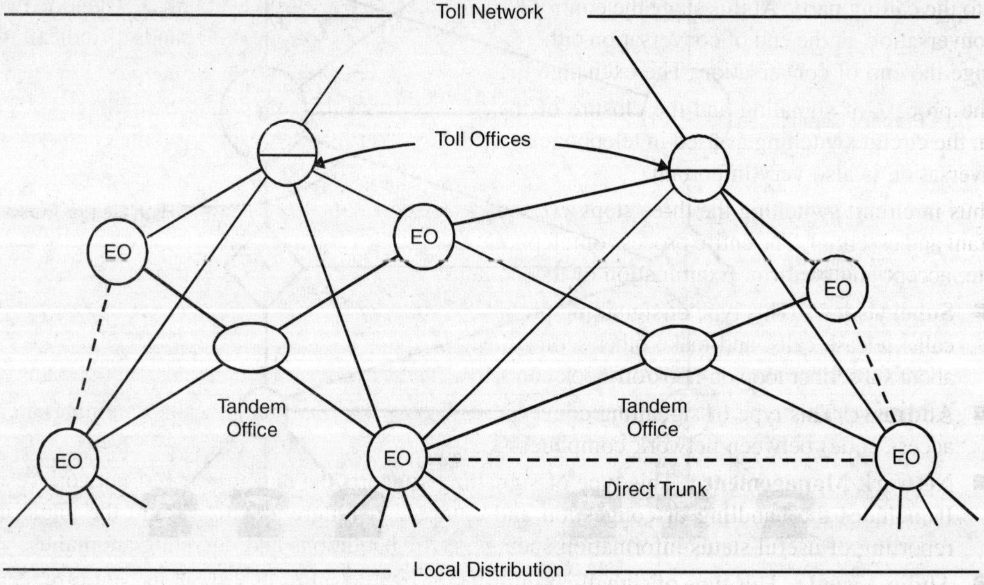

Fig. 3.15. Exchange Area Network.

3.6. COMMON CHANNEL SIGNALLING SYSTEM EVOLUTION OF SS7

Common Channel Signalling System No. 7 (SS-7) is a global standard for telecommunications defined by the International Telecommunication Union (ITU).

3.6.1. Signalling

The exchange of information between the users and the telephone system is called **signaling**. Signaling is required to establish end to end path (connection) between the calling and the called users through the telephone system before users communicate voice messages. Figure 3.16 shows full signaling between the calling user, the telephone system and the called user.

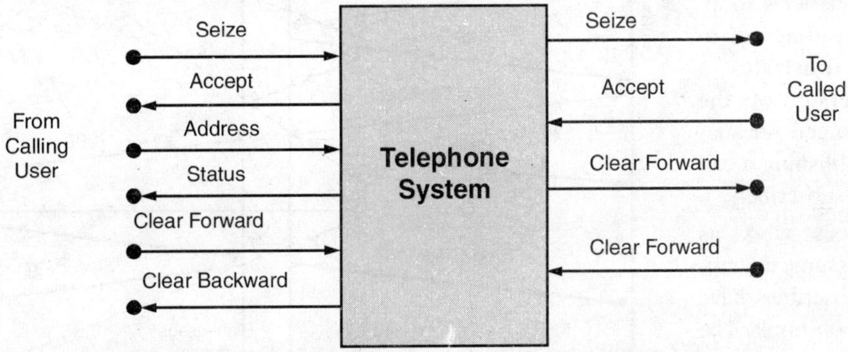

Fig. 3.16. Signaling exchange between users and telephone system

When the calling user lifts the hand set a circuit through the telephone exchange gets closed indicating that the particular user intend to make a call. The user gets a dial tone indicating that the exchange is ready to serve. These two steps correspond to seizure and accept signaling. The called number (address) is passed to the exchange which in turn checks whether the called line is free. If free, the exchange seizes the called party by sending ringing signal. Ringing and response corresponds to seizure and accept signaling between the exchange and the called party. When the called party is busy or responds to ringing, the exchange gives status to the calling party. At this stage the contro! closes the relevant cross point contact. Then the two users start conversation. at the end of conversation either or both the parties replace their handsets indicating to the exchange the end of conversation. The exchange then clears all connections on both sides.

The process of signaling and the closure of the path between the called and the calling party clearly explain the circuit switching as used in telephone system. Clearing/release of lines and the contact at the end of conversation is also very important.

Thus in circuit switching the three steps viz. call set-up, message transfer and the call release are very important and essential. The entire process of telephone communication needs service primitives like request, indicate, accept and confirm. Examination of the signaling diagram carefully would clarify these steps.

- **Supervisory :** This type of signalling provides the necessary control and status signals to establish calls, release calls, and make other service features possible. It includes the informing of exchanges about subscriber loop on-hook/off-hook conditions, and providing information about the status of calls.
- **Address** : This type of signalling conveys addressing information (subscriber number, area code, access code) between network components.
- **Network Management :** This type of signalling supports the management of network resources. It includes the handling of congestion and component failure situations, and the gathering and reporting of useful status information such as traffic conditions and operating anomalies.
- **Audio-Visual :** This type of signalling informs the calling subscriber about the status of a call, and alerts the called subscriber about a waiting call.

Two categories of signalling are discussed below :

1. Subscriber signalling and
2. Interexchange signalling.

3.6.2. Subscriber Signalling

Subscriber signalling refers to the signals exchanged between a subscriber and a local exchange via the subscriber's loop. The sequence diagram in Figure 3.17 illustrates (a simplified version of) the signals exchanged for successfully establishing a call between two subscribers.

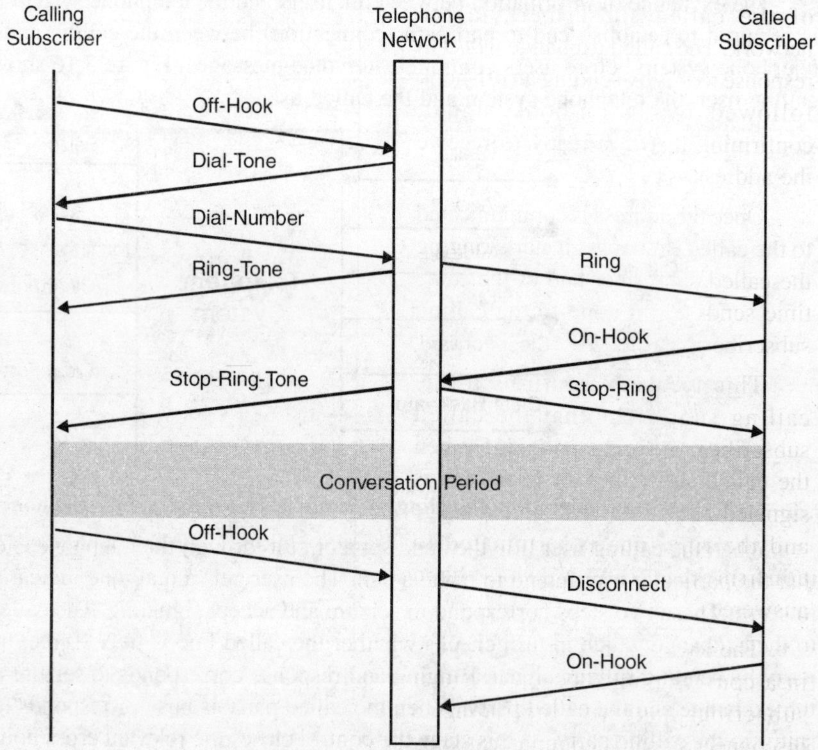

Fig. 3.17. Sample Subscriber Signalling Scenario.

This process works as follows. We assume that initially both subscribers have their phones on-hook. The calling subscriber sends an off-hook signal to the local exchange by lifting the receiver. The switch responds by activating an audible dial tone over the subscriber loop, informing the subscriber that a number may be dialed now. The subscriber dials the other party's number, and each dialed digit is signaled to the local exchange. This number is used as an address to route the call through the network and finally arrives at the local exchange of the called party. This local exchange applies a ringing signal to the called subscriber's loop. At the same time, a confirmation is sent back to the calling subscriber's local exchange, which in turn applies a ring tone to the calling subscriber's local loop. This serves as an audible feedback that the number is dialed. When the called subscriber lifts the receiver, an off-hook signal is sent back to its local exchange. This causes the local exchange to stop the ringing, and is propagated back to the calling subscriber's exchange which in turn stops the ring tone. The network connects the two parties and they may engage in a conversation. Either subscriber can terminate the call by pressing the hook switch. This sends an on-hook signal to the local exchange, which in turn terminates the call and informs the other subscriber accordingly.

3.6.3. Interexchange Signalling

Interexchange signalling refers to the signals exchanged between two or more network exchanges in order to handle calls. The sequence diagram in Figure 3.18 illustrates (a simplified version of) the signals exchanged between two local exchanges via the trunk lines which may involve zero or more tandem exchanges. The scenario is the same as that of Figure 3.10 except that here we are looking inside the network.

This process works as follows. After the calling subscriber has dialed the number of the called subscriber and this address has been provided to the calling exchange, the latter conveys a connect signal to the local exchange of the called subscriber via the network. The called exchange *winks* in response to this by sending an off-hook followed by an on-hook signal, confirming that it is ready to receive the address.

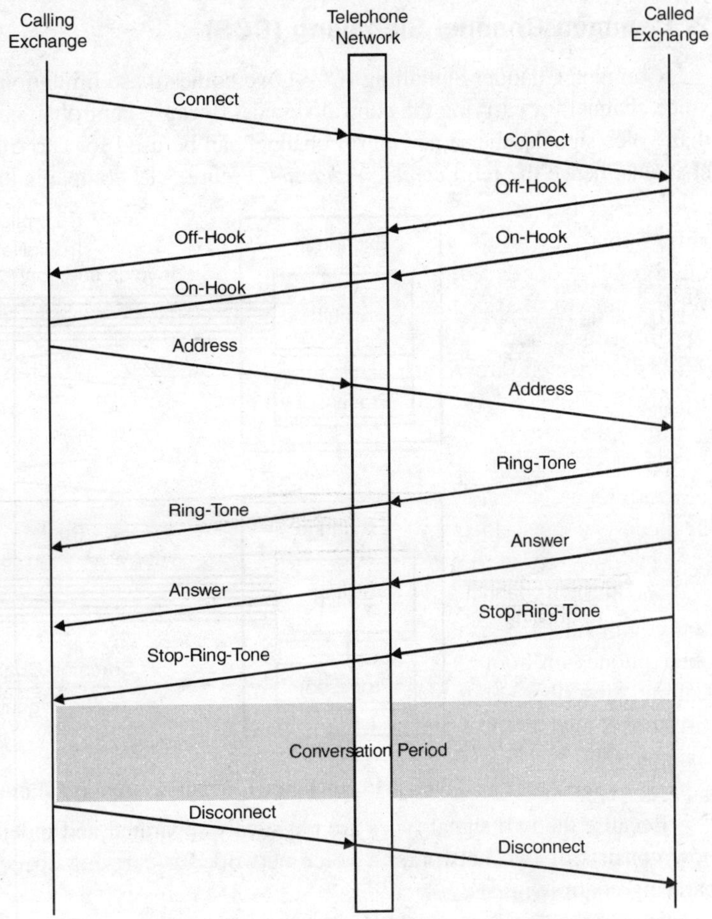

Fig. 3.18. Sample Interexchange Signalling Scenario.

Once the address is communicated to the called exchange, it starts ringing the called subscriber and at the same time sends a ring tone to the calling subscriber's phone.

This feedback confirms to the calling subscriber that the called subscriber's phone is ringing. When the called subscriber answers, this is signaled back to the calling exchange and the ring tone is terminated to confirm that the call has been answered.

The two parties can now engage in a conversation. As soon as either subscriber hangs up the phone, a disconnect signal is communicated by its local exchange to the local exchange of the other party and the call is terminated.

3.6.4. Common Channel Signalling

There are two general methods for conveying control signals between the components of a network :

1. Inchannel signalling
2. Common channel signalling.

1. Inchannel signalling

With **inchannel signalling** Inchannel the control signals occupy the same channel as the voice signal. It comes in two forms :

(*i*) **in-band signalling** uses audio tones for conveying its signals,

(*ii*) **out-of-band signalling** reserves a narrow band within the voice band for conveying control signals.

Inchannel signalling has the *advantage* of using the same trunk lines and equipment for carrying control signals as for voice signals.

It has the disadvantage of being very limited both in speed and richness of signals that can be conveyed.

2. Common Channel Signalling (CCS)

Common Channel Signalling (CCS) overcomes these limitations by using a channel separate from the voice channel for carrying the control signals. Because control signals have a lower bandwidth requirement than voice signals, the same control channel can be used for carrying the control signals of multiple voice channels, hence the term *common channel*. Figure 3.19 compares inchannel signalling to CCS.

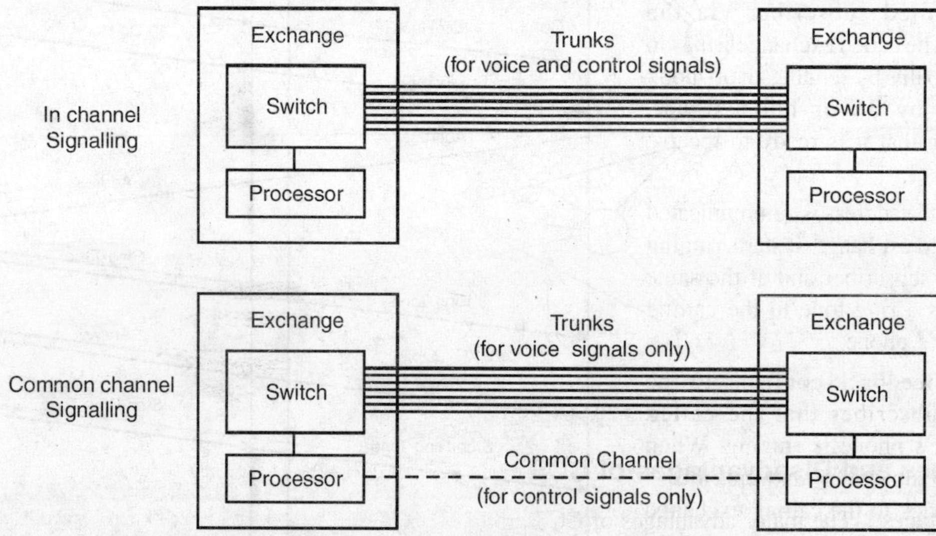

Fig. 3.19. Inchannel Versus Common Channel Signalling.

Because the two signal types are physically separated and independently handled, the overall network now consists of two networks: a voice network for carrying speech, and a distributed data network for carrying control signals.

The relationship of these two networks results in two modes of operation :

1. Associated
2. Non-associated.

In the **associated mode** the common channel is closely associated with the trunks so that the control signals for a call follow the same route as the voice signal.

In the **non-associated mode** the exchanges act as the Signalling Points (SP) as before, but additional **Signal Transfer Points** (STPs) are introduced to implement a signalling network of different topology to the voice network. As a result, the control signals no longer necessarily follow the same route as the voice signals. This produces two independent networks with links at the switching points. Figure 3.20 compares the two modes.

The non-associated mode, while considerably more complex, offers all the flexibility of data networks for routing the control signals. For example, it is now possible to route call signalling information to a central point where information can be looked up in a database to obtain a service profile for handling the call.

The Common Channel Interoffice Signalling (CCIS) system in North America which is based on the CCITT Signalling System Number 6 (SS6) is a living example of common channel signalling. CCIS supports both the associated and the nonassociated mode of operation. It uses packet switching for its control signalling.

Each packet consists of a set of Signal Units (SUs) for conveying signalling information.

The limited line speed of CCIS and SS6 (4.8 kbps and 2.4 kbps, respectively) and the limited size of the SUs has lead to the development of a replacement for SS6. This is discussed in the next section.

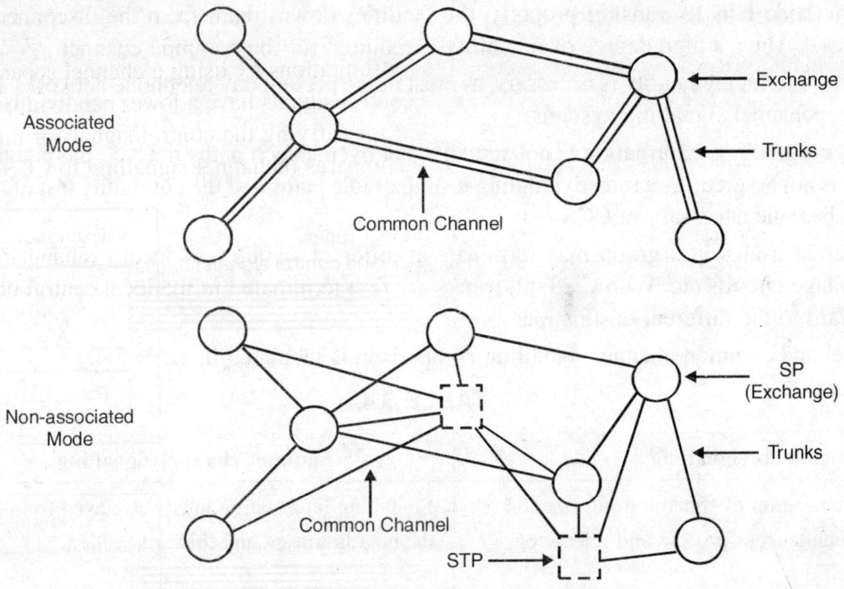

Fig. 3.20. CCS Modes of Operation.

Advantages and Disadvantages of CCS

Advantages : The major advantages of CCS are listed below:

1. For each associated trunk group, only one set of signalling facility is required. The channel used for CCS need not be associated with any particular group. Figure 3.19 shows a CCS network that is disassociated from the message network structure.
2. The introduction of SPC switching machines and CCS provides efficient routing procedure.
3. CCS allows for signalling at any time in the entire duration of a call, not only at the beginning.
4. CCS removes most of the signalling costs associated with inter office trunks.
5. Information can be exchanged between processor at high speed.
6. The CCS provides acceptable quality for network related signalling tones such as DTMF, MF and SF achieved.
7. As there is no need of line signalling equipment on every function, considerable cost savings can be achieved.
8. The D1 channel bank use 1 bit per time slot for signalling and 7 bits for voice which provides signalling rate 8 kbps. D2 channel provides higher data rate which use 1 bit in every sixth frame for signalling. This is referred as "robbed bit signalling."
   When CCS is utilized, the associated T carrier system no longer needs to carry signalling information on a per channel basis and a full 8 bits of voice can be transmitted in every time slot of every frame.
9. As separate channels are used for voice and control. There is no chance of mutual interference and the error rate is very low.
10. CCS enables more services to the subscribers. A signalling link operating at 64 k bit/s normally provides signalling up to 1000 or 1500 speech circuits.

Disadvantages of CCS

Some disadvantages of CCS are listed below:

1. The CCS network is basically a store and forward network. So, in a established circuit, the signalling information's are stored, processed and then forwarded to next node. This cause additional overhead and disconnect to the continuity.

2. If one node fails to transmit properly, the facilities downstream from the disconnect will not be released. Thus, a high degree of reliability is required for the common channel.
3. Proper interfacing facility is necessary, as most of the present day telephone networks are equipment with inchannel signalling systems.
4. As the signalling information is not actually sent over speech paths in CCS, the integrity of speech path is not assured. As a remedy, routing testing of idle paths and the continuity test of an established path become necessary in CCS.
5. Different trunks in a group may terminate at different switch, say local exchange, other foreign exchange circuits etc. With CCS, all trunks are first terminated to the local central office and then forward to the different destination.

In channel and common channel signalling comparison is tabulated in Table 3.4.

TABLE 3.4.

Inchannel signalling	Common channel signalling
Automatic propagation of signalling information enables the simultaneous process and release of associated facilities.	Signalling information must be relayed from one node to the next in a store and forward fashion.
Integrity of speech and signalling are maintained as they are on the same path.	Special equipments should be provided for the integrity as they are travelling on separate path.
Separate signalling equipment is required for each trunk and hence is expensive.	Only one set of signalling equipment is required for a whole group of trunk circuits and there fore CCS is economical.
Transfer of information such as address digits is from common control network originating office—Voice channel—receiving office—common control network.	Transfer of information is directly between control elements (processors of SPC systems).
Trunks are held up during signalling.	Trunks are not required for signalling.

3.6.5. Signalling System Number 7

The CCS7 (*i.e.* SS7 or C7) is a global standard for telecommunication defined by ITU-T. The standard defines the procedures and protocols by which network elements in the PSTN transfers information over a digital signalling network to effect wireless (cellular) and wireless call setup, routing and control.

CCITT's Signalling System Number 7 (SS7) was designed with the requirements of digital telephone networks in mind. As a common channel signalling standard, it is suitable for use with a wide range of circuit-switched digital networks. Like its predecessor, SS6, it uses packet switching, but is primarily designed to work with 64 kbps digital channels of modern exchanges. However, it can also operate over analog channels of lower speeds, as well as digital channels of higher speeds.

SS7 is a very large and complex system. It is the signalling system of choice for ISDN. The CCITT Q.700 series of recommendations which is comprised of numerous parts is entirely devoted to the definition of SS7.

Since SS7 is a data network (similar to X.25, but designed for the specific application of signalling), it is useful to compare its protocol architecture to the OSI reference model. This is illustrated in Figure 3.18.

Each level is separately discussed below.

OSI Layer	SS7 Level		Purpose
Higher layers		Operations and Maintenance Applications Part	Provides network management functions for the operation, administration, and maintenance of the network.
Network	4	User Parts	Deals with signalling data contents.
		SCCP	Provide additional network services.
	3	Signalling Network Functions	Routing of signalling data through a set of signal transfer points.
Data Link	2	Signalling Link Control	Reliable, ordered delivery of signal units over a signalling data link.
Physical	1	Signalling Data Link	Transmission of signalling data bits over the physical channel.

Fig. 3.21. SS7 Protocol Architecture.

The CCS-7 signalling was introduced in India with the commissioning of new technology switches in major cities by the end of 1995–96. These new technology switches are manufactured by multinational companies such as LM Ericsson of Sweden, Siemens of Germany, AT & T of the U.S.A., Alcatel of France, and Fijitsu of Japan.

3.6.5.1. Purpose of SS7 Network

1. An interesting feature of SS7 is that, it is a prerequisite for introduction of the following new services :
   (*a*) The SS7 provides the internal control and network intelligence essential to an ISDN.
   (*b*) Intelligent Network (IN) which uses network elements such as SSP, SCP and intelligent peripheral (IP), for achieving improved subscriber substation.
   (*c*) Personal Communication Systems (PCS) which uses SS7 to provide personalized voice, data, image and video communication services that can be accessed regardless of location network and time utilizing advanced microcell technology. It enables personalized billing, personalized numbering (one number for one person at any place at any time) and time and location independent charging.
2. The primary purpose of SS7 was to access remote database to lookup. This lookup process has several benefits because the carriers do not have to maintain a full database at each switching node, but they know how to get the remote data base end find the information quickly.
3. SS7 achieves, increased revenue generation, additional call charging capabilities, variety of calling-card services.
4. SS7, in integration with SPC systems enables quick and efficient call setup and teardown across the network in less than one second. Also, this integration provides for better supervision, monitoring and billing systems integration.
5. SS7 enables full use of the channel for the talk path because the signalling is done out of band on its own separate channel.
6. SS7 achieves enhanced call features such as call forwarding, calling party name/ number display and three way calling. Also with SS7, efficient and secured worldwide communication is possible. It also provides, toll free and toll wire line services.
7. It handles the rerouting of network by using automatic protection switching services such as those found in SONET.
8. SS7 uses packet switching concept. Hence SS7 network is capable of preventing the misrouted calls, duplication of call requests and lost packets (requests for service).

3.6.5.2. Features of SS7

Introduction of this form of signalling is essential for the development of telecom networks in any country because of the following features:

1. Internationally standardized by the ITU.
2. SS7 is suitable for any transmission medium *i.e.*, can be operated over both terrestrial and satellite links.
3. Even though SS7 is optimized to work with digital SPC exchanges utilizing 64 kbps digital channels, it is suitable for operation over analog channels.
4. SS7 is suitable for various communication services such as telephony, text, data, images and video.
5. Transport mechanism is application independent.
6. High performance and flexibility along with a future oriented concept which meets new requirements.
7. High reliability for message transfer.
8. Processor friendly structure of the messages (signal units in multiples of 8 bits).
9. SS7 networks are much faster, efficient in the call setup and teardown process.
10. Flexible structure which accommodates all the technical advancement in telecommunication systems.

3.6.5.3. SS7 Network Architecture

The SS7 network is built out of the essential components interconnected by signalling links. SS7 messages are exchanged between network elements over 56 or 64 kbps bidirectional channels called signalling links. In this section, the essential components used in the signalling network, signalling link types and the basic SS7 architecture are described.

3.6.6. Signalling Points of SS7 Network

There are three essential components used in SS7 network. They are SSP, STP and SCP. Fig. 3.21 (*a*) shows the symbols that are used to depict these three key elements of any SS7 network. STP and SCP's are customarily deployed in pairs. While elements of a pair are not generally co-located, they work redundantly to perform the same logical function. When drawing complex network diagram, these may be depicted as a single element for simplicity as shown in Fig. 3.22 (*b*).

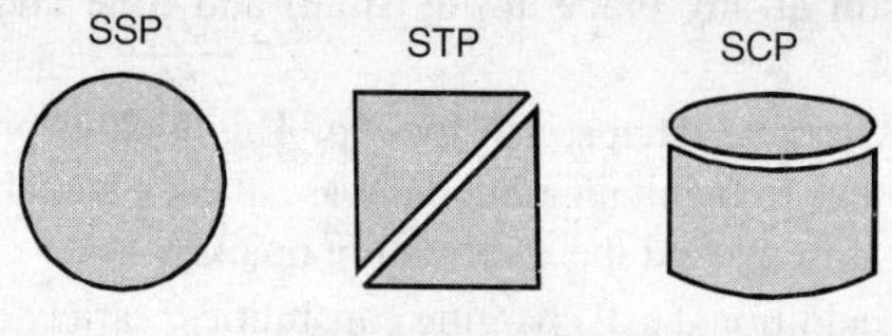

(*a*) Symbols of basic components

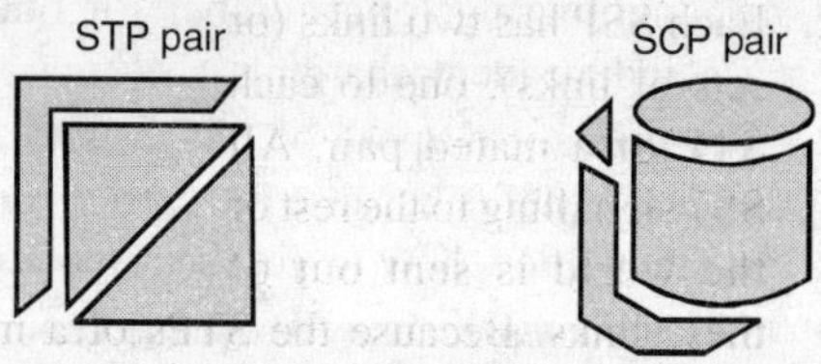

(*b*) Pair of components

Fig. 3.22. Components of SS7 Network.

Signal Switching Points (SSP's)

SSP's are telephone switches (end offices or tandems) equipped with SS7 capable software and terminating signalling links. An SSP sends signalling message to other SSP's to setup manage, and release voice circuits required completing a call.

An SSP may also send a query message to a centralized database (an SCP) to determine how to route a call. Sometimes SSP also referred as service switching point.

Signal Transfer Point (STPS)

STPS are the packet switches of the SS7 network. An STP routes each incoming message to an outgoing signalling link based on routing information contained in the SS7 message. Because it acts as a network hub, an STP provides improved utilization of the SS7 network by eliminating the need for direct links between signalling points.

An STP may perform global title translation, a procedure by which the destination signalling point is determined from digit present in the signalling message. An STP can also acts as a "fire wall" to screen SS7 messages exchanged with other networks.

Signal Control Points (SCP's)

SCP's are databases that provide information necessary for advanced call processing capabilities. An SCP sends a response to the originating SSP containing the routing number(s) associated with the dialed number. An alternate routing number may be used by the SSP if the prime number is busy or the call is unanswered within a specified time.

SCP's and STP's are generally deployed in mated pair configurations in separate physical locations to ensure network wide service in the event of isolated failure.

3.6.7. Basic Signalling Architecture

The essential components STP, SSP and SCP are combined together to form the architecture of SS7 network. Figure 3.23 shows a simple example of how the basic elements of SS7 network are deployed to form two interconnected networks.

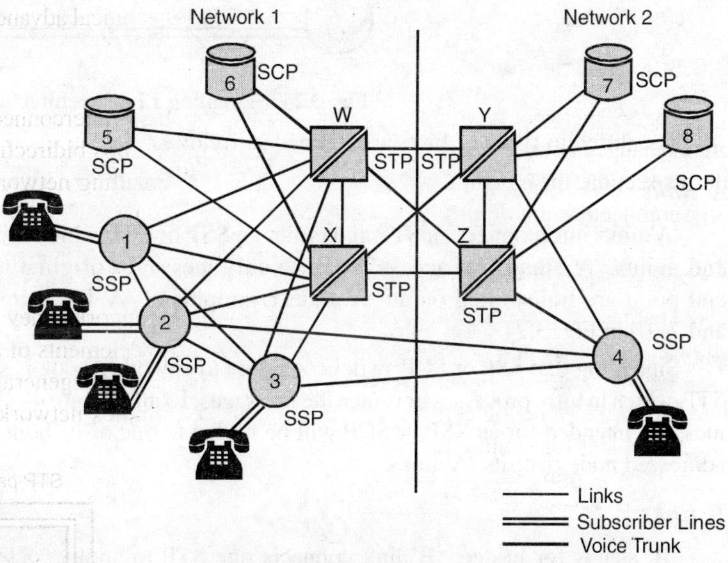

Fig. 3.23. Typical SS7 Network.

1. **Mated pair of STPs.** STPs W and X are referred as a matched pair of STPs because both perform identical functions. They are redundant. Similarly Y and Z form a matched pair.
2. Each SSP has two links (or sets of links), one to each STP of a mated pair. All SS7 signalling to the rest of the world is sent out of these links. Because the STPs of a mated pair are redundant, messages sent over either link (to either STP) will be treated equivalently.
3. The STPs of a matched pair are joined by a link (or set of links).
4. **Quad.** Two mated pairs of STPs interconnected by four links (or set of links) are referred as a quad.
5. **Matched pairs of SCPs.** SCPs are usually (though not always) deployed in pairs. As with STPs, the SCPs of a pair are intended to function identically. Pairs of SCPs are referred as matched pairs of SCPs. Note that they are not directly joined by a pair of links.

3.6.8. SS7 Signalling Link Types

Signalling links are logically organized by link type (A through F) according to their use in the SS7 signalling network. SS7 signalling links are characterized according to their use in the signalling network. Virtually all links are identical in that they are 56 kbps or 64 kbps bidirectional data links that support the same lower layers of the protocol.

Figure 3.24 shows the signalling link architecture.

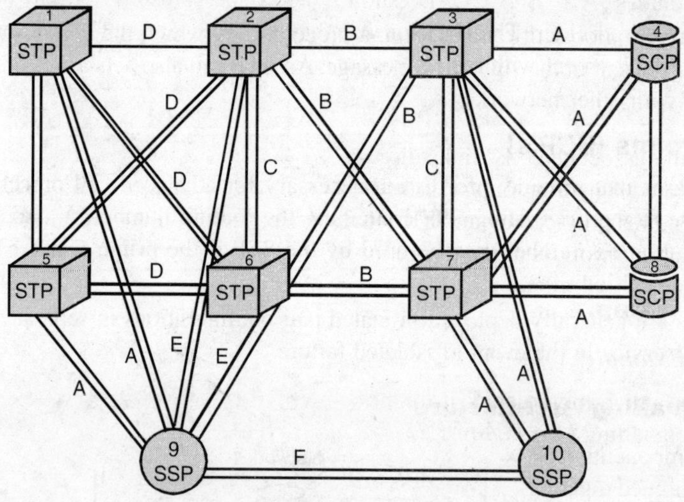

Fig. 3.24. Signalling Link Architecture.

The different types of links are defined as follows :

A links

'A' links interconnect an STP and either an SSP or SCP which are collectively referred to as signalling end points. 'A' stands for access, because only messages originating from or destined to the signalling end point are transmitted on an 'A' link. Examples of 'A' links are 1–9, 3–10, 3–4, 3–8, 4–7, 5–9, 7–8 and 7–10 in Fig. 3.24.

Signalling that SSP or SCP switches to send to any other node is sent on either of it's a links to its home STP, which in turn, processes or routes the messages. Examples are 4–7–10 (or) 8–7–10 in Fig. 3.21. Similarly, message intended for an SSP or SCP will be routed to one of its home STP which will forward them to the addressed node over its 'A' links.

B links

'B' stands for bridge. 'B' link connects one STP to another STP. Bridge describes the quad of links interconnecting peer pairs of STPs. 'B' links, or D links interconnecting two mated pairs of STPs are referred as B/D links.

C links

'C' stands for cross. A cross (C) links connects, STPs performing identical functions into a matched pair. A 'C' link is used only when an STP has no other route available to a destination signalling point due to link failure. Thus the 'C' links are used to enhance the reliability of the signalling network.

D links

The link 'D' denotes diagonal and describes the quad of links interconnecting mated pairs of STPs at different hierarchical levels. Because, there is no clear hierarchy associated with a connection between networks, interconnecting links are referred as B, D or B/D links.

E links

'E' link stands for extended. This link connects an SSP to an alternate STP. 'E' link provide an alternate signalling path if an SSP's home STP cannot be reached via an A link.

Hence 'E' links provide backup connectivity to the SS7 network. In Fig. 3.24, 9–2 and 9–6 are alternative E links. A network includes "A", "B", "C", "D" and E links. But E links may or may not be deployed at the discretion of the network provider. The decision of whether or not to deploy "E" links can be made by comparing the cost of deployment with the improvement in reliability.

F links

"F" stands for fully associated. "F" links directly connect two signalling end points (SSP's and SCPs). "F" links allow associated signalling only. F links are not usually used in networks with STPs. Because they bypass the security features provided by an STP, F links are not generally deployed between networks. Their use within an individual network is at the discretion of the network provider.

3.6.9. SS-7 Call Setup

Basic Call Setup Example

Before going into much more detail, it might be helpful to look at several basic calls and the way in which they use SS7 signalling (see *Figure 3.25*).

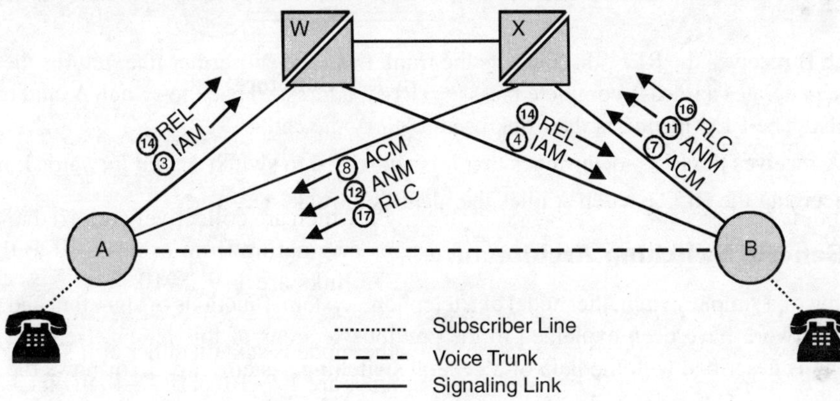

Fig. 3.25. Call Setup Example.

In this example, a subscriber on switch A places a call to a subscriber on switch B.

1. Switch A analyzes the dialed digits and determines that it needs to send the call to switch B.
2. Switch A selects an idle trunk between itself and switch B and formulates an initial address message (IAM), the basic message necessary to initiate a call. The IAM is addressed to switch B. It identifies the initiating switch (switch A), the destination switch (switch B), the trunk selected, the calling and called numbers, as well as other information beyond the scope of this example.
3. Switch A picks one of its A links (e.g., AW) and transmits the message over the link for routing to switch B.
4. STP W receives a message, inspects its routing label, and determines that it is to be routed to switch B. It transmits the message on link BW.
5. Switch B receives the message. On analyzing the message, it determines that it serves the called number and that the called number is idle.
6. Switch B formulates an address complete message (ACM), which indicates that the IAM has reached its proper destination. The message identifies the recipient switch (A), the sending switch (B), and the selected trunk.
7. Switch B picks one of it's A links (e.g., BX) and transmits the ACM over the link for routing to switch A. At the same time, it completes the call path in the backwards direction (towards switch A), sends a ringing tone over that trunk towards switch A, and rings the line of the called subscriber.

8. STP X receives the message, inspects its routing label, and determines that it is to be routed to switch A. It transmits the message on link AX.
9. On receiving the ACM, switch A connects the calling subscriber line to the selected trunk in the backwards direction (so that the caller can hear the ringing sent by switch B).
10. When the called subscriber picks up the phone, switch B formulates an answer message (ANM), identifying the intended recipient switch (A), the sending switch (B), and the selected trunk.
11. Switch B selects the same A link it used to transmit the ACM (link BX) and sends the ANM. By this time, the trunk also must be connected to the called line in both directions (to allow conversation).
12. STP X recognizes that the ANM is addressed to switch A and forwards it over link AX.
13. Switch A ensures that the calling subscriber is connected to the outgoing trunk (in both directions) and that conversation can take place.
14. If the calling subscriber hangs up first (following the conversation), switch A will generate a release message (REL) addressed to switch B, identifying the trunk associated with the call. It sends the message on link AW.
15. STP W receives the REL, determines that it is addressed to switch B, and forwards it using link WB.
16. Switch B receives the REL, disconnects the trunk from the subscriber line, returns the trunk to idle status, generates a release complete message (RLC) addressed back to switch A, and transmits it on link BX. The RLC identifies the trunk used to carry the call.
17. STP X receives the RLC, determines that it is addressed to switch A, and forwards it over link AX.
18. On receiving the RLC, switch A idles the identified trunk.

3.6.10. A Generic Switching Architecture

The working of a simple switch, the model of a telephone system, functions of signaling and the operation of a telephone network have been explained in the previous sections of this lesson. Here the operation of circuit switching is described with the help of a general switching system. Fig. 3.26 shows the diagram of a

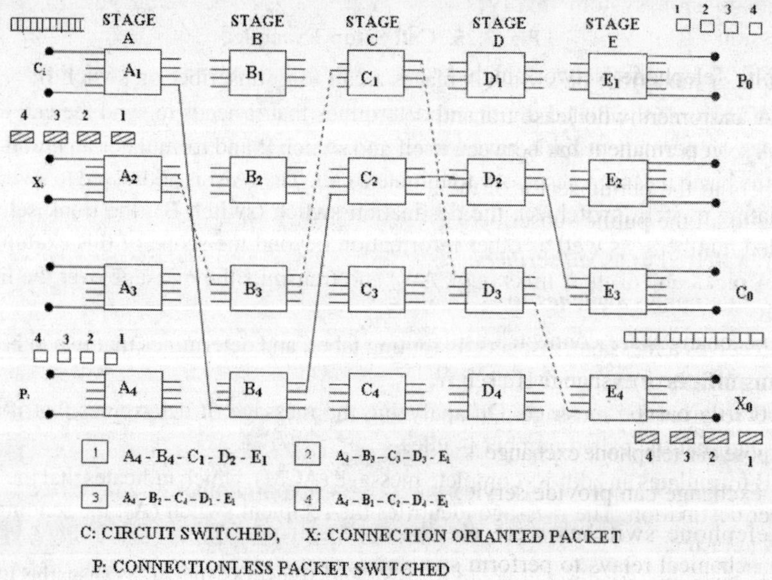

Fig. 3.26. Switching Schemes

general switching system. In fact this diagram will be used to explain various switching schemes such as circuit switching, connectionless packet switching and connection oriented packet switching. Only circuit switching is discussed in this section.

The architecture in the figure has five stages namely *A*, *B*, *C*, *D* and *E* of switching and each stage has a group of four switches namely 1, 2, 3 and 4 (A_1, A_2, A_3 and A_4 at the first stage). Users are distributed in four groups. The calling users are connected to the first stage and similarly the called users are connected to the fifth stage. For convenience the users are shown separately connected as calling users/input lines and called users/output lines. Actually the same user when makes a call is the calling user and becomes the called user when receives the call. In a telephone system the user works in a half duplex mode and is connected to the network with two wires. Internally the network actually works on four wires.

A circuit switched user, C_i wants to make a call to another circuit switched user C_0 in another country. C_i seizes the switch A_1 and on knowing the called user number, the network finds and specifies the end to end path, say, A_1-B_4-C_2-D_2-E_3. All these intermediate switches and the interconnecting links are simultaneously switched on and the entire circuit (path) from C_i to C_0 is dedicated to the call between them for the entire duration. The message from C_i flows to C_0 as if a direct pipe has been laid between the two users. This pipe is withdrawn at the end of the call. The probability that all the switches and the links will be working without any defect at the time and duration of the call is rather low. The cost of dedication the entire path under such a constraint is rather quite high. It may be noted that a single failure anywhere in the circuit renders the call aborted. Except for the call set up delay, the communication in this mode of switching is real time and the quality of service (QoS) is excellent because there is no more processing is taking place in the network and as such there is no interference after the call has been set up. The users once get the pipe are the owners of the pipe for the duration of the call. Circuit switching is considered very reliable. The reliability, the QoS and the real time communication demand high price.

Summary

- The telephone network had its beginnings in the late 1876s.
- The transmission system involves analog transmission (voice communication) and digital transmission.
- The Public Telephone Network includes mainly two types of subscribers: Private and Public
- All those customers who lease transmission media, equipments and services from the telephone companies on permanent basis have known as **private subscribers.**
- Subscribers to the public sector of the telephone network share facilities and equipments that are available to all the public subscribers to the network.
- Users are referred to as ***subscribers.***
- The system employs ***switches*** to facilitate connections between subscribers.
- The two exchanges are connected using a set of interexchange lines called **trunk lines.**
- **Switching offices** or exchange is a central location where subscribers are interconnected, permanently or temporarily basis.
- The purpose of a telephone exchange is to make a path for a call to be completed between two subscribers.
- Manual exchange can provide service only to minimum number of subscriber.
- Early telephone switches were **electromechanical**; they are so named because they use electromechanical relays to perform switching functions.
- An automated switching system is a system of switches, sensors and devices that allows subscribers to give instructions directly to the switch without having to go through an operator.

- The numbering plan is used to identify the subscribers connected in a telecommunication network.
- A **dial plan** establishes the expected number and pattern of digits for a telephone number.
- Each country decides what rules to follow when issuing telephone numbers. Such a numbering plan is called national numbering plan.
- An *open* dialing plan is one in which there different dialing arrangements for local are and long distance telephone calls.
- A *closed numbering* plan is one in which the subscriber's number is a standard length, and a *closed dialing* plan is one in which the subscriber's number is used for all calls, even in the same area.
- Common Channel Signalling System No. 7 (SS-7) is a global standard for telecommuni-cations defined by the International Telecommunication Union (ITU).
- Signalling refers to the exchange of control information between the components of a network (telephones, switches, etc.) in order to establish, manage, and disconnect calls.
- Subscriber signalling refers to the signals exchanged between a subscriber and a local exchange via the subscriber's loop.
- Interexchange signalling refers to the signals exchanged between two or more network exchanges in order to handle calls.
- In the **associated mode** the common channel is closely associated with the trunks so that the control signals for a call follow the same route as the voice signal.
- In the **nonassociated mode** the exchanges act as the Signalling Points (SP).
- The CCS7 (*i.e.* SS7 or C7) is a global standard for telecommunication defined by ITU-T.

Review Questions

1. Explain how the call is set up using SS7 signaling network.
2. Briefly describe interoffice trunk, tandem trunk and tandem switches.
3. Explain the telephone switching hierarchy in details.
4. Describe local central office telephone exchanges and operated assisted local exchange in details.
5. Explain the working of automated central office switches and exchanges.
6. What is meant by SS7 signalling points ? List and describe the three SS7 signalling points.
7. How is multiplexing of telephone channels achieved ? What is FDM hierarchy ?
8. Explain formation of groups, super group, master groups and radio channels.
9. Define composite baseband signal and describe the process of formation of radio channels.
10. Discuss the public telephone network.
11. Explain the typical telephone circuit route for the medium distance call.
12. Discuss the role of switching offices in the PSTN services.
13. List the end instruments and its use.
14. What are the primary functions of transmission systems?
15. Define local loop and trunks.
16. Discuss the subscriber line circuit in brief.
17. What are the two types of subscribers to the public telephone network? Briefly discuss them.
18. What is the use of multiplexing? List the various multiplexing techniques.
19. What are the limitations of manual exchanges?
20. With a block diagram, explain the functions of a step by step switching system.
21. Explain the routing of a local call with a suitable example and diagram.

22. List the disadvantage of strowger switching system.
23. What are the unique features of cross bar switches?
24. What is the basic principle of cross bar? With necessary diagrams explain the operation.
25. What are the different forms of signalling?
26. What is an automated central office switch?
27. Briefly describe the North American Telephone Switching hierarchy.
28. What are the purposes of telephone network signalling function?
29. Distinguish Inchannel signalling and CCS?
30. What are Inband and Outband Signalling?
31. List the advantages and disadvantages of Inband Signalling?
32. Draw the CCS Signalling message formats.
33. What is STP, SSP and SCP
34. Draw SST network architecture.
35. What are the signalling link types?
36. Explain with necessary diagrams, each type of the voice frequency (VF) signalling.
37. Explain architecture of T1 link and DS-1 frame format in relation to the PCM signalling.
38. What are the three ways of implementing CCS? Explain each types of signalling with neat diagrams.
39. List the advantages and disadvantages of CCS.
40. List the purpose and features of SS7.
41. Explain all the signalling link types of SS7 link architecture.
42. Explain with neat diagram the protocol architecture of SS7.
43. Name three types of signalling units used in SS7. With neat diagrams explain each fields associated with the signalling units.
44. Discuss the telephone numbering plan and its importance.
45. Discuss the various ITU recommendations in numbering plan.
46. Explain the dialing plans for national numbering plan.
47. Write short notes on:
    (*a*) Instruments
    (*b*) Local loop
    (*c*) Trunks circuits
    (*d*) Exchanges
    (*e*) Automated central office
    (*f*) Subscriber line circuit
    (*g*) Line finder and allotter
    (*h*) Telephone numbering plan
    (*i*) Recommendation-E64
    (*j*) Open and close dialing plans
    (*k*) Subscriber signalling
    (*l*) Interexchange signalling
    (*m*) Multiplexing
    (*n*) FDM
    (*o*) WDM
    (*p*) SDM

❑❑❑

III

UNIT

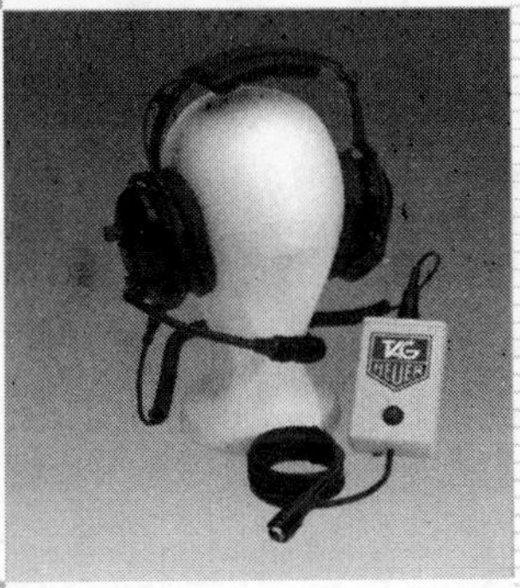

➤ Multiplexing of Telephone Channels

SYLLABUS

Multiplexing of Telephone Channels : Introduction, Time Division Multiplexing, T1 Digital Carrier, Digital Hierarchy, Digital Carrier Line Encoding, T Carrier Systems, Digital Carrier Frame Synchronization, Bit Versus Word Interleaving, Statistical TDM, Codecs and Combo Chips, Frequency Division Multiplexing, FDM Hierarchy, Composite Baseband Signal, Formation of Mastergroup, Wavelength Division Multiplexing.

CHAPTER

4

MULTIPLEXING OF TELEPHONE CHANNELS

INSIDE THIS CHAPTER

4.1. INTRODUCTION : MULTIPLEXING

One requirement of a data communications system is for it to transmit as many intelligent signals as possible in a fixed period using a single-communications channel. **Multiplexing** is a technique which makes it possible to cram a number of logical channels (each capable of supporting an independent connection) into the same physical channel or line. Multiplexing may be defined as any process of sending of a number of separate signals together, over the same cable or bearer, simultaneously and without interference.

As we know that usage of data and telecommunication is increasing day by day, so the traffic also. We can accommodate to add individual lines each time for a new channel or we may use transmissions lines, with the higher bandwidth (fiber optics or satellite microwaves), so that it can carry more numbers of signals.

In other words, we can define multiplexing is the set of techniques that allows the simultaneous transmission of multiple signals across a single data link.

The objective of multiplexing should be obvious:

(*i*) To reduce costs by better utilizing the capacity of a line.

(*ii*) To reduce complexity.

Multiplexing is an operation in which a number of independent signals are combined into a composite signal for transmission over a communication channel. The signals must be combined in a way so that they can be separated at the receiving end without interfering one another. The separation of the composite signal into the original signals is called **demultiplexing.** The transmission media may be metallic wire, a coaxial cable, optical fiber, a terrestrial microwave radio system or a satellite system, refer Fig. 4.1.

The rate of data transmission is measured in the number of bits per second (bps) transmitted. The bps rate is often confused with the baud rate. Baud refers to the rate at which a modulated signal between two devices changes in 1 second. For example, if the signal between two modems changes frequency or phase at a rate of 2,400 times per second, the baud is 2,400. If we use a modulation method in which a single modulation change carries one bit, the 2,400 baud is also 2,400 bits per second. Using more sophisticated modulation methods, several bits of information can bits of data are transmitted with each modulation change, the data transfer rate is 4,800 bits per second at 2,400 baud. The data signals being transmitted are normally multiplexed to increase the transmission rate of data over the communications channel or to increase the efficiency of the channel by allowing multiple users of the same channel.

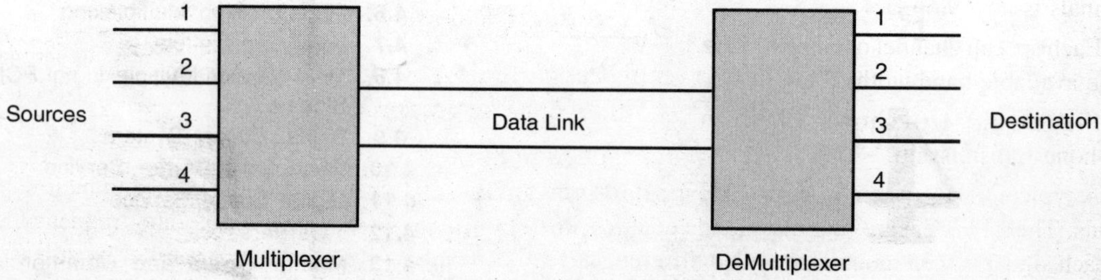

Fig. 4.1. Concept of Multiplexing/Demultiplexing.

In a multiplexed system, many lines (say *n* lines) share the bandwidth of one link (Physical path) as shown in Fig. 3.27.

TIPS

Multiplexing is an operation in which a number of independent signals are combined into a composite signal for transmission over a communication channel.

In telecommunication, the multiplexing means the use of one telecommunication line to handle several channels of voice or data. The best example of multiplexing is our TV cable. We select a particular channel using remote control from the cable, which carries many channels.

The individual channels entering and leaving the terminal station is called baseband channels. A link, which carries multiplex signal, is called broad band channel.

Types of Multiplexing

There are many domains in which multiplexing can be achieved. They are time, space, phase, frequency and wavelength. However, the most common techniques of multiplexing are as follows:

1. Frequency division multiplexing (FDM) (mainly used for Analog communication).
2. Time division multiplexing (TDM) (mainly used for digital communication discussed in Chapter 5).
3. Wave division multiplexing (WDM) (mainly used for Analog communication).
4. Space Division Multiplexing (SDM).

4.2. FREQUENCY DIVISION MULTIPLEXING (FDM)

FDM is a broad band analog transmission technique in which multiple signals are transmitted over a single cable simultaneously

Frequency Division Multiplexing (FDM) is an analog technique where each communications channel is assigned a carrier frequency. To separate the channels, a guard-band would be used. This is to ensure that the channels do not interfere with each other.

For example, if we had our 3 terminals each requiring a bandwidth of 3 kHz and a 300 Hz guard-band, Terminal 1 would be assigned the lowest frequency channel 0-3 kHz, Terminal 2 would be assigned the next frequency channel 3.3 kHz-6.3 kHz and Terminal 3 would be assigned the final frequency channel 6.6 kHz-9.6 kHz.

The frequencies are stacked on top of each other and many frequencies can be sent at once. The downside is that the overall line bandwidth increases. Individual terminal requirement were 3 kHz bandwidth each, in the above example: the bandwidth to transmit all 3 terminals is now 9.6 kHz.

Each speech channel occupies 4 kHz of the available bandwidth.

Figure 4.2 (*b*) shows FDM in telephone transmission.

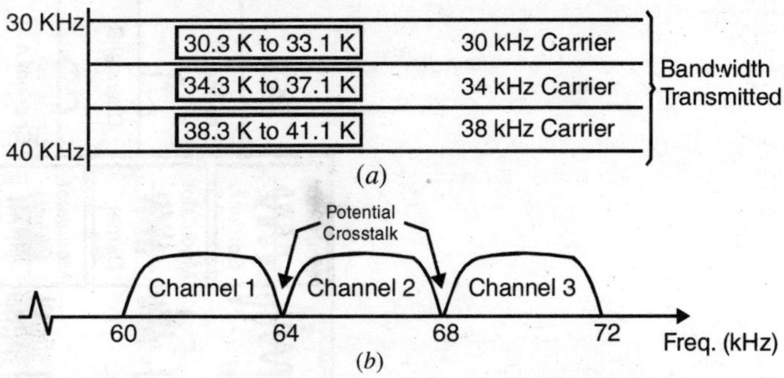

Fig. 4.2. FDM in Telephone Transmission.

A typical FDM system is shown in Fig. 4.3 (*a* and *b*). The message signals are assumed to be low-pass signals. The LPF's ensure that the signals entering the modulators are bandlimited to some frequency, say, W. Each signal is then modulated with a different carrier. The frequencies of the carriers differ by at least so that the modulated signals are separated in frequency domain. The BPF's are used to restrict the modulated signals to their prescribed ranges. The outputs of the BPF's are added into a composite signal and sent through the communication channel.

At the receiver end, the composite signal is passed through different BPF's so that the modulated message signals are separated. Then demodulation follows.

TIPS

Frequency Division Multiplexing (FDM) is an analog technique where each communications channel is assigned a carrier frequency.

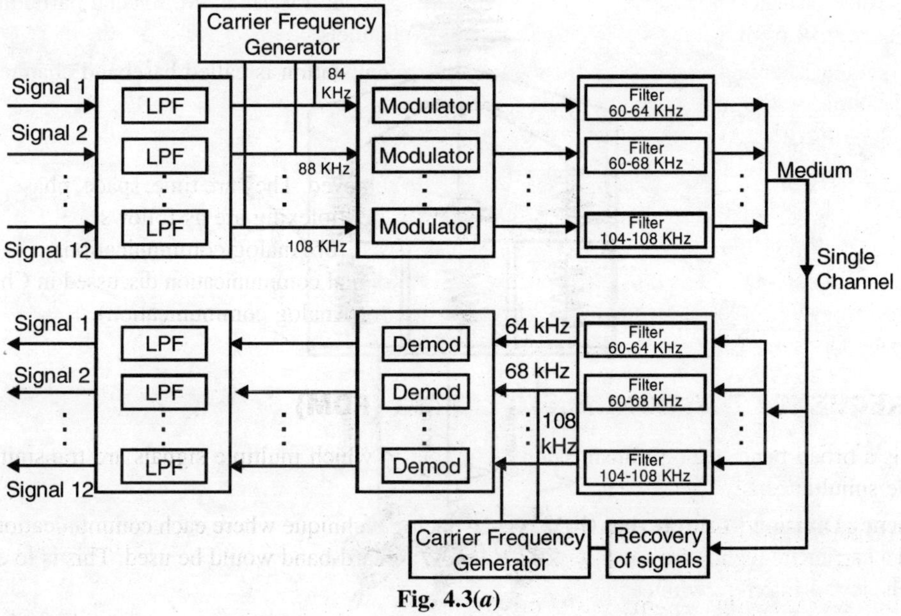

Fig. 4.3(*a*)

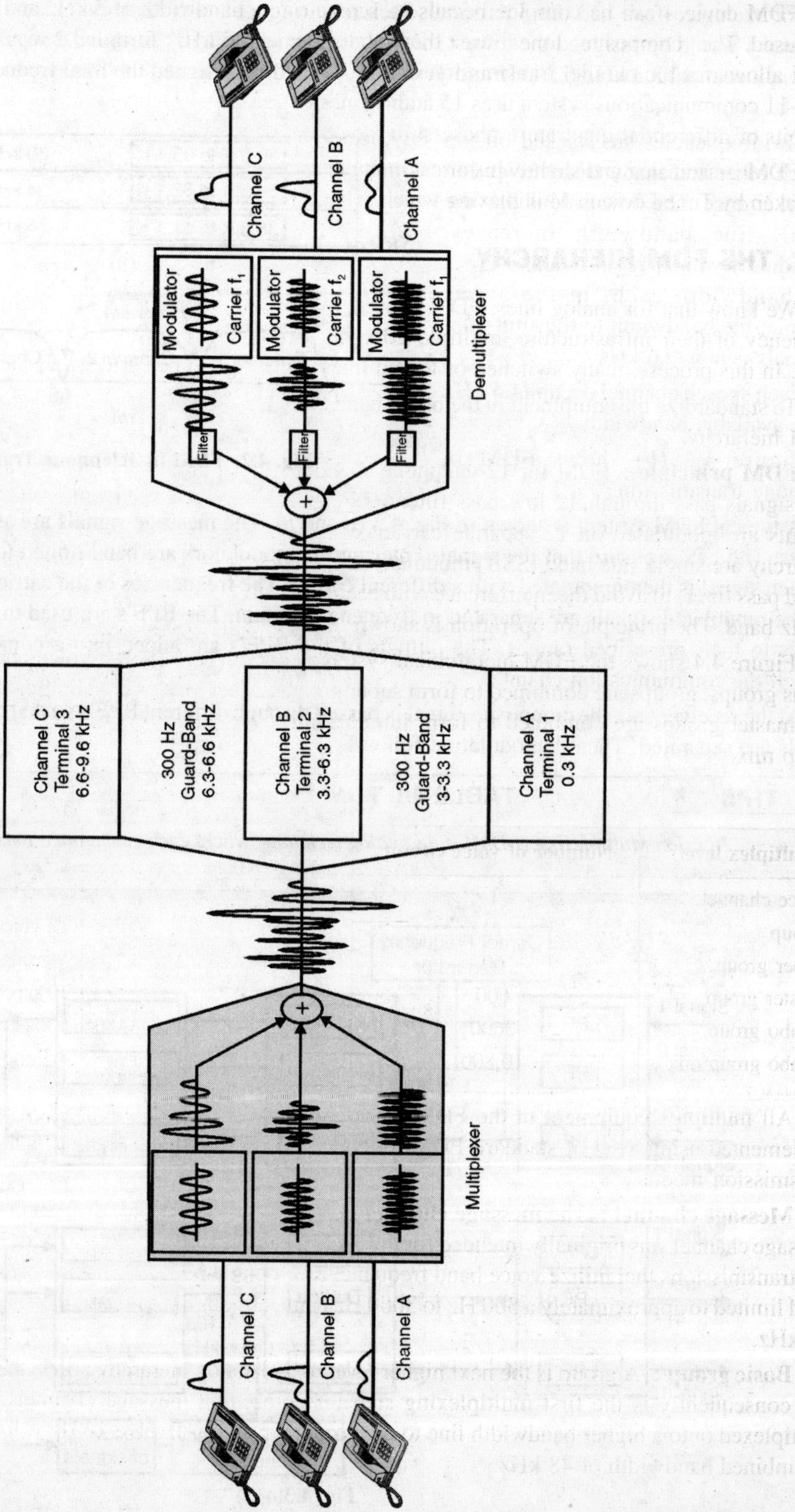

Fig. 4.3. (*b*) A typical FDM System.

FDM devices can be complex because a separate modulator/demodulator circuit is required for each tone used. The composite tones are then modulated to a single carrier frequency for radio transmission. FDM allows for the parallel transmission of data over a single communications channel. For example, the Link-11 communications system uses 15 audio tones to transmit 30 bits of parallel data. Each tone transmits two bits of differential quadrature phase-shift keyed data.

FDM is an analog and slightly historical multiplexing technique. It is prone to noise problems and has been overtaken by Time Division Multiplexing which is better suited for digital data.

4.3. THE FDM HIERARCHY

We know that for analog lines, FDM is used. Telephone companies traditionally used to maximize the efficiency of their infrastructure, multiplexed signals from lower bandwidth lines onto higher bandwidth lines. In this process, many switched or leased lines can be combined into fewer but bigger channels.

To standardize the equipment in the broad band transmission system CCITT recommended the following FDM hierarchy.

FDM principle : FDM for 12 telephone channels (group multiplex level) is considered for example. The signals pass through 12 low pass filters (LPF) to remove any high frequency components. The LPF outputs are modulated on 12 separate carrier signals separated by 4 kHz. The modulations in all FDM hierarchy are single side band (SSB) modulation. The output of each of the 12 modulations must be filtered (band pass filter) to avoid the interference with each other. BPF are used to restrict each signal to the allocated 4 kHz band. The principle of operation is shown in Fig. 4.3 (*a*).

Figure 4.4 shows the FDM hierarchical system. As the figure shows, voice channels are combined to forms groups, groups are combined to form super groups, super groups are combined to form master groups and master groups are combined to form jumbo groups and jumbo groups are combined to form jumbo group mix.

TABLE 4.1. FDM Hierarchical System.

Multiplex level	Number of voice circuits	Frequency band (kHz)	Formation
Voice channel	1	0–4	—
Group	12	60–108	12 voice circuits
Super group	60	312–552	5 group (12 × 5 = 60)
Master group	600	564–3084	10 super group (60 × 10 = 600)
Jumbo group	3600	564–17548	6 master group
Jumbo group mix	10,800	3000–6000	3 Jumbo groups

All multiplex equipment in the FDM hierarchy uses SSB modulation. Each level of the hierarchy is implemented using a set of standard FDM modules. The multiplex equipment is independent of particular transmission media.

Message channel : The **message channel** is the basic building block of the FDM hierarchy. The basic message channel was originally intended for the analog transmission of voice signals, although it now includes any transmissions that utilize voice band frequency from 0 to 4 kHz. The basic voice band circuit is actually band limited to approximately a 300 Hz to 3000 Hz band, although for practical consideration it is considered a 4 kHz.

Basic group : A group is the next higher level in the analog hierarchy above the basic message channel and consequently is the first multiplexing step for combining message channels. 12 voice channels are multiplexed onto a higher bandwidth line to create a basic group. Twelve 4 kHz voice band channels occupy a combined bandwidth of 48 kHz.

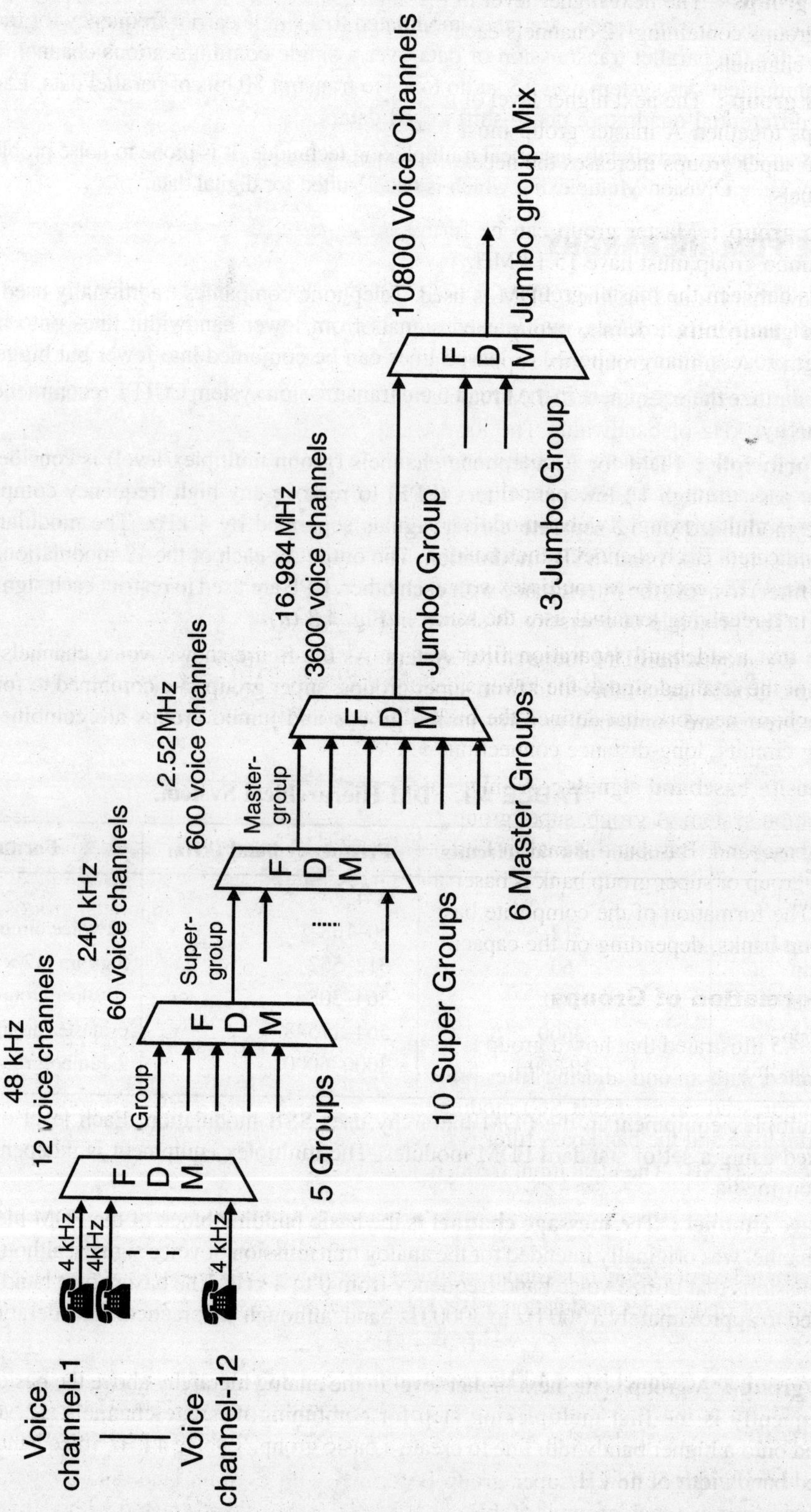

Fig. 4.4. The FDM Hierarchical System.

Super groups : The next higher level in the analog hierarchy is the super group, which is formed by FDM five groups containing 12 channels each. A super group has a bandwidth of 240 kHz and supports up to 60 voice channels.

Master group : The next higher level of multiplexing is the master group, which is formed by FDM 10 super groups together. A master group must have 2.40 kHz of bandwidth, but the need for guard bands between the super groups increases the necessary bandwidth to 2.52 MHz Master groups support up to 600 voice channels.

Jumbo group : Master group can be further multiplexed in master group banks to create a jumbo group. A jumbo group must have 15.12 MHz (6 × 2.52 MHz) but is augmented to 16.984 MHz to allow for guard bands between the master groups.

Jumbo group mix : Jumbo group can be further multiplexed, which is formed by 3 FDM Jumbo groups together. A Jumbo group mix support up to 10800 voice channels.

All multiplex equipment in the FDM hierarchy used SSB modulation. Thus, every voice circuit required approximately 4 kHz of bandwidth. The lowest level building block in the hierarchy is a channel group consisting of 12 voice channels. A channel group multiplex uses a total bandwidth of 48 kHz. Figure 4.5 shows a block diagram of an A5 channel group multiplexer, the most common A –type channel bank used for first' level multiplexing. Twelve modulators using 12 separate carriers generate 12 double-sidebands signals as indicated. Each channel is then bandpass filtered to select only the lower sideband of each double-sideband signal. The composite multiplex signal is produced by super posing the filter outputs. Demultiplex equipment in a receiving terminal uses the same basic processing in reverse order.

Notice that a sideband separation filter not only removes the upper sideband but also restricts the bandwidth of the retained signal: the lower sideband. These filters therefore represented a basic point in the analog telephone network that defined the bandwidth of a voice circuit. Since FDM was used on all long-haul analog circuits, long-distance connections provided somewhat less than 4 kHz of bandwidth.

Composite baseband signals : A single message channel is baseband *i.e.*, the modulating signal in a communication system. A group, super group or signals prior to modulation of the final carrier is known as composite baseband. Baseband signal in Fig. 4.3, the output of a channel bank is baseband. And, also the output of a group or super group bank is baseband. The final output of the FDM multiplexer is the composite baseband. The formation of the composite base band signal can include channels, groups, super group and master group banks, depending on the capacity of the systems.

4.3.1. Formation of Groups

A Fig. 4.5 illustrated that how a group is formed with an A-type channel bank. Each voice band channel is band limited with an anti aliasing filter prior to modulating the channel carrier. The combination of the sideband suppressed carrier (SSB-SC) modulator. A balanced modulator is a double side band suppressed carrier modulation and the band pass filter is tuned to the difference between the carrier and the input voice band frequencies (LSB). The ideal input frequencies for the channel banks are determined from the following relation,

$$Fc = 112 - 4n \text{ kHz}$$

where, f_c = channel carrier frequencies and n = number of channels from 1 to 12. Therefore, for channel 1, a 0 kHz to 4 kHz bands of frequencies modulation a 108 kHz carrier. The output of a channel band pass filter is

$$F_{\text{out}} = (f_c - 4 \text{ kHz}) \text{ to } f_c$$

Earn voice band channel has a 4 kHz bandwidth.

Super Group

Figure 4.6 illustrates how a super group is formed with a group bank and combining network. The frequency spectrum for each group is 60 kHz to 108 kHz. Each group is mixed with a different group carrier

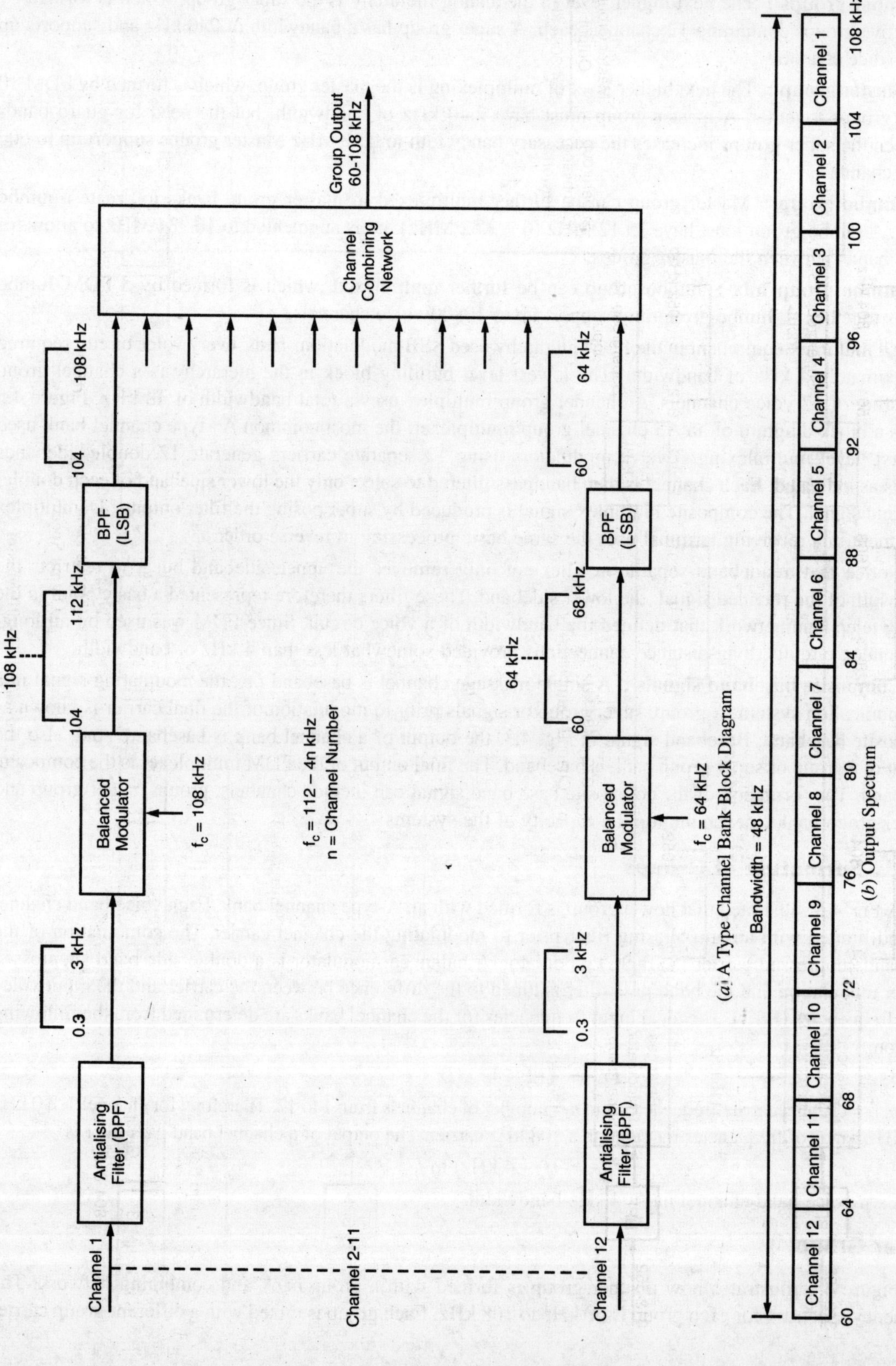

Fig. 4.5. A block diagram of A-Type channel group multiplexer.

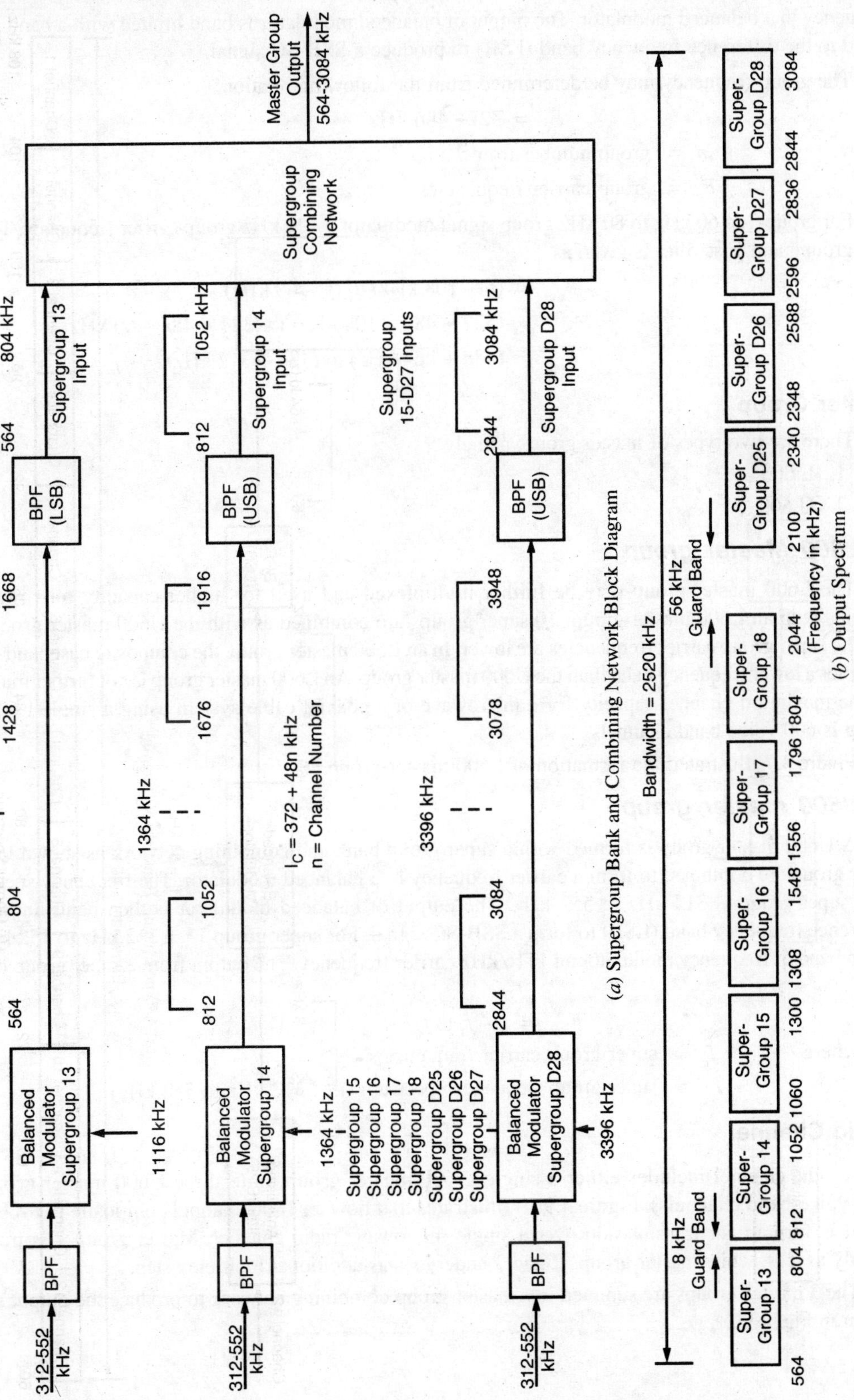

(*a*) Supergroup Bank and Combining Network Block Diagram

(*b*) Output Spectrum

Fig. 4.6. Formation of L600 Master Group.

frequency in a balanced modulator. The output of balanced modulator is band limited with a band pass filter tuned to the difference frequency band (LSB) to produce a SSB-SC signal.

The group frequency may be determined from the following relation:

$$F_c = 327 + 48n \text{ kHz}$$

where, n = group number from 1 to 5

F_c = group carrier frequencies

For group-1, a 60 kHz to 80 kHz group signal modulator a 420 kHz group carrier frequency. The output of a group band pass filter is given as

$$F_{out} = (f_c - 108 \text{ kHz}) \text{ to } (f_c - 60 \text{ kHz})$$

$$= (327 + 48n - 108 \text{ kHz}) \text{ to } (327 + 48n - 60 \text{ kHz})$$

$$F_{out} = (48n + 264 \text{ kHz}) \text{ to } (48n + 312 \text{ kHz})$$

Master Group

There are two types of master group, namely :

1. L 600
2. U 600

1. L600 Master group

The L600 master group may be further multiplexed and used for higher capacity microwave radio systems with an L600 master group: 10 super groups are combined as with the U600 master group, except that the super group carrier frequencies are lower. In an L600 master group, the composite baseband spectrum occupies a lower frequency band than the U600 master group. An L600 master group is not further multiplexed. So the maximum channel capacity for a microwave or a coaxial cable system using a single L600 master group is 600 voice band channels.

Figure 4.7 illustrated the formation of L600 master group.

2. U600 master group

A U600 master group is formed with a super group bank and combining network as shown in Fig. 4.8 super group are combined to form a carrier frequency in a balanced modulator. The frequency spectrum for each super group is 312 kHz to 552 kHz. The output of balanced modulator is then band limited to the difference frequency band (LSB) to form a SSB-SC signal. For super group 13, a 312 kHz to 552 kHz super group band of frequency modulator an 1116 kHz carrier frequency. The output from a super group band pass filter is

$$F_{out} = f_c - f_s \text{ to } f_c$$

where f_c = super group carrier frequencies

F_c = super group frequency spectrum (*i.e.*, 312 kHz to 552 kHz)

Radio Channel

A radio channel includes either a single L 600 master group up to three U600 master groups (*i.e.*, 1800 voice-band channels). Figure 4.9(*a*) illustrates that how an 1800 channel composite FDM baseband signal is formed for transmission over a single microwave radio channel. Master group-1 is transmitted directly as it is, while master groups 2 and 3 undergoes as additional multiplex step.

The 3 master groups are summed in a master group combining network to produce the output spectrum shown in Fig. 4.9(*b*).

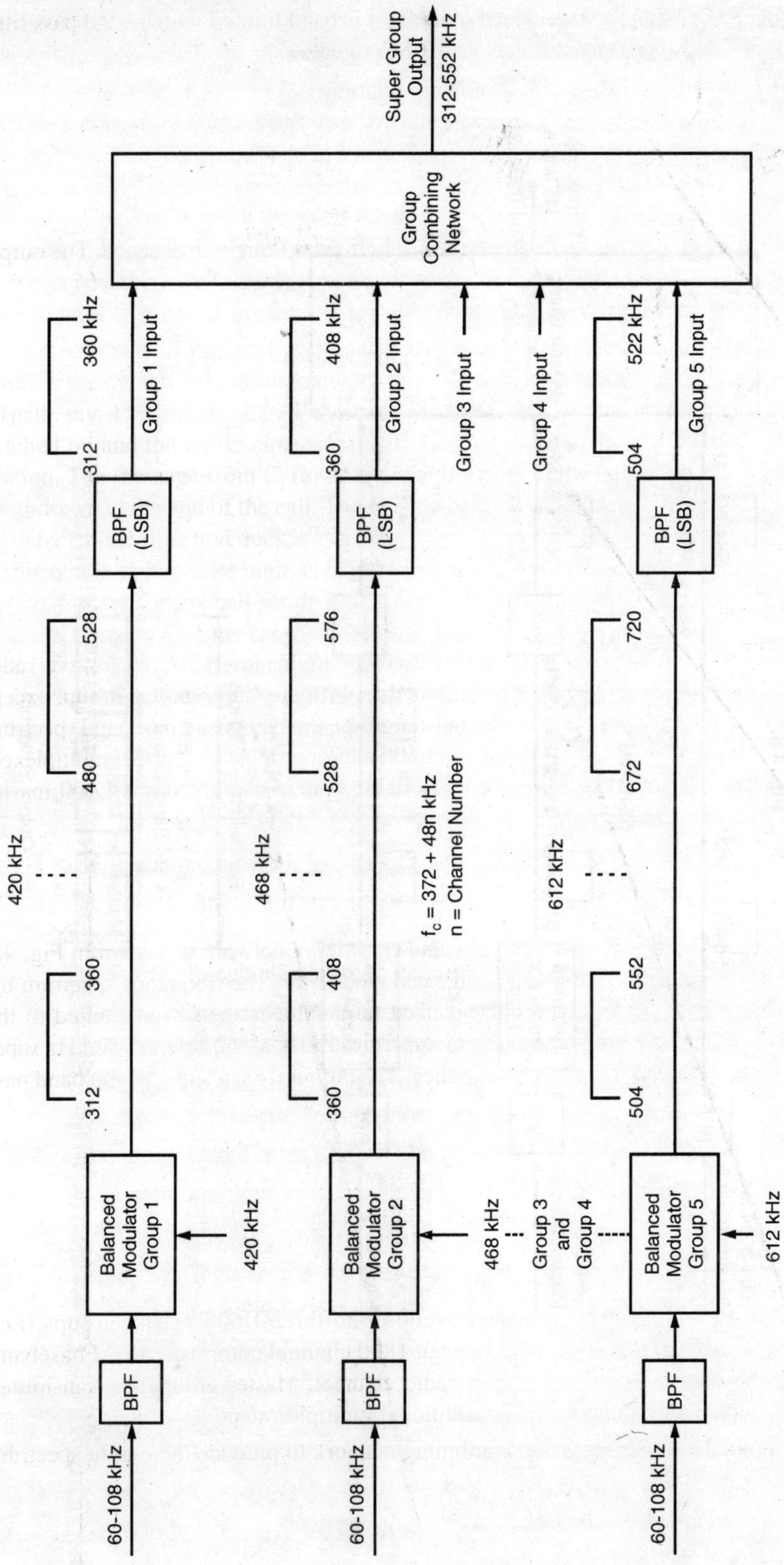

(a) Group Bank and Combining Network Block Diagram

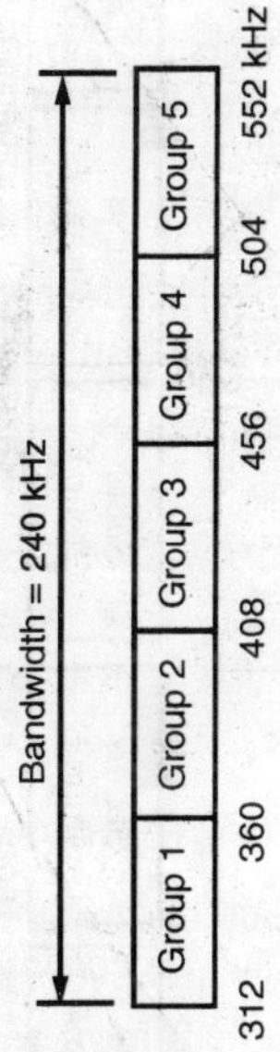

(b) Output Spectrum

Fig. 4.7.

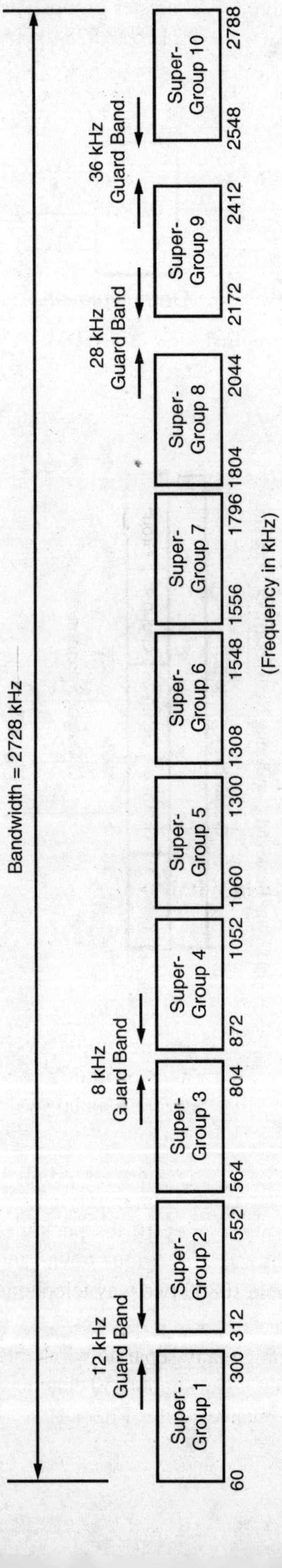

Fig. 4.8.

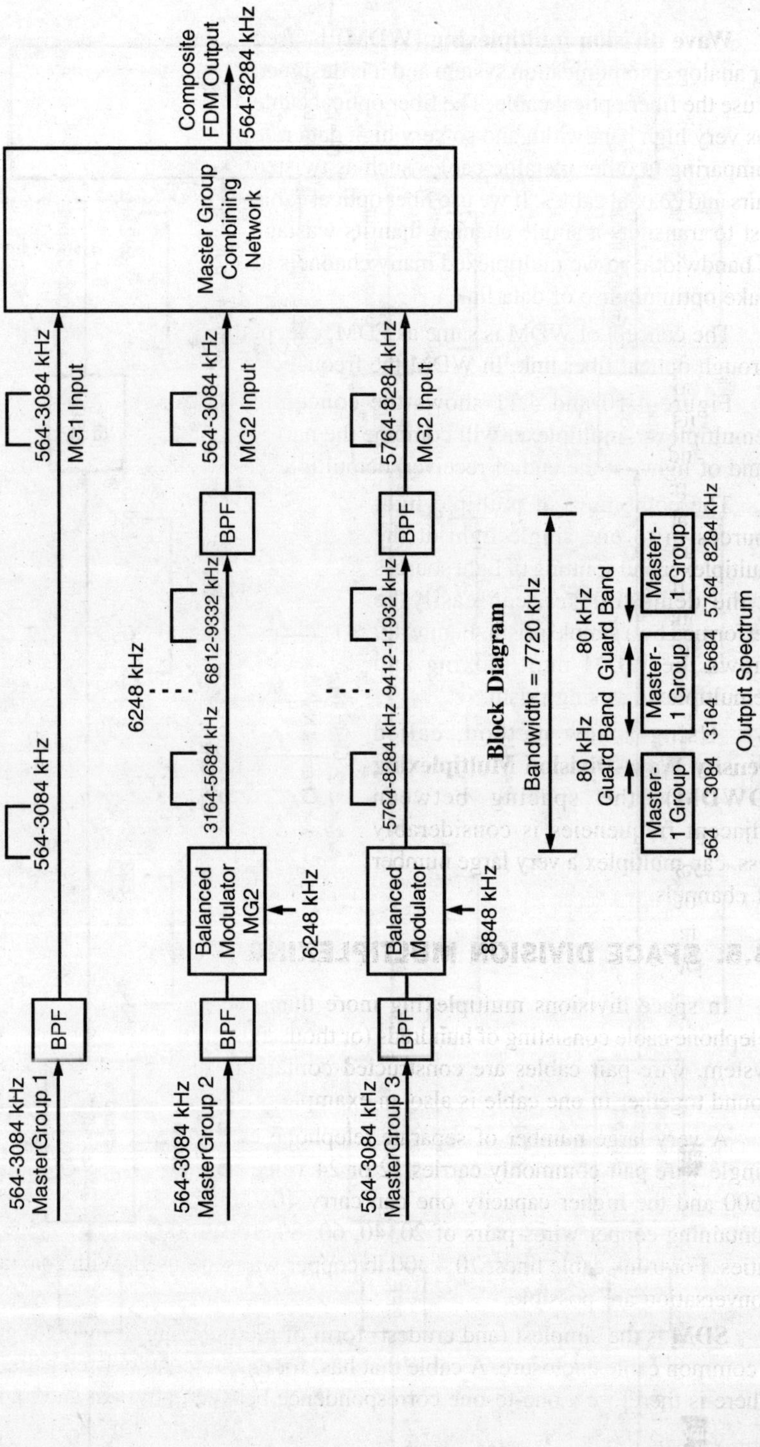

Fig. 4.9. Three-Mastergroup Radio Channel

The system shown in Fig. 4.9 may be increased from 180 voice channel directly to master group. The additional 312 kHz to 552 kHz super group extends the composite output spectrum from 312 kHz to 8284 kHz.

4.4. WAVE-DIVISION MULTIPLEXING (WDM)

Wave division multiplexing (WDM) is used for analog communication system and it is designed to use the fiber optical cable. The fiber optical cable has very high bandwidth and so very high data rate comparing to other metallic cables such as twisted pairs and coaxial cables. If we use fiber optical cable just to transmits a single channel than its wastage of bandwidth, so we multiplexed many channels to make optimum use of data link.

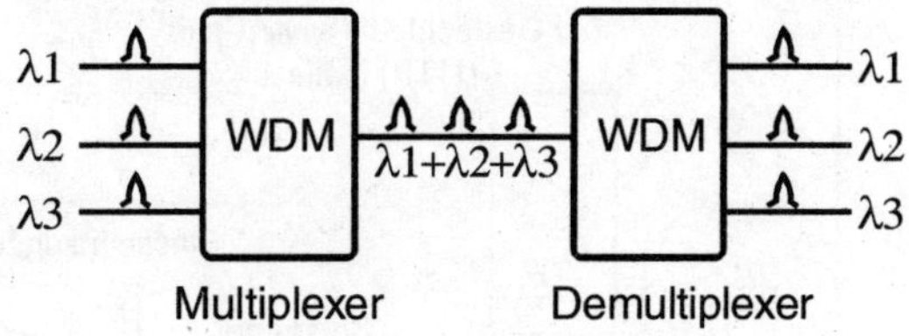

Fig. 4.10. The Conceptual View of a WDM.

The concept of WDM is same as FDM, except that WDM involves optical signals which are transmitted through optical fiber link. In WDM the frequencies are very high that is in optical region.

Figure 4.10 and 4.11 shows the conceptual view of a WDM. We have WDM multiplexer and demultiplexer, multiplexer will combine the narrow bands of light from the various sources to make a wider band of light. At the end of receiver, demultiplexer is used to separate the various signals.

The combining of multiple light sources in to one single light at the multiplexer and splitting of light sources at the demultiplexer, can easily be performed by a simple prism. Figure 4.9 shows the WDM multiplexing and demultiplexing using prism.

Using a new method, called **Density Wave-Division Multiplexing (DWDM)**, the spacing between adjacent frequencies is considerably less, can multiplex a very large number of channels.

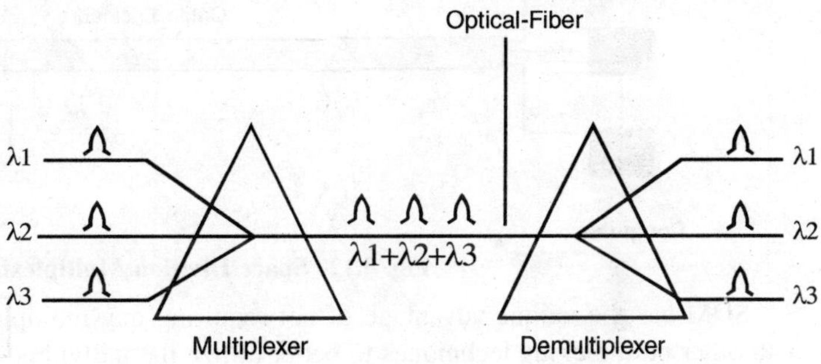

Fig. 4.11. Prisms used in WDM.

4.5. SPACE DIVISION MULTIPLEXING (SDM)

In space divisions multiplexing more than one physical transmission path are grouped together. A telephone cable consisting of hundreds (or thousands) of twisted pair constitutes a space division multiplexed system, wire pair cables are constructed containing many hundreds of wire pairs. Several coaxial tubes bound together in one cable is also an example of space division switching.

A very large number of separate telephone calls can be transmitted together down a coaxial system. Single wire pair commonly carries 12 or 24 voice channels. But one single coaxial tube commonly carries 3600 and the higher capacity one can carry 10800. In India, the telephone cables (5 or 10 lbs per mile) containing copper wires pairs of 20, 40, 60, 80, 100, 1000 and 5 per cent extra used in various towns and cities. For trunk cable lines, 70 – 300 lb copper wires are used. With coaxial cable 10800 two way telephone conversation are possible.

SDM is the simplest (and crudest) form of multiplexing. It involves grouping many separate wires into a common cable enclosure. A cable that has, for example, 50 twisted pairs inside it can support 50 channels. There is therefore a one-to-one correspondence between physical and logical channels refer Fig. 4.12.

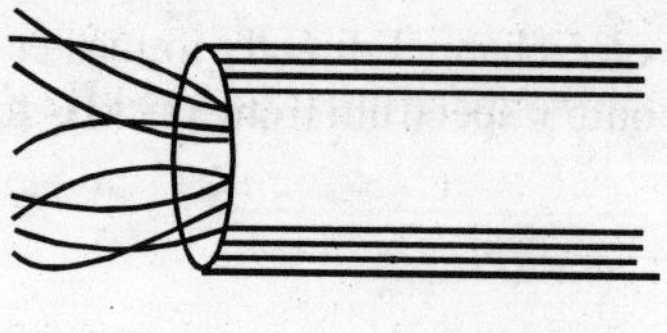

(*a*) Unshielded Twisted-pair (UTP) cable

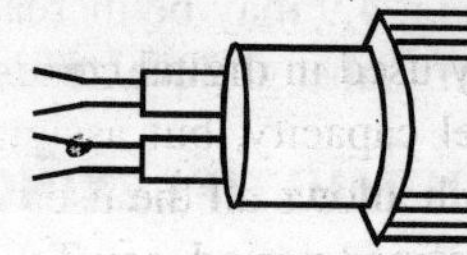

(*b*) Shielded Twisted-pair (STP) cable

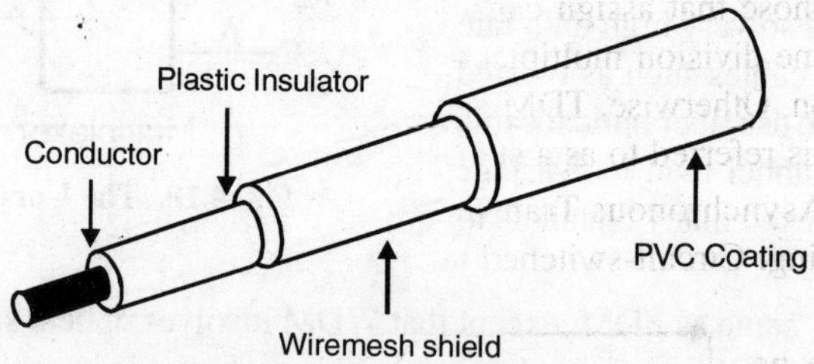

(*c*) Coaxial cable

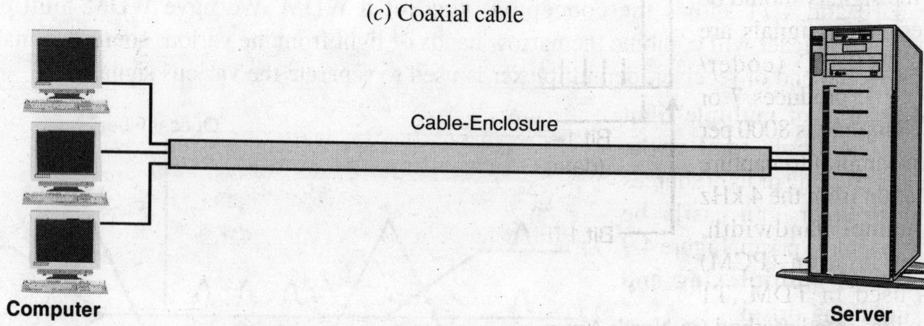

Fig. 4.12. Space Division Multiplexing (SDM).

SDM has the unique advantage of not requiring any multiplexing equipment. It is usually combined with other multiplexing techniques to better utilize the individual physical channels.

The space division switching is not limited to voice frequency circuits alone. Many high capacity transmission systems using either frequency or time division multiplexing can also be space division multiplexed.

4.6. TIME DIVISION MULTIPLEXING

In 1970's TDM was introduced and now favored over FDM. TDM is the sharing of a common transmission medium in time. In TDM, the time available is divided into small slots, and each of them occupied by a piece of one of the signals to be sent. Thus the multiplexing device should scan the input signal in round-robin fashion. TDM is a base band technology in which individual channels of data or voice are interleaved into a stream of framed bits across a communication channel. Figure 4.13 shows the Principle of TDM.

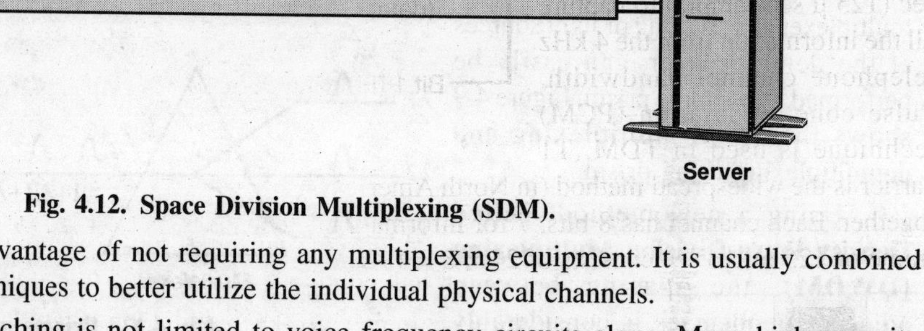

Fig. 4.13. Principle of TDM.

TDM is more commonly used in digital communication systems. **Time-division multiplexing (TDM)** grants each user full channel capacity, but assigns time slots to each user. In simple term TDM can be defined that a technique which allows all the users to share the time period, use the available data link with the full capacity for the predefined period, say T. Transmissions from various sources are interleaved in the time domain.

There are two basic modes of operation for TDM: those that repeatedly assign a portion of the transmission capacity to each source and those that assign capacity only as it is needed. The first form of operation is referred to as synchronous time division multiplexing (STDM) when necessary to; distinguish it from the "as-needed" mode of operation. Otherwise, TDM is generally understood to imply the synchronous variety. The as-needed form of TDM is referred to as a statistical time division multiplexing.

Multiplexing (ATDM), Asynchronous Transfer Mode (ATM), statistical time division multiplexing (Stat-Mux), or packet switching. Circuit-switched telephone networks use STDM, whereas data networks typically use ATDM.

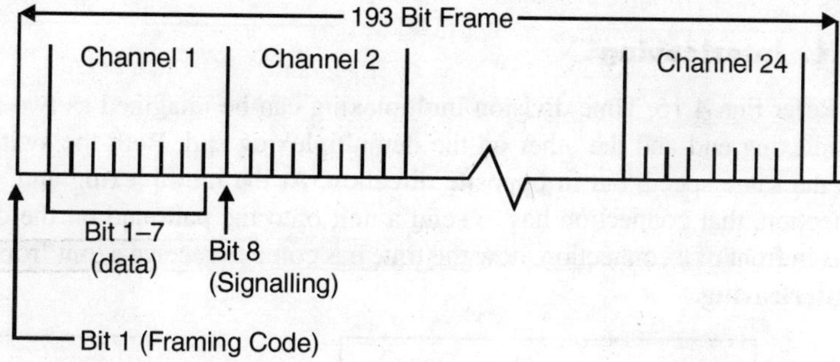

Fig. 4.14.

If analog signals are to be multiplexed, the signals should be sampled. The analog signals are digitized by CODEC (coder/decoder) device. It produces 7 or 8 bit number. Sampling is 8000 per sec (125 μ sec/sample) to capture all the information from the 4 kHz telephone channel bandwidth. Pulse cone modulation (PCM) technique is used in TDM. T1 carrier is the widespread method (in North America and Japan) T1 consists of 24 voice channels multiplexed together. Each channel has 8 bits, 7 for information and 1 for signalling, as shown in Fig. 4.14. The output frame of 125 micro sec consists of $24 \times 8 = 192$ bits plus one extra bit for framing.

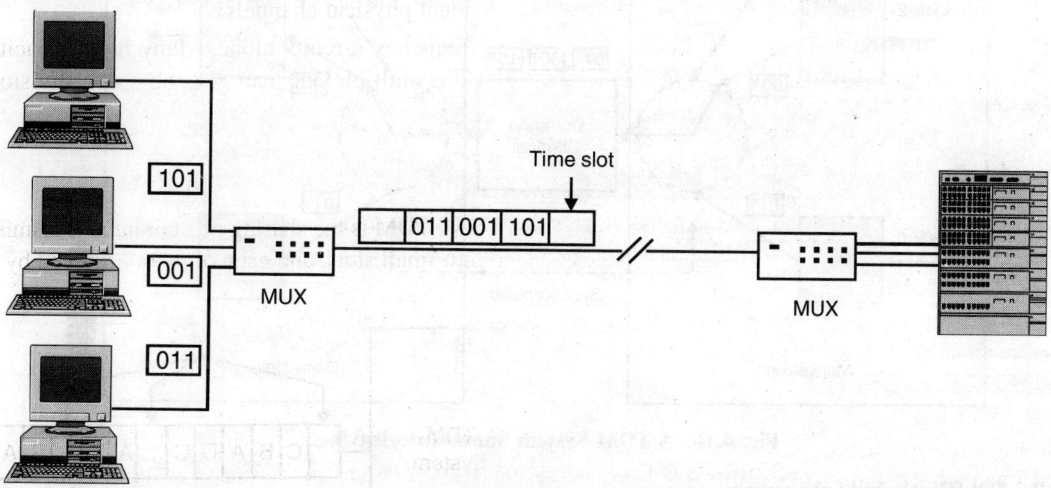

Fig. 4.15. A TDM system.

Pulse Code Modulation is the most prevalent encoding technique used for TDM digital signals. Two or more voice channels are sampled and converted to PCM codes and then time multiplexed onto a single transmission line. Each user is connected to a time-division multiplexer. Data signals from the user are fed to the time-division multiplexer buffer, and the time slots are rotated among the users and scanned for data. The data from each user can be in the form of bits, bytes, or blocks. The data signals from all users are compiled into frames for transmission on a single, high-speed communications channel. Transmit and receive frames are used for half-duplex communications. Transmit frames are sent and a receive time slot is enabled for return information. In this manner, a single carrier frequency and modem may be used to transmit and receive information at a fairly high rate of speed. Since time slots are preset and assigned, if a user has no data to transmit, the time slot is wasted.

TDM requires special multiplexing/demultiplexing hardware (MUX) at either end of the line (see Fig. 4.15). Because the channels are spread across time, some means of initial synchronization is also needed. Basically, the receiving end needs to know which time slot belongs to the first channel when the connection is established, and can work everything else out from this reference point.

Each channel has a time slice assigned to it whether the terminal is being used or not. Again, to the send and receiving stations, it appears as if there is a single line connecting them. All lines originate in one location and end in one location.

4.6.1. Interleaving

Refer Fig. 4.16, Time division multiplexing can be imagined as two fast rotating switches, one on the multiplexing end and the other on the demultiplexing end. Both the switches are synchronized and rotate with the same speed, but in opposite direction. At the multiplexing end, as the switch opens in front of a connection, that connection has to send a unit onto the path and on the demultiplexing end, as the switch opens in front of a connection, now the time has come to receive a unit from the path. This process is referred as **Interleaving.**

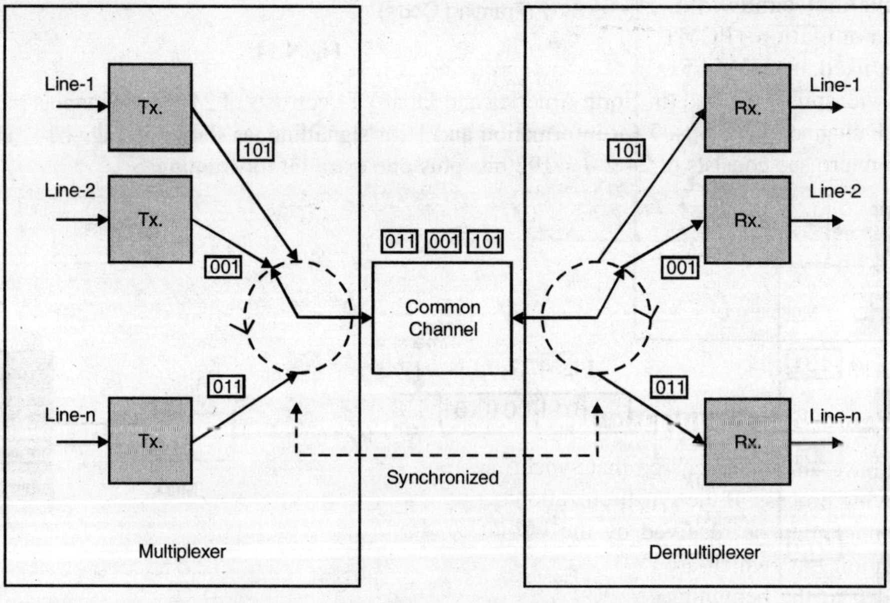

Fig. 4.16. A TDM System Shows Interleaving.

4.6.2. Bit/Byte Interleaving

Time division multiplexing can be either bit by bit or byte by byte basis, known as bit interleaving and byte interleaving respectively. Bit interleaving also known as digit interleaving, used in North American Digital Hierarchy (T-1 system). Byte interleaving also known as word interleaving, used in Digital services (DS) and SONET (synchronous optical network system).

Figures 4.16 and 4.17 shows these two schemes. It can be seen from the example shown in Fig. 4.17, that we have three data, A, B and C. When one bit of each data is being is sent then it is referred as bit interleaving and when we sent 4 bits (word/byte) of A, 4 bits of B and 4 bits of C then it is referred as byte/ word interleaving shown in Fig. 4.17. Reverse process is done at the receiver end to recover the data.

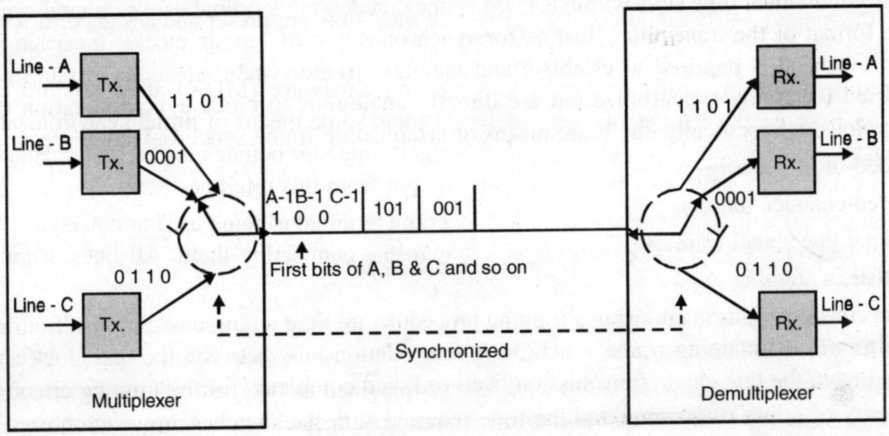

Fig. 4.17 (*a*) Bit interleaving.

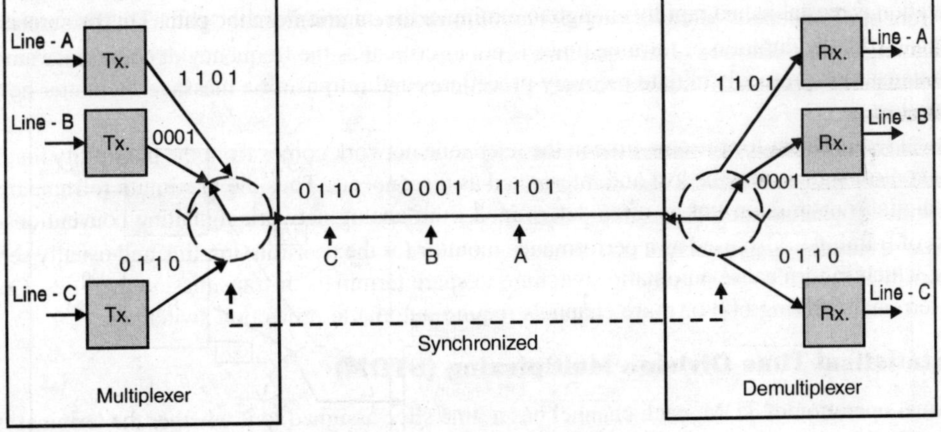

Fig. 4.17 (*b*) Byte interleaving.

4.6.3. Frame Synchronization

As we have already discussed that synchronization between multiplexer and demultiplexer is must in the interleaving process. If the synchronization between multiplexer and demultiplexer is not there than a bit of one channel may be received by the wrong channel. To avoid this problem, we use one or more synchronization bits, which are usually added to the beginning of the each frame. These synchronizing bits follow a pattern, frame to frame, that allows the demultiplexer to synchronize with the incoming stream so that it can separate the time slots accurately. Mostly,

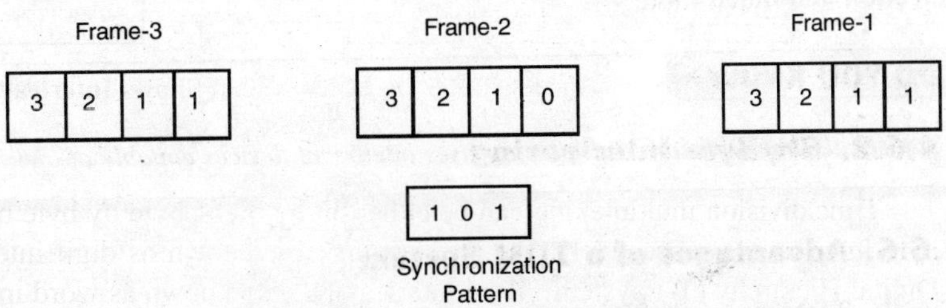

Fig. 4.18. Synchronization Bits Pattern.

this synchronization information consists of one bit per frame, alternating between 1 and 0, as shown in Fig. 4.18.

4.6.4. Framing

To identify individual time slots within a TDM frame, a receiving terminal uses a counter synchronized to the frame format of the transmitter. Just as for synchronization of sample clocks, a certain amount of transmission overhead is required to establish and maintain frame synchronization. In fact, most of the techniques used for frame synchronization are directly analogous to clock synchronization techniques discussed previously. Specifically, the basic means of establishing frame synchronization are;

1. Added-digit framing
2. Added-channel framing
3. Unique line signal framing
4. Statistical framing

The main considerations in choosing a framing procedure are time required to establish framing, effects of channel errors in maintaining frame synchronization, relationships between the line clock and sample clocks derived from the line clock, transmission overhead, and complexity of the framing circuitry.

The severity of a loss of framing and the time required to reestablish synchronization depend on the nature of the traffic. Since a loss of framing implies a loss of data on all channels, the mean time between misframes must be as long as possible. For voice traffic in frequent misframes can be tolerated if frame synchronization is reestablished rapidly enough to minimize the duration of the "glitch" in the output speech.

For data traffic the duration of reframe times is not as critical as the frequency of occurrence since most data communications protocols initiate recovery procedures and retransmit a message no matter how many data bits are lost.

A critical requirement for reframe time in the telephone network comes from the possibility that various in-channel control signals may be lost and interpreted as disconnects. Thus the maximum reframe time on a particular digital transmission link is often determined from analog network signalling conventions.

A loss of framing is also used as a performance monitor for the transmission link and usually sets alarm conditions, which in turn cause automatic switching to spare terminals or transmission facilities. Generally speaking, terminals serving 600 or more channels require automatic protection switching.

4.6.5. Statistical Time Division Multiplexing (STDM)

In normal operation of TDM, each channel has a time slice assigned to it whether the terminal is being used or not. In case when there is no data to transmit in assigned slot of time than it is wastage of time as well as bandwidth. So an another type of TDM, known as **Statistical Time Division Multiplexing** uses intelligent devices capable of identifying when a terminal is idle. They allocate time only to lines when required. This means that more lines can be connected to a transmission medium as this device statistically compensates for normal idle time in data communication lines. Newer STDM units provide additional capabilities such as data compression, line priority, mixed speed lines, host port sharing, network port control, automatic speed detection and much more.

DO YOU KNOW►

Statistical Time Division Multiplexing *uses intelligent devices capable of identifying when a terminal is idle.*

4.6.6. Advantages of a TDM System

1. TDM is more efficient,

2. More easier to operate,
3. Less complex and
4. Less expensive than FDM
5. Its ability to handle devices with varying speeds, its effectiveness when used with devices that transmit data almost continuously, and its simple implementation.

4.6.7. Time Division Multiplex Loops and Rings

It is mentioned that TDM is not as acquiescent to applications with distributed sources and sinks of traffic as is FDM. In this section a particular form of a TDM network is described that is quite useful in interconnecting distributed nodes. The basic structure of interest is referred to as a TDM loop or TDM ring and is shown in Fig. 4.19.

Basically, a *TDM ring* is configured as a series of unidirectional (two-wire) links arranged to form a closed circuit or loop. Each node of the network is implemented with two fundamental operational features. First, each node acts as a regenerative repeater merely to recover the incoming bit stream and retransmit it. Second, the network nodes recognize the TDM frame structure and communicate on the loop by removing and inserting data into specific time slots assigned to each node. As indicated in Fig. 4.19, a full-duplex connection can be established between any two nodes by assigning a single time slot or channel for a connection. One node insets information into the assigned time slot that propagates around the loop to the second node (all intervening nodes merely repeat the data in the particular time slot). The destination node removes data as the assigned time slot passes by and inserts return data in the process. The return data propagates around the loop to the original node where it is removed and replaced by new data, and so forth.

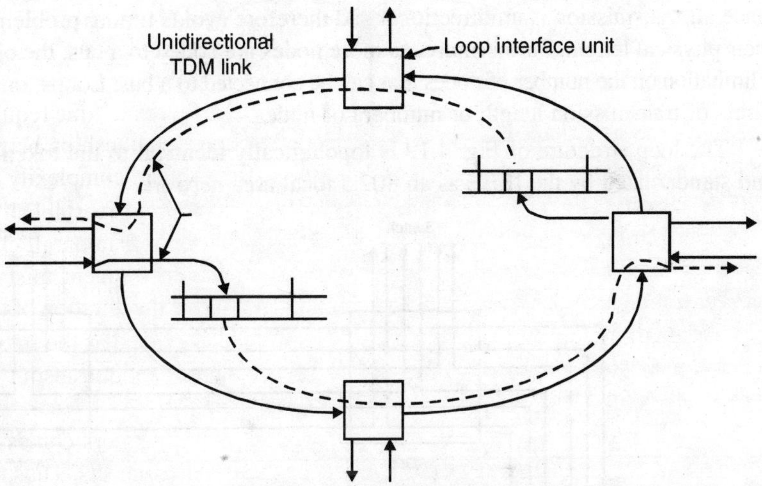

Fig. 4.19. TDM Loop.

Since other time slots are not involved with the particular connection shown, they are free to be used for other connections involving arbitrary pairs of nodes. Hence a TDM Loop with C time slots per frame can support C simultaneous full-duplex connections. If, as channels become available, they are reassigned to different pairs of nodes, the transmission facilities can be highly utilized with high concentration factors and provide low blocking probabilities between all nodes. Thus a fundamental attraction of a loop network is that the transmission capacity can be assigned dynamically to meet changing traffic patterns. In contrast, if a star network with a centralized switching node is used to interconnect the nodes with four-wire links, many of the links to particular nodes would be underutilized since they cannot be shared as in a loop configuration.

Another feature of the loop-connected network is the ease with which it can be reconfigured to accommodate new nodes in the network. A new access node is merely inserted into the nearest link of the network and the new node has complete connectivity to all other nodes by way of the TDM channels. In contrast, a star structured network requires transmission to the central node and expansion of the centralized switching facilities.

The ability to reassign channels to arbitrary pairs of nodes in a TDM loop implies that the loop is much more than a multiplexer. It is, in fact, a distributed transmission and switching system. The switching capabilities come about almost as a by-product of TDM transmission. TDM loops represent the epitome of integrated transmission and switching.

DO YOU KNOW

TDM loops have been used within computer complexes to provide high capacity and high interconnectivity between processors, memories and peripherals.

The loop structure in this application is sometimes more attractive than more conventional bus structures since all transmission is unidirectional and therefore avoids timing problems on bidirectional uses that limit their physical length. Furthermore, as more nodes are added to a bus, the electrical loading increases, causing a limitation on the number of nodes that can be connected to a bus. Loops, on the other hand, have no inherent limits of transmission length or numbers of nodes.

The loop structure of Fig. 4.19 is topologically identical to the token-passing ring developed by IBM and standardized by the IEEE as an 802.5 local area network.

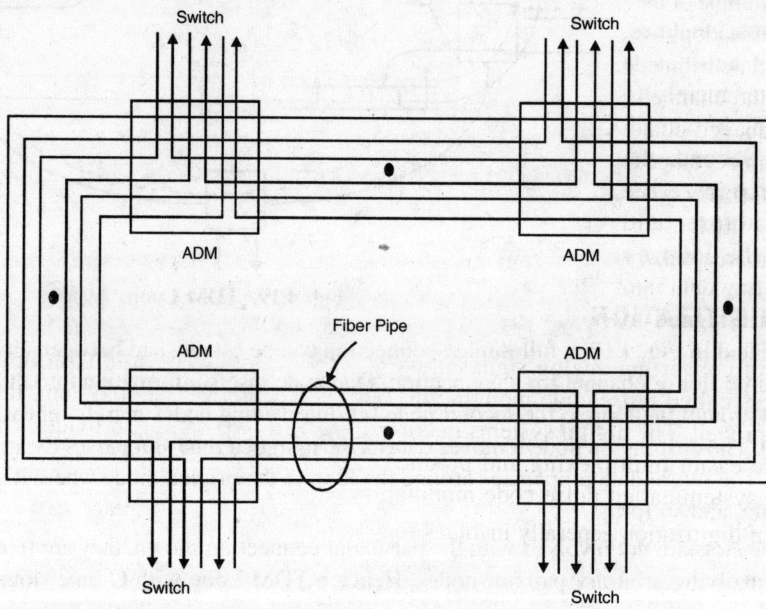

Fig. 4.20. Functional Mesh, Fiber Loop and ADM's.

However a token-passing ring operates differently than a TDM loop in that there is only one channel. When a node on a ring becomes active, it uses the entire capacity of the outgoing link until it is through sending its message. In contrast, a node on a loop uses only specific time slots in the TDM structure, allowing other nodes to be simultaneously connected, using other time slots. In essence, a TDM loop is a distributed-circuit switch and an 802.5 ring is a distributed-packet switch. A particularly attractive use of a loop with high-bandwidth links is shown in Fig. 4.19. This figure illustrates the use of add-drop multiplexers (ADMs) that access whatever bandwidth is needed at a local node but pass the rest on to other nodes.

In typical applications the amount of bandwidth allocated to each node is quasi-static. It is changed only in response to macroscopic changes in traffic patterns, possibly as a function of the time of day. This basic operation is generally often referred to as a cross-connect function as opposed to a switching function, which involves call-by-call reconfigurations.

An important point to note about Fig. 4.20 is the ability to utilize a general-purpose physical topology but define an arbitrary functional topology on top of it.

One obvious limitation of a loop is its vulnerability to failures of any link or node.

The effect of a node failure can be minimized by having bypass capabilities included in each node. When bypassed, a node becomes merely a regenerative repeater, as on T-carrier transmission links. Link failures can be circumvented by providing alternate facilities. Figure 4.21 shows one particular structure using a second, reverse-direction loop to provide backup capabilities in the case of failures. When fully operational, the network can use the reverse loop as a separate, independent network for traffic as needed. Whenever a failure occurs, the nodes adjacent to the break establish a new loop by connecting the forward path to the reverse path at both places. Hence all nodes continue to have full connectivity to any node on the new loop.

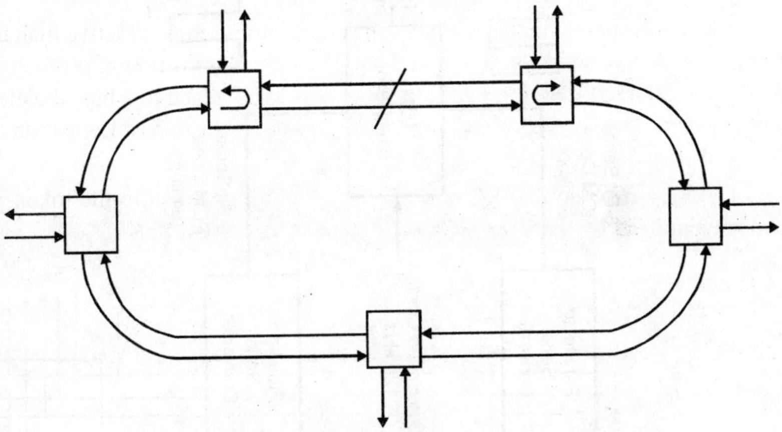

Fig. 4.21. Use of Reverse Loops.

A particular example of the use of the dual reverse loops for both protection and distributed queued access to the channels is the distributed queued dual-bus (DQDB) system developed by QPSX in Australia and standardized by the IEEE as an 802.6 metropolitan area network.

4.7. VOICE DIGITIZATION

The process of converting analog signal into digital signals is called **digitization**. It involves the modulation process called pulse code modulation. With digital transmission system, the quality of the system can be improved. The digital systems in comparison with the analog, provides a better switching interface, enhance easier multiplexing and produce clear signals. To convert analog signals to digital signals, a coding system called pulse code modulation (PCM) is used.

The process of digitization generally involves four steps as shown in Fig. 4.22.

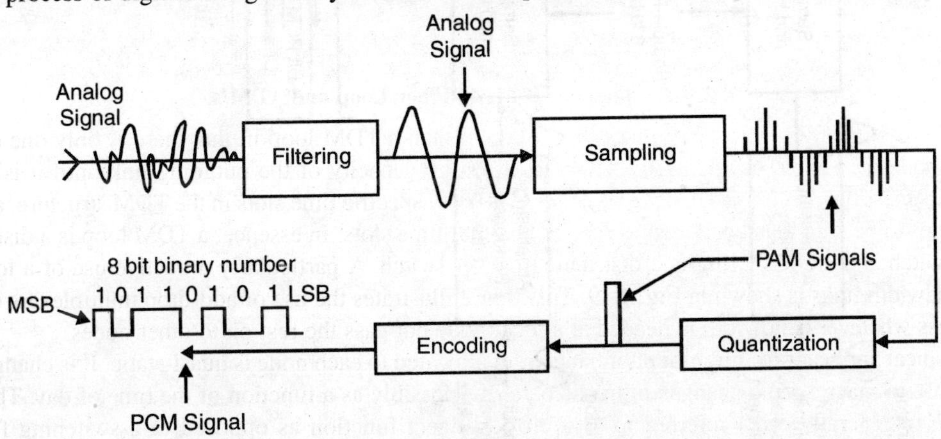

Fig. 4.22. A Simple Block Diagram of PCM Encoder.

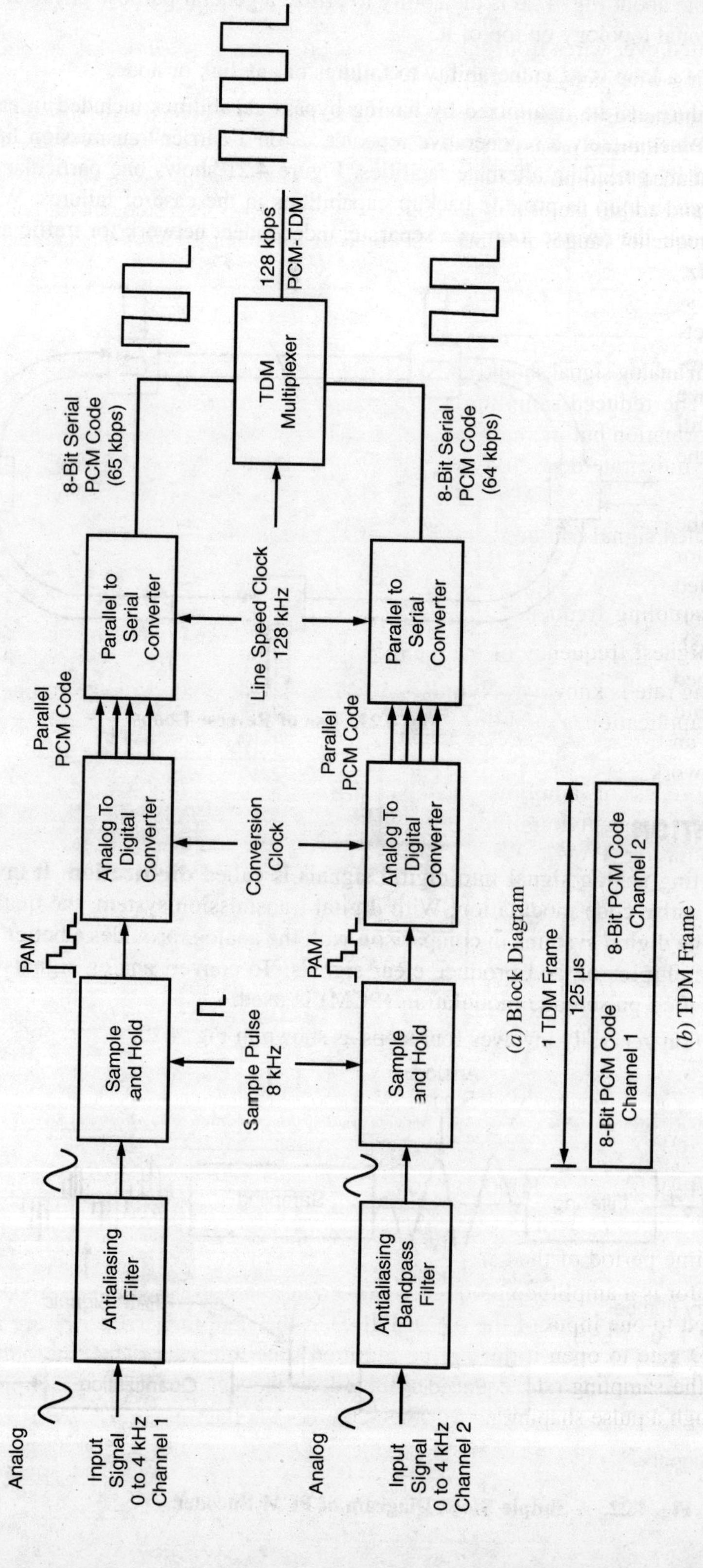

Fig 4.23. Two Channel PCM-TDM System

4.7.1. Filtering

When signal is transmitted over wires from subscriber to local exchange, electrical noise and crosstalk disturbances from neighboring wire paths also will be picked up.

The signal also gets attenuated as the distance increases. Thus the speech signal transmitted is combined with noise and attenuated. By filtering process, the lower frequencies are filtered out to remove electrical noise which may also be induced from power lines. The higher upper frequencies are filtered out because they require additional bits and add to the cost of a digital transmission system. Frequencies below 300 Hz and above 3400 Hz (voice frequency range) are filtered from the analog signal. Thus the actual bandwidth of the filtered signal is 3100 Hz.

4.7.2. Sampling

The sampled signal from analog signal should carry sufficient information so as to receive at the receiver with minimum distortion. The reduced sampling rate contains less information. The higher sampling rate contains sufficient information but increase in rate extends the bandwidth required for transmitting the sampled signal. The Nyquist rate derived by Harry Nyquist in 1933 gave engineering compromise for the sampling rate.

By principle, the sampled signal can be recovered exactly when

$$f_s \geq (2)\,(f_m) \qquad \text{...(4.1)}$$

where f_s = sampling frequency

f_m = highest frequency of the band limited signal.

This minimum sampling rate is known as **Nyquist rate**. The speech signal is in the frequency range of 300 to 3400 Hz. For this application, a sampling rate of 8000 samples per second is almost a worldwide standard.

Figure 4.23 shows the process of sampling. It can be realized that the sampling process is equivalent to amplitude modulation of a constant amplitude pulse train. Hence this technique is normally referred to as pulse amplitude modulation (PAM).

The sampling frequency

$$f_s = 2\,(f_m) \qquad \text{...(4.2)}$$

$$f_s \approx 8000 \text{ Hz} \qquad \text{...(4.3)}$$

As

$$f_s = \frac{1}{T} \qquad \text{...(4.4)}$$

$$T_s = \frac{1}{f_s}$$

$$T_s = \frac{1}{8000}$$

$$T_s = 125 \ \mu \text{ sec} \qquad \text{...(4.5)}$$

where, T_s = time period of the sampling.

In PAM sample generator is a amplitude modulator. In simple version, it contains a AND gate. The signal to be converted is fed to one input of the AND. Pulses at the sampling frequency are applied to the other input of the AND gate to open it during the required time intervals. Thus, the output of the gate consists of pulses at the sampling rate, equal in amplitude to the signal voltage at each instant. The pulses are then passed through a pulse shaping network which gives the flat tops.

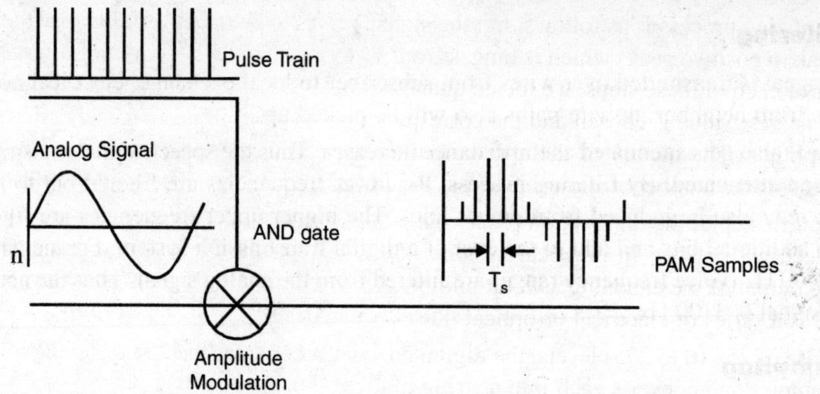

Fig. 4.24. Generation of Pulse Amplitude Modulation Samples.

4.7.3. Quantization and Encoding

Instead of sending a pulse train capable of continuously varying one of the parameters, the PCM technique produces a series of binary code which represents the approximate amplitude of the signal sample at that instant. This process of approximation is called signal quantization. Thus, the combined operations of sampling and quantizing generate a quantized PAM waveform.

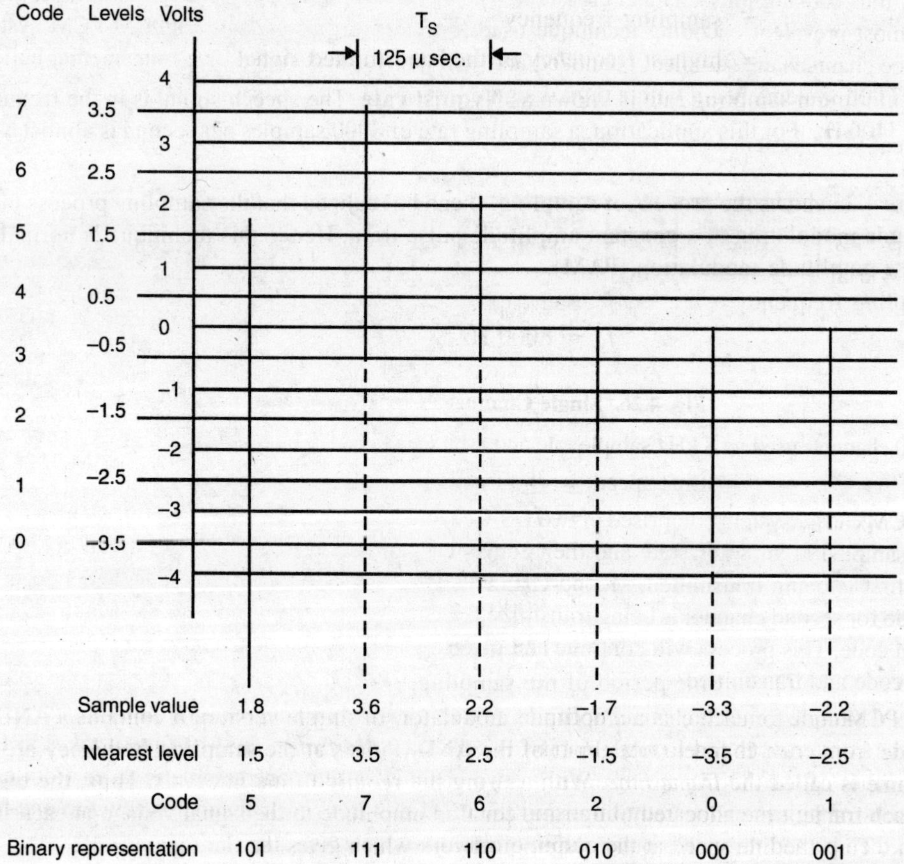

Fig. 4.25. Quantization With Number of Levels 8.

The quantizing process is as follows. Let the signal $m(t)$ is to be quantized. Identify the signal excursion (negative peak to positive peak) which is ranges from V_L to V_H. Divide the total range into M equal intervals. In the center of each of these steps, we locate quantization levels. If the sampled amplitude is in a particular interval, the nearest value is chosen and the corresponding binary digit is approximated as equivalent to the sampled amplitude. The quality of the approximation may be improved by reducing the size of the steps, thereby increasing the number of allowable levels. Figure 4.25 shows the process of quantization and encoding.

In the fourth step of the digitization process, the quantized samples are encoded into a digital bit stream (series of electrical pulses). In practical system, 256 levels can be used to obtain quality of signal. This results in the output of the encoding as 8 bit number. This binary number (or 8 bit word) is transmitted over the network as a series of electrical or optical pulses.

In the case of 256 (0 to 255) levels, the digital encoder recognizes the 255 different voltage levels of the quantized samples and converts each into a string of eight bits (1*s* and 0*s*).

The series of pulses is called a digital bit stream. The PCM process require a 64 kbps channel to encode a 4 kHz audio input signal because 8000 samples/sec × 8 bit/words = 64000 bps.

4.8. TIME DIVISION MULTIPLEXING OF PCM SIGNALS

A PAM or PPM can be employed for TDM. These methods are now obsolete as the transmitted pulses dispersed due to attenuation and delay distortion. They spread in time and interfere with the pulses of adjacent channels, thus causing inter-channel crosstalk. The PCM technique used in TDM eliminates this problem. PCM is most prevalent encoding technique used for TDM digital signals. In TDM-PCM system, two or more voice channels are sampled, quantized and encoded to PCM codes. The fundamental building blocks for most TDM systems begin with a digital signal level 0 (DS-0).

Typical PCM system for TDM transmission is shown in Fig. 4.26.

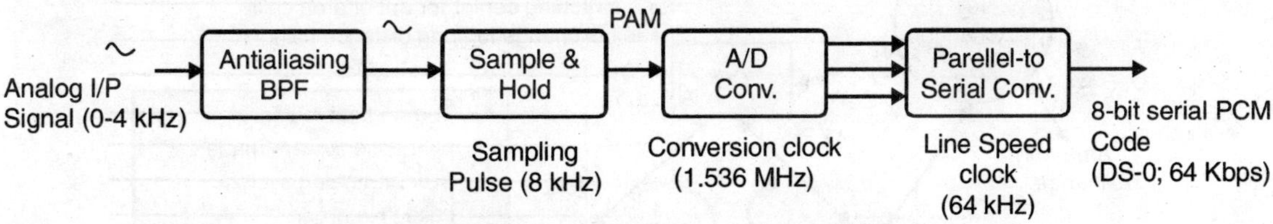

Fig. 4.26. Single Channel PCM Transmission system.

DS-0 channels used an 8 kHz sample rate and 8-bit PCM code, which resulted to a 64 kbps PCM line speed,

$$\text{Line speed} = (800 \text{ samples/sec}) \times (8 \text{ bits/sample}) = 64 \text{ kbps.}$$

A PCM carrier system comprised of two DS-0 channels may be time-division multiplexed. Each channel's input is sampled at an 8 kHz rate and then converted to an eight-bit PCM code. While the PCM code for channel first is being transmitted, second channel's is sampled and converted to a PCM code. Now while PCM code for second channel is being transmitted, the next sample is taken from first channel and converted to a PCM code. This process will continue and samples are taken alternatively from each channel, converted to PCM code and transmitted.

The PCM code for each channel occupies a fixed time slots within the total TDM time frame. One 8-bit PCM code from each channel (total 16 bits) is called a *TDM frame*, and the time it takes to transmit one TDM frame is called the *frame time*. With a two channel system, one sample is taken from each channel during each frame; time allocated to transmit the PCM bits from each channel is equal to one-half the total fame time. Thus the line speed at the output of multiplexer is

$$\text{Line speed} = (2\text{-channels/frame}) \times (8000 \text{ samples/ sec}) \times (8 \text{ bits/sample}) \quad ...(4.6)$$

$$\text{Line speed} = 2 \times 64 \text{ kbps} = 128 \text{ kbps}$$

4.9. DIGITAL CARRIER SYSTEMS

Telephone companies make uses of multiplexing long distance calls onto high- speed trunks. International standards bodies issue multiplexing rates that from the digital transmission hierarchy. Digital carrier is a digital signalling represents a telecommunications service. The base level in the hierarchy is the equivalent of a single 64 kbps digitized voice. The hierarchy differs from region to region.

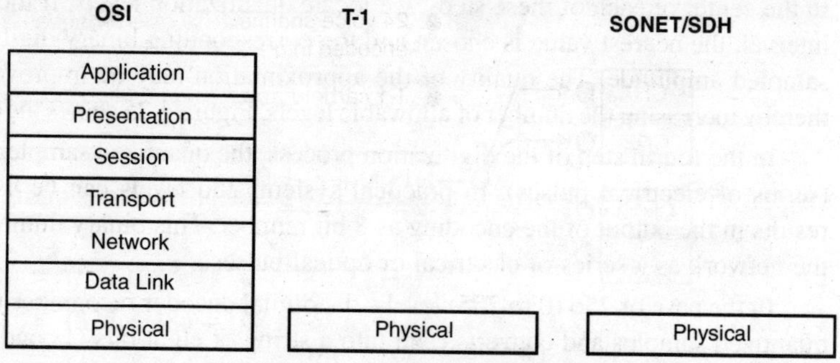

Fig. 4.27. Comparison of the layer.

Digital service in North America defines a four level transmission hierarchy called T-carrier, range from T_1, T_2, T_3 and T_4. In Europe and South America, there is five-level transmission hierarchy called **E-carrier**, range from E_1, E_2, E_3, E_4 and E_5. Both systems are using PCM to encode an analog signal in digital form. The signal is sampled 8000 times per second, and each sample value is encoded in an 8-bit value. The signal transmission uses TDM. SONET and SDH are intended to provide a specification for taking advantage of high-speed digital transmission capability of optical fiber.

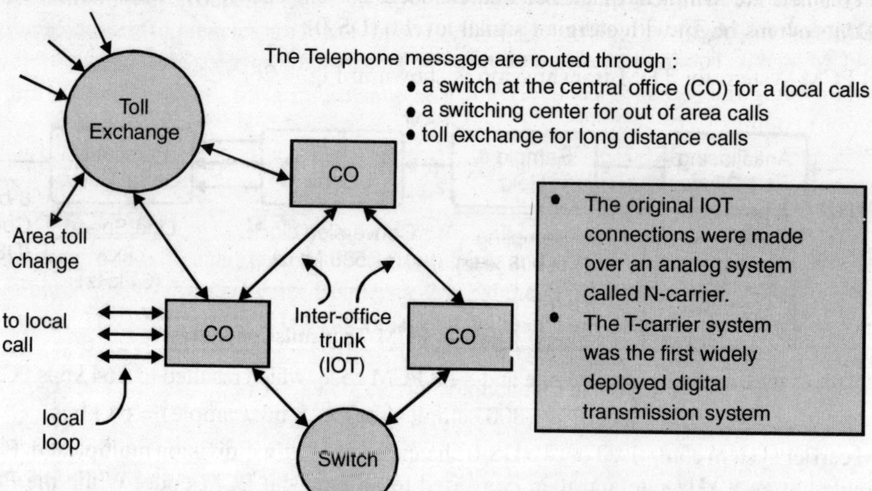

Fig. 4.28. Organization of telephone services.

T-1 and SONET/SDH standard define the rates and formats, the physical layer, network and the network operational criteria.

The non-local calls are routed between central offices through inter-office trunks (IOT). The original IOT connections were made an analog system called the N-carrier. The N-carrier system deployed an analog transmission, allowing 12 voice calls to share the same line. The L-carrier system was designed in the late 1960s for handling much higher capacity. The system can carry up to 6000 calls on the coaxial cable. However, the repeater was needed to improve signal periodically along the line. The system was very expensive for investment.

The new digital transmission called T-carrier system was later developed. The T-1 version can handle 24 calls on two pairs of copper wire.

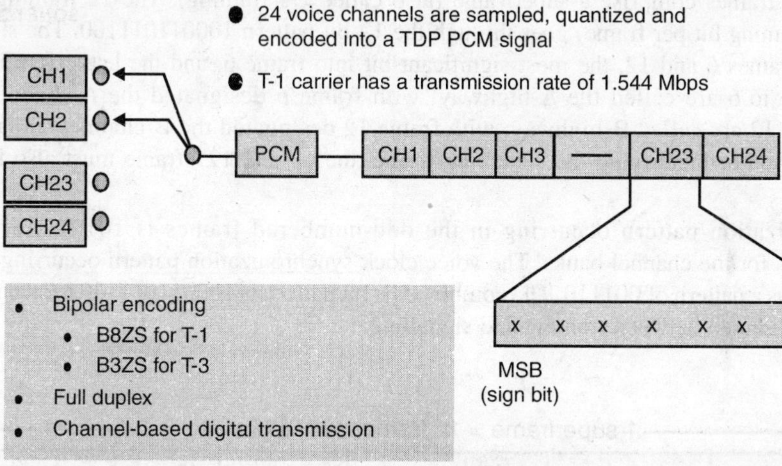

Fig. 4.29. T-1 carrier system.

Frame

The Bell system has established time division multiplexed PCM systems in North America. The transmission system is known as the T-carrier systems. The T1 carrier is the fundamental building block of the multiplexing hierarchy. The T1 carrier consists of 24 voice channels that are sampled, digitized and encoded to a PCM signal. The transmission rate onto the T1 carrier is 1.544 Mbps. The destructions T1 and DS1 refer to the circuit and signal respectively. Pulse trains in T carrier rely on the B8ZS (Bipolar with 8 Zero Substitution) coding.

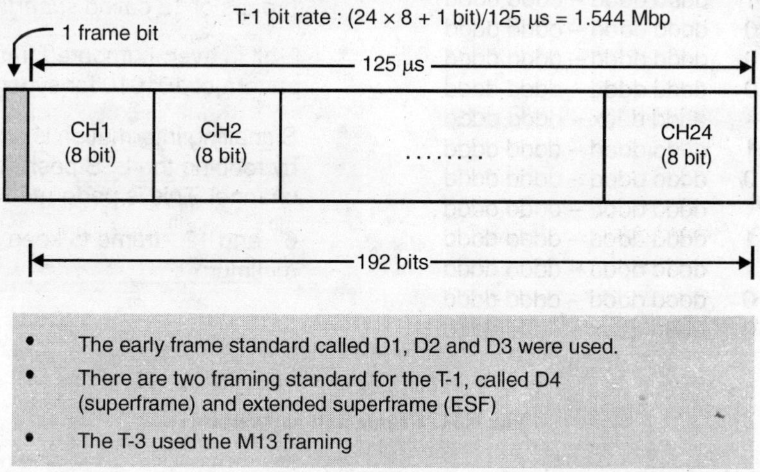

Fig. 4.30. T-1 frame.

A DS1 frame contains 193 bits divided into 24 slots (one for each voice channel) of 8 bits and one extra bit called a framing bit (f-bit), which is used for synchronizing, as illustrated in Fig. 4.29.

Eight-bit voice samples are taken each of 24 channels at a rate of 8000 times per second (64 Kbps rate).

To support this speed, T1 must transmit a DS1 frame every 1/8000 of a second (125 μs). This yields the bit rate of (24 × 8 + 1) bits/125 μs= 1.544 Mbps.

Superframe

12 consecutive frames comprise a superframe (also called D4 framing). The 12 framing bits in this superframe (one framing bit per frame) goes through the 12-bit pattern 100011011100. The signalling bits are substituted in frames 6 and 12, the most significant bit into frame 6, and the least significant bit into frame 12. Frames 1 to 6 are called the A-highway; with frame 6 designated the A-channel signalling frame. Frames 7 to 12 are called B-highway; with frame 12 designated the B-channel signalling frame. Therefore, in addition to identifying the signalling frame, the 6th and 12th frame must also be positively identified.

The synchronization pattern occurring in the odd-numbered frames is 101010. These are the synchronization bits for the channel banks. The voice clock synchronization pattern occurring in the even-numbered frames has a pattern of 001110. The combined Fs bit pattern is 100011011100 for the superframes. It is call the S bit is shared between framing and signalling.

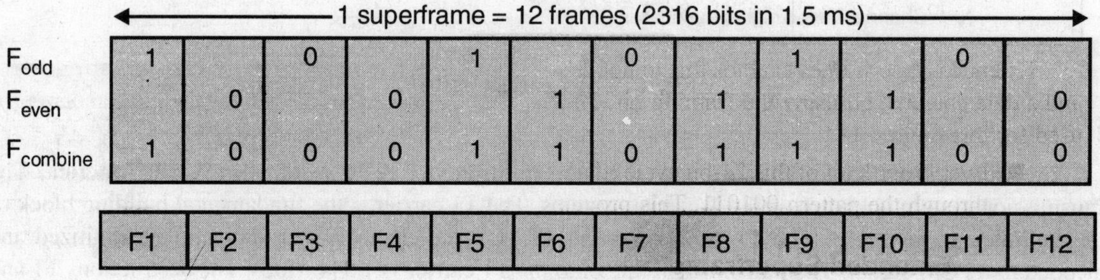

Frame #	F	data (192 bits)
1	1	dddd dddd – dddd dddd
2	0	dddd dddd – dddd dddd
3	0	dddd dddd – dddd dddd
4	0	dddd dddd – dddd dddd
5	1	dddd dddx – dddd dddd
6	1	dddd dddd – dddd dddd
7	0	dddd dddd – dddd dddd
8	1	dddd dddd – dddd dddd
9	1	dddd dddd – dddd dddd
10	1	dddd dddd – dddd dddd
11	0	dddd dddd – dddd dddd
12	0	dddd dddx dddd dddd

- T-1 carrier frames are transmitted in groups of 12 called *superframes*
- F-bit in even-numbered frame has a pattern of 101010 for synchronization
- Signalling information is accomplished by robbing the LSB position of each channel. This is performed only in the 6th and 12th frame to keep distortion minimum

Fig. 4.31. Frame and superframe.

Extended Superframe

The extended superframe format consists of 24 frames of 193 bit, totalling 4632 bits, of which 24 are framing bits. The extended superframe occupies 3 ms.

$$\left[\frac{1}{1.544\ \text{Mbits/s}}\right]\left[\frac{193\ \text{bits}}{\text{Frame}}\right](24\ \text{frames}) = 3\ \text{ms}$$

TABLE 4.2. Extended Superframe Format

Frame number	Framing bit	Frame number	Framing bit
1	C	13	C
2	F	14	F
3	F	15	F
4	S = 0	16	S = 0
5	C	17	C
6	F	18	F
7	F	19	F
8	S = 0	20	S = 1
9	C	21	C
10	F	22	F
11	F	23	F
12	S = 1	24	S = 1

Extended Superframes extends this to not only provided frame synchronization, but also error detection and a data channel, all using the framing bit. The value in every 193rd bit (in bits 193, 386, and so on) are used for three purposes :

- Every fourth bit of this 24-bit cycle (that is, the framing bits for frames 4, 8, 12, 16, 20, and 24) goes through the pattern 001011. This provides the frame synchronization.

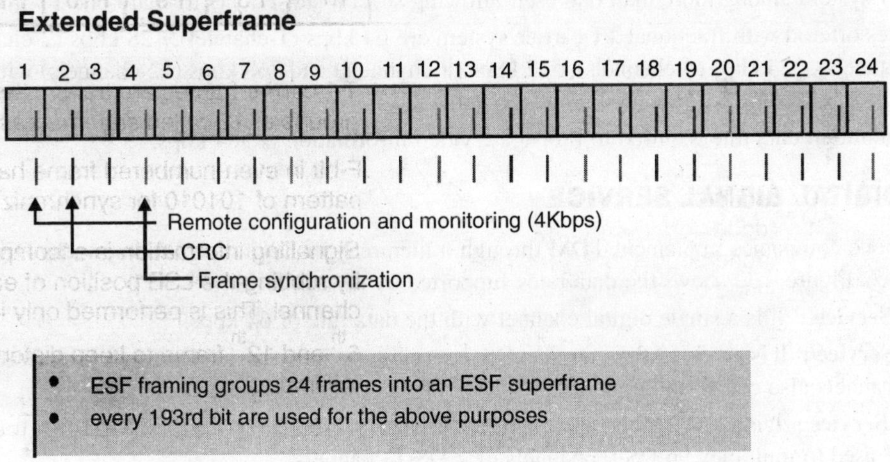

Fig. 4.32. Extended Superframe.

- The framing bits for frames 2, 6, 10, …, 22).are used to send a 6 bit CRC, generated from the data in the previous 24 frames. This provides error detection. The receiving CSU can then track the error rate and generate an alarm if it gets too high. This error checking is done constantly while the links is in service and for any type of data.
- The remaining framing bits (for frames 1, 3, 5, 7,…, 23) provide a 4kbps supervisory data channel that is used for other functions such as remote configuration and monitoring of CSUs.

The eighth bit in every channel of frames 6, 12, 18 and 24 is used for signalling between central offices.

4.10. FRACTIONAL T CARRIER SERVICE

A fractional T1 is one or more channels bundled together and sold to a customer as a set. This allows a consumer to purchase less than a full T1's bandwidth at a lower cost. Like the channelized T1, individual channels can be voice or data and a CSU/DSU is used to split the channels, however less than the full set of 24 channels is available to the customer for use.

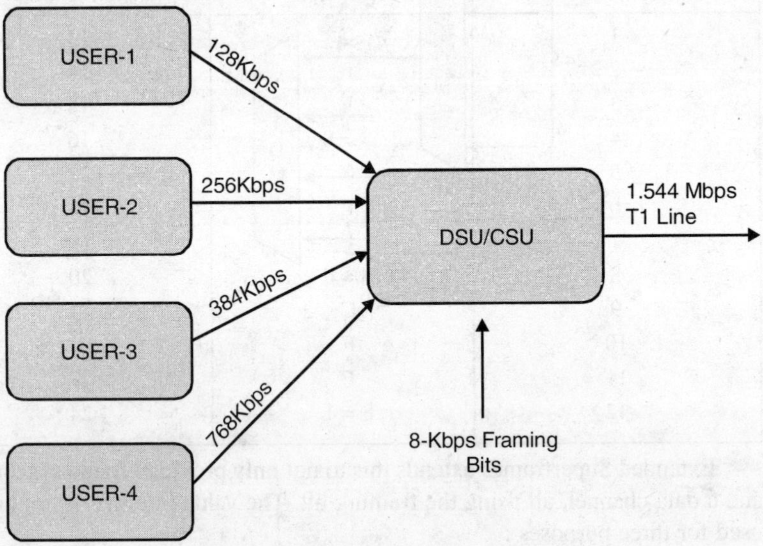

Fig. 4.33. Fractional T1 Carrier System.

Fractional T-carrier emerged because standard T1 carriers provide a higher capacity (higher bit rate) than most users require. Fractional T1 carrier system distributes the channels (*i.e.,* bits) in a standard T1 system among more than one user, allowing several subscribers to share one T1 line.

Bit rates offered with fractional T1 carrier system are 64 kbps (1-channel), 128 kbps (2-channels) 256 kbps (4-channels), 384 kbps (6-channels), 512 kbps (8-channels) and 768 kbps (12-channels) with 384 kbps (1/4 T1) and 768 kbps (1/2 T1) being the most common.

The minimum data rate required to propagate video information is 384 kbps.

4.11. DIGITAL SIGNAL SERVICE

Telephone companies implement TDM through a hierarchy of Digital signal, known as Digital Signal (DS) Service. Figure 4.32 shows the data rates supported by at each level.

DS-0 Service : It is a single digital channel with the data rate of 64 kbps.

DS-1 Service : It is a 1.544 Mbps service. DS-1 service as single service for 1.544 Mbps transmissions. 24 DS-0 channels also can be multiplexed.

DS-2: Service : It is a 6.312 Mbps service. It can be used as a single service for 6.312 Mbps transmissions or it can be used to multiplex 96 DS-0 channels or 4 DS-1 channels.

DS-3 Service : It is a 44.376 mbps service. DS-3 can be used as a single service for 44.376 Mbps transmissions. It may be used to multiplex 7 DS-2 channels, 28 DS-1 channels, or 672 DS-0 channels.

DS-4 Service : It is a 274.176 Mbps service. DS-4 may be used to multiplex 6 dDS-3 channels, 42 DS-2 channels or 4032 dS-0 channels.

DS-5 Service : It is a 560.160 Mbps service. DS-5 may be used to multiplex 2 DS-4 channels, 12 DS-3 channels, 88 DS-2 channels, 362 DS-1 channels or 875 DS-0 channels.

Telephone companies implement digital services (DS-0 to DS-5) by using t lines. The capacities of these lines matched to the data rates of the digital services. DS-0 has been defined as a basis for reference purposes, it actually does not offer as a service.

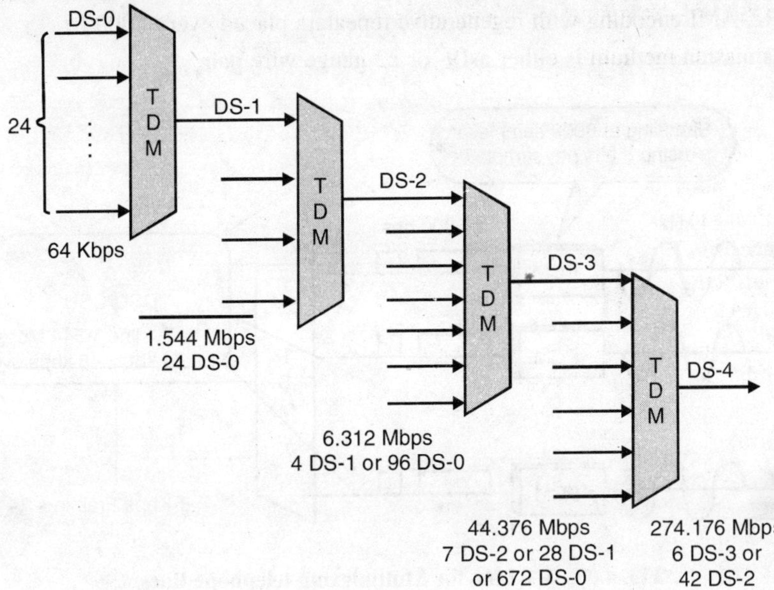

Fig. 4.34. Digital Signal Service Using TDM Hierarchy.

DS-1, DS-2, DS-3, DS-4 and DS-5 digital signals may be placed on T-1, T-2, T-3, T-4, and T-5 lines respectively. Table 4.3, lists the digital signals, their bit rates, channel capacities and services offered for the line types.

TABLE 4.3. The Digital Signals, Their Bit Rates, Channel Capacities and Services Offered.

DS	Line	Rate (Mbps)	Voice Channels	Services
DS-1	T-1	1.544	24	Voice or Data
DS-2	T-2	6.312	96	Voice or Data Picture phone
Ds-3	T-3	44.736	672	Voice or Data, Picture Phone, Broadcast
DS-4	T-4	274.176	4032	Same as T-3
DS-5	T-5	560.160	8064	Same as T-3

TIPS ▶

Telephone companies implement TDM through a hierarchy of digital signals, known as digital signal (DS) service.

4.12. T-CARRIERS

T-Carriers are used for the transmission of PCM-encoded time-division multiplexed digital signals. It utilized special line-encoded signals and metallic cables that have been conditioned to meet the relatively high bandwidths required for high-speed digital transmission

T1 and T1C Carrier System

1. Utilizes PCM and TDM techniques to provide short-haul transmission of 24 VB signals (48 VB signals for T1C)

2. Range from about 5 to 50 miles
3. Use BPRZ-AMI encoding with regenerative repeaters placed every 6000 ft
4. The transmission medium is either a 19- or 22-gauge wire pair.

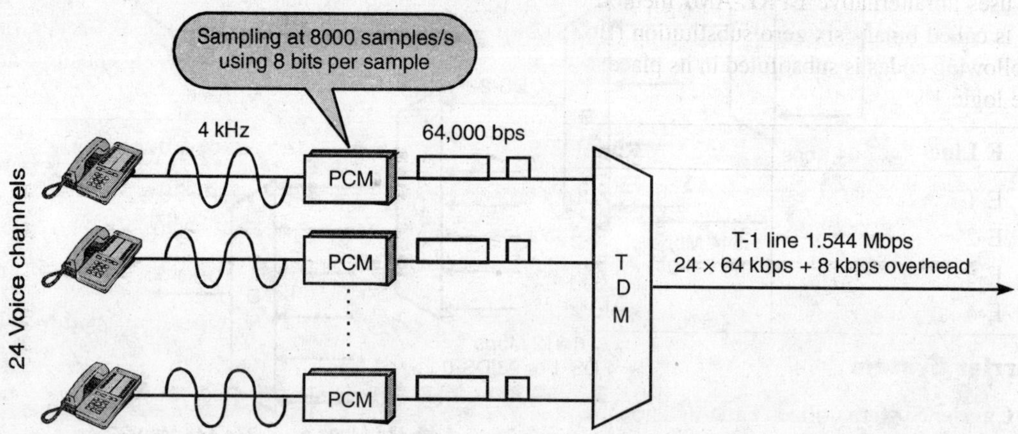

Fig. 4.35. T-1 Line for Multiplexing telephone lines.

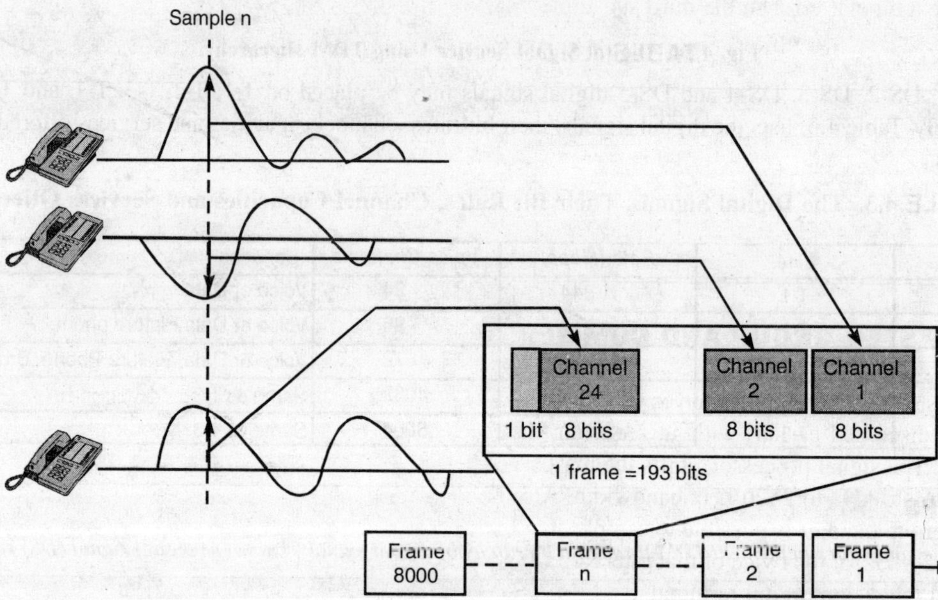

Fig. 4.36. T-1 frame structure.

Line Rate of the T1 Carrier

$$\left(\frac{8 \text{ bits}}{\text{channel}} \times \frac{24 \text{ channels}}{\text{frame}}\right) + \frac{1 \text{ framing bit}}{\text{frame}} = 193 \frac{\text{bits}}{\text{frame}}$$

$$193 \frac{\text{bits}}{\text{frame}} \times 8000 \frac{\text{frames}}{\text{seconds}} = 1.544 \text{ Mbps}$$

T2 Carrier System

T2 Carrier System utilizes PCM to time division multiplex 96 VB channels into a single 6.312 Mbps data signal for transmission up to 500miles over a special LOCAP (low capacitance) cable. T2 Carrier System uses an alternative BPRZ-AMI method of ensuring that ample transitions occur in the data. This method is called binary six zero substitution (B6ZS). With B6ZS, whenever six consecutive 0's occur, one of the following codes is substituted in its place: 0 - + 0 + - or 0 + - 0 - +. The + and – represents positive and negative logic 1's.

E Line	Rate (Mbps)	Voice Channels
E-1	2.048	30
E-2	8.448	120
E-3	34.368	480
E-4	139.264	1920

T3 Carrier System

T3 Carrier System utilizes Time-division multiplexes 672 PCM-encoded voice channels for transmission over a single metallic cable. The encoding technique used with T3 carriers is binary three zero substitution (B3ZS). There are four substitution patterns used with B3ZS: 00-, -0-, 00+, and +0+. The pattern chosen should cause a bipolar error in the third substitute bit.

TABLE 4.4. DS and T lines rates

Service	Line	Rate (Mbps)	Voice Channels
DS-1	T-1	1.544	24
DS-2	T-2	6.312	96
DS-3	T-3	44.736	672
DS-4	T-4	274.176	4032

4.13. MASTER GROUP AND COMMERCIAL TELEVISION

The master group terminal receives voice band channels which are being frequency division multiplexed, as we have discussed earlier, without requiring that each voice band channel be demultiplexed to voice frequencies. The signal processor, shifts the master group signals frequencies, from a 564 kHz to 3084 kHz bandwidth to a 0 kHz to 2520 kHz bandwidth. Also dc restorations for the television signal. After shifting the master group band, it is possible to sample at a 5.1 MHz rate. The sampling rate for the commercial TV signal is 10.2 MHz, at the twice of that rate. Refer Fig. 4.37, which shows a block diagram of a master group /commercial television digital terminal.

A 9-bit PCM code is used to digitize each sample of the master group or television signal, to meet the transmission requirements. The output of the digital processor is approximately 46 Mbps for the master group and 92 Mbps for the TV signal.

The digital terminal shown in Fig. 4.36 has three main functions, which are as follows :

1. Conversion of the parallel data from the output of the encode to the serial data,
2. Flame synchronizing bits are inserted, and
3. It converts the serial binary signal to form more suitable for transmission.

As there is no 92 Mbps line speed in the digital hierarchy, the 92 Mbps digital must be split into two 46 Mbps signals for the commercial TV terminal.

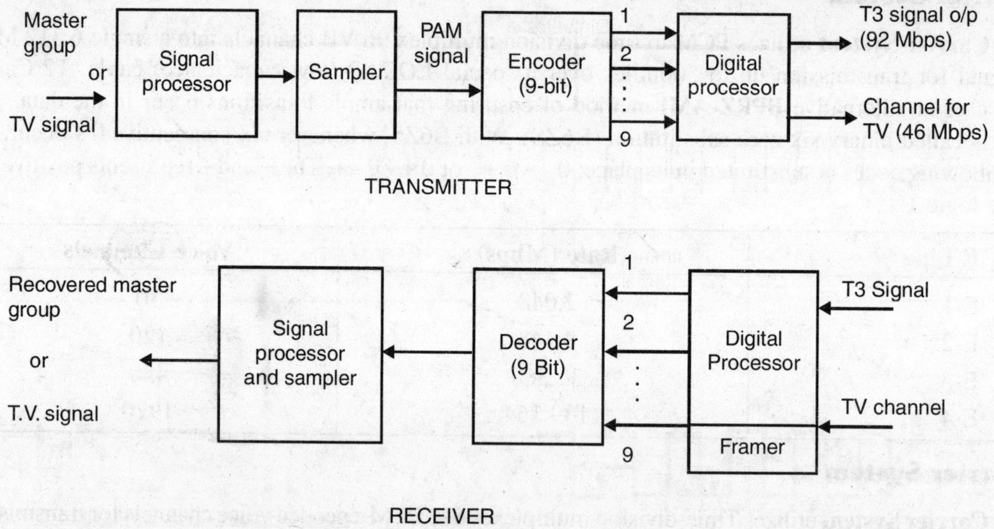

Fig. 4.37. Block Diagram of a Master Group/Commercial TV Digital Terminal.

4.14. PICTURE PHONE TERMINAL

A picture phone is used to transmit the low quality video signals for the use of non-dedicated subscribers. This picture phone signal is encoded into T2 capacity of 6.312 Mbps, which is substantially less than that for commercial network broadcast signals, so that cost may be reduced and makes the service affordable. At the same time, it allows the transmission of adequate detail and contrast resolution to satisfy the average picture phone subscriber.

Picture phone service is ideally suited to a differential PCM code in which only difference of the magnitude between the present sample and the previous sample is encoded and transmitted. By using differential PCM, we need to transmit substantially fewer bits than encoding the actual sample, so that result in saving of bandwidth.

4.15. DATA TERMINAL

Data terminal are designed to transmit the signals other than voice. In most of the cases the data rates generated by each individual subscriber are substantially less than the data rate capacities of digital lines. Therefore, it seems only logical that terminals be designed that transmit data signal from several sources over the same digital line.

A method of coding the data is shown in Fig. 4.38. This method is more efficient, it codes the transition time. In the coding format shown in Fig. 4.38 is a uses three bit code to identify when transition occurs on the data and whether the transition is from a '1' to '0' or vice-versa. The first bit is referred as address bit. When there is no transition it indicates logic '1' and when transition did occur, it indicates logic '0'. The second bit indicates whether the transition occurred during the first half (0) or during the second half (1) of bit period. The third bit indicates the direction of the transition; a '1' for this bit indicates a 0-1 transition; and a '0' indicates a 1-0 transition.

Consequently; where there are no transition in the data, a second of all '1' is transmitted. Although the transmissions of only address bit would be sufficient, but however, the sign bit provides a degree of error protection and limit error propagations.

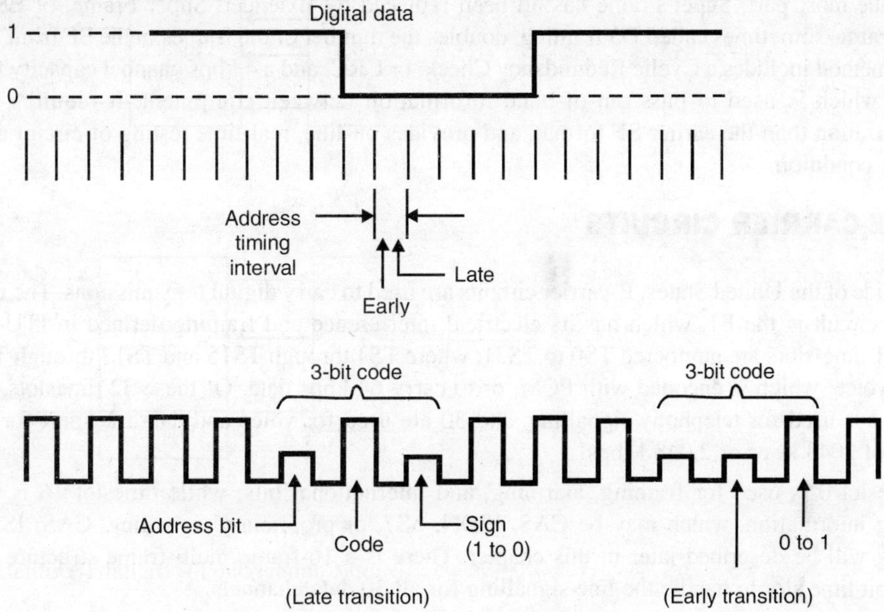

Fig. 4.38. Data Coding Format.

The advantage of using a coded format rather than the original data is that coded data are more efficiently substituted for voice in analog system. Without this coding format, transmitting a 250 kbps data signal requires the same bandwidth as would be required to transmit 60 voice channels with analog multiplexing. When we use this coded format, a 50 kbps data signal displaces three 64 kbps PCM-encoded channels and a 250 kbps data stream displaces only 12 voice-band channels.

4.16. FRAME SYNCHRONIZATION

Framing, or **frame synchronization**, is the process by which incoming frame alignment signals, *i.e.,* distinctive bit sequences, are distinguished from data bits, permitting the data bits within the frame to be extracted for decoding or retransmission. A T1 time slot is an 8-bit segment for each DS0. A frame consists of 24 time slots (one for each DS0 in the T1) and an additional framing bit, for a total of 193 bits.

The framing bit does not contain any user data; however, it is used by the transmitting and receiving equipment for the actual synchronization of the incoming data with the receiver. Within a bit stream, framing bits are used to determine the beginning or end of a frame. They occur at specified positions in the frame, do not carry information, and are commonly repetitive. If the transmission is temporarily interrupted, or a bit slip event occurs, the receiver must re-synchronize. T1s use two types of framing: Super Frame (SF) and Extended Super Frame (ESF), with the latter being the most prevalent method implemented.

Super Frame (SF) is an older framing standard for T1s that is also referred to as D4 or even D3/D4 framing. In order to determine where each channel is located in the stream of data being received, each set of 24 channels is aligned in a frame. The frame is 192 bits long and is terminated with the framing bit (the 193rd bit), which is used to find the end of the frame. To ensure that the framing bit will be located by receiving equipment, a specific pattern is sent on this bit. Equipment will search for a bit which has the correct pattern, and will align its framing based on that bit. The pattern sent is 12 bits long, so every group of 12 frames is called a Super Frame. The pattern used in the 193rd bit is 1000 1101 1100.

For the most part, Super Frame has all been replaced by Extended Super Frame, or ESF. **Extended Super Frame**, sometimes called D5 framing, doubles the number of the frames in the SF from 12 to 24. This framing method includes a Cyclic Redundancy Check, or CRC, and a 4 kbps channel capacity for a data link channel, which is used to pass out-of-band information between equipment. It requires less frequent synchronization than the earlier SF format, and provides on-line, real-time testing of circuit capability and operating condition.

4.17. E-CARRIER CIRCUITS

Outside of the United States, E-carrier circuits are used to carry digital transmissions. The most common E-carrier circuit is the E1, which has its electrical interference and framing defined in ITU-T G.703 and G.704. E1 time-slots are numbered TS0 to TS31, where TS1 through TS15 and TS17 through TS31 are used to carry voice, which is encoded with PCM, or to carry 64 kbps data. Of these 32 timeslots, 1 is used for framing, 1 is used for telephony signalling, and 30 are used for voice and data transport for an aggregate capacity of 2048 kbps or 2.048 Mbps.

Timeslot 0 is used for framing, alarming, and international bits; while timeslot 16 is used to carry signalling information, which may be CAS, ISDN, SS7, or proprietary signalling. CAS, ISDN, and SS7 signalling will be described later in this chapter. There is a 16-frame multi-frame structure that allows a single 8 bit time slot to handle the line signalling for all 30 data channels.

The E1 standard also allows for optional CRC-4 error-checking and is most commonly used in Europe, parts of Asia, South and Central America and Africa.

4.18. CODECS

Codec is a generic term that refers to the coding and decoding function performed by a device. It converts analog signals to digital codes and digital codes to analog signals.

A codec is a large-scale integration (LSI) chip designed for use in the telecommunication industry for central office switches, digital handsets; private branch exchanges (PBXs), voice store- forward systems and digital echo suppressors. The codec is applicable for any purpose that needs the digitizing of analog signals, such as in a PCM-TDM carrier system.

Functions of Codec

The codec performs following functions:

1. Analog sampling.
2. Analog to digital and digital to analog conversion.
3. Digital companding.

Recently developed codecs are referred as combo-chips because they combine codec and filter functions in the same LSI package. The filters (input and output) perform the following functions :

1. Noise rejection
2. Band limiting,
3. Antialiasing
4. Reconstruction of analog audio waveforms after decoding.

4.19. COMBO-CHIPS

Essentially, a combo chip replaces the older codec and filter chip combination. A combo-chip can provide the analog to digital conversion and transmit and receive filtering necessary to interface a full-duplex *i.e.,* four wire voice telephone circuit to the PCM highway of a TDM carrier system.

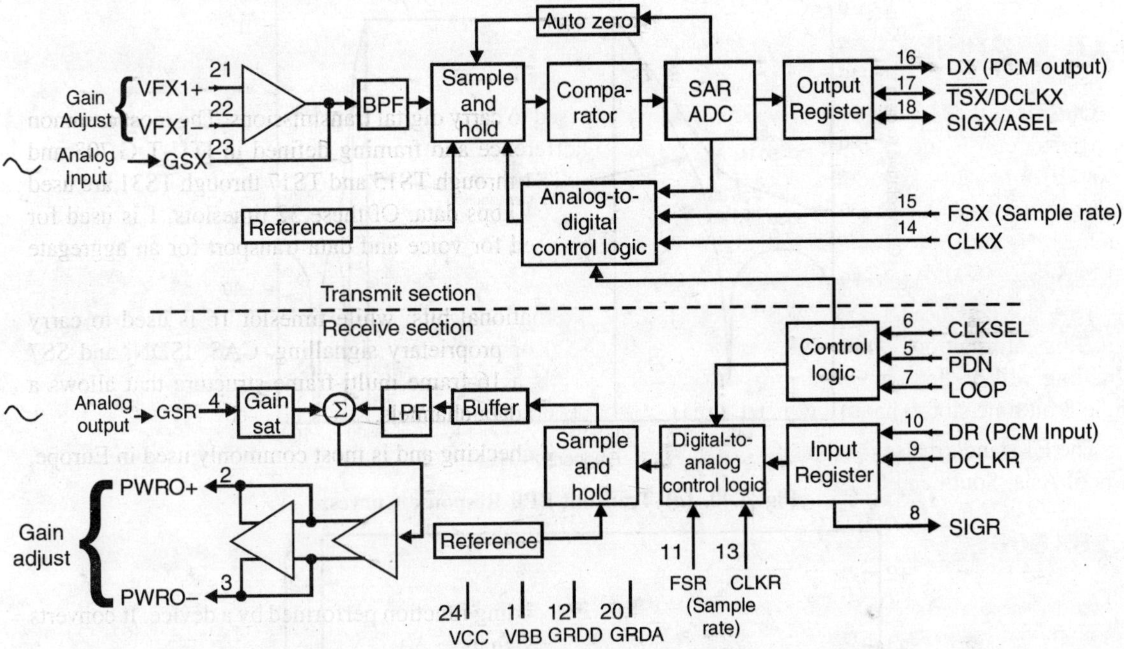

Fig. 4.39. A Simple Block Diagram of Combo-Chip.

Functions of Combo-Chip

A combo- chip performs the following functions :

1. Band pass filtering of the analog signals before encoding and after decoding.
2. Encoding and decoding of voice signals.
3. Encoding and decoding of signalling and supervision information.
4. Digital companding.

Figure 4.38 shows the simple block diagram of combo-chip. The frequency response curves for transmit band pass filter and receive band pass filter shown in Fig. 4.40 (*a*) and (*b*).

There are two types of modes of operation performed in combo-chip. They are as follows :

1. Fixed-data-rate mode
2. Variable-data-rate mode

Both the modes of operation in combo-chip are discussed in brief here :

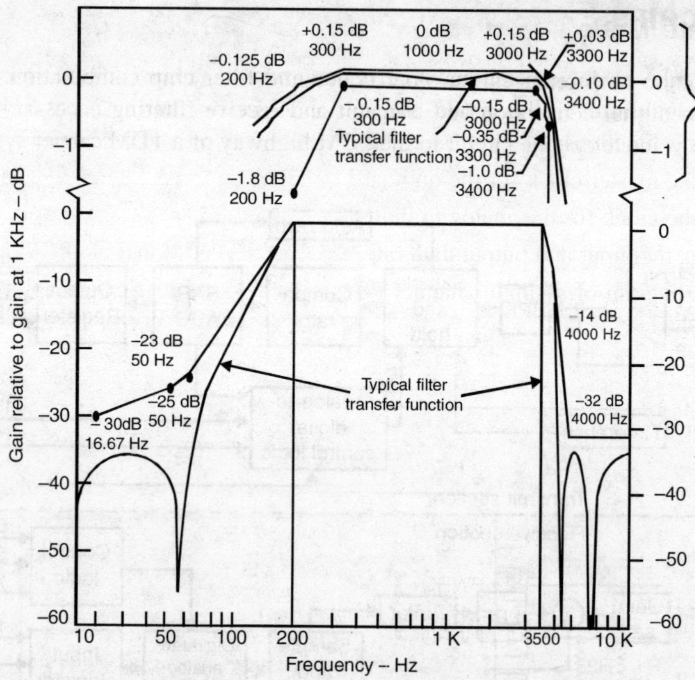

Fig. 4.40. (*a*) Transmit BPF Response Curves.

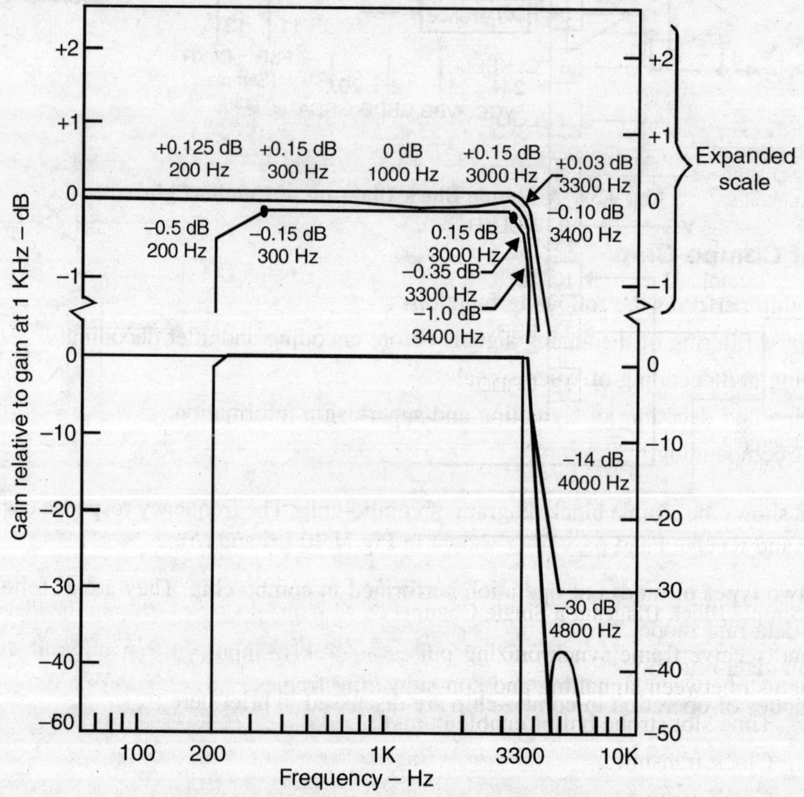

Fig. 4.40. (*b*) Receive LPF Response Curves.

1. Fixed-Data-Rate Mode

In the fixed-data-rate mode, transmit and receive data must be either 1.536 Mbps or 2.048 Mbps— the same as the master clock rate. In this mode, the master transmits and receives clocks on a combo-chip performs the following functions:

- It provides the master clock for the on board switched capacity filter.
- It provides the clock for the analog to digital and digital to analog convertors.
- It determines the input and output data rate between the codec and the PCM highway.

A simple block diagram of a single channel PCM system using a combo-chip in the fixed- data-rate mode shown in Fig. 4.41.

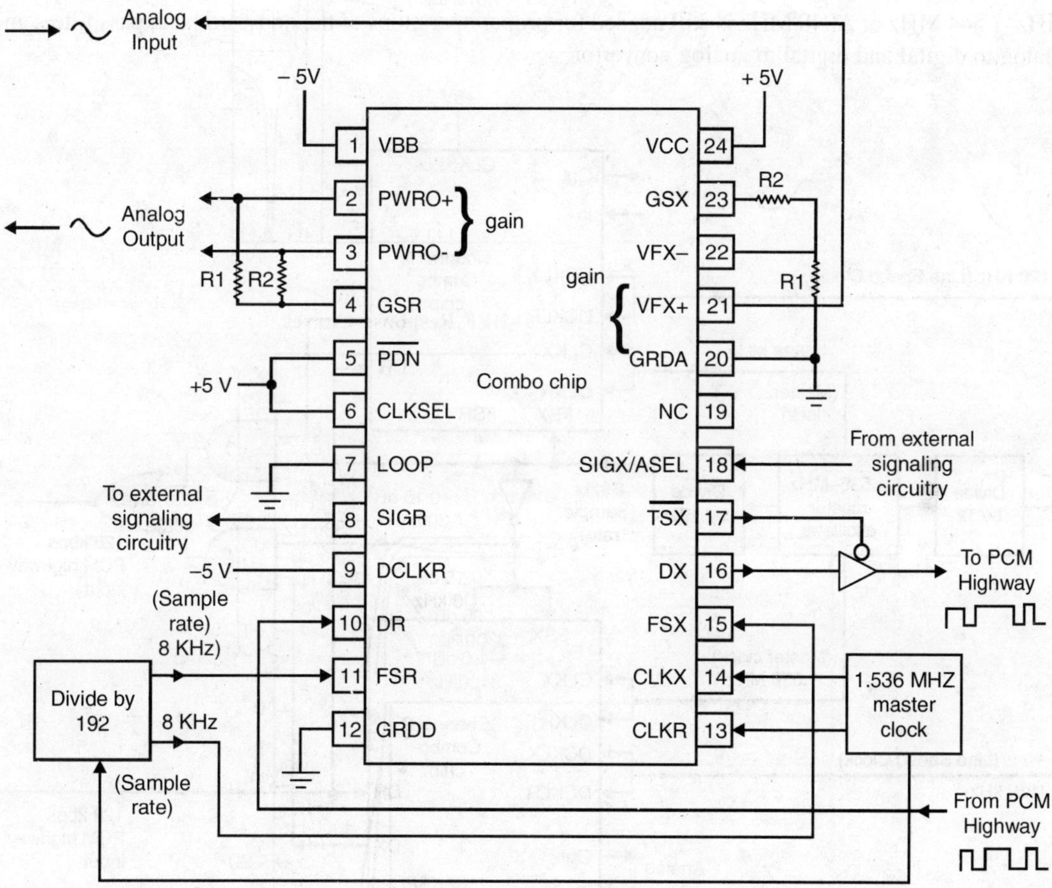

Fig. 4.41. A Simple Block Diagram a Single Channel PCM Using a Combo-Chip in the Fixed Data Rate.

Transmit and receive frame synchronizing pulses are 8 kHz input set transmit and receive sampling rates and distinguish between signalling and non-signalling frames.

TSX	Time slot strobe buffer enable output
CLK X	Clock transmit
CLK R	Clock receive

DX Data o/p

DR Data input

In the fixed data rate mode, data are input and output for a single channel in short bursts and that's why, it is known as *burst mode*.

- The inputs and output bits rates from the codec are equal to the master clock frequency *i.e.,* 1.536 Mbps.
- The DX and DR are enabled only 1/24 of the total frame time *i.e.,* 125 μs.

2. Variable-Data-Rate Mode

Variable-data-rate mode provides the ability to vary the frequency of transmit and receive bit clocks. It allows for a flexible data input and output clock frequency. In this mode, a master clock frequency of 1.536 MHz, 1.544 MHz or 2.048 MHz is still needed for proper operations of the on board band pass filters and the analog to digital and digital to analog convertor.

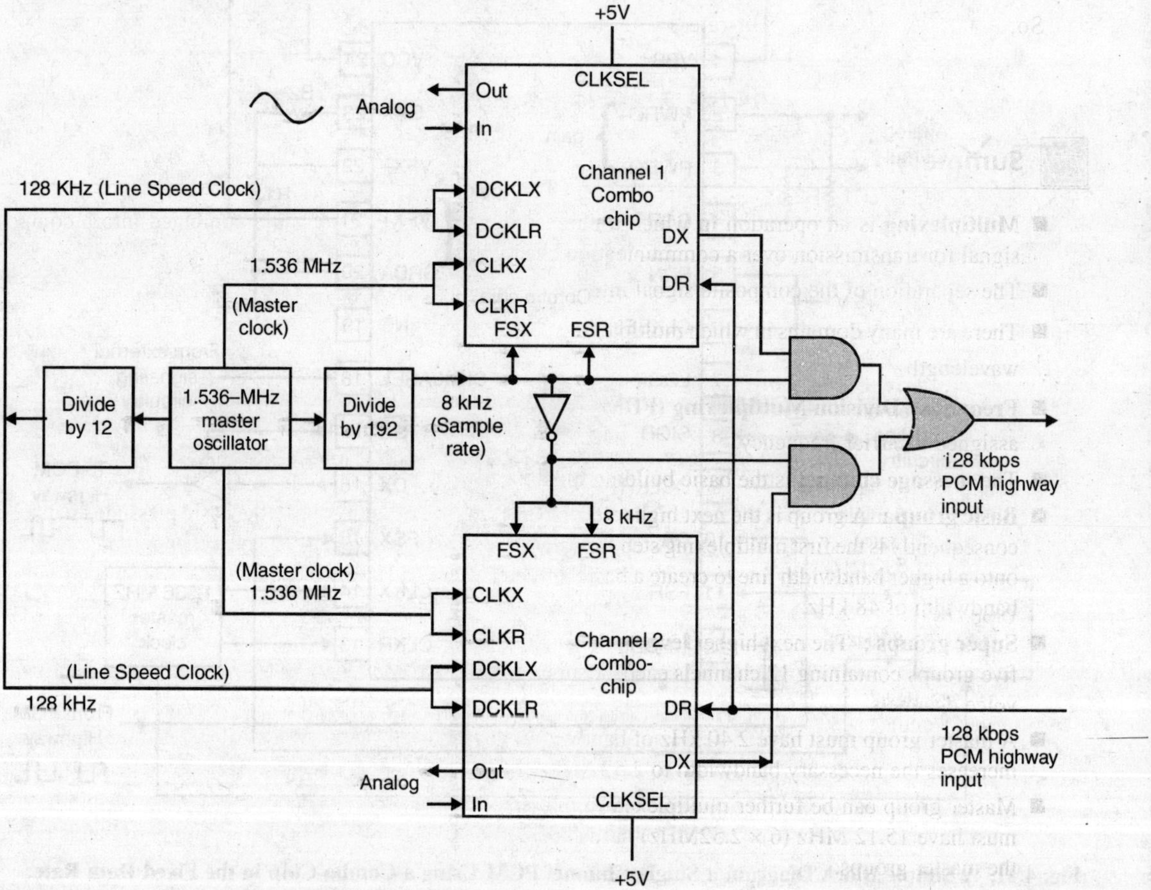

Fig. 4.42. Simple Block Diagram for a Two-Channel-PCM-TDM System.

However in the variable data rate mode, DCL KR and DCL KX becomes the data clock for receive and transmit PCM highways, respectively.

This mode of operation also known as the *shift register mode*. When FSX (Frame synchronizing pulses transmit) is high, data rate are transmitted on to the PCM highway on the next eight consecutive positive transition of DCLKX. Similarly, which FSR (Frame synchronizing receive pulse) is high, data rate from the PCM highway are clocked into the codec on the next eight consecutive negative transitions of DCL KR.

Figure 4.42 shows the block diagram for a two channel-PCM-TDM system using a combo- chip in the variable-data-rate mode with a master clock frequency of 1.536 MHz, a sample rate of 8 kHz, and a transmit and receive data rate is 128 kbps.

If the sample rate is 8 kHz the frame time is 125 μs. Therefore one-eight bit PCM word for each channel is transmitted and/or received during each 125 μs frames. Now for 16 bits to occur in 125 μs, a 128 kHz transmit and receive data clock is required.

$$t_b = \left(\frac{1 \text{ channel}}{8 \text{ bits}}\right) \times \left(\frac{1 \text{ frame}}{2 \text{ channels}}\right) \times \left(\frac{125\ \mu\text{s}}{\text{Frame}}\right) = \frac{125\ \mu\text{s}}{16 \text{ bits}}$$

$$t_b = 7.8125\ \mu\text{s/bit}$$

So,

$$\text{Bit-rate} = \frac{1}{t_b} = \frac{1}{7.8125\ \mu\text{s}}$$

$$\text{Bit-rate} = 128 \text{ kbps.}$$

Summary

- **Multiplexing** is an operation in which a number of independent signals are combined into a composite signal for transmission over a communication channel.
- The separation of the composite signal into the original signals is called **demultiplexing.**
- There are many domains in which multiplexing can be achieved. They are time, space, phase, frequency and wavelength.
- **Frequency Division Multiplexing (FDM)** is an analog technique where each communications channel is assigned a carrier frequency.
- The **message channel** is the basic building block of the FDM hierarchy.
- **Basic group :** A group is the next higher level in the analog hierarchy above the basic message channel and consequently is the first multiplexing step for combining message channels. 12 voice channels are multiplexed onto a higher bandwidth line to create a basic group. Twelve 4 kHz voice band channels occupy a combined bandwidth of 48 kHz.
- **Super groups :** The next higher level in the analog hierarchy is the super group, which is formed by FDM five groups containing 12 channels each. A super group has a bandwidth of 240 kHz and supports up to 60 voice channels.
- A master group must have 2.40 kHz of bandwidth, but the need for guard bands between the super groups increases the necessary bandwidth to 2.52 MHz Master groups support up to 600 voice channels.
- Master group can be further multiplexed in master group banks to create a jumbo group. A jumbo group must have 15.12 MHz (6 × 2.52MHz) but is augmented to 16.984 MHz to allow for guard bands between the master groups.
- Jumbo group can be further multiplexed, which is formed by 3 FDM Jumbo groups together. A Jumbo group mix support up to 10800 voice channels.
- A single message channel is baseband i.e the modulating signal in a communication system. A group, super group or signals prior to modulation of the final carrier is known as **composite baseband**.
- A radio channel includes either a single L 600 master group up to three U600 master groups (*i.e.,* 1800 voice-band channels).

- In space divisions multiplexing more than one physical transmission path are grouped together.
- TDM is the sharing of a common transmission medium in time. In TDM, the time available is divided into small slots, and each of them occupied by a piece of one of the signals to be sent.
- **Time-division multiplexing (TDM)** grants each user full channel capacity, but assigns time slots to each user.
- Time division multiplexing can be either bit by bit or byte by byte basis, known as bit interleaving and byte interleaving respectively. Bit interleaving also known as digit interleaving.
- To identify individual time slots within a TDM frame, a receiving terminal uses a counter synchronized to the frame format of the transmitter.
- **Statistical Time Division Multiplexing** uses intelligent devices capable of identifying when a terminal is idle.
- Basically, a *TDM ring* is configured as a series of unidirectional (two-wire) links arranged to form a closed circuit or loop.
- TDM loops have been used within computer complexes to provide high capacity and high interconnectivity between processors, memories and peripherals.
- The process of converting analog signal into digital signals is called **digitization**.

Review Questions

1. Explain in detail about CODECS?
2. What is meant by TDM ? Describe briefly about statistical and synchronous TDM.
3. Explain CODECS and COMBO chips in detail.
4. Explain AT and T's FDM in detail.
5. Explain in detail about CCITT time division multiplexing.
6. Explain CODEC and COMBO chips in detail.
7. Assume a data stream is made of ten os (zeros). Encode this stream using following encoding schemes :
   (*i*) Polar NRZ-1
   (*ii*) Monchester
   (*iii*) Differential Monchester
   (*iv*) HDB3.
8. Explain CCIT time division multiplexing in detail.
9. Explain T-1 carrier system in detail.
10. Draw and explain the AT&T's FDM hierarchy. Four digital data channels each transmitting 1 Mbps use a satellite channel of 1 MHz. Design an appropriate configuration using FDM.
11. In which type of multiplexing are CODECS and Combo chips used. What is the difference between the two of them ? Explain the transmitter and receiver sections of the CODECS with the help of block diagrams.
12. Write short note on T Carriers.
13. Describe the DS hierarchy. How are T lines related to DS service ? A DS-0 signal has a data rate of 64 Kbps. Where does this number come from ?

❑❑❑

IV

UNIT

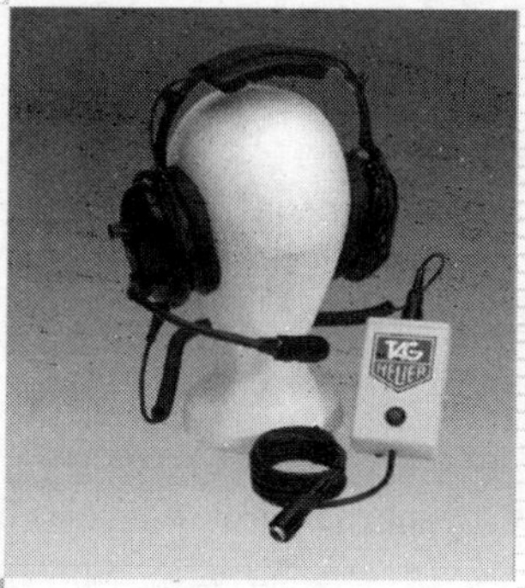

- ➤ Introduction to Data Communication
- ➤ Interface Standards
- ➤ Transmission Media

SYLLABUS

Data Communications : Components, Protocols and Standards, Standards Organizations, Line Configuration, Topology, Transmission Mode, Digital Signals, Digital to Digital Encoding, Digital Data Transmission, DTE-DCE Interface, Interface Standards, Modems, Cable Modem, Transmission Media—Guided and Unguided, Transmission Impairment, Performance, Wavelength and Shannon Capacity.

CHAPTER 5

INTRODUCTION TO DATA COMMUNICATION

INSIDE THIS CHAPTER

5.1. INTRODUCTION TO COMMUNICATION

People can communicate with other people. This is one unmistakable characteristic that points out the difference between humans and other animals (i.e. the ability to communicate). During the early days, communication among humans was limited by distance. People communicated by natural senses of hearing and sight or by written words. Actual words could be conveyed although the process was very slow. As the distance increased beyond the normal range of vision and hearing it became very difficult or near to impossible to communicate using the above methods. People searched for better ways of communicating faster and over long distances. Smoke signals and flag signaling were used to send messages. These were used for centuries until the nineteenth century when electrical signals came in use. **In 1869, the telephone was invented.** This marked a real breakthrough. At that time, it seemed like a miracle to be able to talk to someone so many miles away as easily and quickly as they were standing next to you. The arrival of wireless communication nearly three decades after the Telephone, completed the breakthrough.

Communication is the process of conveying information at a distance or it is the basic process of exchanging information. In other words, communication is the science of transmitting, receiving and processing of information. Electrical communication is a process in which the message or information is transmitted from one point to another point or from one person to another in the form of electrical signal, through some communication link.

DO YOU KNOW►

Communication *is the process of conveying information at a distance or it is the basic process of exchanging information.*

The electronic equipments or devices which are used for the purpose of communication are known as communication equipments and integrating of different communication equipments for the purpose of communication form a **communication system.**

5.2. ELECTRICAL COMMUNICATION SYSTEM

In communication, the physical information, such as sounds, pictures, words etc., is converted into an electrical message called signal and this electrical signal is being conveyed at the far place, and there, these electrical signals are reconverted into physical message.

Figure 5.1 shows the simple block diagram of electrical communication system and it has three main basic components,

1. Transmitter
2. Communication channel or transmission media
3. Receiver

Transmitter

Transmitter is the sub-system that takes the information signal and processes it prior to transmission. The transmitter modulates the information onto a carrier signal, amplifies the signal and broadcasts it over the channel.

Communication Channel

Communication Channel is the medium which transports the modulated signal to the receiver. Air acts as the channel for broadcasts like radio may also be a wiring system like cable TV or the Internet.

Receiver

Receiver is the sub-system that takes in the transmitted signal from the channel and processes it to retrieve the information signal. The receiver must be able to discriminate the signal from other signals which may use the same channel (called tuning), amplify the signal for processing and demodulate (remove the carrier) to retrieve the information. It also, then processes the information for reception (for example, broadcast on a loudspeaker).

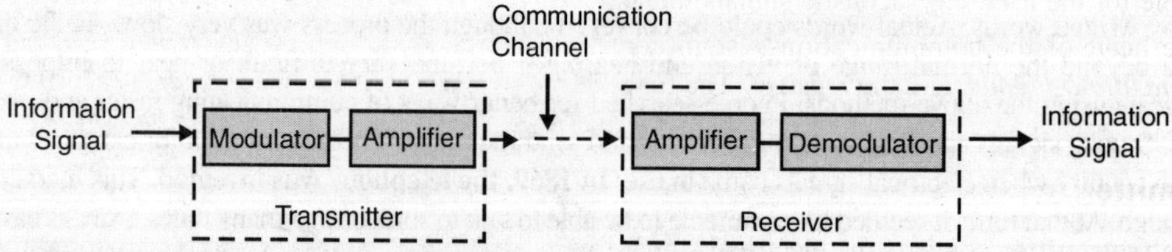

Fig. 5.1. Simple block diagram of communication system.

5.3. TYPES OF COMMUNICATION SYSTEMS

Broadly we classify the communication system in two types depending upon the type of the signal which is used in the communication system for processing, namely

1. Analog Communication System
2. Digital Communication System

5.3.1. Analog Communication System

We have described an electrical communication system in rather broad terms based on the implicit assumption that the message signal is a continuous time-varying waveform. We refer to such continuous-time signal waveforms as ***analog signals*** and to the corresponding information sources that produce such signals as ***analog sources***.

Analog signals can be transmitted directly via carrier modulation over the communication channel and demodulated accordingly at the receiver. We call such a communication system an ***analog communication system.***

In general, an analog communication system can be represented by the functional block diagram shown in Fig. 5.2. The information generated by the source may be of the form of voice (speech source), a picture (image source), or plain text.

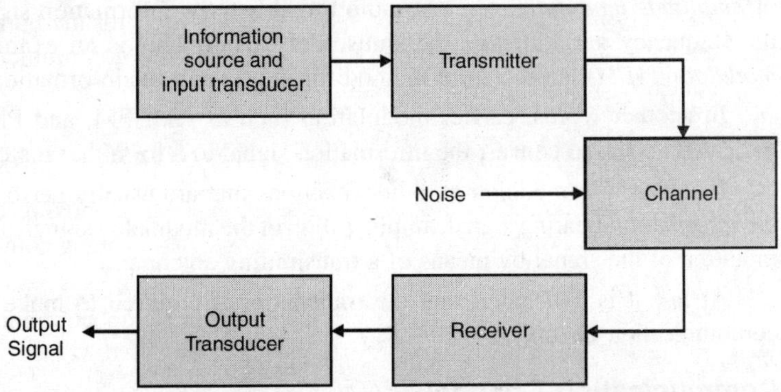

Fig. 5.2. Functional block diagram of an analog communication system.

Information Source

The communication system exists to convey a message. This message comes from the **information source**, which originates it, in the sense of selecting one message from a group of messages. In other words we can say that the function of information source is to provide required message that is to be transmitted. An essential feature of any source that generates information is that its output is described in probabilistic terms; *i.e.*, the output of a source is not deterministic. Otherwise, there would be no need to transmit the message.

Transducer

A **transducer** is usually required to convert the output of a source into an electrical signal that is suitable for transmission. For example, a microphone serves as the transducer that converts an acoustic speech signal into an electrical signal, and a video camera converts an image into an electrical signal. At the destination, a similar transducer is required to convert the electrical signals that are received into a form that is suitable for the user; e.g., acoustic signals, images, etc.

The heart of the communication system consists of three basic parts, namely, the *transmitter,* the *communication channel,* and the *receiver.* The functions performed by these three elements are described as follows:

Transmitter

The **transmitter** converts the electrical signal into a form that is suitable for transmission through the physical channel or transmission medium. For example, in radio and TV broadcast, the Federal Communications Commission (FCC) specifies the frequency range for each transmitting station. Hence, the transmitter must translate the information signal to be transmitted into the appropriate frequency range that matches the frequency allocation assigned to the transmitter. Thus, signals transmitted by multiple radio stations do not interfere with one another. Similar, functions are performed in telephone communication systems where the electrical speech signals from many users are transmitted over the same wire.

In general, the transmitter performs the matching of the message signal to the channel by a process called *modulation.* Usually, modulation involves the use of the information signal to systematically vary either the amplitude, frequency, or phase of a sinusoidal carrier. For example, in AM radio broadcast, the information signal that is transmitted is contained in the amplitude variations of the sinusoidal carrier, which is the center frequency in the frequency band allocated to the radio transmitting station. This is an example of *amplitude modulation.* In FM radio broadcast, the information signal that is transmitted is contained in the frequency variations of the sinusoidal carrier. This is an example of *frequency modulation. Phase modulation* (*PM*) is yet a third method for impressing the information signal on a sinusoidal carrier.

In another words, carrier modulation such as AM, FM, and PM is performed at the transmitter, as indicated above, to convert the information signal to a form that matches the characteristics of the channel.

In addition to modulation, other functions that are usually performed at the transmitter are filtering of the information-bearing signal, amplification of the modulated signal, and in the case of wireless transmission, radiation of the signal by means of a transmitting antenna.

At last it is concluded that the transmitter is required to make the suitable for conduction over the communication channel.

Communication Channel

The **communications channel** is the physical medium that is used to send the signal from the transmitter to the receiver. There are two types of channels, namely point to point channel and broadcast channels. In point to point channel usually employ a variety of physical media, including wire lines, optical fiber cables, and wireless (microwave radio) such as telephone channels. On the other hand, in broadcast channels provide a capability where many receiving stations can be reached simultaneously from a single transmitter.

In wireless transmission, the channel is usually the atmosphere (free space). Examples of broadcast channels are satellite communications, television broadcasting, radio broadcasting, etc.

The words "*tele*", "*phon*", and "*graph*" are derived from Greek.

Tele – means 'at a distance'

Phon – means sound or speech

Graph - means writing or drawing

Therefore, *telecommunication* means communication at a distance. This can be done through wires called transmission lines or through atmosphere by a radio link. Other examples include:

Telephone – speaking at a distance

Television – seeing at a distance

Telegraph – writing at a distance

Complex telephone networks are built using a lot of devices and methods, some of which will be discussed later in this book.

Whatever the physical medium for signal transmission, the essential feature is that the transmitted signal is distorted in a random manner by a variety of possible mechanisms. During the transmission and reception process the signal get corrupted due to noise inducted in the system.

Noise may be defined as unwanted signals which tend to interfere with the required signals. The most common form of signal degradation comes in the form of additive noise, which is generated at the front end of the receiver, where signal amplification is performed. This noise is often called *thermal noise.*

In wireless transmission, additional additive disturbances are man-made noise, and atmospheric noise picked up by a receiving antenna. Automobile ignition noise is an example of man-made noise, and electrical lightning discharges from thunderstorms is an example of atmospheric noise. Interference from other users of the channel is another form of additive noise that often arises in both wireless and wire line communication systems.

Receiver

The function of the **receiver** is to recover the message signal contained in the received signal. If the message signal is transmitted by carrier modulation, the receiver performs *carrier demodulation* in order to extract the message from the sinusoidal carrier. Since, the signal demodulation is performed in the presence of additive noise and possibly other signal distortion, the demodulated message signal is generally degraded to some extent by the presence of these distortions in the received signal. Other than performing the primary function of signal demodulation, the receiver also performs a number of peripheral functions, including signal filtering and noise suppression.

Destination is the final stage which is used to convert an electrical signal in to its original form by using the output transducer.

DO YOU KNOW▶

Analog signals can be transmitted directly via carrier modulation over the communication channel and demodulated accordingly at the receiver. We call such a communication system an analog communication system.

5.3.2. Digital Communication System

An analog source output may be converted into a digital form and the message can be transmitted via digital modulation and demodulated as a digital signal at the receiver. The communication system which uses the digital signals as input is known as **digital communication system.**

There are some main advantages to transmitting an analog signal by means of digital modulation.

1. The most important reason is that signal fidelity is better controlled through digital transmission than analog transmission.
2. In particular, digital transmission allows us to regenerate the digital signal in long-distance transmission, thus eliminating effects of noise at each regeneration point.
3. In contrast, the noise added in analog transmission is amplified along with the signal when amplifiers are used periodically to boost the signal level in long-distance transmission.
4. Another reason for choosing digital transmission over analog is that the analog message signal may be highly redundant. With digital processing, redundancy may be removed prior to modulation, thus conserving channel bandwidth.
5. Yet another reason may be that digital communication systems are often cheaper to implement.

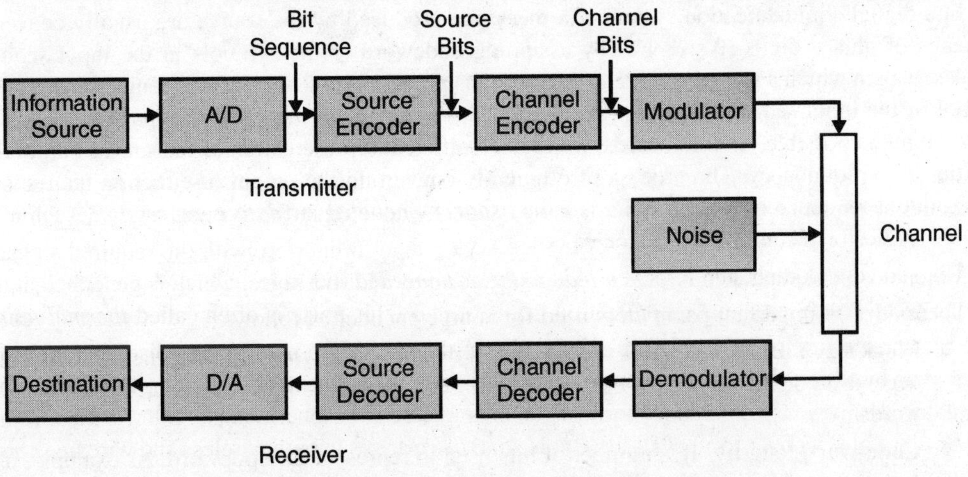

Fig. 5.3. Typical block diagram of a digital communication system.

In some applications, the information to be transmitted is inherently digital; e.g., in the form of computer data, etc. In such cases, the information source that generates the data is called a ***discrete (digital) source***.

In a digital communication system, the functional operations performed at the transmitter and receiver must be expanded to include message signal discretization at the transmitter and message signal synthesis or interpolation at the receiver. Additional functions include redundancy removal, and channel coding and decoding.

Figure 5.3 shows a typical block diagram of a digital communication system.

Information Source

The source output may be either an analog signal, such as audio or video signal, or a digital signal, such as the output of a computer which is discrete in time and has a finite number of output characters. The analog information is converted to an equivalent digital signal by sampling and quantization that process is known as **Analog to Digital conversion (ADC).** The A/D produces analog-to-digital conversion (for analog source) including sampling and quantizing.

Sampling : Pulse amplitude modulation (PAM) : Converting the continuous input to discrete sequence.

Quantization : Pulse code modulation (PCM) : Encoding each quantized sample into a *digital word* including uniform and nonuniform quantization.

The discrete information sources are characterized by source alphabet, symbol rate, source alphabet probabilities and probabilistic dependence of symbols in a sequence.

The *source alphabet* will have letters, digits, and special characters available from the information source.

The *symbol rate* is the rate at which the information sources the symbols/characters and its unit is symbol/sec.

Source alphabet occurs with its own probabilities. Each source alphabet from a source has independent reoccurrence rate in the sequence. A parameter called source entropy may be defined as the average information contents per symbol in a long message. It is expressed in bits/symbol. The source information rate is given by the product of source entropy and symbol rate expressed in bits per second. It represents the minimum number of bits per sec on an average to represent the information from the source. That is to say, source information rate representations the minimum average data rate needed to convey the information from source to destinations.

Source Encoding/Decoding

In a digital communication system, the messages produced by the source are usually converted into a sequence of binary digits (0's & 1's) by assigning codeword to the symbols in the input sequence. The simplest way in which a source coder can perform this process is to assign a fixed length binary code to each symbol of the input sequence. Ideally, we would like to represent the source output (message) by as few binary digits as possible. In other words, we seek an efficient representation of the source output that results in little or no redundancy. The process of efficiently converting the output of either an analog or a digital source into a sequence of binary digits is called ***source encoding* or *data compression,*** such as Huffman coding, JPEG (for picture), MPEG (for video).

The source-encoding step removes *redundant* or *unneeded* messages.

There are few important parameters used for source encoder as follows :

- **Block size :** Block size gives the number of distinct code-words that the encoder can represent and depends upon the number of bits in the codeword. As an example, the block size of 2^5 *i.e.*, 32 code-words.
- **Code-word length :** It is number of bits used to represent the codeword. As example, if 5 bits are assigned to each code-word, then code-word length will be 5 bits.

- **Average data rate :** It is the output bits per second from the source encoder. It may be the long term data rate that can be achieved by the encoder. Actually, the source encoder assigns many numbers of bits to each input symbol and that's why the data rate is higher than the symbol rate. For example, if we consider that the symbols are given to the source encoder and the length of code-word is 5 bits, then the output data rate from the source encoder would be,

$$\text{Data rate} = \text{Codeword length} \times \text{Symbol rate}$$

$$\text{Data rate} = 10 \times 5 = 50 \text{ bits/sec}$$

The coding efficiency is the rate of actual data rate to minimum achievable data rate.

The **source decoder** at the receiver converts the binary output of the channel decoder in to a symbol sequence.

Channel Encoder/Decoder

The sequence of binary digits from the source encoder, which we call the *information sequence,* is passed to the *channel encoder.* The purpose of the channel encoder is to introduce, in a controlled manner, some redundancy in the binary information sequence which can be used at the receiver to overcome the effects of noise and interference encountered in the transmission of the signal through the channel. Thus, the added redundancy serves to increase the reliability of the received data and improves the fidelity of the received signal. In effect, redundancy in the information sequence aids the receiver in decoding the desired information sequence. Channel coding is used to counteract channel impairments such as noise, fading and jamming by selectively introducing redundancies in the transmitted data. Through channel coding, communication system can obtain coding gain which is the decrease of signal to noise ratio (SNR) to get the same Bit error rate (BER) performance.

The channel encoder provides a different type of communication security than that provided by the encryptor. By adding some known redundancy to the message. Error control is achieved by adding extra bits to the output of the source encoded data. The extra redundancy bits added do not convey any information but provides detection and correction of errors in the message bits at the receiver. The coding and decoding process at the encoder and decoder needs the memory for processing of binary data. Typical channel codes: *block codes, convolutional codes.*

There are some important parameters that a channel encoder must possess.

(*i*) The coding rate which is depend upon the extra bits added by the channel encoder.

(*ii*) Coding techniques used.

(*iii*) Coding efficiency.

(*iv*) Error detection/correction capability.

(*v*) Complexity of encoder and decoder.

Digital Modulator/Demodulator

As we have already discussed about the need of modulation in communication system. Analog modulation methods are employed in analog communication system such as amplitude, frequency and phase modulation. A bit sequence cannot be transmitted from channel (fiber, cable, atmosphere and other transmission medium) directly. We must use the bit sequence to generate appropriate digital waveform, which is suitable to transmission medium. For example *Pulse waveform* (baseband signal) can directly transmitted through *cable*. For wireless, the baseband signals must be shift to a frequency band (RF signal), which is suitable to transmit through atmosphere.

In digital communication system, digital modulation techniques are used such as amplitude shift keying (ASK), frequency shift keying (FSK), phase shift keying (PSK) and minimum shift keying (MSK).

The output of channel encoder is a digital signal composed of symbols. These digital symbols can be transmitted directly on a physical channel after passing through a line coding system (such as NRZ, RZ, Manchester, Differential Manchester). This type of transmission is referred to as base band modulation because the base band signal has been coded and transmitted at the base band frequency only. However, for wireless communication this digital signal is carrier modulated using sinusoidal as carrier signal. Carrier modulation enables efficient transmission of the signal by minimizing the effect of channel noise.

At the receiving end, we use digital demodulator to converts modulated signals in to binary bits.

Communication Channel

The **communications channel** is the physical medium that is used to send the signal from the transmitter to the receiver. There are two types of channels, namely point to point channel and broadcast channels. In point to point channel usually employ a variety of physical media, including wire lines, optical fiber cables, and wireless (microwave radio) such as telephone channels. On the other hand, in broadcast channels provide a capability where many receiving stations can be reached simultaneously from a single transmitter. In wireless transmission, the channel is usually the atmosphere (free space). Examples of broadcast channels are satellite communications, television broadcasting, radio broadcasting etc.

Whatever the physical medium for signal transmission, the essential feature is that the transmitted signal is distorted in a random manner by a variety of possible mechanisms. During the transmission and reception process the signal get corrupted due to noise inducted in the system.

Noise may be defined as unwanted signals which tend to interfere with the required signals. The most common form of signal degradation comes in the form of additive noise, which is generated at the front end of the receiver, where signal amplification is performed. This noise is often called *thermal noise.*

In wireless transmission, additional additive disturbances are man-made noise, and atmospheric noise picked up by a receiving antenna. Automobile ignition noise is an example of man-made noise, and electrical lightning discharges from thunderstorms is an example of atmospheric noise. Interference from other users of the channel is another form of additive noise that often arises in both wireless and wire line communication systems.

DO YOU KNOW▸

In a digital communication system, the functional operations performed at the transmitter and receiver must be expanded to include message signal discretization at the transmitter and message signal synthesis or interpolation at the receiver.

5.4. COMPARISON OF DIGITAL AND ANALOG COMMUNICATION SYSTEMS

Analog communication involves transmission of a continuous of values over a defined interval of time. Processing of analog information is generally not easy and it is more prominent to the noise. Further, there is no facility of error detection/correction in the analog system. Digital communications overcome these problems very effectively. We now discuss the advantages and few disadvantages of digital communication systems.

5.4.1. Advantages Digital Communication

(*i*) Reduction of noise is possible in digital communication system.

(*ii*) Reduction of distortion and other impairments is possible in digital communication system.

(*iii*) Error detection and correction is possible in digital communication system that will reduce the probabilities of error by employing channel encoder.

(*iv*) Regeneration of signal is easier than analog communication system.

(*v*) Signal processing and storage is easier due to use of digital signal processing techniques and digital computer.

(*vi*) Security of data is possible due to encryption and decryption and spread spectrum coding provides antijamming facility.

(*vii*) Data compression and image processing is possible in digital communication system.

(*viii*) Large amount of noise interference may be tolerated.

(*ix*) Use of VLSI technology may be possible.

(*x*) Handling of multimedia information such as integration of voice, video and data is easier.

(*xi*) There are many signal processing algorithms can be used to increase transmission rate and reliability of digital communication system.

(*xii*) Digital circuits are more reliable and can be produced at lower cost than analog circuits, such as DSP, FPGA and ASIC.

(*xiii*) Digital signal is less subject to distortion and interference than analog signal.

(*xiv*) The digital signal is more easer to "reborn" or regenerated, compared to analog signals because:

- Digital signal has only two states (0 or 1)
- On the contrary, analog signal are not two-state signal.

It can take an infinite variety of amplitude.

5.4.2. Disadvantages of Digital communication

1. Digital communication requires more bandwidth than analog communication.
2. It needs synchronization.

5.5. LIMITATIONS OF COMMUNICATION SYSTEM

The primary limitations of information transmission by electrical means are bandwidth and noise.

1. Bandwidth Limitation
2. Noise Limitation

5.5.1. Bandwidth Limitation

The **bandwidth** can be defined as the frequency range or the band of frequency needed for a particular transmission. The concept of bandwidth applies to the both systems and signals. If the signal changes frequently with time, its frequency contents extends over a wide range and we may say that the signal has large bandwidth. In the same way, the ability of a system to follow signal variations is reflected in its usable frequency response or transmission bandwidth. Every communication system has a finite bandwidth that restricts the rate of signal variations. The bandwidth is allocated by regulatory agencies. Such type of regulation is required to avoid interference among the signals of the same frequency. However, that allocated bandwidth for a given transmission may not be adequate to convey the entire information.

According to information theory, greater the bandwidth of a transmission of the system, the more is the information that can be transmitted. Now we know that to transmit video signal in Television broadcasting needs bandwidth in MHz (India it is 7 MHz), while the audio signal has slow variations and it requires less bandwidth (approximately 150 kHz). For example, when we listen music in AM and FM radio, the performance of FM radio is better than AM radio, why? The reason is related to the bandwidth used in AM and FM radio. The full amount of information available to the human ear is contained in the range of frequency from 0Hz to 15 kHz. The maximum modulating frequency is only 5 kHz in AM transmission whereas it is 15 kHz in FM. Therefore, an AM radio receiver cannot reproduce all the information contained in the music because this would be needed bandwidth of 30 kHz. On the other hand, the bandwidth allocated to FM transmission is 200 kHz. So that FM receiver can easily reproduce the transmitted information. This is why FM has a better fidelity than AM and it is preferred to listen to a FM receiver.

Thus bandwidth is a fundamental limitation of any communication system.

DO YOU KNOW▶

The bandwidth can be defined as the frequency range or the band of frequency needed for a particular transmission.

5.5.2. Noise Limitation

Noise is another limitation of any communication system. It can be defined as an unwanted form of energy which tends to interfere with the transmission and reception of the desired signal in a communication system. Some sort of noise is always present in the communication system. Noise cannot be removed completely but it can be minimized by using some noise limitation circuits.

DO YOU KNOW▶

Noise is defined as an unwanted form of energy which tends to interfere with the transmission and reception of the desired signal in a communication system.

5.6. DIGITAL RADIO

In a digital radio, the digital pulses modulate an analog carrier between two or more points. The digital transmission system needs a transmission channel may a guided media such as metallic parallel lines, coaxial cables, optical fiber, or unguided media *i.e.* wireless transmission.

Digital modulation is sometimes called **digital radio** because digitally modulated signals can be propagated through earth's atmosphere and used in wireless communication systems.

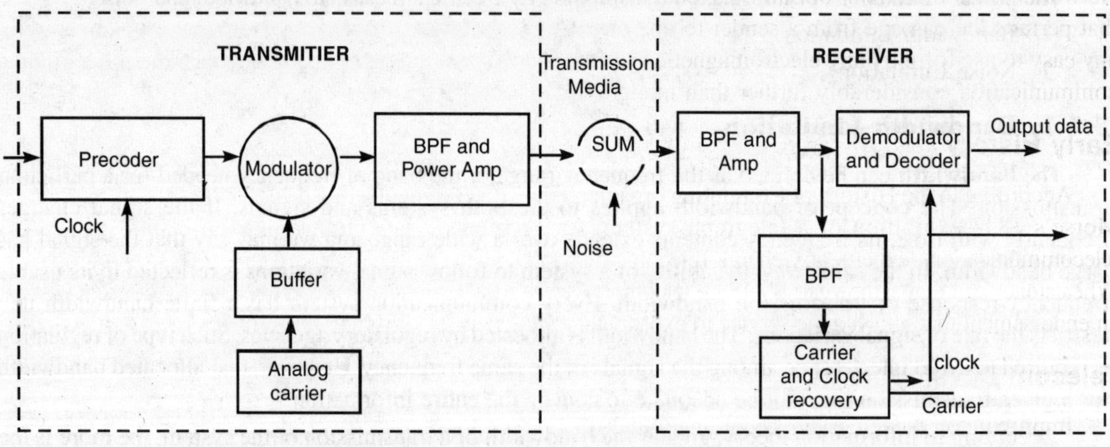

Fig. 5.4. A simple block diagram of Digital Radio.

5.7. INTRODUCTION : DATA COMMUNICATIONS

What is Data Communications?

The advancement in technology around the world has occurred at very high rate, transforming the telecommunication industry into an extremely dynamic and highly sophisticated field. Due the advent of

VLSI chips and the accompanying low cost microprocessors, microcontrollers, computers have dramatically increased need for the exchange of digital information.

Data Communications is the transfer of data or information between a source and a receiver. The source transmits the data and the receiver receives it. The actual generation of the information is not part of Data Communications nor is the resulting action of the information at the receiver.

Data Communication is interested in the transfer of data, the method of transfer and the preservation of the data during the transfer process.

In the data communication world, **data** means the information stored in digital form. A single unit of data is **datum** and the data is plural. *Data Communications* is the process of transferring the binary (digital) information between two or more points.

The purpose of Data Communications is to provide the rules and regulations that allow computers with different disk operating systems, languages, cabling and locations to share resources.

The rules and regulations are called protocols and standards in Data Communications.

5.8. HISTORY OF DATA COMMUNICATION

Though data communication may seem a relatively new innovation, the history of data connections dates to the early 19th century. Through an array of technological developments that enjoyed commercial, government and military contributions, data communication evolved from simple wired connections to a robust global exchange of information.

Definition

Before studying the history of data communication, it is first important to understand exactly what constitutes this form of information exchange. Because the history of data communication extends well beyond modern computer connections and wireless devices, some might find the communication channel's earliest roots surprising. According to Dr. Carl Rebman, professor of Data Communications and Computer Networks at the University of San Diego, lexicographers have defined data communication as "Any process that permits the passage from a sender to one or more receivers of information of any nature, delivered in any easy to use form by any electromagnetic system." This definition helps historians trace the roots of data communication considerably further than modern digital equipment.

Early History

According to the History of Computing organization, data communication has its earliest roots in Samuel Morse's 1837 exhibition of a telegraph system. An account of data communication history posted by telecommunications experts at General Telecom, LLC also points to a telegraph patent that inventor Charles Wheatstone filed that same year. By 1843, telegraph service had become adopted by the Great Western Railway, an endorsement that allowed the service to expand across the nation.

Telecom Developments

Improving on the telegraph, according to the History of Computing, Alexander Graham Bell introduced the telephone in 1876.

Though standard telephone lines did not carry data traffic until nearly a hundred years later, the development of early telecommunications—coupled with an 1895 invention by Guglielmo Marconi, the radio—laid the groundwork for numerous subsequent developments in communication technology. In 1947, Bell Labs introduced the transistor, a device that found integration in myriad subsequent electronic products. The U.S. government expanded on these technologies in 1958 with its launch of a communications-oriented satellite, and the first facsimile transmission over standard telephone lines occurred four years later.

Computer Connections

After the first fax transmission in 1962, the modulation of data into sound for transmission across telephone lines spread in popularity for several years.

Though modulation/demodulation, or modem, technology continued to carry slower data traffic for the remainder of the 20th century, according to the History of Computing, the 1969 development of Internet Protocol (IP) marked a significant milestone in data communication history.

Within the following decades, early packet communication technologies like Asynchronous Transfer Mode (ATM), Frame Relay and Integrated Services Digital Network (ISDN) emerged as a viable solution for commercial and high-end residential data needs. By 1991, more than 1 million servers had come online using Internet Protocol technology, and the World Wide Web emerged as the primary component of the Internet by the mid-1990s.

Wireless

As wired data communication expanded, a separate form of data exchange that required no wires experienced a concurrent development. According to the wireless communication reference website WirelessCommunication.nl, wireless technologies developed rapidly during World War I. By 1921, the first commercial radios voice broadcast, a communication channel that transmitted data in the form of modulated radio waves, had taken place and police dispatchers had adopted the technology. In 1974, the Federal Communication Commission (FCC) began allocating wireless spectrums for wireless communication, and wireless companies began integrating radio packet data as early as 1984. This technology evolved into the various digital wireless packet data protocols used into the 21st century.

Three main developments that led to the growth of data communications systems:

- Large-scale integration of circuits reduced the cost and size of terminals and communication equipments.
- Developments of software systems made establishment of communication networks easy and
- Competition among providers of transmission facilities reduced the cost of data circuits

Summary of History of Data Communication

- 1838: Samuel Morse & Alfred Veil Invent Morse Code Telegraph System
- 1876: Alexander Graham Bell invented Telephone
- 1910:Howard Krum developed Start/Stop Synchronisation
- 1930: Development of ASCII Transmission Code
- 1945: Allied Governments develop the First Large Computer
- 1950: IBM releases its first computer IBM 710
- 1960: IBM releases the First Commercial Computer IBM 360

Main Contributors of Data Communication

Transmission Technology

- Packet Switching Technology
- Internet
  - 1967: ARPANET by Advanced Research

Project Agency (ARPA) of U.S.

- 1975: TCP/IP protocol

- LAN Technology
  - DIX-Ethernet & IEEE 802 Networks
- **WAN :** 1976: ISO releases HDLC & CCITT releases X.25 (PSPDN)
- In 70's & 80's main thrust in Wide Area Networking (WAN) was to put *Data on Voice Circuits* using Modem & on ISDN lines
- In 90's the trend is reverse. Major Efforts were on putting *Voice OverData* using:
  - Voice Over Frame Relay
  - Voice Over Internet
  - Voice Over ATM etc

5.9. STANDARD ORGANIZATION FOR DATA COMMUNICATION

The importance of standards in the field of communication cannot be overstressed. Standards enable equipment from different manufacturer and with different operating characteristics to become components of the same network. Standards also enable different networks in different geographical locations (e.g., different countries and continents) to be interconnected.

Standards are developed by national and international organizations established for this exact purpose. We will discuss a number of important standards developed by various organizations as following:

1. The **International Standards Organization** (ISO) is a voluntary organization with representations from national standards organizations of member countries, major vendors and end-users. ISO is active in many areas of science and technology, including information technology. ISO standards are published as *ISO serial-no* (e.g., ISO 8632).
2. The **International Telecommunication Union-Telecommunication Sector (ITU-T)** formerly the Consultative Committee for International Telegraph and Telephone (CCITT), is one of four permanent part of the International Telecommunication Union based in Geneva, Switzerland. It is a standards organization devoted to data and telecommunication, with representations from governments, major vendors, telecommunication carriers, and the scientific community. The ITU-T has developed three sets of specifications as following :
   (*i*) **The V series :** It is for modem interfacing and data transmission over telephone lines.
   (*ii*) **The X series :** It is for data transmission over public digital networks, E-mail and directory services.
   (*iii*) **The I & Q series :** For Integrated services Digital Network (ISDN) and its extension Broadband ISDN.

   The ITU-T is separated into 14 study groups that prepare the recommendations for the defined topics. These are revised and republished every four years. ITU-T standards are very influential in the field of telecommunications and are adhered to by most manufacturers and carriers.
3. The **Institute of Electrical and Electronic Engineers** (IEEE) is a US standards organization with members throughout the world. IEEE is active in many electric, electronics, computer, and communication related areas. The IEEE standards for local area networks are widely adopted. IEEE standards are published as *IEEE serial-no* (e.g., IEEE 908). The IEEE works closely with ANSI (American National Standards Institute) to develop communications and information processing standards with the ultimate object of advancing theory, creativity and product quality in any field related to electrical engineering.
4. The **American National Standards Institute (ANSI)** is a completely non profit private organization comprised of manufactures and users of data processing equipment and services. It is the official standards agency for the United States and is the US voting representative for the ISO.

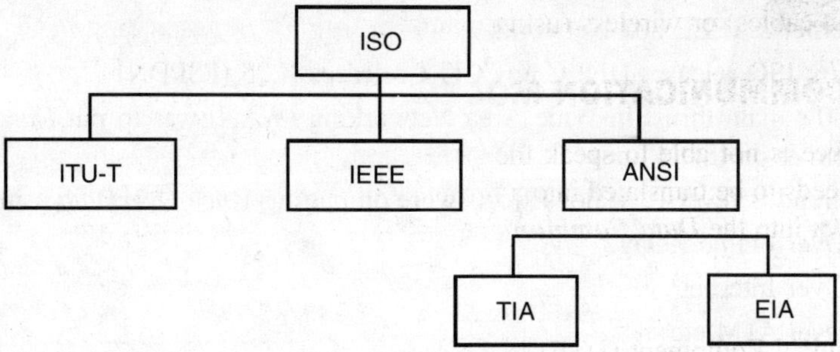

Fig. 5.5. Standard Organization for data communication.

5. The **Electronic Industries Association** (EIA) is a US trade association best known for its EIA-232 standard. EIA is responsible for developing the RS (recommended standards) series of standards for data and telecommunications. EIA activities include standards development, increasing public awareness and lobbying.
6. The **Telecommunication Industry Association (TIA)** is the major trade association in the information technology and communications industry. The TIA represents manufactures of communications and information technology products and service providers for the global marketplace through its core competencies.
7. The **European Computer Manufacturers Association** (ECMA) is a standards organization involved in the area of computer engineering and related technologies. ECMA directly cooperates with ISO and CCITT.

In addition to these organizations, and because of their global market influence, large vendors occasionally succeed in establishing their products as *de facto* standards.

5.10. DATA COMMUNICATION COMPONENTS

The basic purpose of a **data communications circuit** is to transfer digital information from one place to another using electronic circuit.

It provides a transmission path between source and destination. Thus data communications can be summarized as the transmission, reception and processing of digital information.

A **station** is simply an endpoint where subscribers get access to the circuit; sometimes it may be called a **node**, which is the location of computers, computer terminals and other digital computing equipment.

We have as many types of data communication circuits as there are types of data communication equipments. These data communication circuits employing electronic communications equipment and facilities to interconnect digital computer equipment.

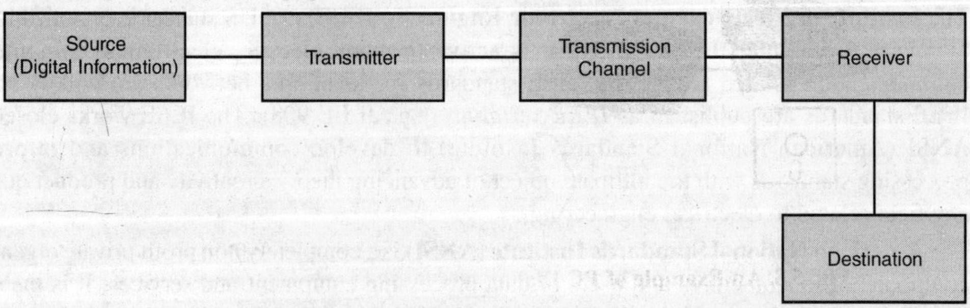

Fig. 5.6. Data communication circuit block diagram.

Communication facilities are the path of transmission from one station to another station either physically (using transmission cables) or wireless (using microwave, IR etc) transmission media.

5.11. DATA COMMUNICATION MODEL

Often the source is not able to speak the same communication language as the medium and the data communications needs to be translated into a form that the medium can understand. Data communication is further broken down into the ***Data Communication Model*** which consists of 5 parts as shown in figure :

1. DTE

The Data Terminal Equipment (DTE) is the source and receiver's equipment which is generating the data stream.

Examples of DTE devices are: PCs, terminals, end user devices.

2. DCE

The interface between the source and the medium, and the medium and the destination or receiver is called the Data Communication Equipment (DCE) and is a physical piece of equipment. Its job is to provide the interface from the DTE device to the medium. Examples of DCE devices are modems, channel banks, and routers.

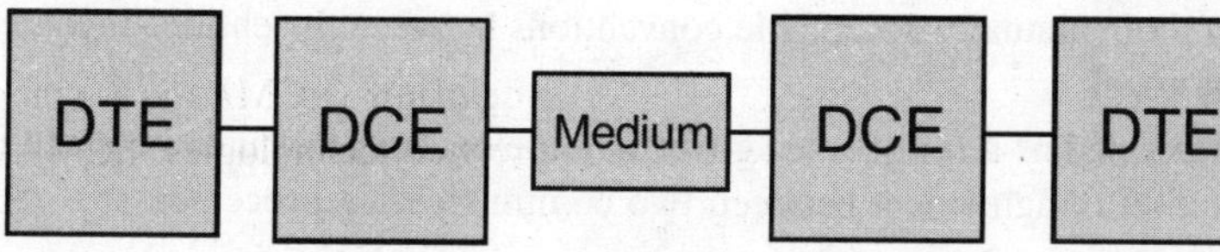

Fig. 5.7. Data Communication Model

An example of this would be your PC dialing into Internet Service Provider (ISP):

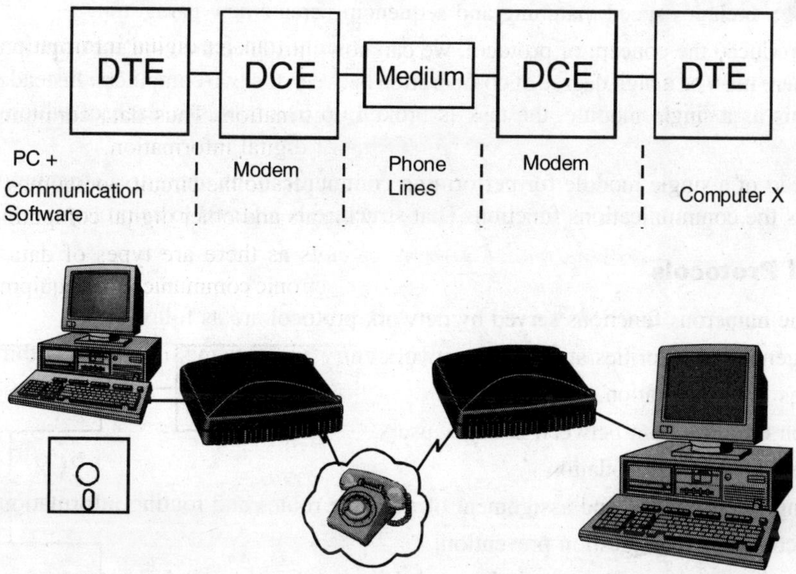

Fig. 5.8. An Example of PC Dialing into internet service provider (ISP).

The output of the DTE device (the PC) is an RS232 signal which is a digital communication consisting of logic ones and zeros.

The medium (the phone line) is made for carrying analog communications consisting of voice transmission.

The output of the PC and the requirements of the phone line are mutually incompatible. The phone line doesn't understand digital signals and the PC doesn't understand analog transmission.

The DCE device (the modem) provides the conversion (the interface) so that digital data can communicate across the analog medium.

5.12. PROTOCOLS

A protocol is used for communication between entities in different systems. The terms 'entity' and 'system' are used in a very general sense. Examples of entities are user application programs, file transfer package, database management systems, electronic mail facilities, and terminals.

Examples of systems are computers, terminals, and remote sensors. Note that in some cases the entity and the system in which it resides are coextensive (*e.g.* terminal).

In general, an entity is anything capable of sending or receiving information and a system is physically distinct object that contains one or more entities.

For two entities to communicate successfully, they must 'speak the same language'.

What is communicated?

How it is communicated and when it is communicated?

Must conform to some mutually acceptable conventions between the entities involved. The conventions are referred to as a protocol.

Protocol may be defined as a set of rules governing the exchange of data between two entities. It also may be conceived as a set of agreement between two communicating processes.

The key elements of protocol are:

- **Syntax:** Includes such things as data format and signal levels.
- **Semantics:** Includes control information for co-ordination and error handling.
- **Timing:** Includes speed matching and sequencing create new paragraph.

Having introduced the concept of protocol, we can now introduce the concept of protocol architecture. It is clear that there must be a high degree of co-operation between the two computers. Instead of implementing the logic for this as a single module, the task is broken up into subtasks, each of which is implemented separately.

Thus, instead of a single module for performing communications, there is a structured set of modules that implements the communications function. That structure is referred to as protocol architecture.

Functions of Protocols

Some of the numerous functions served by network protocol are as follows:

- Management of priorities at both the network entry and transmission levels within the network.
- Process synchronization.
- Session establishment between network users.
- Means for protocol validation.
- Routing establishment and assignment of message routes and routing information.
- Flow control and congestion prevention.
- Sequencing—sequenced transmission and delivery of messages.
- Addressing of network components and users.
- Efficient network resources utilization.

- Resource management, monitoring and protection.
- Layered transparency between networks users and nodes.
- Reliable message transmission, including error, control and recovery.
- Testing of network resources, such as links and routes.
- Security and privacy.
- Optional packet switching through message segmenting and pipelining.
- Session termination between network users.
- Orderly exchange of data messages.

Type of Protocols

There are two types of network services:

1. Connection-oriented Service
2. Connectionless Service

Two distinct techniques are used in data communications to transfer data. Each has its own advantages and disadvantages. Computer networks communicate using protocols, which defines the procedures that the system involved in the communication process will use. Protocol may be mainly classified as either **connection-oriented** method or the **connectionless** method.

1. Connection-oriented services

With a **connection oriented protocol,** a logical connection is established between the end points (for example virtual circuit) before to the transmission of data. It requires a session connection be established before any data can be sent.

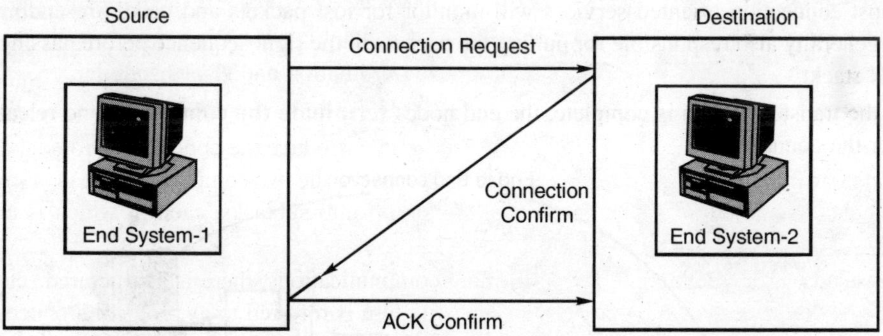

Fig. 5.9 (*a*). Set up phase

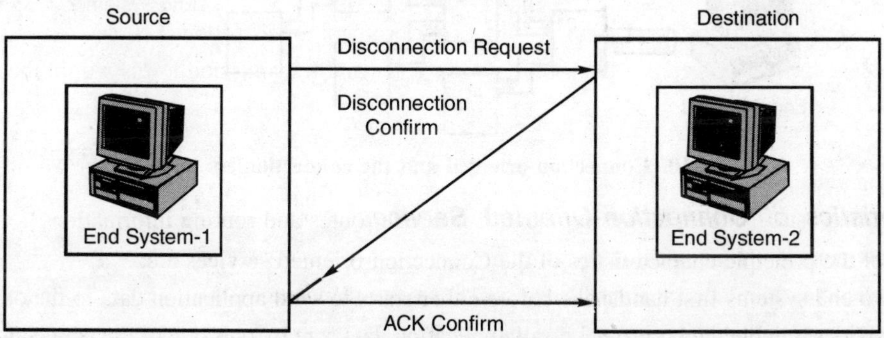

Fig. 5.9 (*b*). Disconnection Phase

Connection oriented protocol is similar to making a standard telephone call. There is a sequence of actions and acknowledgements, such as setting up the call, establishing the connection, and then disconnecting the connection. This method is often called a "reliable" network service. The high degree of reliability is achieved by using a rigid set of procedures for establishing the connection, transferring the data, acknowledging the data and then releasing the connection. It can guarantee that data will arrive in the same order.

Connection-oriented services set up virtual links between end systems through a network, as shown in Fig. 5.9. Note that the packet on the left is assigned the virtual circuit number 01. As it moves through the network, routers quickly send it through virtual circuit 01. If data are lost or damaged, the destination station requests that they may be sent again.

In connection-oriented service, communication proceeds through three phases:

1. Connection establishment
2. Data transfer
3. Connection release.

Connection-oriented service involves three phases: connection establishment, data transfer, and connection termination.

During **connection establishment**, the end nodes or stations may reserve resources for the connection. The end nodes also may negotiate and establish certain criteria for the transfer, such as a window size used in TCP connections. This resource reservation is one of the things exploited in some denial of service (DOS) attacks. An attacking system will send many requests for establishing a connection but then will never complete the connection. The attacked computer is then left with resources allocated for many never-completed connections. Then, when an end node tries to complete an actual connection, there are not enough resources for the valid connection.

The **data transfer phase** occurs when the actual data is transmitted over the connection. During data transfer, most connection-oriented services will monitor for lost packets and handle resending them. The protocol is generally also responsible for putting the packets in the right sequence before passing the data up the protocol stack.

When the transfer of data is complete, the end nodes **terminate the connection** and release resources reserved for the connection.

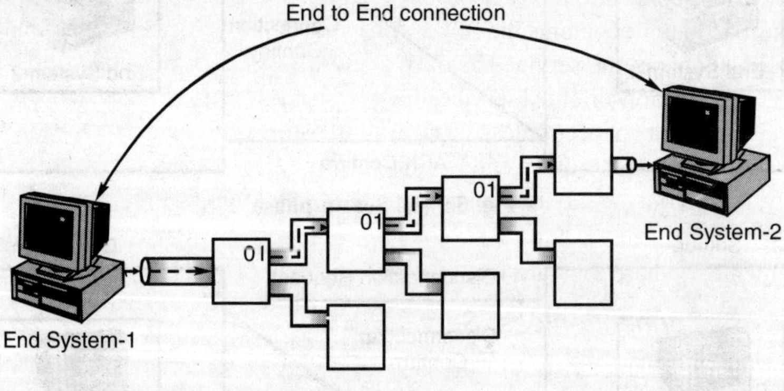

Fig. 5.10. Connection-oriented and the connectionless services

Characteristics of Connection-Oriented Services

Some of the principle characteristics of the Connection-oriented services are:

1. Two end systems first handshake before either starts to send application data to the other.
2. Provides reliable data transfer i.e., all application data sent by one side of the connection arrives at the other side of the connection in order and without any gaps.

3. Provides flow-control, i.e., it makes sure that neither end of a connection overwhelms the buffer in the other end of the connection by sending so many packets too fast.
4. Provides congestion control, i.e., regulates the amount of data that an application can send into the network.
5. Provides some means of error control (i.e., Error detection and Error correction)
6. Whenever a connection is no longer is needed, a specific handshake drop the connection.

Applications of Connection-Oriented Services

1. Transmission Control Protocol (TCP) is a connection-oriented protocol.
2. A Wireless Area Network (WAN) service that uses the connection-oriented model is frame relay. The service provider's sets up PVC (Permanent virtual circuit) through the network as required or requested by the customer.
3. ATM (asynchronous transfer mode) is another networking technology that uses the connection-oriented virtual circuit approach.
4. File transfer and remote terminal protocol will not tolerate loss of data, and require the packets to remain in ordered. For this kind of application a connection-oriented services needed.

TIPS

With a connection oriented protocol, a logical connection is established between the end points (for example virtual circuit) before to the transmission of data.

2. Connectionless Services

A connectionless service does not require a session connection between sender and receiver. The sender simply starts sending packets (called datagram) to the destination. This service does not have the reliability of the connection-oriented method, but it is useful for periodic burst transfers. Neither system must maintain state information for the systems that they send transmission to or receive transmission from. A connectionless network provides minimal services.

A good analogy for a connectionless service is the process of sending letters through the postal system. Each transmission (the "letter") contains the full destination address and is processed independent of related messages. As described above, the service has only to ensure that each reaches its host within certain time parameters. Unlike a connection-oriented service, the system has free reign on what happens enroute between the sender and receiver. with a connectionless service, all transmissions are independently routed, and perhaps re-assembled in some order at the other end — the service in between has no inherent responsibility for ensuring ordinarily, it need only assure that each transmission gets delivered from its source to its destination.

Connectionless protocol are usually described as stateless because the end points have no protocol defined way to remember where they are in a conversation of message exchanges.

Characteristics of Connectionless Service

Some of the principle characteristics of the Connectionless services are:

1. No handshaking require.
2. No guarantee of reliable data transfer.
3. No flow control or congestion control.
4. A message can be delayed to ensure another arrives first.
5. A message can be handed off to a trusted third party in the distribution network.

6. A message can be intercepted by a third party, copied or logged, and passed on to the intended receiver,
7. Widely different channels of communication can be used for transmitting messages.

Applications of Connection-Less services

1. UDP (User Datagram Protocol) is a connectionless network protocol.
2. LANs operate as connectionless systems. A computer attached to a network can start transmitting frames as soon as it has access to the network. It does not need to set up a connection with the destination system ahead of time.
3. For video conferencing and voice on internet protocol.
4. The Internet is one big connectionless packet network in which all packet deliveries are handled by IP. However, TCP adds connection-oriented services on top of IP. TCP provides all the upper-level connection-oriented session requirements to ensure that data is delivered properly.

TIPS

A connectionless service does not require a session connection between sender and receiver.

5.13. STANDARDS AND STANDARD ORGANISATIONS

The importance of standards in the field of communication cannot be overstressed. Standards enable equipment from different manufacturer and with different operating characteristics to become components of the same network. Standards also enable different networks in different geographical locations (e.g., different countries and continents) to be interconnected.

A **standard** is an object or procedure considered by an authority or by general consent as a basis of comparison. Data communication standards are guidelines that have been normally accepted by the data communication industry. Data communications standards are not laws; rather, they are simply suggested methods of implementing procedures and accomplishing results. The guidelines outline procedures and configurations of equipments that help ensure an orderly transfer of information between two or more computers or data communication equipments.

Basically there are two types of standards as follows:

(*i*) Open system
(*ii*) Closed or Proprietary system

Open System

In open system standards, any company can manufacture compatible equipments or software's; however royalty must be paid to the original company. An example of an open system standard is IBM's personal computer. The open system has many advantages like customer choice, compability between vendors and competition by smaller companies.

It has few disadvantages include less product control and increased difficulty acquiring agreement between companies for changes or updates.

Closed or Proprietary System

Closed or proprietary standards system are generally manufactured and controlled by one company. Others companies are not permitted to manufacture equipments or software are using this standard. Closed system has advantages like, monopoly, tight control, and easier consensus.

This standard has got few disadvantages; higher price, high investment, lack of choice and reduced customer protection against the manufacturer going out of business.

An example of a closed standard is Apple Macintosh computers.

5.14. LINE CONFIGURATION

Line configuration means the way two or more communication devices attach to a link.

A Link is the physical communication pathway that transfers data from one device to another.

For communication to occur, two devices must be connected in same way to the same link at the same time.

There are two possible line configurations.

1. Point-to-Point.
2. Multipoint.

1. Point-to-Point

A point-to-point line configuration provides a dedicated link between two devices. The entire capacity of the channel is reserved for transmission between those two devices. Most point-to-point line configurations use an actual length of wire or cable to connect the two ends, but other options, such as microwave or satellite links, are also possible as shown in Fig. 5.11. When you change television channels by infrared remote control, you are establishing a point-to-point line configuration between the remote control and the television's control system.

Fig. 5.11. Point-to-point line configurations

2. Multipoint

A multipoint or broadcasting line configuration is one in which more than two specific devices share a single link as shown in Fig. 5.12. In a multipoint environment, the capacity of the channel is shared, either spatially or temporarily. If several devices can use the link simultaneously, it is a spatially shared line configuration. If users must take turns, it is a time-shared line configuration

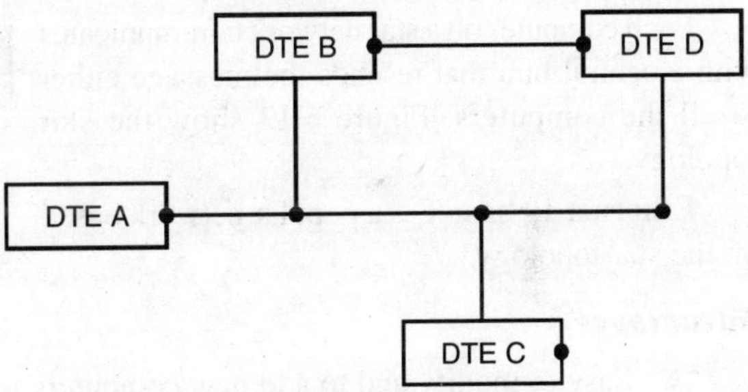

Fig. 5.12. A multipoint line configuration

5.15. PHYSICAL TOPOLOGY

The **physical topology** defines how devices are connected to make a network. For the sake of simplicity, the network wiring is symbolized as a single line between the two devices. In actuality, it may be a twisted pair of wires, a coaxial cable, and an optical fiber. Right now, we're merely focusing on the "shape" of the network, technically known as its *topology*.

We have seen that a topology is essentially a stable geometric arrangement of computers in a network. If you want to select a topology for networking, the following points should be considered.

- Application S/W and protocols.
- Types of data communicating devices.
- Geographic scope of the network.
- Cost.
- Reliability.

If we want to include more devices (sometimes called *nodes*) on this network, we have several options of network configuration to choose from.

Devices can be connected any or combinations of one or more configurations of the following:

1. Star topology
2. Ring topology
3. Bus topology
4. Mesh topology
5. Hybrid topology
6. Tree topology
7. Cellular topology

1. Star Topology

The primary feature of the star topology is that each station is radially connected to a central node through a direct point to point connection. With devices known as "gateways" at branching points in the network, data flow can be restricted between nodes, allowing for private communication between specific groups of nodes. This addresses some of the speed and security issues of the simple bus topology. However, those branches could easily be cut off from the rest of the "star" network if one of the gateways were to fail.

Star topology networks require a central point of connection between media segment. These central points are referred to as Hubs. Hubs are special repeaters that overcome the electromechanical limitations of a media.

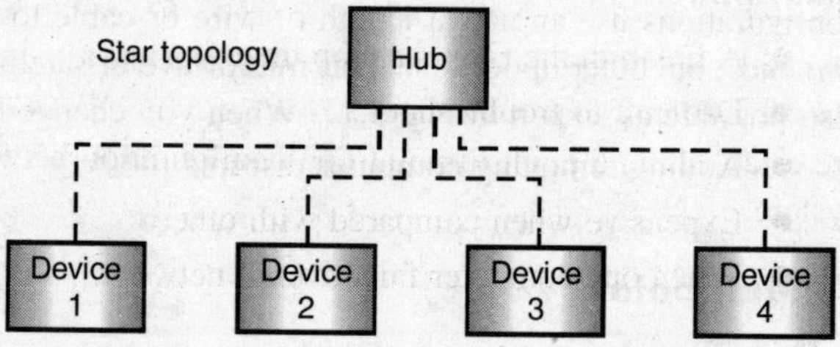

Fig. 5.13.

Each computer on a star network communicates with a central hub that resends the message either to all the computers. Figure 5.13 show the star topology.

Ethernet 10 base T is a popular network based on the star topology.

Advantages

- Easy to modify and to add new computers without disturbing the rest of the network.
- Less expensive than mesh topology.
- Each device needs only one link and one port.
- Easy to install and configure.
- Easy to diagnose network faults.
- Single computer failure does not affect the network.
- Ordinary telephone cables can be used.

Disadvantages

- More cabling is required as compare to others.
- Failure of the central hub brings the entire network down.

2. Ring Topology

In ring topology, each device has a dedicated point-to-point line configuration only with two devices on either side of it. A signal is passed along the ring in one direction, from device to device until it reaches its destination.

With the ring topology, adjacent devices are interconnected in a closed loop configuration as shown in Fig. 5.14. This topology provides the best reliability with the least amount of wiring. Since each node has two connection points to the ring, a single break in any part of the ring doesn't affect the integrity of the network. The devices, however, must be designed with this topology in mind. Also, the network must be interrupted to install or remove nodes. As with bus topology, ring networks are *broadcast* by nature.

Each device in the ring has a repeater. When the devices receive the signal intended for the other node, it just regenerates the bits and passes them along. Ring network passes a **token**. A token is a short message with the electronic address of the receiver. Each network interface card is given a unique electronic address, which is used to identify the computer on the network.

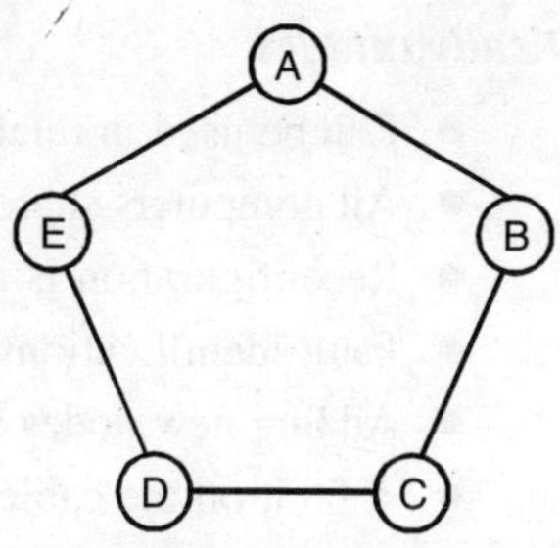

Fig. 5.14. Ring Topology

Advantages

- Easy to install and reconfigure.
- Adding/deleting the new device is easy as only two connections have to be adjusted.
- Fault isolation is simplified.
- No terminators required.

Disadvantages

- A break in the ring can stop the transmission the entire network.
- Difficult to troubleshoot.
- Adding/removing computer disrupts the entire network.
- Expensive when compared with other topologies.
- When one computer fails overall network disrupts.

3. Bus Topology

A bus topology is multipoint. One long cable acts as a backbone to link all the devices in the network. Nodes are connected to the bus cable by drop lines and taps. A drop line is a line running between the device and main cable. A tap is connector that splices into the main cable. There is a limit on the number of taps used and distance between the taps.

A linear bus topology is a multipoint circuit configuration where individual devices are connected by a common, shared communication channel (transmission media) as shown in Fig. 5.15. Very simple to install and maintain. Nodes can be easily added or removed with minimal wiring changes.

On the other hand, the one bus network must handle *all* communication signals from *all* nodes.

This is known as ***broadcast*** networking, and is analogous to a group of people talking to each other over a single telephone connection, where only one person can talk at a time (limiting data exchange rates), and everyone can hear everyone else when they talk (which can be a data security issue).

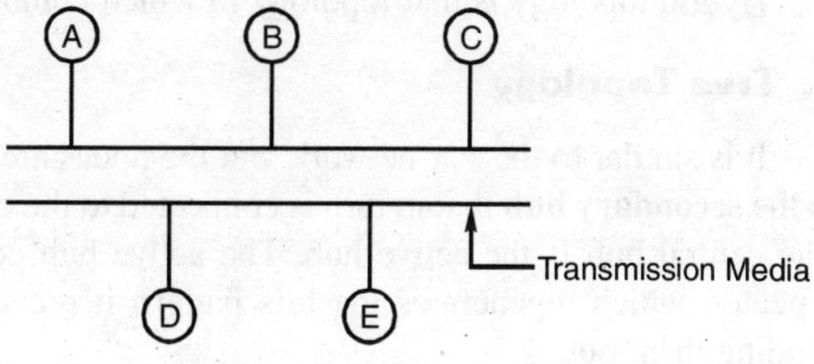

Fig. 5.15. Bus Topology.

Also, a break in the bus wiring can lead to nodes being isolated in groups.

Advantages

- Simple, reliable and easy to use.
- Easy to installation and cheaper than when compared with others.
- Less cabling.

Disadvantages

- Can be used in relatively small networks.
- All computers share the same bus.
- Reconfiguration is difficult.
- Fault identifications is difficult.
- Adding new nodes is difficult.
- A fault on the cable stops all transmission.

4. Mesh topology

In mesh topology each and every computer is connected to each other with direct point to point link.

A fully connected mesh network therefore has n (n–1/2) physical channels to link n devices.

To accommodate these, every device on the network must have n – 1 input/output parts.

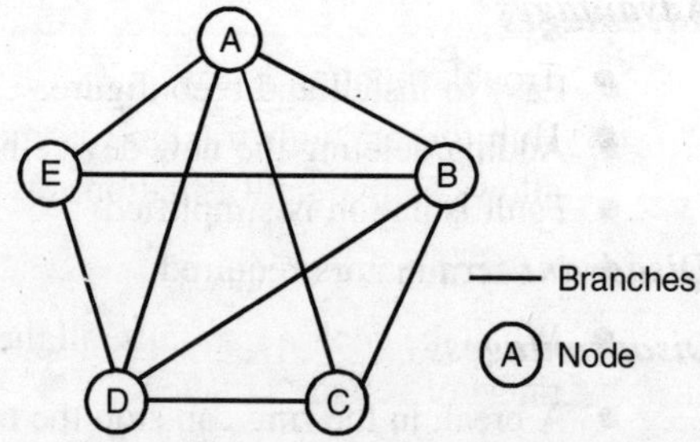

Fig. 5.16. Mesh Topology

Advantages

- Use of dedicated links eliminates the traffic problems.
- It is robust, *i.e.* if one link becomes unusable it does not incapacitate the entire system.
- Privacy is maintained since the message travels along the dedicated lines.
- Point-to-point link makes fault identification and fault isolation easy.

Disadvantages

- The amount of cabling required is high.
- The number if I/O ports required is high.

5. Hybrid Topology

Hybrid topology is that topology in which combination of two or more topology is used.

6. Tree Topology

It is similar to the star network, but the nodes are connected to the **secondary hub** that in turn is connected to the central hub. The central hub is the active hub. The active hub contains the repeater, which regenerates the bits pattern it receives before sending them out.

The secondary hub can be either active or passive.

A passive hub provides a simple physical connection between the attached devices.

Advantages and **Disadvantages** of the tree are same as that of the star network. Also, the addition of the secondary hub allows more devices to be attached to the central hub. It also allow the network to isolate priorities communication from different computers.

The tree topology structure is shown in the Fig. 5.17.

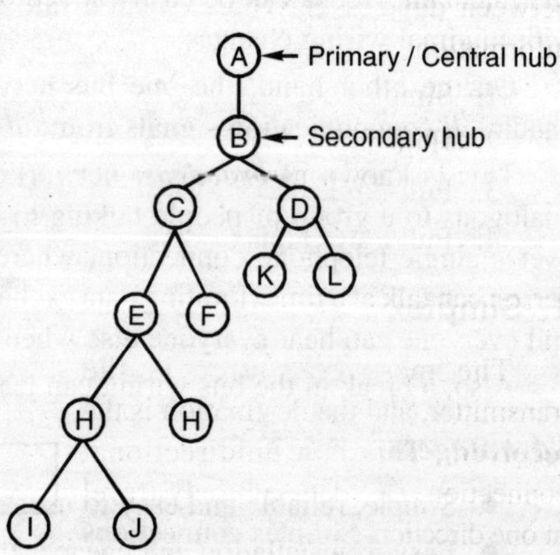

Fig. 5.17. Tree Topology

7. Cellular Topology

The cellular topology is applicable only in case of wireless media that does not require cable connection.

In wireless media, each point transmits in a certain geographical area called a **cell**. Each cell represents a portion of the total network area. Devices that are in the cell communicate through a central hub. Hubs in different cells are interconnected. They route data across the network and provide a complete network infrastructure.

The data is transmitted in the **cellular digital packet data (CDPD)** format.

Advantages

- Troubleshooting is easy.
- Hub-to-hub fault tracking is more complicated, but allows simple fault isolation.

Disadvantages

- When the central hub fails, all the units in the assigned range of cell are affected.

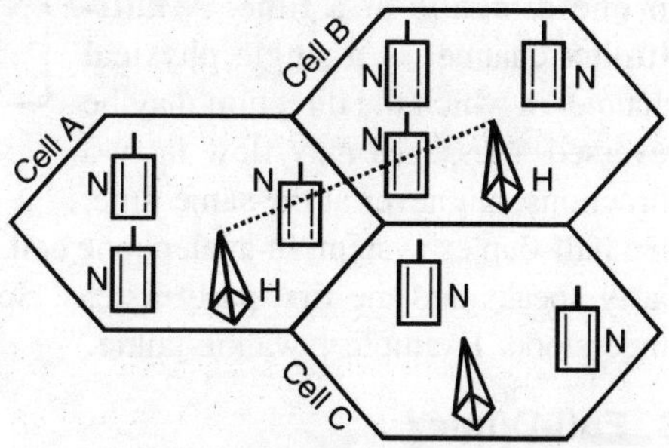

Fig. 5.18. Cellular Topology

5.16. COMMUNICATION MODE

A communications channel is a pathway over which information can be conveyed. It may be defined by a physical wire that connects communicating devices, or by a radio, laser, or other radiated energy source that has no obvious physical presence. Information sent through a communications channel has a source from which the information originates, and a destination to which the information is delivered. Although information originates from a single source, there may be more than one destination, depending upon how many receive stations are linked to the channel and how much energy the transmitted signal possesses.

In a digital communications channel, the information is represented by individual data bits, which may be encapsulated into multibit message units. A byte, which consists of eight bits, is an example of a message unit that may be conveyed through a digital communications channel. A collection of bytes may itself be grouped into a frame or other higher-level message unit. Such multiple levels of encapsulation facilitate the handling of messages in a complex data communications network.

Any communications channel has a direction associated with it, communication modes or connections between devices may be classified into three categories depending upon the direction of the flow of information:

1. Simplex
2. Half-duplex
3. Full-duplex
4. Full/Full Duplex

1. Simplex

The message source is the transmitter, and the destination is the receiver. This is a unidirectional connection, *i.e.*, data can only travel in one direction. **Simplex connections** are useful in situations where a device only receives or only sends data.

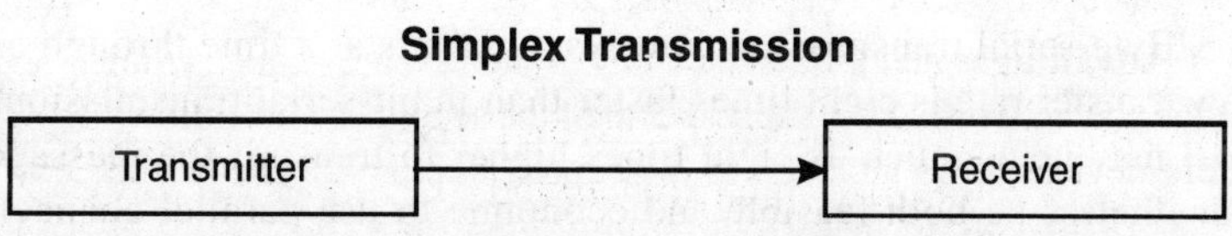

Fig. 5.19. Simplex transmission.

For example, a printer or radio station is a simplex channel because it always transmits the signal to its listeners and never allows them to transmit back.

2. Half-Duplex

This is a bidirectional connection, with the restriction that data can travel in one direction at a time. A **half-duplex** channel is a single physical channel in which the direction may be reversed. Messages may flow in two directions, but never at the same time, in a half-duplex system. In a telephone call, one party speaks while the other listens. After a pause, the other party speaks and the first party listens. Speaking simultaneously results in garbled sound that cannot be understood. Example : walkie-talkie.

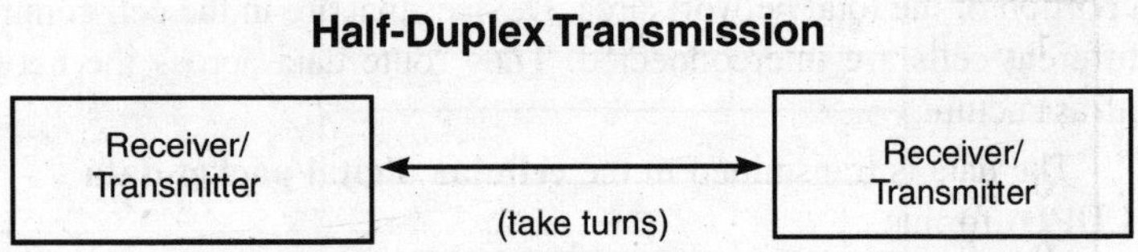

Fig. 5.20. Half-Duplex transmission.

3. Full-Duplex

This is a bidirectional connection in which data can travel in both directions at once. A **full-duplex connection** is really consists of equivalent to two simplex connections in opposite directions a forward channel and a reverse channel, linking the same points.

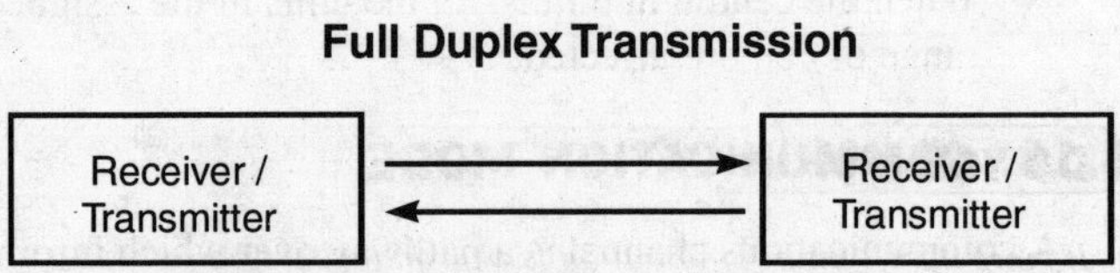

Fig. 5.21. Full-Duplex transmission.

A full-duplex channel allows simultaneous message exchange in both directions. The transmission rate of the reverse channel may be slower if it is used only for flow control of the forward channel. Examples- mobile phone, telephone etc.

4. Full/Full Duplex

In the **full/ full duplex** mode transmission is possible in both the directions simultaneously but not between the same two stations. For example, one station is transmitting to a second station and may be receiving from a third station at the same time. This mode is applicable to multipoint circuits.

5.17. TRANSMISSION MODE

Data communications over a medium can be divided into two basic types :

1. Parallel data communication or Parallel transmission mode
2. Serial data communication or Serial transmission mode

5.17.1. Parallel Transmission Mode

If more than 1 bit of information is transmitted over the data transmission medium at a time then it is considered a **Parallel Communication**.

Byte-serial transmission conveys eight bits at a time through eight parallel channels. Although the raw transfer rate is eight times faster than in bit-serial transmission, eight channels are needed, and the cost may be as much as eight times higher to transmit the message. When distances are short, it may nonetheless be both feasible and economic to use parallel channels in return for high data rates. The popular Centronics printer interface is a case where byte-serial transmission is used as shown in Fig. 5.22. As another example, it is common practice to use a 16-bit-wide data bus to transfer data between a microprocessor and memory chips; this provides the equivalent of 16 parallel channels. On the other hand, when communicating with a timesharing system over a modem, only a single channel is available, and bit-serial transmission is required.

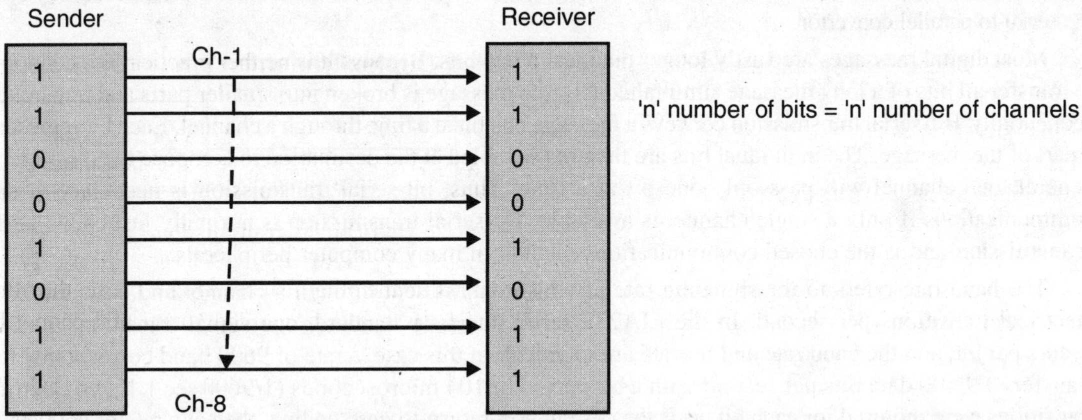

Fig. 5.22. Parallel transmission mode.

DO YOU KNOW▶

If more than 1 bit of information is transmitted over the data transmission medium at a time then it is considered a Parallel Communication.

5.17.2. Serial Transmission Mode

The physical connection determines how many bits (1's or 0's) can be transmitted at a single instance of time. If only 1 bit of information can be transmitted over the data transmission medium at a time then it is considered a **Serial Communication**. Normally, within the system the data is moving in parallel transmission

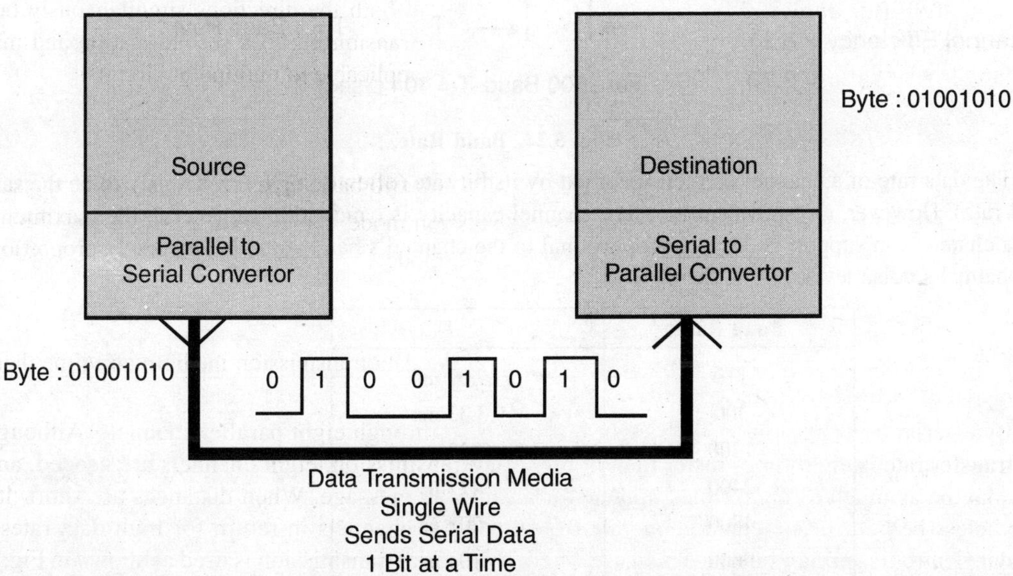

Fig. 5.23. Serial transmission.

mode. Now to transfer these data from one system to another system over a single transmission channel, we must convert parallel data into serial data by using the parallel to serial convertor at source end. Similarly at the destination, we receive data in serial mode (1 bit at a time), so it should be converted back to parallel form by using the serial to parallel convertor.

Most digital messages are vastly longer than just a few bits. Because it is neither practical nor economic to transfer all bits of a long message simultaneously, the message is broken into smaller parts and transmitted sequentially. Bit-serial transmission conveys a message one bit at a time through a channel. Each bit represents a part of the message. The individual bits are then reassembled at the destination to compose the message. In general, one channel will pass only one bit at a time. Thus, bit-serial transmission is necessary in data communications if only a single channel is available. Bit-serial transmission is normally just called **serial transmission** and is the chosen communications method in many computer peripherals.

The baud rate refers to the signaling rate at which data is sent through a channel and is measured in electrical transitions per second. In the EIA232 serial interface standard, one signal transition, at most, occurs per bit, and the baud rate and bit rate are identical. In this case, a rate of 9600 baud corresponds to a transfer of 9,600 data bits per second with a bit period of 104 microseconds (1/9600 sec.). If two electrical transitions were required for each bit, as is the case in non-return-to-zero coding, then at a rate of 9600 baud, only 4800 bits per second could be conveyed. The channel efficiency is the number of bits of useful information passed through the channel per second. It does not include framing, formatting, and error detecting bits that may be added to the information bits before a message is transmitted, and will always be less than one.

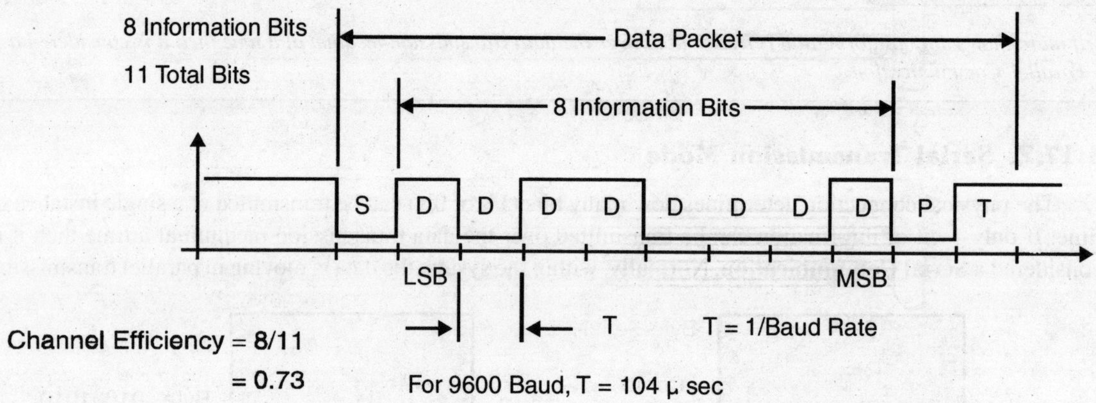

Fig. 5.24. Baud Rate.

The data rate of a channel is often specified by its bit rate (often thought erroneously to be the same as baud rate). However, an equivalent measure channel capacity is bandwidth. In general, the maximum data rate a channel can support is directly proportional to the channel's bandwidth and inversely proportional to the channel's noise level.

Baud Rate	T
110	9.09 ms
300	3.33 ms
1200	833 μs
2400	417 μs
4800	208 μs
9600	104 μs
19200	52 μs

A **communications protocol** is an agreed-upon convention that defines the order and meaning of bits in a serial transmission. It may also specify a procedure for exchanging messages. A protocol will define how many data bits compose a message unit, the framing and formatting bits, any error-detecting bits that may be added and other information that governs control of the communications hardware. Channel efficiency is determined by the protocol design rather than by digital hardware considerations. Note that there is a tradeoff between channel efficiency and reliability—protocols that provide greater immunity to noise by adding error-detecting and -correcting codes must necessarily become less efficient.

DO YOU KNOW

If only 1 bit of information can be transmitted over the data transmission medium at a time then it is considered a Serial Communication.

5.17.3. Comparison of Parallel and Serial Transmission Modes

The transfer rate comparison is relative to serial versus parallel. Serial data transfer is used for high transfer rates over long distances. Data can be transferred at much faster rates over long distances using serial methods than using parallel data transfer. At shorter distances, typically less than 15 feet, parallel data transfers are used.

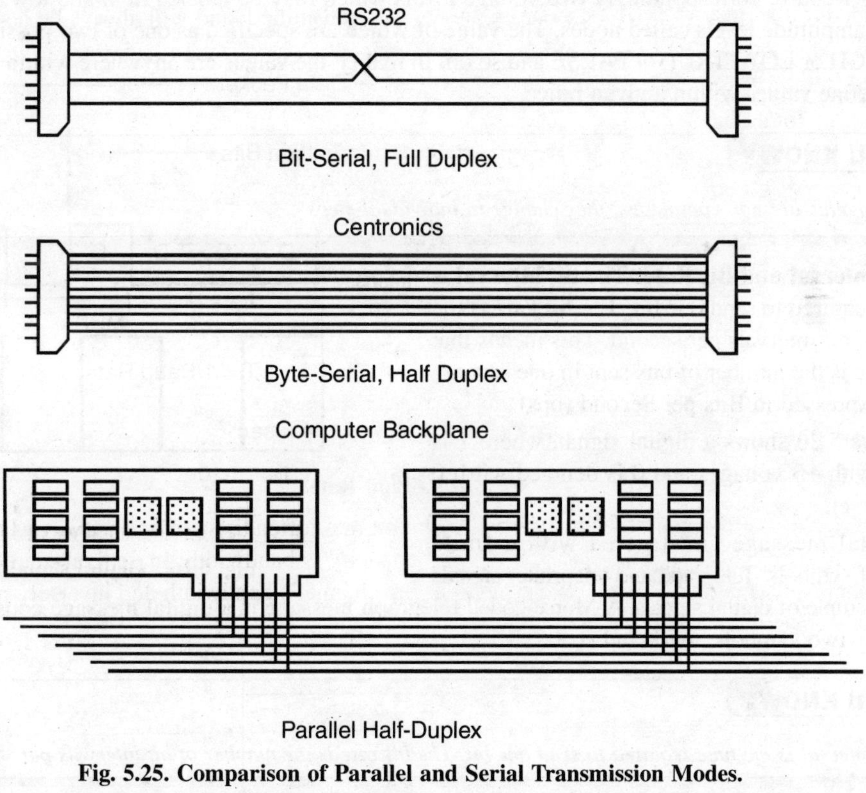

Fig. 5.25. Comparison of Parallel and Serial Transmission Modes.

A personal computer uses 64 bit high speed parallel address and data busses to transfer data within the confines of the computer. If 64 bit high speed data transfers were attempted for long distances, cable characteristics would cause the signals to lead or lag each other and all the bits would not arrive at the exact same time. The bits would arrive staggered and the data would be corrupted.

Communications	Advantages	Disadvantages
Parallel	1. Fast Transfer Rates	2. Short distances only 3. More number of channels 4. Complex and costly
Serial	1. Long Distances 2. Single channel 3. Simple	1. Slow transfer rates

Serial data communication only transfers one bit at a time and does not have the timing problems (race conditions) of parallel data communications.

5.18. DIGITAL SIGNAL

Digital signals are non-continuous, they change in individual steps. They consist of pulses or digits with discrete levels or values. The value of each pulse is constant, but there is an abrupt change from one digit to the next. Digital signals are may be defined as signals whose both dependent variable and independent variables are discrete in nature. A digital signal comprises pulses occurring at discrete interval of time.

However, generally, we use binary number system to represent a digital signal. In binary system, we use two digits 1 and 0, correspondingly two voltage levels which may be labeled high and low. Digital signals have two amplitude levels called nodes. The value of which are specified as one of two possibilities such as 1 or 0, HIGH or LOW, TRUE or FALSE and so on. In reality, the values are anywhere within specific ranges and we define values within a given range.

DO YOU KNOW▶

Digital signals are non-continuous, they change in individual steps.

Bit Interval and Bit Rate: The **bit interval** is the time required to send one bit. The **bit rate** is the number of bit intervals per second. This means that the bit rate is the number of bits sent in one second, usually expressed in Bits per Second (bps)

Figure 5.26 shows a digital signal, where 1 is denoted with +5 voltages and 0 is denoted wit h 0 voltage level.

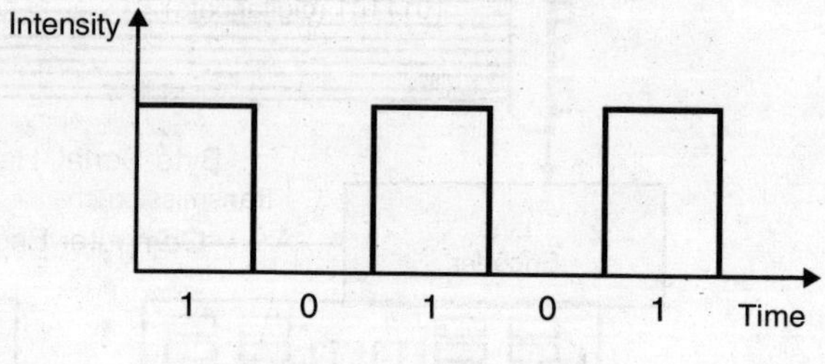

Abrupt Amplitude Variations

Fig. 5.26. Digital signal.

Digital message constructed with a finite number of symbols. Telegraph and teleprinter signals are the example of digital signals. A Morse coded telegraph message is a digital message constructed from a set of only two symbols, mark and space.

DO YOU KNOW▶

The bit interval is the time required to send one bit. The bit rate is the number of bit intervals per second.

5.19. DIGITAL TO DIGITAL ENCODING (LINE CODING)

When digital data is to be transmitted from one point to another point (for example, one computer to another computer), the digital data is to convert in to digital signal, so that it can be transmitted. This process of converting digital data in to digital signal is known as **line encoding**.

Line encoding is the method used to represent the digital information on the media. A pattern, that uses either voltage or current, is used to represent the 1s and 0s of the digital signal on the transmission link.

Refer Fig. 5.28, from the sender computer the digital data is given to the block, Line encoders to convert the stream, of bits (0s and 1s) in to digital signal and then it is transmitted over the line or transmission media or channel (may be physically connected through wire or using wireless method). At the receiving end the digital signal is again converted in to digital data by using line decoder block and finally digital data is given to the receiver computer.

Important terms used in line encoding:

1. Signal element

A signal element carries data element and it is the shortest unit (time wire) of a digital signal.

2. Data element (bit)

Data element is the smallest entity that can represent a piece of information known a bit.

"Data elements are what we need to send;
Signal elements are what we can send.
Data elements are being carried and signal
Element are the carriers"

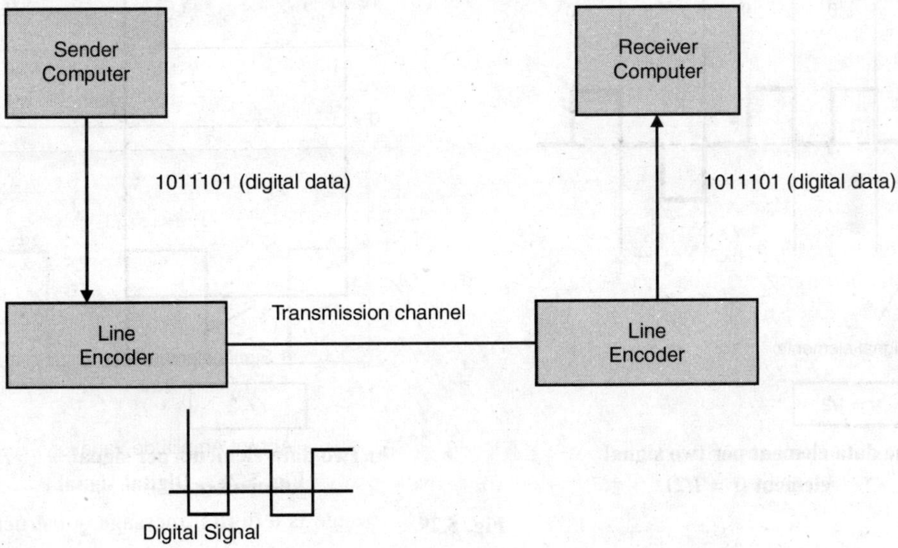

Fig. 5.27. Block diagram of line coding.

3. Data rate or bit rate

The **data rate** is defined the number of data element (or bits) sent in one second. Its unit is bps (Bits per second).

4. Signal rate

The **signal rate** is the number of signal element sent in one second. Its unit is baud. It is prefer in data communication to increase the data rate while decreasing the signal rate. Speed of transmission can be increase by increasing the data rate whereas decreasing the signal rate, decrease the bandwidth requirement.

We define a ratio '*r*' which is the number of data element carried by each digital signal element. It is the ratio of number of data elements to the number of signal elements. For example, a data element is represented

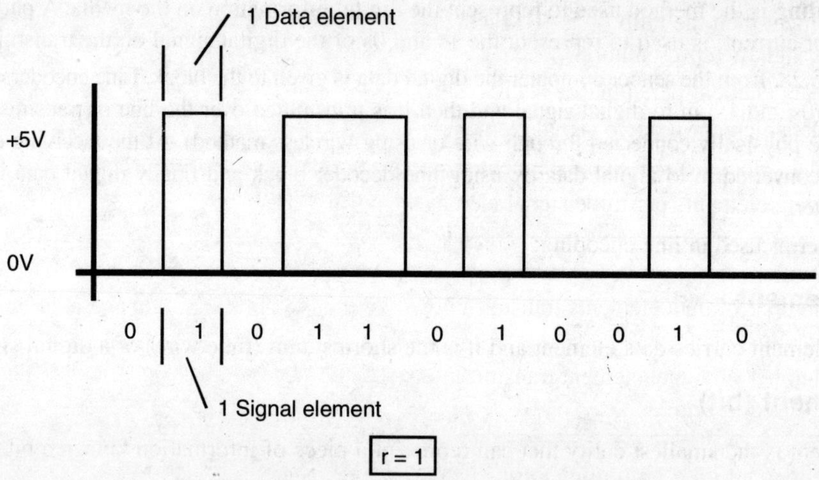

Fig. 5.28. One Data Element per signal (r = 1)

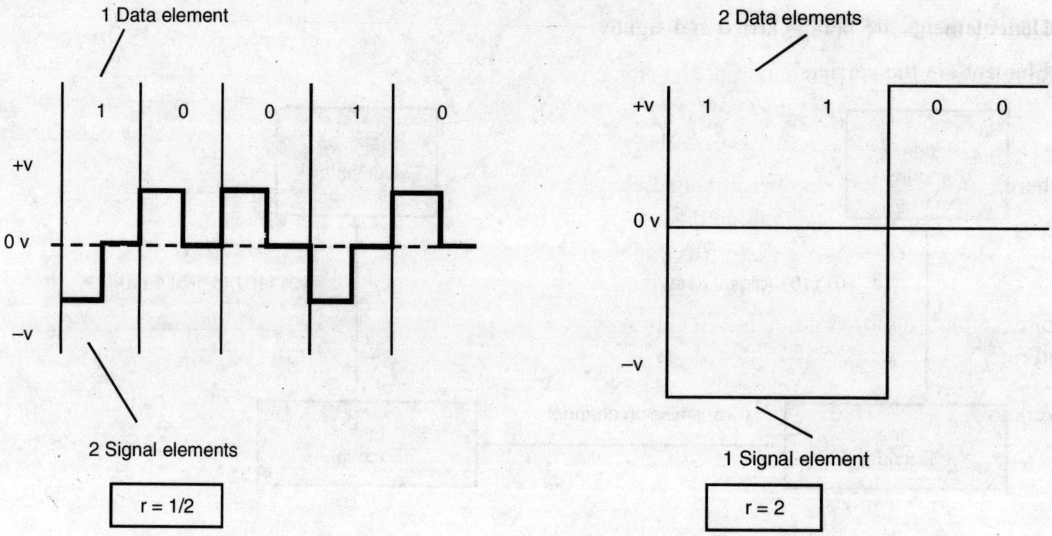

(*a*) One data element per two signal element (r = 1/2)

(*b*) Two data elements per signal element (r = 2)

Fig. 5.29.

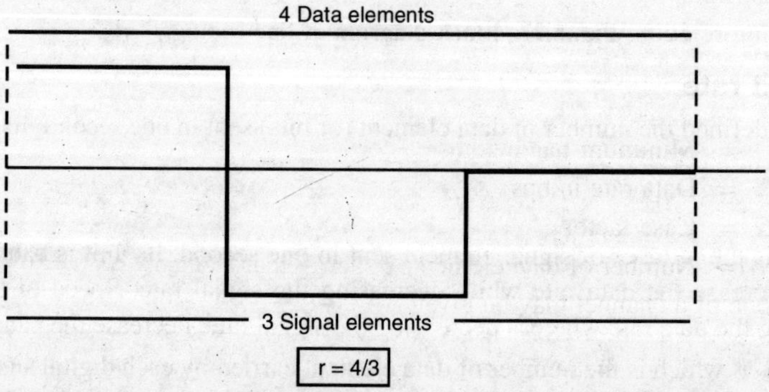

Fig. 5.30. Four data elements per three signal element (r = 4/3)

by a signal element then the ratio 'r' is '1'. Figures 5.28, 5.29 (*a*) and (*b*) and Fig. 5.30 shows the different types of ratio's signals representation.

A. One data element per signal ($r = 1$)
B. One data element per two signal element ($r = 1/2$)
C. Two data elements per signal element ($r = 2$)
D. Four data elements per three signal element ($r = 4/3$)

TIPS

- The number of data elements transmit in one second is referred as data and measured in *bits per second* (bps).
- The number of signal element transmit in one second is referred as signal rate or baud rate and it is measured in *baud*.

It is pretended in data communication that to increase the data rate which decreasing the signal rate. The requirement of bandwidth the data communication channel is also depend upon the ratio (r).

5. Relationship between Data Rate and Signal Rate

The relationship between data rate and signal rate is given by,

$$S = C \times N \times \frac{1}{r} \qquad ...(5.2)$$

where, S = Number of signal elements
N = Data rate in bits.
C = Case factor (its values lies between 0 and 1)
r = No.'s of data element / each signal element

For example; suppose bit rate is 100 kbps, $r = 2$ and value of case factor (C) is 1. Find the average value of baud rate.

We know,

$$S = C \times N \times \frac{1}{r}$$

$$S = 1 \times 100000 \times \frac{1}{2}$$

$$S = 50 \text{ kbaud}$$

6. Bandwidth

Baud rate determine the bandwidth. Although the actual bandwidth of a digital signal is infinite but the effective bandwidth is finite.

By using following relation, the minimum bandwidth can be determined,

$$B_{min} = C \times N \times \frac{1}{r} \qquad ...(5.3)$$

where, B_{min} = Minimum bandwidth required
N = Data rate in bps
C = Case factor
r = Number of data elements / each signal element

Maximum data rate of a channel may be given by

$$N_{max} = \frac{1}{C} \times B \times r \qquad ...(5.4)$$

where, N_{max} = maximum data rate

B = Bandwidth

According to Nyquist theorem, maximum data of a channel is given as

$$N_{max} = 2 \times B \times \log_2 (L) \quad \text{...(5.5)}$$

A signal with L levels actually can carry $\log_2 (L)$ bits per level, if $C = ½$ and $r = 1$;

$$N_{max} = \frac{1}{C} \times B \times r = 2 \times B \times \log_2 (L)$$

7. DC Component

There are some lines encoding schemes leaves residual **direct current (dc) components**. A dc signal has zero frequency so it cannot travel through a media and it is very difficult to handle dc components by channel.

The dc component is undesirable due to following two reasons :

(1) If the signal has to pass through a system, say transformer that does not allow dc component to pass through, so the signal is distorted and that may cause errors in the output signal.

(2) The dc component has extra energy lies on the line and it is waste full.

For example, we have two line encoding schemes as shown in Fig. 5.31 (*a*) and (*b*). In Fig. 5.31 (*a*), line coding has dc component because its average value is not zero, *i.e.* the positive voltages are not cancelled by the negative voltages. This waveform does not pass through the transformer properly. In Fig. 5.31 (*b*) the average value is zero, than means there is no dc components in this case. This waveform can pass through the transformer properly.

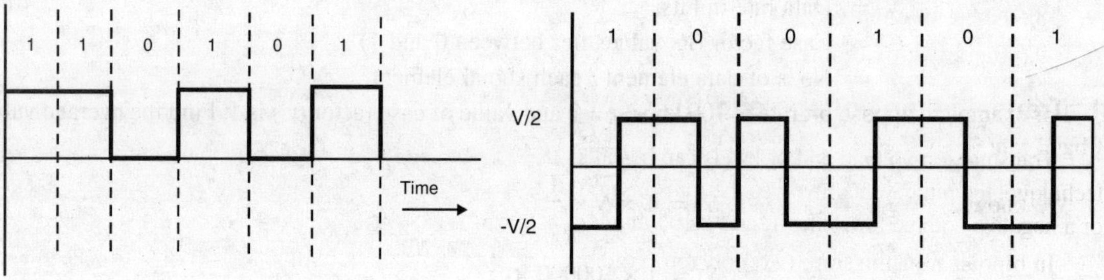

(*a*) a signal with dc components (*b*) a signal without dc components

Fig. 5.31.

8. Base-Line Wandering

A long string of either logic 1's or logic 0's in NRZ produces a condition in which a receiver may loss

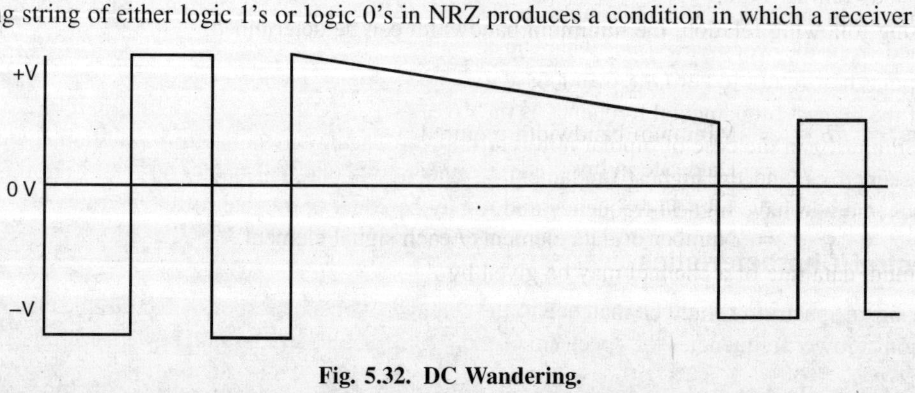

Fig. 5.32. DC Wandering.

its amplitude reference for optimum discrimination between received 1's and 0's. This state is known as **dc wandering or base line wandering**. This problem may also arise when there is a significant imbalance in the number of 1's and 0's transmitted. Figure 5.32 shows how dc wandering is produced from a long string of successive logic 1's. It is observed that after a long string of 1's; there is likely that 1 may be changed to 0 (1 to 0 errors). Similarly, when a long string of 0's increases the chances of that 0 may be changed to 1 (0 to 1 errors).

DC components and DC wandering or base line wandering are short coming of few of the line encoding schemes.

5.19.1. Factors Affecting the Choice of Line Coding Scheme

Line coding is used to convert a sequence of bits in to a digital signal. There are several line codes that are used for the electrical representation of binary signal symbols 1 and 0. Each representation has its own advantages and disadvantages. The most important criteria for choosing a particular waveform are based on spectral characteristics, timing context, bandwidth, noise immunity etc.

These important factors are discussed as follows :

1. Transmission voltage and dc components
2. Bandwidth consideration
3. Spectral characteristics
4. Duty cycle
5. Clock and framing bit recovery
6. Power efficiency
7. Error detection
8. Noise Immunity
9. Bandwidth Compressive

1. Transmission Voltage and DC Components

Transmission voltage and or levels can be categories as being either unipolar or bipolar. In this transmission technique, the binary data involves the transmission of only a single non-zero voltage level (e.g., either a positive or a negative voltage for logic '1' and zero voltage ground for logic 0).

In bipolar transmission, two non zero voltages are involved (for example, a positive voltage for a logic '1' and an equal magnitude negative voltage for a logic '0' or vice-versa. Considering an equal probability of the occurrence of logic 1 and 0, symmetrical bipolar voltage reduces the average power by 50%.

2. Bandwidth Consideration

To determine the maximum bandwidth required to propagate a line-encoded digital signal. We must determine the highest fundamental frequency in the signal. The highest fundamental frequency may be determined from the worst-case (*i.e.,* where fastest transitions occur) binary bit sequences. For example, with unipolar non-return to zero, the worst case condition is an alternating frequency 1 and 0 sequence, the period of the highest fundamental frequency is the time of two bits that is equal to one-half of the bit rate (bit rate/2). Similarity, in the case of bipolar non-return to zero, again the worst-case condition is an alternating 1 and 0 sequences, and the highest fundamental frequency is one-half of the bit rate. Like this, each line coding scheme will have highest frequency and it may be equal or one half of the bit rate.

3. Spectral Characteristics

The signal spectrum should be matched to the channel frequency response. Now, if the channel has high attenuation at lower frequencies the spectrum of the signal should have a small power spectral density in this range to avoid excessive distortion in the signal. It is also important to note that it is desirable to have zero power spectral density (PSD) at dc, *i.e.,* ω = 0, unable the system to be AC coupled. Transformers and AC coupling are used at the repeaters. Significant lower in low frequency components causes dc wander in the pulse stream when AC coupling is used. The AC coupling is required since the DC path provide by the cable pairs between the repeaters sites are used to transmit the power required to operate the repeaters.

4. Duty Cycle

The duty cycle of a binary pulse may be used to categorize the type of transmission. If this binary pulse is maintained for the entire bit time, this is called non-return to zero. If the active time of binary pulse is less than 100% of the bit time, it is called return to zero (RZ).

5. Clock and Framing Bit Recovery

Bit synchronization is the most essential requirement in any digital communication system. To recover and maintain clock and framing synchronization from the received data, there should be enough transition in the data waveform. In unipolar NRZ and bipolar NRZ, encoding a long string of 1s and 0s generation a data signal void transition and therefore, is inadequate for clock recovery. In bipolar RZ, a transition occurs in each bit position regard less of whether the bit is a '0' or a '1'. This BPRZ is the best encoding scheme for clock recovery.

6. Power Efficiency

Power efficiency means the requirement of power for the transmission should be as small as possible, for a given bandwidth and a specified detection error probability.

7. Error Detection and Correction Capability

It is preferred that line encoding scheme should have some means to detect the error and preferably correct the detected errors. With unipolar NRZ, BPNRZ, unipolar RZ and Bipolar RZ encoding, there is no means to determine that the received data have the error or not. However, in BPRZ- AMI (Bipolar Return to zero-Alternate Mark Inversion) encoding, an error in any bit will cause a bipolar violation (*i.e.*, the reception of two or more consecutive logic 1s with the same polarity). Therefore, BPRZ-AMI has a build-in error detection mechanism.

8. Noise Immunity

The transmitted symbols (1s and 0s) should be protected against noise, in order to reduce the error in transmission. The bipolar signals are having more immunity against noise, them the unipolar signals. However more power is to be transmitted.

9. Bandwidth Compression

There are some line encoding scheme such as multilevel codes increase the efficiency of BW utilization by using more number of levels so that in a bit interval essentially 2 bit may be transmitted.

5.19.2. Common Line Coding Schemes

There are following types of techniques or schemes used for Line Encoding (refer to Fig. 5.33)

1. Unipolar—NRZ (Non-return to zero)
2. Polar :
   - NRZ- L (Non-Return to zero-level)
   - NRZ-I (Non-Return to zero-Inversion)

- R Z (Return to zero)
- Biphase—Manchester and Differential Manchester

3. Bipolar :
   - AMI (Alternate Mark Inversion)
   - Pseudoternary
4. Multilevel
5. Multi-transition

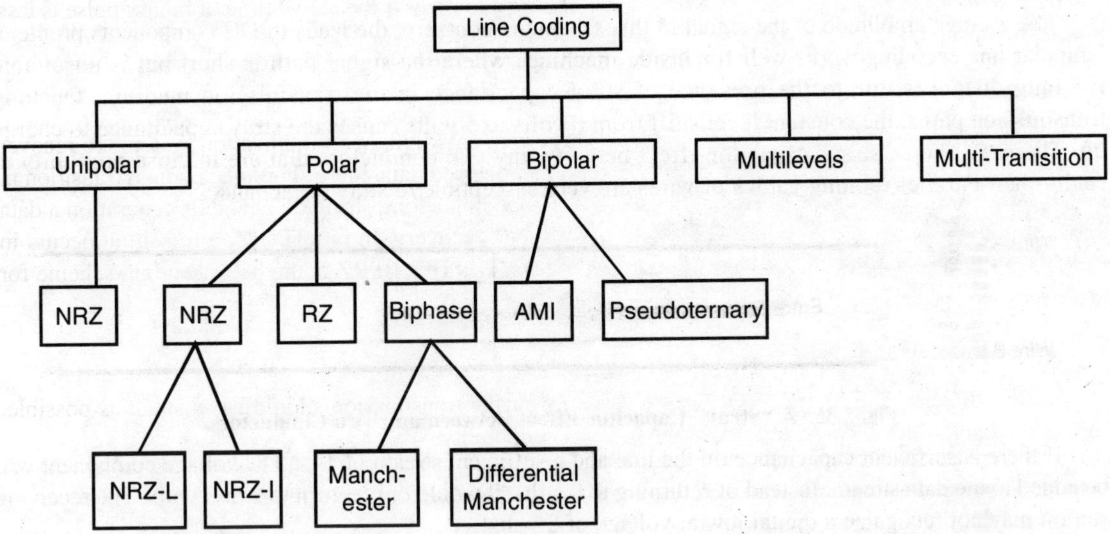

Fig. 5.33. Classifications of Line Coding Scheme.

5.19.3. Level Encoding

The simplest form of line coding uses a different signal level to encode each discrete symbol transmitted. Communications engineers commonly refer to the unbalanced code as a unipolar code and the balanced code as a polar code.

5.19.3.1. Unipolar (Non-Return to Zero)

- **Unipolar encoding** technique is the simplest technique to represent the digital signal with only one level of amplitude.
- One level of amplitude mean one voltage level (say V) stands for binary 1 and another level (say 0 V) stands for binary 0.
- In this scheme signal level is only one side of the reference level (0 V level) normally above the reference level that means having a single polarity only.

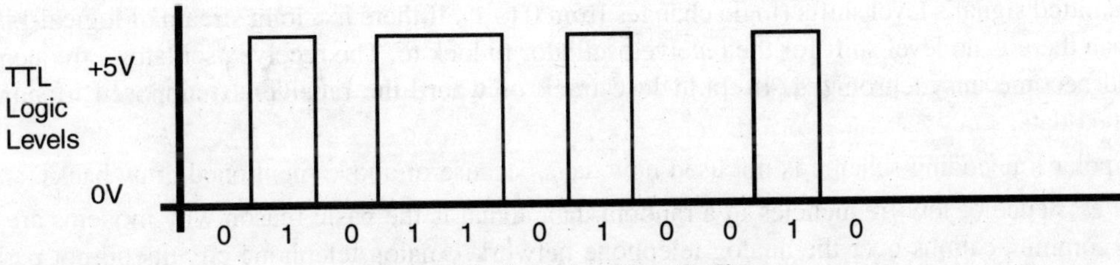

Fig. 5.34. Unipolar Line Coding.

Figure 5.34 shows the representation of the digital data. A common example of Unipolar line encoding is the logic levels used in computers and digital logic. Logic High (1) is represented by +5V and a logic Low (0) is represented by 0 V.

This unipolar line encoding also referred as NRZ (Non-Return to Zero) scheme. A unipolar scheme was designed as NRZ scheme in which, the positive voltage defines the binary '1' and binary '0' for zero voltage.

It is called NRZ because the signal does not return to zero during the bit interval.

This scheme has two major draw backs.

1. DC Components

The average amplitude of the signal in this scheme is non-zero, the leads to DC components problem. Unipolar line encoding works well for inside machines where the signal path is short but is unsuitable for long distances due to the presence of stray capacitance in the transmission medium. On long transmission paths, the constant level shift from 0 volts to 5 volts causes the stray capacitance to charge up. There will be a "stray" capacitor effect between any two conductors that are in close proximity to each other. Parallel running cables or wires are very susceptible to stray capacitance.

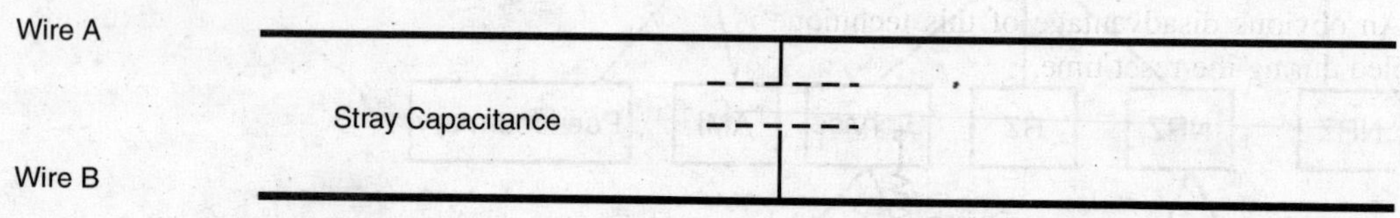

Fig. 5.35. A "Stray" Capacitor Effect Between any Two Conductors.

If there is sufficient capacitance on the line and a sufficient stream of 1s, a DC voltage component will be added to the data stream. Instead of returning to 0 volts, it would only return to 2 or 3 volts. The receiving station may not recognize a digital low at voltage of 2 volts!

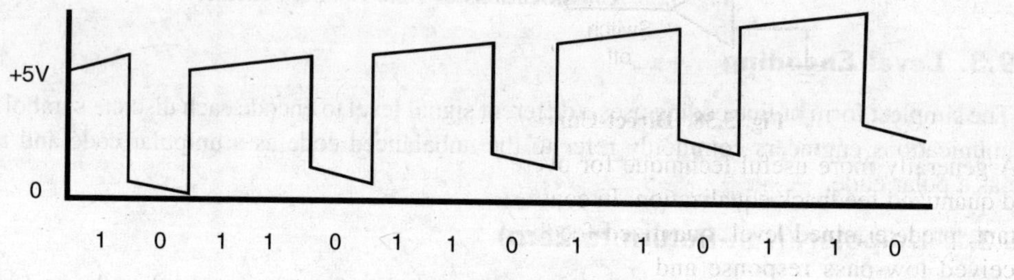

Fig. 5.36. DC Voltage Component.

2. Lack of Synchronization

If the data consists of a long sequence of 1's and 0's, and then there is no change in the signal level during this period, so that leads to synchronization problem. Unipolar line encoding can have synchronization problems between the transmitter and receiver's clock oscillator. The receiver's clock oscillator locks on to the transmitted signal's level shifts (logic changes from 0 to 1). If there is a long stream of logical 1s or 0s in a row than there is no level shift for the receive oscillator to lock to. The receive oscillator's frequency may drift and become unsynchronized. It could lose track of where the receiver is supposed to sample the transmitted data.

Unipolar line coding scheme is not used now days because of above mentioned draw backs.

The existence of low frequencies in a random data signal is the basic reason why modems are needed for data communications over the analog telephone network. Analog telephone circuits do not pass direct current. It is also the reason that NRZ coding is not often used for long-distance transmission. Direct-current wander is not unique to data transmission systems. It is a phenomenon that must be reconciled in television receivers, radar receivers, or radiation detectors.

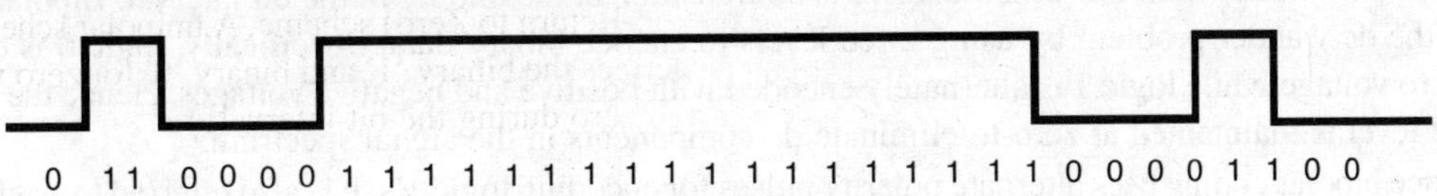

Fig. 5.37. Unipolar NRZ Signal.

One technique of offsetting dc wander is referred to as dc or baseline restoration. As illustrated in Figure 5.38, dc restoration involves passing received pulses through a capacitor, detecting them, and then removing the charge on the capacitor before the next pulse arrives. Charge on the capacitor is removed by driving the voltage to a specific threshold (0 V in Figure 5.37) and then removing the driving voltage when threshold is reached. Since all charge on the capacitor is removed after each pulse, the baseline or decision reference level is constant at the beginning of a signal interval.

An obvious disadvantage of this technique is that the signal input must have zero amplitude or be disabled during the reset time.

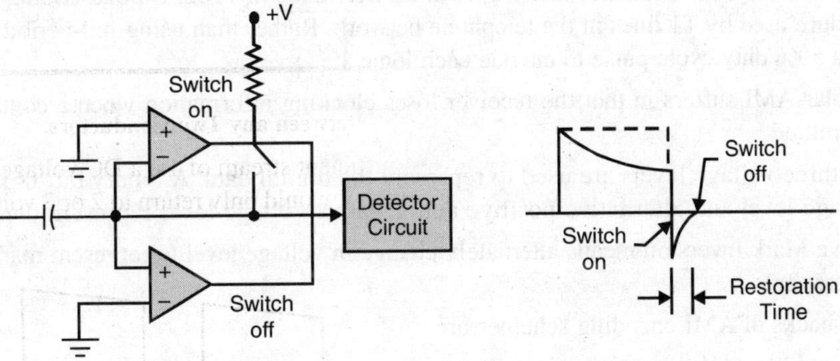

Fig. 5.38. Direct-Current Restoration for Unipolar Pulses.

A generally more useful technique for overcoming baseline wander is to use decision feedback, also called quantized feedback equalization. In contrast to dc restoration, which drives the capacitor voltage to a constant, predetermined level, quantized feedback compensates for dc wander by locally generating the unreceived low-pass response and adding it to the received signal. To accomplish this, the original data stream is reconstructed. As shown in Figure 5.39, the reconstructed data stream is passed through a low-pass filter that generates a pulse equal to the tail or droops characteristic of the channel. The feedback signal adds to the received signal to eliminate the droop (intersymbol interference). Using a frequency-domain analysis, the feedback response is complimentary to the channel response. Quantized feedback is used in ISDN basic rate line interfaces.

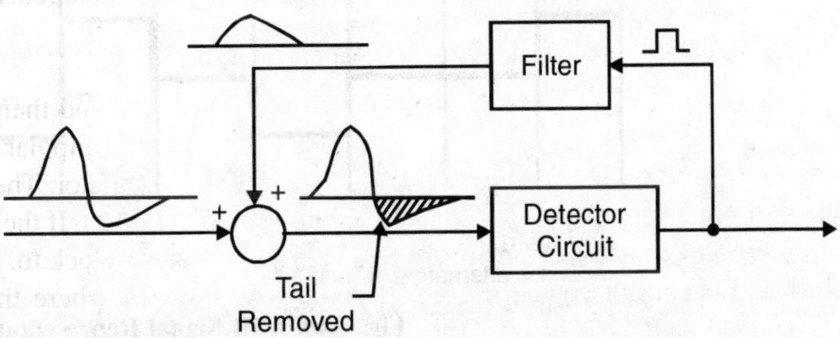

Fig. 5.39. Decision Feedback Equalization.

5.19.3.2. Bipolar Encoding Scheme

Bipolar encoding scheme utilizing the three voltage levels to represent digital data. These voltage levels are +v, 0v and –v. in this scheme, there is no transition at the middle of the bit interval. Bipolar coding solves the dc wander problem by using three levels to encode binary data. Specifically, logic 0 is encoded with zero voltage while logic 1 is alternately encoded with positive and negative voltages. Hence the average voltage level is maintained at zero-to eliminate dc components in the signal spectrum.

Since bipolar coding uses alternate polarity pulses for encoding logic 1's, it is also referred to as **alternate mark inversion** (AMI).

The binary bit '0' bit is represented by zero level and binary bit '1' is represented by alternative positive and negative voltage level. There are three encoding schemes under bipolar encoding scheme.

(1) AMI (Alternate Mark Inversion)
(2) Pseudoternary (variation of AMI)
(3) Twinned binary codes

(1) Alternative Mark Inversion (AMI)

The most widely used of the multilevel binary codes is the Bipolar-AMI code. The Bipolar-AMI code was first used in the PCM based T1-carrier system by Bell Labs in 1962. Bipolar coding is the basic line-coding procedure used by T1 lines in the telephone network. Rather than using full-period pulses, however, T1 lines use a 50% duty cycle pulse to encode each logic 1.

The Bipolar-AMI suffers in that the receiver loses clocking information when a continuous stream of zeros is transmitted.

In AMI, three voltage levels are used to represent the digital data. A binary bit '0' is represented by the zero voltage level and alternative positive and negative voltage level represent the binary bit '1'.

Alternative Mark Inversion means alternately change in voltage level to represent mark (*i.e.,* binary bit '1' from telegraphy).

The drawbacks of AMI encoding schemes are :

- Problem of synchronization
- DC component occurs whenever there is a long string of binary bit '0'.

Figure 5.40 shown the representation of the digital data in AMI encoding scheme.

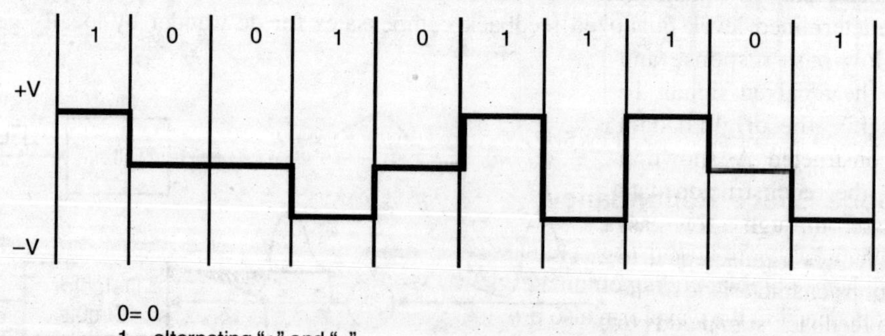

Fig. 5.40. AMI Signal Representation.

(2) Pseudoternary

This is a variation of the AMI, in this scheme binary bit '1' is represent with the zero level and binary bit '0' is represent with alternative +v and –v to voltage level.

It is used for long distance communication.

Figure 5.41 shown the representation of the digital data in Pseudoternary encoding scheme for the digital data 1001011101.

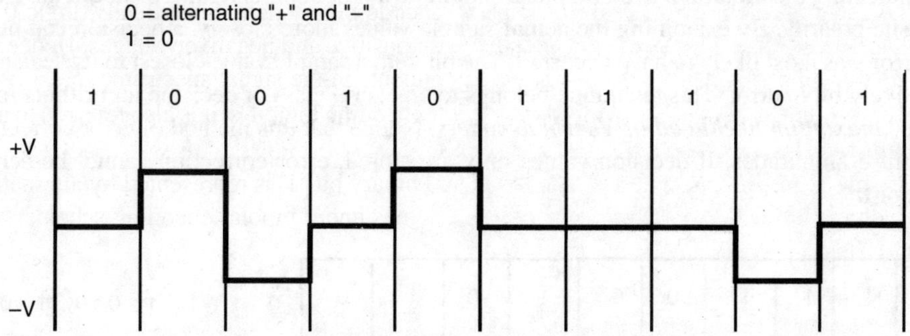

Fig. 5.41. Pseudoternary Encoding Scheme.

(3) Twinned Binary Codes

To eliminate the DC component from the spectrum and to keep from increasing the symbol rate, the **twinned binary codes** can be used. A twinned binary code increases the number of available symbols by increasing the number of signal levels.

The main disadvantage of twinned binary codes is a slight loss in noise immunity.

$0 \Rightarrow 1$ transmits "+"
$1 \Rightarrow 0$ transmits "–"
$0 \Rightarrow 0$ transmits "0"
$1 \Rightarrow 1$ transmits "0"

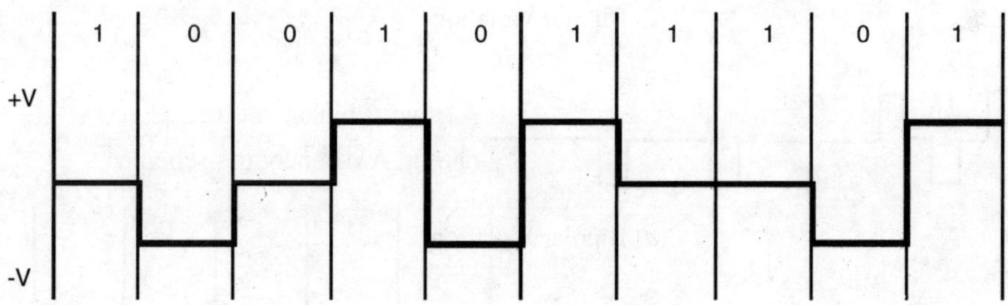

Fig. 5.42. Twinned Binary Codes.

5.19.3.3. Code Space Redundancy

In quintessence, bipolar coding uses a ternary code space but only two of the levels during any particular signal interval. Hence bipolar coding eliminates dc wander with an inefficient and redundant use of the code space. The redundancy in the waveform also provides other benefits. The most important additional benefit is the opportunity to monitor the quality of the line with no knowledge of the nature of the traffic being transmitted. Since pulses on the line are supposed to alternate in polarity, the detection of two successive pulses of one polarity implies an error. This error condition is known as a **bipolar violation**. No single error can occur without a bipolar violation.

Hence the bipolar code inherently provides a form of line code parity. The terminals of T1 lines are designed to monitor the frequency of occurrence of bipolar violations, and if the frequency of occurrence exceeds some threshold, an alarm is set.

In T-carrier systems, bipolar violations are used merely to detect channel errors. By adding some rather sophisticated detection circuitry, the same redundancy can be used for correcting errors in addition to detecting them. Whenever a bipolar violation is detected, an error has occurred in one of the bits between and including the pulses indicating the violation. Either a pulse should be a 0 or an intervening 0 should have been a pulse of the opposite polarity. By examining the actual sample values more closely, a decision can be made as to where the error was most likely to have occurred. The bit with a sample value closest to its decision threshold is the most likely bit in error. This technique belongs to a general class of decision algorithms for redundant signals called ***maximum likelihood** or **Viterbi decoders***. Notice that this method of error correction requires storage of pulse amplitudes. If decision values only are stored, error correction cannot be achieved (only error detection).

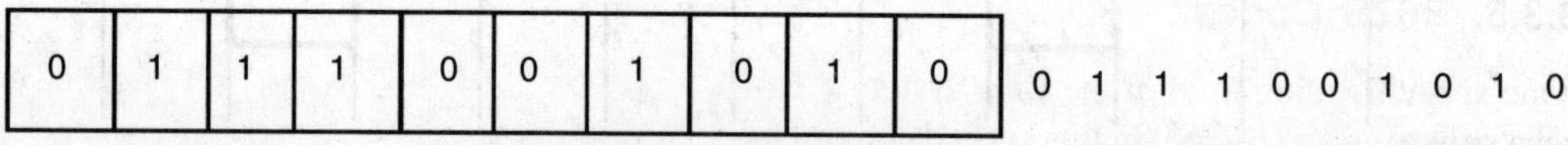

(*a*) Binary Code

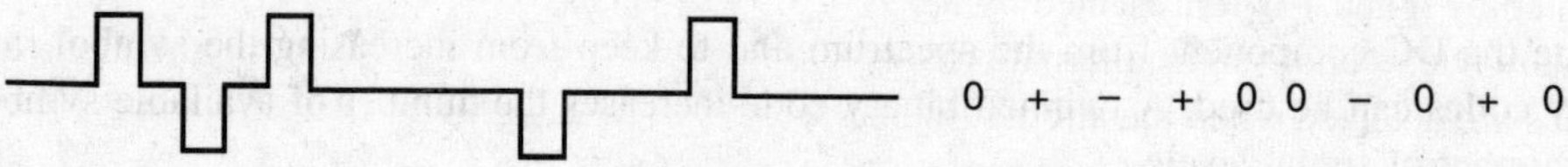

(*b*) Normal Bipolar

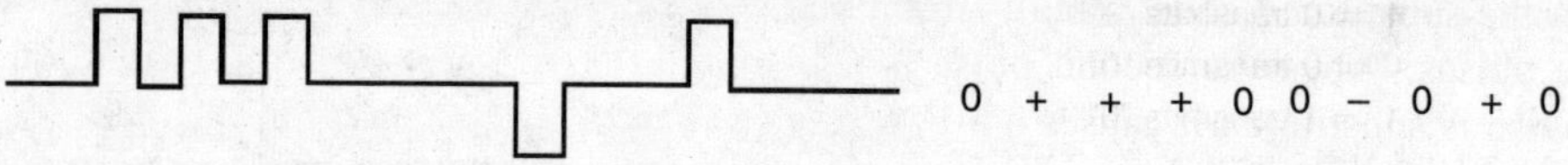

(*c*) Bipolar Violation

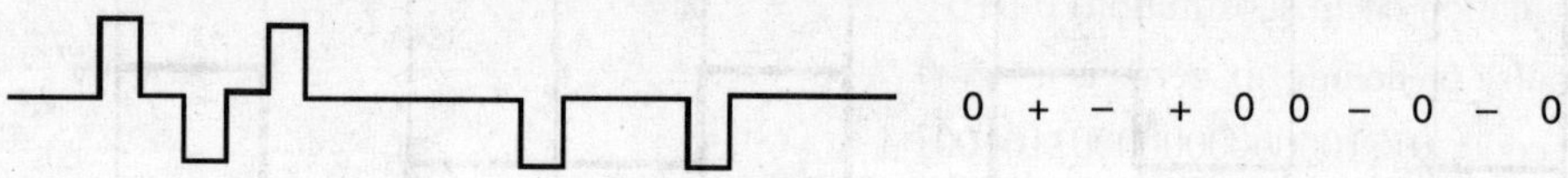

(*d*) Bipolar Violation

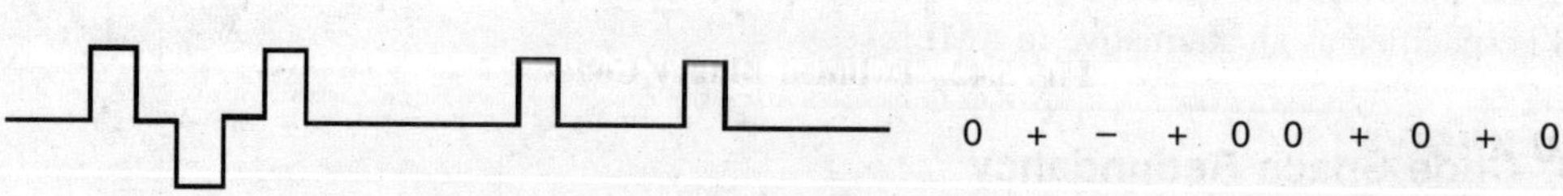

(*e*) Bipolar Violation

Fig. 5.43. Code Representations of Binary Violation.

An additional application of the unused code space in bipolar coding is to purposely insert bipolar violations to signify special situations such as time division multiplexing framing marks, alarm conditions, or special codes to increase the timing content of the line signals. Since bipolar violations are not normally part of the source data, these special situations are easily recognized. Of course, the ability to monitor the quality of the line is compromised when bipolar violations occur for reasons other than channel errors.

Figure 5.43 shows an example of binary violation.

5.19.3.4. N-Zero Substitution

The technique used to solve the pulse density problem is to use special coding to ensure that sufficient 1's are always present to provide accurate timing. The special code pattern is inserted into the transmitted bit stream. At the receiver, the presence of the special code is detected and the signal adjusted to the original information.

One way of ensuring that there are large numbers of 1's in any transmitted bit stream is to detect long strings of 0's and substitute for them code patterns containing mostly ones. At the receiver these code patterns are recognized and the inserted pattern is replaced with 0's. Such algorithms are called binary N-zero substitution schemes or BNZS schemes.

5.19.3.5. B6ZS Codes

One special coding used in the Bell System's T2 transmission lines (the DS2 signal) is called **binary six-zero substitution (B6ZS).** In this special coding, the receiver detects where the substitution has occurred by detecting a bipolar violation.

In a normal bipolar code, the pulses representing 1s are alternately positive and negative around zero. Zero is represented by 0 and 1 is represented by a + or – sign, which indicates the polarity of the pulse. For the normal bipolar code, + and – signs alternate in the code with 0s placed as they occur in the input system. Bipolar violations are detected when the alternating pattern of + and – signs is interrupted; that is if two + or two – signs occur in sequence.

In the B6ZS code, whenever a string of six 0s occurs, a special code, is substituted. The code pattern substituted depends on the polarity of the 1 pulse immediately preceding the string of six 0s. Note also that in the substituted code, a built- in bipolar violation is caused by the substitution of the second and fifth pulses. When the second pulse is substituted it causes a +0+ or –0– violation, and when the fifth pulse is substituted it causes a –0– or +0+ violation. Here are specific examples:

1. The 1 pulse preceding six zeros is a + :
   Binary code: 0101000000110000001001
   Representation (with substitution) 0-0+0+ –0– + – +0+ –0– + –00+
2. The 1 pulse preceding six zeros is a – :
   Binary code: 0111000000000001101001
   Representation (with substitution): 0- + - 0 - + 0 + - 0 - + 0+ - + - + + 0 0 -

The B3ZS and B6ZS codes are specified for the DS3 and DS2 North American standard rates, respectively, and B8ZS is specified as an alternative to AMI for the DS1 and DS1C rates.

DO YOU KNOW▶

B6ZS Rules

1 Pulse Polarity	Special Code
+	0 + – 0 – +
–	0 – + 0 + –

5.19.3.6. HDB3

A somewhat different special code substitution is recommended by the CCITT in Europe. It is also a BNZS code called *high-density bipolar three-bit substitution* (HDB3). This technique replaces strings of four 0s with the substitute code shown in below :

DO YOU KNOW ►

HDB3 Rules

1 Pulse Polarity	Special Code *Even*	Special Code *Odd*
+	000 +	– 00– +
–	000 –	+ 00 +

Because four 0s are replaced (greater than 3), the code is called HDB3. Its substitution rules depend on how many 1s have occurred in the PCM bit stream since the last substitution as well as the polarity of the 1 pulse that is just ahead of the string of zeros.

For examples:

1. Plus, odd; minus, odd:

   Binary code: 01011<u>0000</u>1010<u>10000</u>

   Substitution : 0 + 0 – + <u>0 0 0 +</u> – 0 + 0 – <u>0 0 0 –</u>

5.19.3.7. Pair Selected Ternary

The BNZS substitution algorithms described in the preceding section are examples of selecting codes in the ternary codes pace to increase the timing content of a binary signal.

Pair selected ternary (PST) 20 is another example.

The PST coding process begins by pairing the binary input data to produce sequences of 2-bit codeword's. These codeword's are then translated in to two ternary digits for transmission. Since there are nine two-digit ternary codes but only four two-digit binary codes, there is considerable flexibility available in selecting the codes. The most useful of the possible coding formats is shown below. This particular format not only ensures a strong timing component but also prevents dc wander by switching modes to maintain a balance between positive and negative pulses. The codes are selected from one column until a single pulse is transmitted. At that time the encoder switches modes and selects codes from the other column until another single pulse (of the opposite polarity) is transmitted.

For example, encode the following binary data stream into a PST line code:

010 011 1010110 0

There are two possible solutions depending on whether the encoder is in the positive or negative mode at the beginning of the sequence:

	01	00	11	10	10	11	00
Case 1 (+ mode) :	0 +	– +	+ –	– 0	+ 0	+ –	– +
Case 2 (– mode) :	– +	– +	+ –	+ 0	– 0	+ –	– +

One potential drawback of the PST coding algorithm is that the binary data stream must be framed into pairs. Hence a PST decoder must recognize and maintain pair boundaries. Recognition of the boundaries is not difficult if random data are being transmitted since a pairwise misframe eventually produces unallowed codes (00, + +, – –). Furthermore, time division multiplex formats typically provide character and pairwise framing automatically.

The power spectrum of a PST line code with equal probabilities for l's and 0's is plotted in Figure 5.44. Also shown is the B6ZS power spectrum and the conventional bipolar power spectrum. An important point to notice in Figure 5.43 is that bipolar coding and its extensions require equal bandwidths. Their only significant difference is that B6ZS and PST have higher energy levels as a result of greater pulse densities. The higher energy levels have the undesirable effect of increasing crosstalk interference in multipair cables.

However, the degradation from the increased crosstalk is somewhat offset by improved accuracy of the recovered sample clock (assuming all signals in the cable are using the same line code).

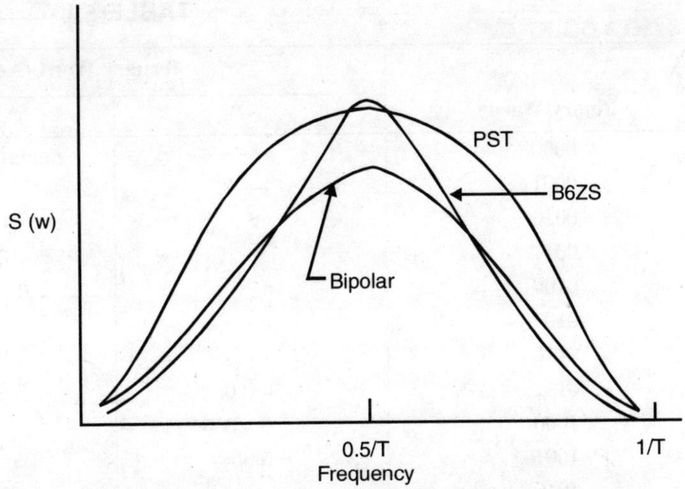

Fig. 5.44. Spectrum of Bipolar, B3ZS, and PST Line Codes for Equally Likely 1's and 0's.

Pair Selected Ternary Encoding Rules :

Binary Input	+ Mode	– Mode
00	– +	– +
01	0 +	0 –
10	+ 0	– 0
11	+ –	+ –

5.19.3.8. Ternary Coding

Since bipolar and PST coding use a ternary code space to transmit binary data, they do not achieve as high an information rate as is possible with more efficient use of the code space. For example, an eight-element ternary code is capable of representing $3^8 = 6561$ different codes. In contrast, 8 bits of binary data produce only $2^8 = 256$ different codes. The previously described line codes do not take advantage of the higher information content of ternary codes; they select codes for their timing content and spectral properties.

One ternary encoding procedure involves mapping successive groups of 4 bits into three ternary (4B3T) digits. Since binary words of 4 bits require only 16 of the 27 possible three-digit ternary codewords, considerable flexibility exists in selecting the ternary codes. Table 5.1 presents one possible encoding procedure. Ternary words in the middle column are balanced in their dc content. Codewords from the first and third columns are selected alternately to maintain dc balance. If more positive pulses than negative pulses have been transmitted, column 1 is selected. When the disparity between positive and negative pulses changes, column 3 is chosen. Notice that the all '0's codeword is not used.

Hence a strong timing content is maintained. Because of the higher information efficiency, however, the ability to monitor performance is sacrificed, and framing is required on three-digit boundaries. 4B3T coding is used on the TI48 span line developed by ITT Telecommunication. This system provides T-carrier transmission for two DS1 signals (48 channels) using a bandwidth that is only 50% greater than a Tl bandwidth (carrying 24 channels).

TABLE 5.1

Binary Word	Ternary Word (Accumulated Disparity)		
	–	0	+
0000	– – –		+ + +
0001	– – 0		+ + 0
0010		– 0 –	+ 0 +
0011		0 – –	0 + +
0100		– – +	+ + –
0101	– + –		+ – +
0110	+ – –		– + +
0111	– 0 0		+ 0 0
1000	0 – 0		0 + 0
1001	0 0 –		0 0 +
1010		0 + –	
1011		0 – +	
1100		+ 0 –	
1101		– 0 +	
1110		+ – 0	
1111		– + 0	

5.19.3.9. Polar Line Encoding

When the digital encoding is symmetrical around 0 Volts, it is called a **Polar Code.**

- Polar line encoding is also called Non-Return to Zero (NRZ).
- Polar line encoding is the simplest pattern that eliminates most of the residual DC problem.
- Polar line encoding scheme uses two voltage levels, one +ve and other is –ve. By using two voltage levels in this scheme reduces the average voltage level on line that means reducing the problem of dc component as observed in unipolar line encoding scheme.
- For example, the voltage level for '0' can be positive and voltage level for '1' can be negative.
- The RS-232D interface uses Polar line encoding.

Types of Polar Line Encoding Scheme

1. Polar NRZ
   - NRZ-L (Non-Return to Zero-Level)
   - NRZ-I (Non-Return to Zero-Inversion)
2. RZ (Return to zero)
3. Bi-phase
   - Manchester
   - Differential Manchester

Polar NRZ (Non-Return to Zero)

- In **polar NRZ** encoding scheme, we use two levels of voltages and the level is always positive or negative. In the scheme also signal level is not returning to zero during the bit interval as observed in unipolar line encoding scheme. There are two types of polar NRZ which are discussed here.

(i) NRZ-L (Non-Return to Zero-Level)

- In **NRZ-L (Non-return to Zero-Level)** the level of the voltage determines the value of the bit and depends upon the type of bit that it represents.
- Generally, it is a –ve logic encoding scheme, where positive level (+ve) represent the '0' binary and negative level (–ve) represent binary '1'. Therefore, the level of the signal depends upon the state of the bit.
- If there is a continuous string of same binary bit (say 1 or 0), if suffers the problem of synchronization.
- Since we use two voltage levels, that reduces the average voltage level on the line and hence problem of dc component is almost eliminated in this scheme.
- Figure 5.45 shows the representation of following digital data using NRZ-L encoding scheme.

For example, 110001100

Representation of binary bits:

To represent binary '1'—— negative voltage (–ve)

To represent binary '0'—— positive voltage (+ve)

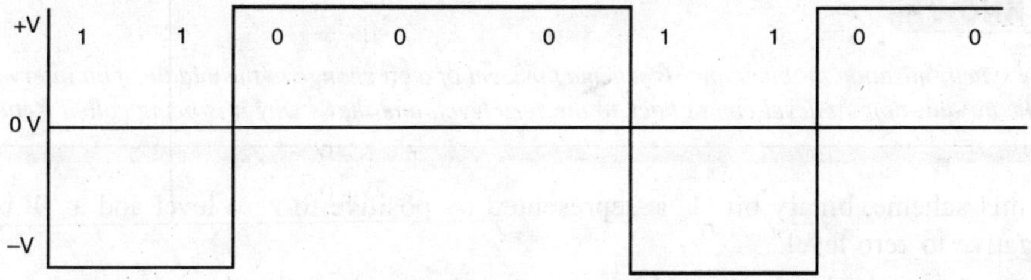

Fig. 5.45. NRZ-L encoding.

(ii) NRZ-I (Non-Return to Zero–Inversion)

- In **NRZ-I (Non-Return to Zero–Inversion)** an inversion of the voltage level represents '1' bit and no change or non-inversion of the voltage level represents '0' bit.

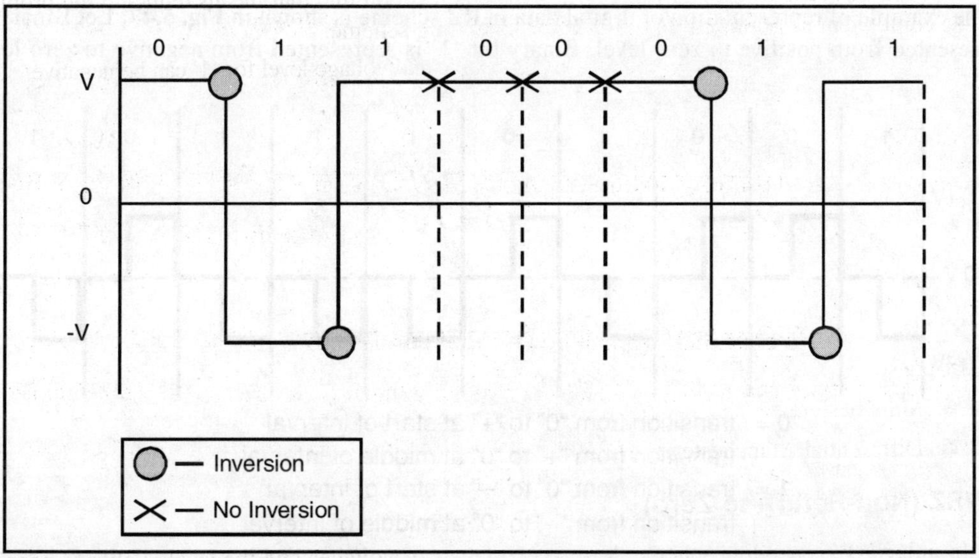

Fig. 5.46. Shows the Representation of the Digital Data, 01100011.

- There is no representation of the bits with +ve or –ve voltage, as it was in polar NRZ-L, but in this scheme change of polarity voltage level will represent the binary bits.
- Synchronization problem is less in this scheme comparing to the NRZ-L, since there is a signal change for binary bit '1'. So, the existence of binary bit '1' in the data stream allows the receiver to synchronization its timer to actual arrival of the signal transmission.
- Only when a long strings of '0' bit present in the data streams, the problem of synchronization will present, because at '0' bit there is no change of level.
- Figure 5.46 shows NRZ-I representation of the digital data, 01100011.

Return to Zero (RZ)

In Return to zero (RZ) line encoding scheme, we have main advantage of synchronization. In NRZ-L, if there is long string of '1' or '0' binary bits, the problem of synchronization occur and in NRZ-I, this problem occurs wherever we have a long string of '0' binary bit. But in RZ scheme, even if there is a long string of '0' binary bit; there is no problem of synchronization.

DO YOU KNOW ▸

To avoid the synchronization problem, in this scheme the level of a bit change at the middle of bit interval, i.e. at the middle of the bit duration the level comes back to the zero level, and that's why it is being called ***Return to zero.***

- In this scheme, binary bit '1' is represented by positive to zero level and a '0' binary bit by negative to zero level.
- It may be noted that in this scheme, we use three voltage levels, +v, –v and zero voltage to represent the digital data. In this scheme, we need two signal change to encode a bit (either 1 or 0), and hence; takes more bandwidth. This is the primary drawback of this scheme.
- There is no problem of dc component and synchronization but it has few drawbacks that it required more bandwidth (twice of Polar NRZ) and system become more complex due to more signal levels.

The example of representation of digital data in RZ scheme is shown in **Fig. 5.47**; Let Binary bit '0' is represented from positive to zero level. Binary bit '1' is represented from negative to zero level.

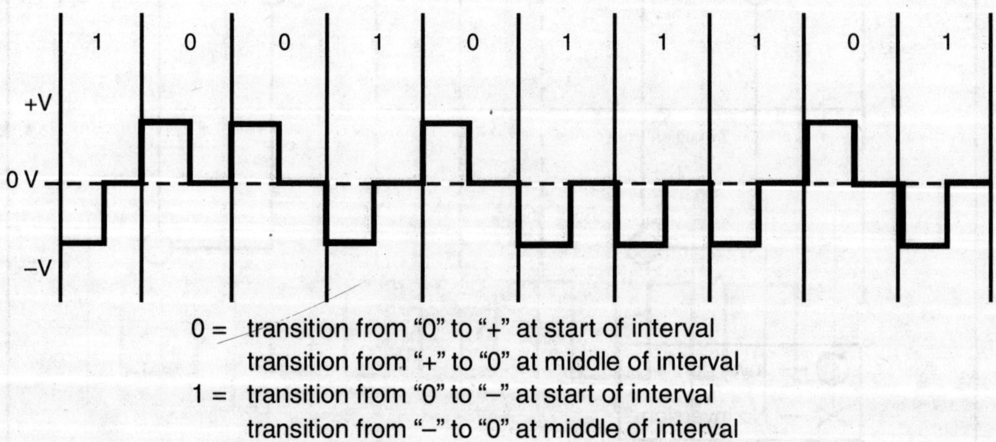

Fig. 5.47. Return to Zero (RZ).

Digital Bi-phase Encoding

Many varieties of line codes achieve strong timing and no dc wander by increasing the bandwidth of the signal while using only two levels for binary data' One of the most common of these codes providing both a strong timing component and no dc wander is the digital bi-phase code, also referred to as "diphase" or a "Manchester" code. Bi-phase encoding scheme is up to some extent it is same as RZ encoding, the only difference is that in Bi-phase signalling, signal does not return to zero, instead it continues to opposite polarity, the middle transition of each bit is used for synchronization. So, there is no problem of synchronization in Bi-phase encoding scheme.

There are two types of bi-phase encoding schemes, which are as follows :

1. Manchester Encoding
2. Differential Manchester Encoding

Manchester Encoding

DO YOU KNOW

Manchester encoding scheme is the combination of RZ and NRZ-L encoding schemes.

In this encoding scheme, the duration of the bit is divided into two halves. The voltage level remains at one level during the first half and moves to the opposite voltage level in the second half.

- The transition at the halfway of the bit provides synchronization.
- In Manchester encoding scheme, a positive to negative transition represent a binary'0' and a negative to positive transition represent the binary '1'.
- In Manchester encoding scheme, the synchronization is achieved at the same level of RZ encoding scheme, with the two voltage levels (+v and –v) only whereas in RZ encoding to achieve synchronization, we have used three levels of voltages.
- Figure 5.48, shown the representation of digital data (1001011101) in Manchester encoding scheme.

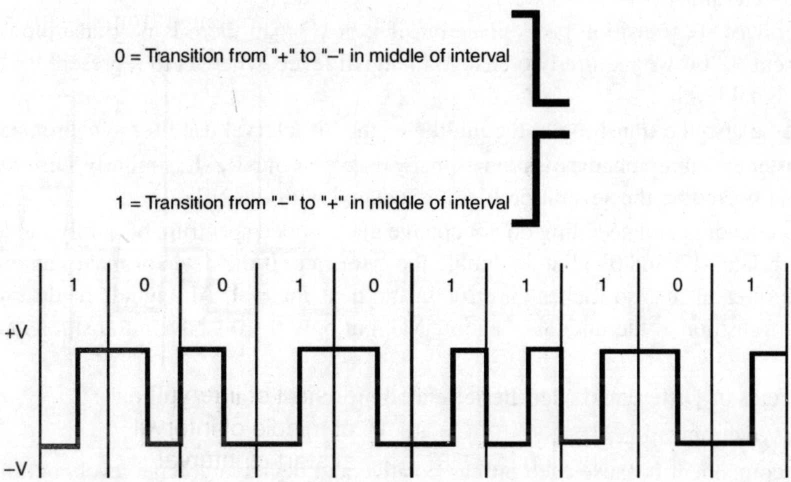

Fig. 5.48. Manchester encoding.

Examination of the frequency spectrum in Figure 5.49 shows that the diphase spectrum is similar to an NRZ spectrum but translated so it is centered about 1/T instead of direct current. Hence digital biphase actually represents digital modulation of a square wave carrier with one cycle per signal interval. Logic 1's cause the square wave to be multiplied by +1 while logic 0's produce multiplication by –1. Diphase is primarily used on shorter links where terminal costs are more significant than bandwidth utilization. The "Ethernet" IEEE 802.3 local area data network uses digital biphase (Manchester) coding.

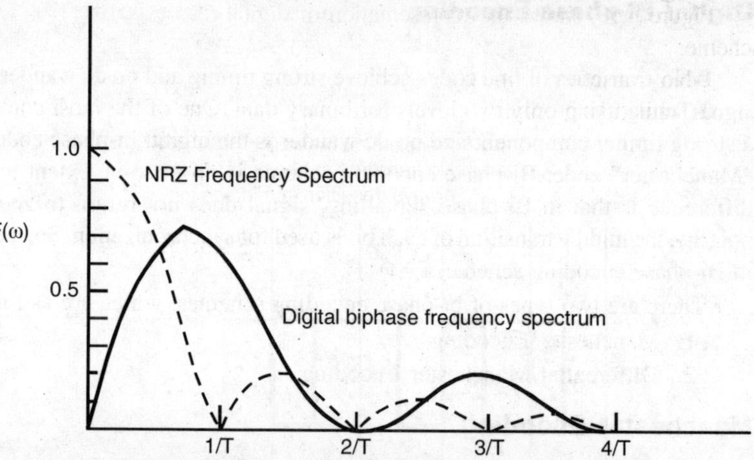

Fig. 5.49. Power spectral density of digital biphase.

Differential Manchester Encoding

Differential Manchester encoding continues the ideas of RZ and NRZ-I. We use the inversion at the middle of the each bit duration for synchronizing purpose but at the same time presence or absence of an additional transition at the beginning of the interval are used to determine the value of bit. One limitation of NRZ and digital biphase signals, as presented up to this point, is that the signal for a 1 is exactly the negative of a signal for a 0. On many transmission media, it may be impossible to determine an absolute polarity or an absolute phase reference.

Hence the decoder may decode all 1's as 0's and vice versa. A common remedy for this ambiguity is to use differential encoding that encodes a 1 as a change of state and encodes a 0 as no change in state. In this manner no absolute reference is necessary to decode the signal. The decoder merely detects the state of each signal interval and compares it to the state of the previous interval. If a change occurred, a 1 is decoded, otherwise a 0 is determined.

If the next bit is '0', transition takes place but if it is '1' than there is no transition of the level. That means to represent '0' bit we require two change in signal levels whereas to represent bit '1', there is only one change of signal level.

In this scheme also the transition at the middle of the bit interval used for synchronization.

As Manchester encoding scheme overcomes many problems of NRZ-L, similarly, Differential Manchester encoding scheme overcome the several problem associated with the NRZ-I.

Differential encoding and decoding do not change the encoded spectrum of purely random data (equally likely and uncorrelated 1's and 0's) but do double the error rate. If the detector makes an error in estimating the state of one interval, it also makes an error in the next interval. All signals of differentially encoded diphase retain a transition at the middle of an interval, but only the 0's have a transition at the beginning of an interval.

The advantages of Differential Manchester encoding schemes are as follows:

(1) Better synchronization

(2) No dc component because each bit has positive and negative voltage level continues.

(3) No base line wandering.

The only problem with this encoding scheme is the signal rate, which is double as compared to NRZ-L and NRZ-I and it required double Bandwidth than the NRZ-L and NRZ-I encoding scheme.

Figure 5.50, shows the representation of digital data (1001011101) in Differential Manchester encoding scheme.

1-No transition

0-Transition presents

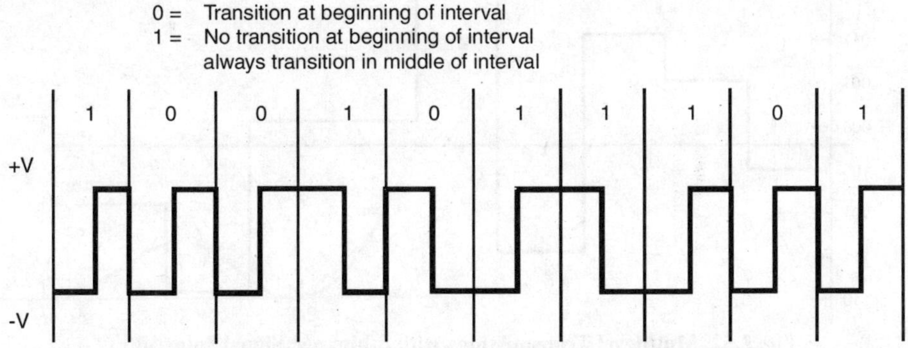

Fig. 5.50. Differential Manchester Encoding Scheme.

5.19.3.10. Coded Mark Inversion

A variety of line codes have evolved similar to the digital biphase code described previously. One of these is referred to as coded mark inversion (CMI) in ITU recommendations. CMI encodes 1's (marks) as an NRZ level opposite to the level of the previous one and 0's as a half-cycle square wave of one particular phase. Figure 5.51 shows a sample encoding for CMI. There is no dc energy in the signal and an abundance of signal transitions exist as in diphase. Furthermore, there is no ambiguity between 1's and 0's. Elimination of the ambiguity actually leads to a major drawback of CMI coding: Its error performance is 3 dB worse than diphase. CMI is the specified interface code for the fourth-level (E4) ITU multiplex signal at 139.264M bps. CMI is also used for electrical interfaces of SONET - STS-3c signal.

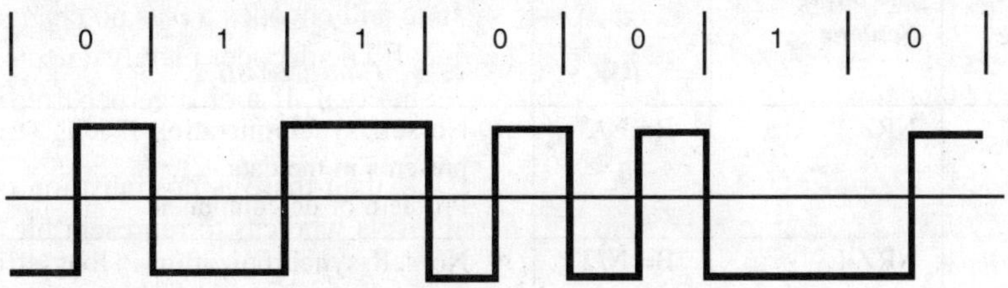

Fig. 5.51. Coded Mark Inversion.

5.19.3.11. Multilevel Signalling

In the line codes discussed so far, two-level (binary) signalling has been assumed. In applications where the bandwidth is limited but higher data rates are desired, the number of levels can be increased while maintaining the same signalling rate. The data rate R achieved by a multilevel system is given as

$$R = \log_2(L)\left[\frac{1}{T}\right] \qquad ...(2.1)$$

where, L = number of levels that can be freely chosen during each interval

T = signaling interval

The signalling rate $1/T$ is often referred to as the symbol rate and is measured in bauds. Within the data communications industry it is common practice to use baud as being synonymous with bit rate. Strictly speaking, however, the bit rate is only equal to the baud rate if binary signalling (1 bit per signal interval) is used. Figure 5.52 show an example of an eight-level transmission format that achieves 3 bits per signal interval (*i.e.*, 3 bits per baud).

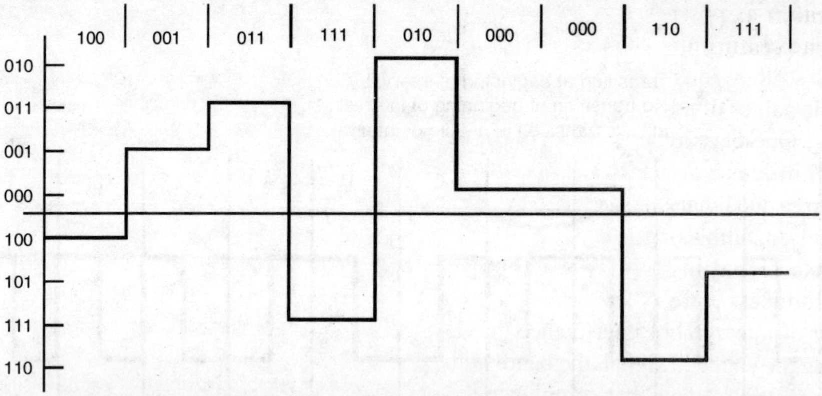

Fig. 5.52. Multilevel Transmissions with 3 bits per Signal Interval.

Multilevel transmission systems achieve greater data rates within a given bandwidth but require much greater signal-to-noise ratios for a given error rate. One aspect of wire line transmission that favors multilevel line coding is the lower baud rate for a given data rate, which in turn reduces the crosstalk. Hence, in crosstalk-limited systems the signal-to-noise ratio penalty of a multilevel line code is not as significant. The TlG T-carrier system developed by AT&T uses a four-level line code at the TlC baud rate (3.152 MHz) to double the capacity of a TlC system (from 48 channels to 96).

The error performance of CMI is 3 dB worse than diphase when bit-by-bit detection is used. The inefficiency rises because for one-half of an interval a 1 looks like a 0. Because CMI has redundancy, some of the inefficiency can be recovered with maximum likelihood (Viterbi) detection.

5.19.4. Comparison of Line Encoding Schemes

S. No.	*Types of Encoding Scheme*	*Encoding Scheme*	*BW*	*Characteristics*
1	Unipolar	NRZ	B= N/2	1. No-self synchronization if long strings of 1 and 0 presents in the data. 2. Problem of dc component.
2	Polar	NRZ-L	B= N/2	1. No self-synchronization if long strings of 1 and 0 presents in the data. 2. Problem of dc component.
3	Polar	NRZ-I	B= N/2	1. No self-synchronization if long strings of 0's presents in the data. 2. Problem of dc component.
4	Polar	RZ	B= N	1. No synchronization problem. 2. Require more bandwidth. 3. More complex. 4. No dc component problem.
5	Biphase	Manchester and Differential manchester	B= N	1. Self-synchronization. 2. Require more bandwidth. 3. More complex. 4. No dc component problem.
6	Bipolar	AMI and Pseudoternary	B= N/2	1. Synchronization problem. 2. Less dc component problem

5.19.5. Partial-Response Signalling

Conventional bandlimiting filters of a digital transmission system are designed to restrict the signal bandwidth as much as possible without spreading individual symbols so they interfere with sample values of adjacent intervals. One class of signalling techniques variously referred to as duo binary, correlative level encoding or partial-response signalling, purposely introduces a prescribed amount of intersymbol interference that is accounted for entire detection circuitry of the receivers. By over filtering the encoded signal, the bandwidth is reduced for a given signalling rate, but the overlapping pulses produce multiple levels that complicate the detection process and increase the signal power requirements for a given error rate.

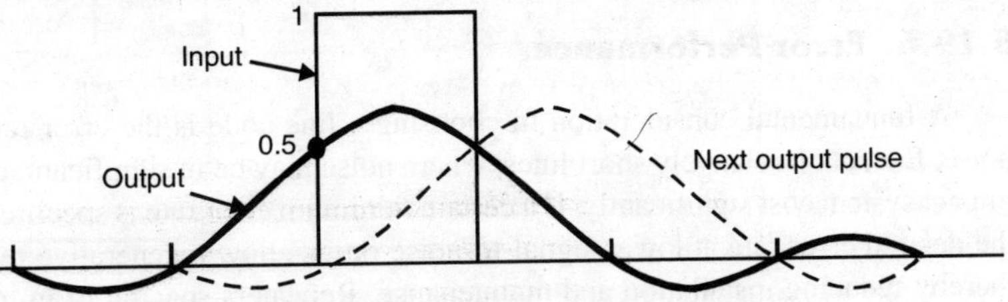

Fig. 5.53. Output Pulse of Partial-Response Channel.

Figure 5.53 shows the pulse response of a typical partial-response system. If the channel is excited by a pulse of duration T, channel filters limit the spectrum such that the main part of the pulse extends across three signal intervals and contributes equally to two sample times. The reason for the term partial response is now apparent: The output only responds to one-half the amplitude of the input.

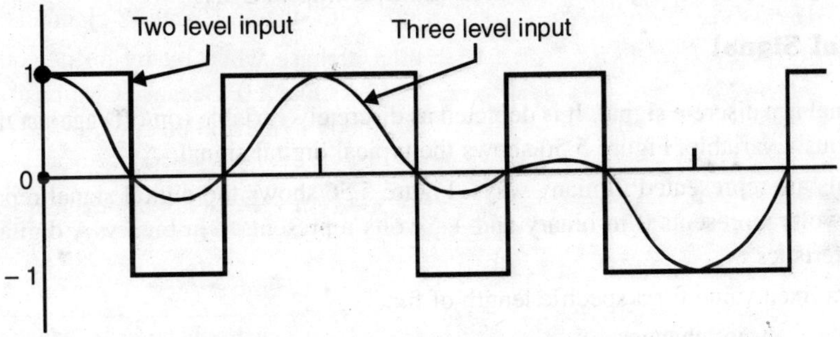

Fig. 5.54. Three Level Partial-response Inputs and Outputs.

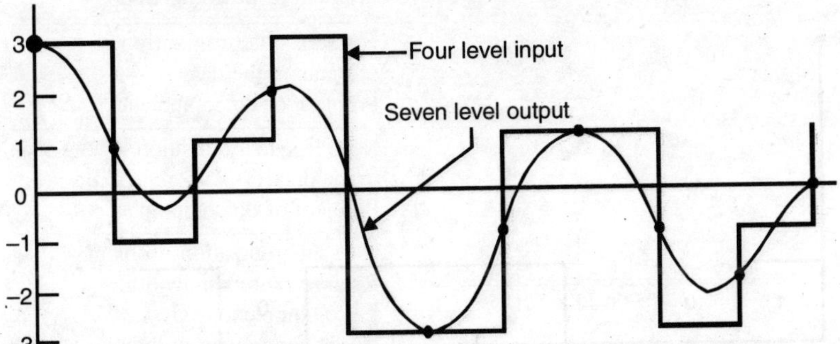

Fig. 5.55. Seven Level Partial-Response Inputs and Outputs.

If the input pulse of Fig. 5.54 is followed by another pulse of the same amplitude, the output will reach full amplitude by virtue of the overlap between pulses. However, if the next input pulse has a negative polarity, the overlap produces zero amplitude at the sample time. Thus, as shown in Figure 5.55, a partial-response system with two level inputs (+1, –1) produces an output with three levels (+1 ,0, –l). In similar fashion, Figure 5.54 shows a system with four input levels (+3, +1, –1, –3) and seven output levels (+ 3,+ 2,+ 1, 0, –1, –2, –3).

5.19.6. Error Performance

A fundamental consideration in choosing a line code is the error rate performance in the presence of noise. Except on relatively short lines, where noise may be in significant, error performance requirement can impact system cost significantly. If a certain minimum error rate is specified, those coding schemes providing the desired error rate at lower signal-to-noise ratios allow regenerative repeaters to be spaced farther apart, thereby reducing installation and maintenance. Repeaters spacing is an important economic factor in wire-line transmission, in optical fiber transmission, and on point-to-point radio links.

5.20. DIGITAL DATA TRANSMISSION

Digital transmission demand has continuously increased since its introduction in 1962 because of its high degree of accuracy. In this section, the advantages of digital transmission and the technical disadvantages of digital transmission are first discussed. There are two basic modes of digital transmission namely asynchronous and synchronous transmission. These two techniques are described in this section. We need to do encoding of digital signals before modulation, referred as line coding. We will be discussing the various techniques of line coding in this chapter and also, the multiplexing technique used for digital signal, referred as Time division multiplexing (TDM).

5.20.1. Digital Signal

A digital signal is a discrete signal. It is depicted as discretely variable (on/off) against the analog signal which is continuously variable. Figure 5.56 shows the typical digital signal.

Digital signals are represented in many ways. Figure 5.56 shows the digital signal represented by two voltages. Here 0 volts represents 0 in binary and + 5 volts represents 1 in binary. A digital signal has the following characteristics :

(*a*) It holds a fixed value for a specific length of time

(*b*) It has sharp, abrupt changes

(*c*) It allows a preset number of values.

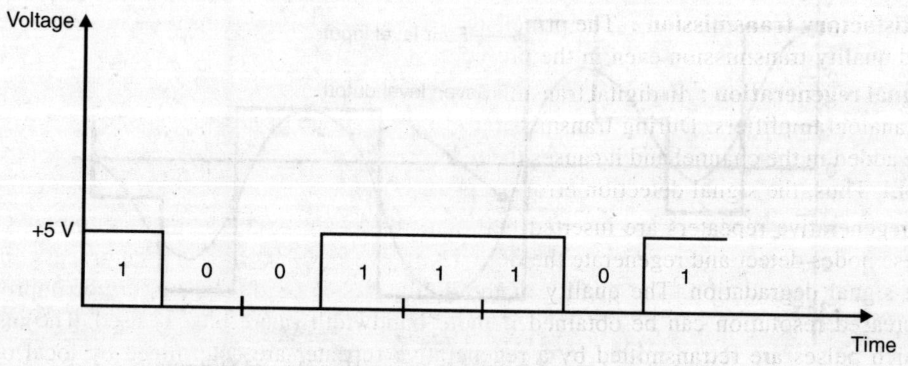

Fig. 5.56. The Digital Signal.

Each pulse (on/off) is known as binary digit (bit) the number of bits transmitted per second is the bit rates of the signal.

Digital signals are non-continuous, they change in individual steps. They consist of pulses or digits with discrete levels or values. The value of each pulse is constant, but there is an abrupt change from one digit to the next. Digital signals are may be defined as signals whose both dependent variable and independent variables are discrete in nature. A digital signal comprises pulses occurring at discrete interval of time.

However, generally, we use binary number system to represent a digital signal. In binary system, we use two digits 1 and 0, correspondingly two voltage levels which may be labelled high and low. Digital signals have two amplitude levels called nodes. The value of which are specified as one of two possibilities such as 1 or 0, HIGH or LOW, TRUE or FALSE and so on. In reality, the values are anywhere within specific ranges and we define values within a given range.

DO YOU KNOW▸

Digital signals are non-continuous, they change in individual steps.

5.20.2. Bit Interval and Bit Rate

The **bit interval** is the time required to send one bit. The **bit rate** is the number of bit intervals per second. This means that the bit rate is the number of bits sent in one second, usually expressed in Bits per Second (bps).

Figure 5.57 shows a digital signal, where 1 is denoted with +5 voltages and 0 is denoted wit h 0 voltage level.

Digital message constructed with a finite number of symbols. Telegraph and teleprinter signals are the example of digital signals. A Morse-coded telegraph message is a digital message constructed from a set of only two symbols, mark and space.

DO YOU KNOW▸

The bit interval is the time required to send one bit. The bit rate is the number of bit intervals per second.

5.20.3. Advantages of Digital Transmission

The following paragraphs emphasize the advantages of digital transmission. The digital transmission use gain and phase equalization to obtain negligible inter symbol interference.

1. **Satisfactory transmission :** The principal advantage of digital transmission is its satisfactory and quality transmission even in the presence of crosstalk and noise.
2. **Signal regeneration :** In digital transmission, it is possible to use regenerative repeaters instead of analog amplifiers. During transmission, small amounts of noise, interference or distortion are added in the channel and it causes the receiver unable to identify the 0*s* and 1*s* of the binary word. Thus, the signal detection error occurs and does not represent the original data exactly.

   If regenerative repeaters are inserted between transmitter and receiver spaced close enough, these nodes detect and regenerate the digital signals. Thus the ıegeneration process eliminates the signal degradation. The quality of the digital transmission can be further improved and increased resolution can be obtained if more bandwidth (mere bits) is used. The instants at which pulses are retransmitted by a regenerative repeater are determined by local oscillator synchronized to the digit rate, which must be extracted from the received wave form.

   Variations in the extracted frequency can cause a periodic variation of the times of the regenerated pulses, which is known as *jitter*. The variation in the times of the regenerated pulses due to changes in propagation time is known as *wander*.

3. **Lower Signal to line noise ratio :** In digital transmission, the noise generated in the terminal and the noise encountered in the transmission line can be separated. By the proper choice of the code (coder-decoder), the quantization noise generated by the terminal equipment can be minimized. The line noise has little effect on the message signal because of the regenerative repeater.

   The signal-to-line noise ratio requirement is lower in digital than in analog transmission, better utilization of noisy media is possible. Digital transmission links provide virtually error free performance at signal to noise ratios of 15.25 dB, depending on the type of line coding or modulation used.
4. Reduction of noise is possible in digital communication system.
5. Reduction of distortion and other impairments is possible in digital communication system.
6. Error detection and correction is possible in digital communication system that will reduce the probabilities of error by employing channel encoder.
7. **Ease of multiplexing :** Time division multiplexing is normally associated only with digital transmission links. The drawback of analog TDM lies in the vulnerability of narrow analog pulses to noise, distortion, crosstalk and intersymbol interference. TDM equipment costs are quite small in the case of digitized signal. Since digitization occurs only at the first level of the TDM hierarchy, high level TDM is more economical. The multiplexing operations of a digital transmission system can be easily integrated into the switching equipment.
8. Signal processing and storage is easier due to use of digital signal processing techniques and digital computer.
9. Security of data is possible due to encryption and decryption and spread spectrum coding provides antijamming facility.
10. Data compression and image processing is possible in digital communication system.
11. Large amount of noise interference may be tolerated.
12. Use of VLSI technology may be possible.
13. Handling of multimedia information such as integration of voice, video and data is easier.
14. There are many signal processing algorithms can be used to increase transmission rate and reliability of digital communication system.
15. Digital circuits are more reliable and can be produced at lower cost than analog circuits, such as DSP, FPGA and ASIC.
16. Digital signal is less subject to distortion and interference than analog signal.
17. The digital signal is more easer to "reborn" or regenerated, compared to analog signals because :
    - Digital signal has only two states (0 or 1).
    - On the contrary, analog signal are not two-state signal.
    - It can take an infinite variety of amplitude.

Other Advantages

Analog transmission system have difficult environments and requires special attention for processing control signals like on-hook/off-hook, coin deposits, clean forward etc. The digital system allows control information independently at less cost. The use of digital subscriber lines (DSL) reduces the signalling costs relative to analog subscriber lines.

The digital transmission uses larger, modern and latest technology. Hence high performance and low cost can be achieved in digital implementations. The time division multiplexer, digital signal processors, basic logic circuits, multiplexing circuits, computer systems are the some of the modern equipments used in digital transmission.

The integration of transmission and switching, performance monitor ability, ease of encryption is the other advantages in digital transmission.

Disadvantages of Digital Transmission

1. Digital communication requires more bandwidth than analog communication.
2. It needs synchronization.
3. Multiplexing difficulties

1. Greater bandwidth : A major disadvantage of digital transmission is the requirement of much greater bandwidth. This is due to the fact that each sample is represented by an 8 bit codeword and each bit is transmitted as a separate discrete pulse. For example, T1 system requires approximately eight times as mush bandwidth as do 24 analog voice channels.

2. Need of synchronization. For efficient reception of digital signal, a timing reference, generally a 'clock' is needed. The clock specifies when to sample the incoming signal to decide which data value was transmitted. The sample clock must be synchronized to the pulse arrival times. For networks when a number of digital transmission links and switches are interconnected, certain synchronization procedures must be established to maintain internal synchronization.

3. Multiplexing difficulties. Since the time of arrival of data is dependent on the distance of travel in TDM, it is not nearly as amenable to applications involving distributed sources and destinations. Also, TDM requires time slot recognition logic as each time division source must duplicate the synchronization. Thus TDM has been primarily used in applications, where all the information sources are centrally located and a single multiplexer controls the occurrence and assignment of time slots.

Other major disadvantage is the incompatibilities with analog facilities.

5.20.4. Pulse Transmission

In digital transmission system, signals propagate as a discrete voltage pulses (positive voltage represents binary 1, and negative voltage represents binary 0). The digital signals are measured in bits per second (bps). In data communications, analog signals are used to transmit information over the telephone system or over radio communication systems.

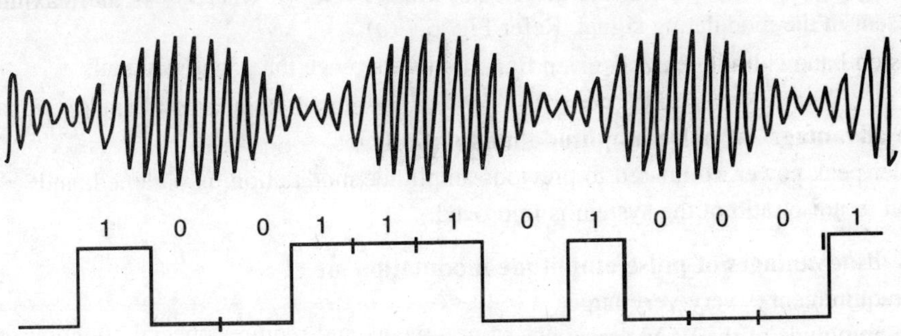

Fig. 5.57. Voice Signal and Digital Signals.

Digital transmission is performed by discrete bundles of energy called pulses. The foundation for this concept arises out of the work done in sampling theory by Henry Nyquist of the Bell Telephone Laboratories. Briefly, Nyquist explained that if a signal, $f(t)$, was band-limited to B Hertz, then the signal could only change at a maximum rate of B Hertz.

He further showed that if a sample of the band-limited signal is taken every 1/2 seconds, no information is lost. The minimum sampling rate of 2 B samples per second is called the **Nyquist sampling rate** and 1/(2 B) is called the Nyquist sampling interval.

A MODEM (modulator/demodulator) converts digital data to analog signals and analog signals can be converted to digital information. This process involves sampling, quantizing. This process is called **digitizing.** Now-a-days, the transmission between central exchanges and long distance sites are done by digitizing voice communication. The analog transmission takes place only between the local exchange or end office and homes. In Figure 5.58 digital to analog to digital conversion processing is shown as block diagram.

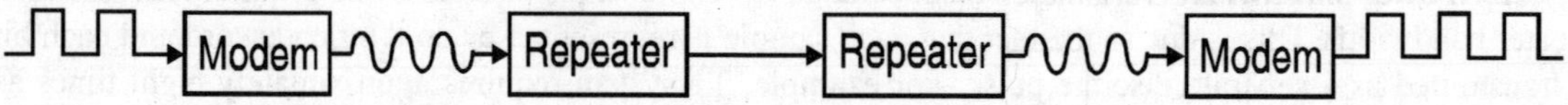

Fig. 5.58. Digital to Analog to Digital Conversion.

Voice converted to digital requires a 64 k bits/sec channel, which happens to be worldwide standard called DS_0 (digital signal, level zero) for transmitting voice calls.

5.20.5. Types of Pulse Modulation System

1. Pulse Amplitude Modulation (PAM)
2. Pulse Width or Duration Modulation (PWM)
3. Pulse Position Modulation (PPM)
4. Pulse Code Modulation (PCM) (Digital in nature)

Advantages of Pulse Modulation System

In pulse modulation system, amplitude limiting techniques may be employed to minimize noise effects, the transmission can be made independent of fading and other difficulties, multiple signals may be transmitted on a single carrier, and Energy is radiated only in short bursts and therefore the consumption of power is low.

Pulse Amplitude Modulation (PAM)

In **Pulse Amplitude Modulation** (PAM), relative amplitudes of successive pulses serve as a measure of signal amplitudes at the corresponding sampling instants. Width and position of the pulse remain constant (fix). Modulating signal is sampled at the basic rate, usually ($1/2f_m$); where f_m is the maximum value of frequency present in the modulating signal. Refer Fig. 6.4 (*a*).

Transmission bandwidth in PAM is given by

$$BW >> f_m \qquad ...(5.1)$$

The main **advantages of pulse amplitude modulation** are:

1. Greater peak power, compared to previous amplitude modulation, is obtained, and
2. Signal to noise ratio of the system is improved.

The main **disadvantages of pulse amplitude modulation** are :

1. BW requirement is very-very large.
2. Since amplitude of the PAM varies therefore noise is more.
3. Also varies the peak power required by the transmitter with modulating signal.

Pulse Duration or Width Modulation (PWM)

The durations or widths of the pulses are varied (from the nominal value) according to the samples. The width of the pulse is varied (around a fixed value) in accordance with the modulating signal whereas amplitude and position remain fixed.

Advantages of PWM

1. Since amplitude is constant, noise is less than PAM.
2. Noise and signal separation is very easy.
3. It does not require the synchronization between transmitter and receiver.

Disadvantages of PWM

1. Large bandwidth needed.
2. Power contents are variables.

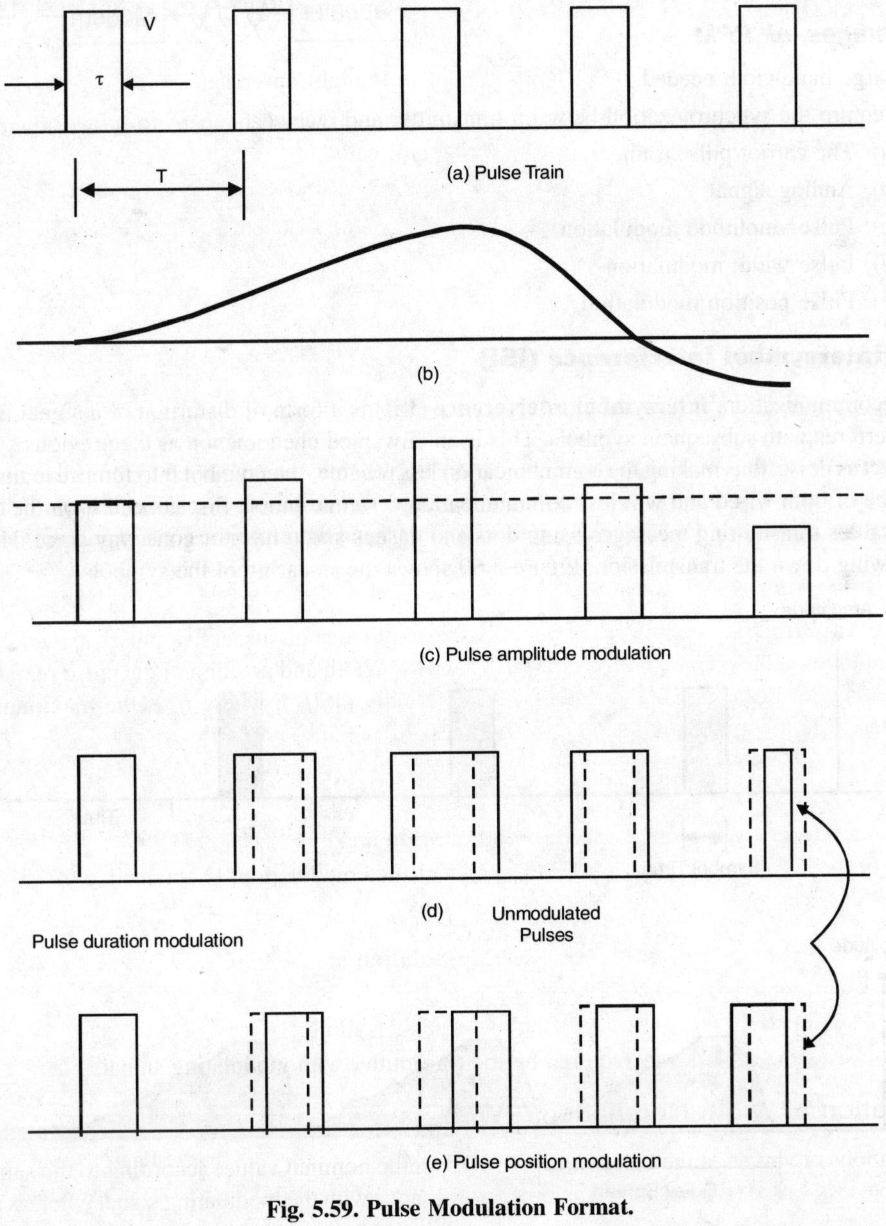

Fig. 5.59. Pulse Modulation Format.

Pulse Position Modulation (PPM)

In **Pulse Position Modulation (PPM),** the amplitude and width of the pulses are kept constant, while the position of each pulse, with reference to the position of a reference pulse, is changed according to the instantaneous sampled value of modulating signal.

Advantages of PPM

1. Since amplitude is constant, noise is less than PAM.
2. Noise and signal separation is very easy.
3. Transmission power for each pulse is same.

Disadvantages of PPM

1. Large bandwidth needed.

Require the synchronization between transmitter and receiver.

(*a*) The carrier pulse train

(*b*) Analog signal

(*c*) Pulse amplitude modulation

(*d*) Pulse width modulation

(*e*) Pulse position modulation

5.20.6. Intersymbol Interference (ISI)

In telecommunication, **intersymbol interference** (ISI) is a form of distortion of a signal in which one symbol interferes with subsequent symbols. This is an unwanted phenomenon as the previous symbols have similar effect as noise, thus making the communication less reliable. Intersymbol interference is an unavoidable consequence of both wired and wireless communication systems. Morse first noticed it on the transatlantic telegraph cables transmitting messages using dots and dashes and it has not gone way since. He handled it by just slowing down the transmission. Figure 5.60 shows the smearing of the symbols.

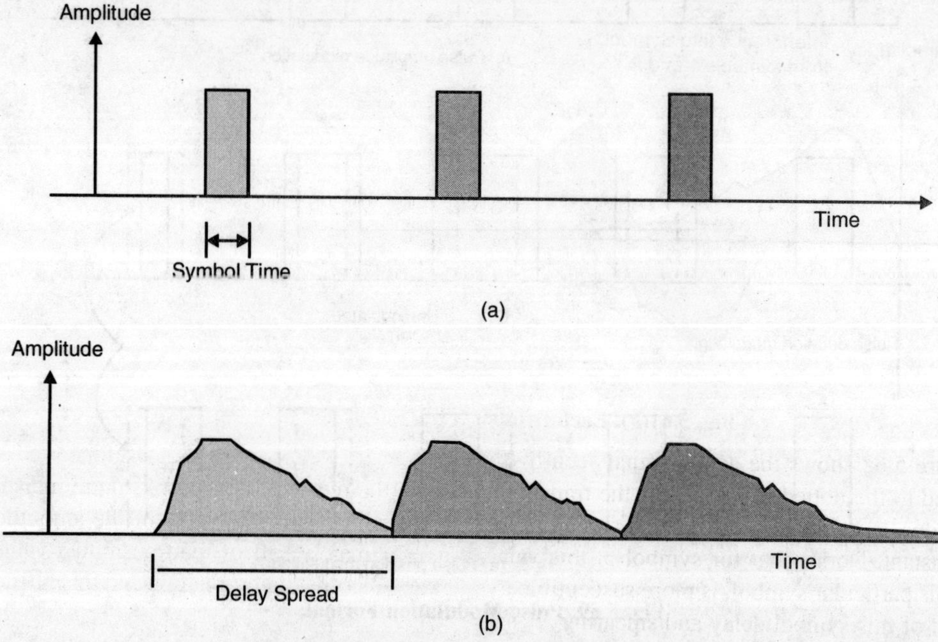

Fig. 5.60. The Smearing of the Symbols.

During these early attempts at transmission, it was noticed is that the received signals tended to get elongated and smeared into each other. A short pulse to represent a dot was received as a much-smeared version of the same thing. The problem appeared to be related to the properties of the medium used and the distance of signal travel. To counter this undesired effect, intermediate repeating stations were established and ways had to be devised to reduce this smearing.

How does this actually affect us now? Figure 5.61 shows a data sequence, 10110, which we wish to send. This sequence is in form of square pulses. Square pulses are nice as an abstraction but in practice they are difficult to create and also require too much bandwidth. So we shape them as shown in dotted line. Advantage of (an arbitrary) shaping at this point is that it reduces bandwidth requirements and can actually be created by the hardware.

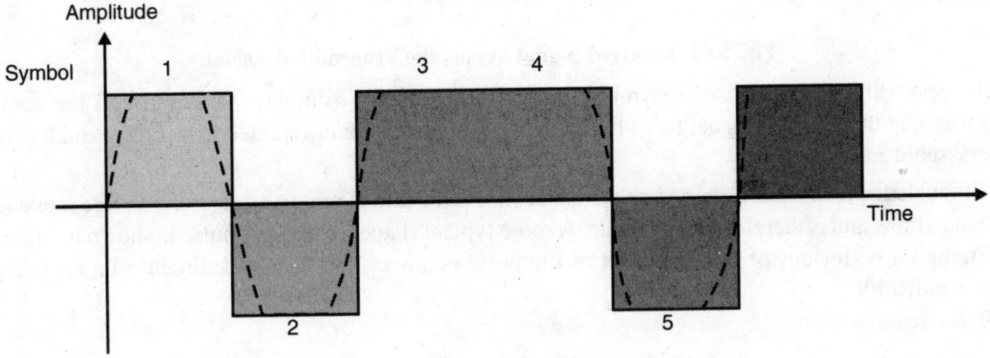

Fig. 5.61(*a*)

Figure 5.61 sequence 101101 to be sent, the dotted line is the shape that is actually sent.

Figure 5.61(*a*) shows each symbol as it is received. We can observe what the transmission medium creates a tail of energy that lasts much longer than intended. The energy from symbols 1 and 2 goes all the way into symbol 3. Each symbol interferes with one or more of the subsequent symbols. The circled areas show areas of interference.

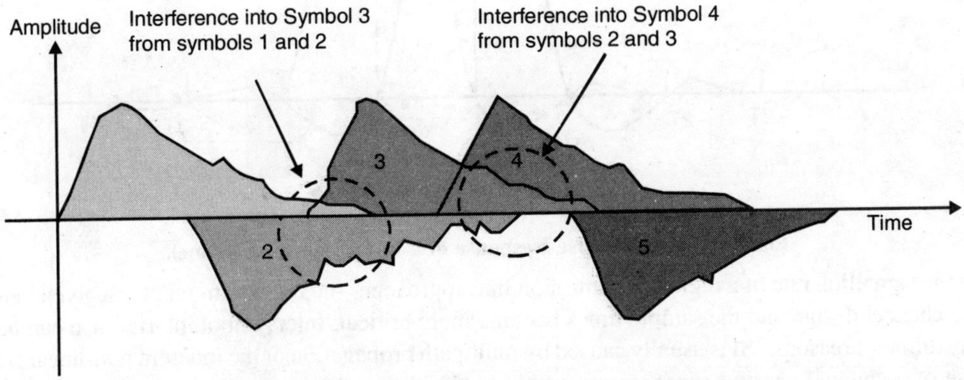

Fig. 5.61(*b*). Each Symbol is Spread by the Medium.

Figure 5.62 shows the actual signal seen by the receiver. It is the sum of all these distorted symbols. Compared to the dotted line that was the transmitted signal, the received signal looks quite indistinct. The receiver does not actually receive this signal; it sees only the little dots, the value of the amplitude at the timing instant. Notice that for symbol 3, this value is approximately half of the transmitted value, which makes this particular symbol, is more susceptible to noise and incorrect interpretation and this phenomena is the result of this symbol delay and smearing.

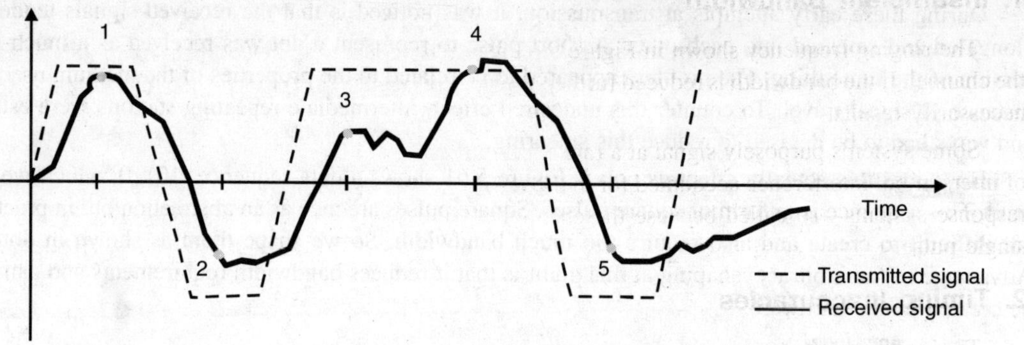

Fig. 5.62. Received Signal Versus the Transmitted Signal.

This spreading and smearing of symbols such that the energy from one symbol affects the next one in such a way that the received signal has a higher probability of being interpreted incorrectly and it is referred as intersymbol interference.

Although discrete, square-shaped pulses are easiest to visualize, preservation of the square shape requires wide bandwidths and is therefore undesirable. A more typical shape for a single pulse is shown in Figure 5.63. The ringing on both sides of the main part of the pulse is a necessary accompaniment to a channel with a limited bandwidth.

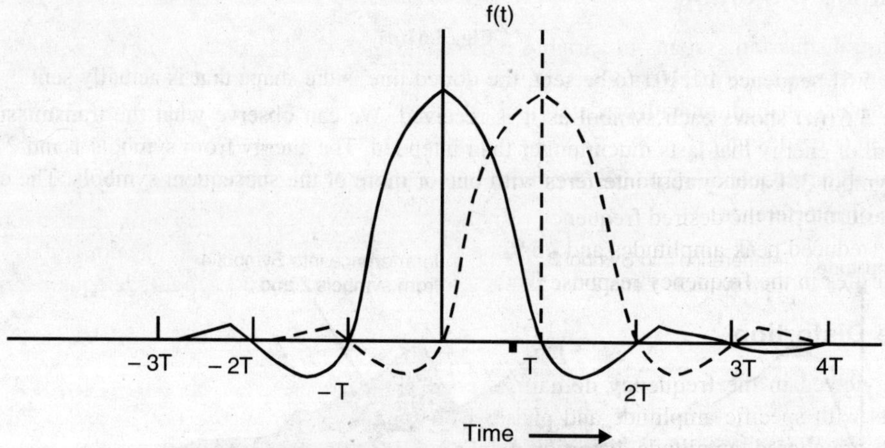

Fig. 5.63. Typical Pulse Response of a Band Limited Channel.

As the signalling rate of a digital transmission line approaches the maximum rate for a given bandwidth, both the channel design and the sample times become more critical. Intersymbol interference can be caused by many different reasons. ISI is usually caused by multipath propagation or the inherent non-linear frequency response of a channel causing successive symbols to "blur" together. It can be caused by filtering effects from hardware or frequency selective fading, from non-linearities and from charging effects.

The main causes of intersymbol interference are :

1. Insufficient Bandwidth
2. Timing Inaccuracies
3. Amplitude Distortion
4. Phase Distortion

1. Insufficient Bandwidth

The ringing frequency shown in Figure 5.63 is exactly equal to the theoretical minimum bandwidth of the channel. If the bandwidth is reduced further, the ringing frequency is reduced and intersymbol interference necessarily results.

Some systems purposely signal at a rate exceeding the Nyquist rate, but do so with prescribed amounts of intersymbol interference accounted for in the receiver. These systems are commonly referred to as partial-response systems—so called because the channel does not fully respond to an input during the time of a single pulse.

2. Timing Inaccuracies

Timing inaccuracies occurring in either the transmitter or the receiver produce intersymbol interference. In the transmitter, timing inaccuracies cause intersymbol interference if the rate of transmission does not conform to the ringing frequency designed into the channel. Timing inaccuracies of this type are insignificant unless extremely sharp filter cutoffs are used while signalling at the Nyquist rate.

Since timing in the receiver is derived from noisy and possibly distorted receive signals, inaccurate sample timing is more likely than inaccurate transmitter timing.

Sensitivity to timing errors is small if the transmission rate is well below the Nyquist rate (e.g., if the transmission rate is equal to or less than the channel bandwidth, as opposed to being equal to the theoretical maximum rate of twice the bandwidth).

3. Amplitude Distortion

Digital transmission systems in variably require filters to band limit transmit spectrums and to reject noise and interference in receivers. Overall, the filters are designed to produce a specific pulse response. When a transmission medium with predetermined characteristics is used, these characteristics can be included in the overall filter design.

However, the frequency response of the channel cannot always be predicted adequately.

A departure from the desired frequency response is referred to as amplitude distortion and causes pulse distortions (reduced peak amplitudes and improper ringing frequencies) in the time domain. Compensation for irregularities in the frequency response of the channel is referred to as amplitude equalization.

4. Phase Distortion

When viewed in the frequency domain, a pulse is represented as the superposition of frequency components with specific amplitude and phase relationships. If the relative amplitudes of the frequency components are altered, amplitude distortion results as above. If the phase relationships of the components are altered, phase distortion occurs. Basically, phase distortion results when the frequency components of a signal experience differing amounts of delay in the transmission link. Compensation of phase distortion is referred to as phase equalization.

The presence of ISI in the system introduces errors in the decision device at the receiver output. Therefore, in the design of the transmitting and receiving filters, the objective is to minimize the effects of ISI, and thereby deliver the digital data to its destination with the smallest error rate possible. Ways to fight intersymbol interference include adaptive equalization and error correcting codes.

5.20.7. Modes of Digital Transmission

The signal is normally transmitted in multiples of a fixed length unit, typically of 8 bits. The bit is represented as a specific voltage level relative to a reference level. A high signal relative to a reference indicates the transmission of a binary 1 while a low signal represents a binary 0 in positive logic.

For the receiving device to decode and interpret this bit pattern correctly, it must be able to determine the following :

(*i*) The reception of digital data involves sampling the incoming signal once per bit time to determine the binary value.

(*ii*) To sample the incoming bits properly, the receiving system should know the arrival time and duration of each bit that it receives.

(*iii*) The start and end of character or byte.

(*iii*) The start and end of each completed message.

The transmission of binary data across a link can be accomplished either in parallel or serial mode. The advantage of serial over parallel transmission is the use of common channel and thus the reduced cost. There are two basic modes of digital transmission for establishing a time base (sample clock) for effective reception of the signal. They are asynchronous and synchronous transmission. Both techniques involve fundamentally different techniques. But these two approaches are common for achieving the desired synchronization.

Asynchronous Transmission

Asynchronous transmission involves separate transmission of groups of bits or characters. In each group, a specific predefined time interval is used for each discrete signal. The transmission times of the group are independent of each other. Hence a sample clock is reestablished for reception of each group.

To alert the receiver for the arrival of a new group, a start bit usually a '0' is added to the beginning of each byte. To indicate the completion of the group, a stop bit usually a '1' is appended to the end of the byte. Thus the size of the group is atleast 10 bits. That is 8 bit contains information, one for start bit and one for stop bit. The transition of each byte is separated by a gap of varying duration. The gap is usually identified by an idle channel or by a stream of additional stop bits. The Fig. 5.64 shows the typical character format.

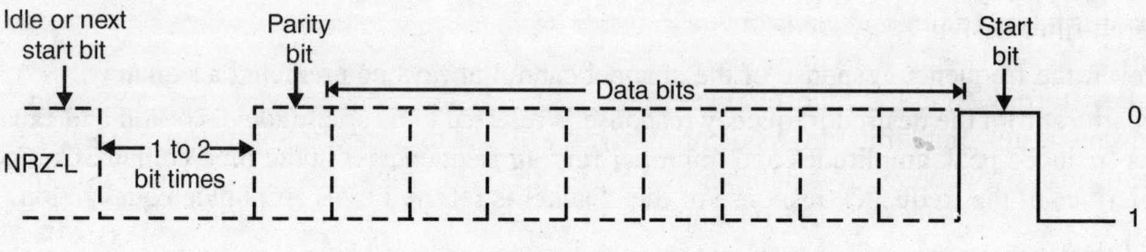

Fig. 5.64. The Typical Character Format.

For error detection, a party bit (odd or even) is added MSB position. Figure 6.10 shows schematic illustration of asynchronous transmission.

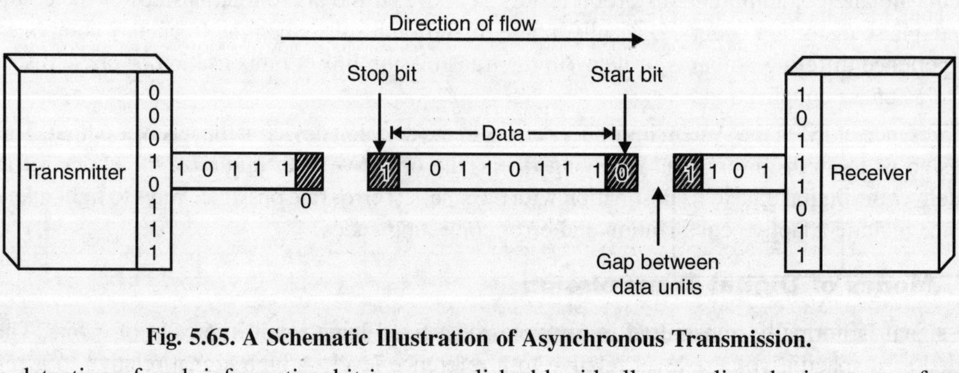

Fig. 5.65. A Schematic Illustration of Asynchronous Transmission.

The detection of each information bit is accomplished by ideally sampling the input waveform at the middle of each signal interval. Asynchronous transmission automatically provides character framing and is inherently flexible in the range of average rates that can be accommodated. If a steady stream of character is sent, the interval between two characters is uniform and equal to the stop element. Asynchronous transmission has been used in voiced data sets (modems) for transmission rates up to 1200 bps.

Disadvantages of Asynchronous Transmission

1. Since the sample time for each information bit is derived from a single start bit, asynchronous system do not perform well in high noise environments. This problem can be eliminated by adding more than one start bit, but it causes the system more complex.
2. **Timing error :** The asynchronous implies a free running clock in the receiver. So any offset in the clock frequency of the receiver causes timing error. If the receiver is 5 percent slower or faster than the transmitter, the sampling of the eight information bit will be displaced by 45 percent.
3. **Framing error :** By timing error, the bit count may be out of alignment. It bit 7 is a 1 and bit 8 is a 0, bit 8 could be mistaken for a start bit. This condition is termed as a framing error, as the character plus starts and stop bits sometimes referred to as a frame. False appearance of start bit due to noise also causes framing error.
4. The major drawback of asynchronous transmission is its poor performance in terms error rates on noisy lines.
5. Asynchronous transmission is simple and cheap but requires an overhead of two to three bits per character. The percentage overhead could be reduced by sending larger blocks of bits between the start and stop bits.
6. The addition of stop and start bits and the insertion gaps into the bit stream made asynchronous transmission slower.

Synchronous Transmission

To achieve greater efficiency, synchronous transmission is used. The main advantage of synchronous transmission is speed. Hence it is more useful in high speed applications like data transfer between computers. For digital signal, synchronous transmission can be accomplished with Manchester or differential Manchester encoding. In Asynchronous transmission, digital signals are sent as separate transmission of groups whereas in synchronous transmission, the digital signals are sent continuously at constant rate. Hence the receiving terminal must establish and maintain a sample clock that is synchronized to the incoming data for an indefinite period of time.

Figure 5.66 shows a typical synchronous frame format. The frame format starts with a preamble called flag to determine the beginning of the block of data. A postamble bit is added at the end of the block of data to identify the end. Other bits are added to the block that conveys control information used in data link procedures. Thus the data block plus, preamble, postamble and control information is called a frame.

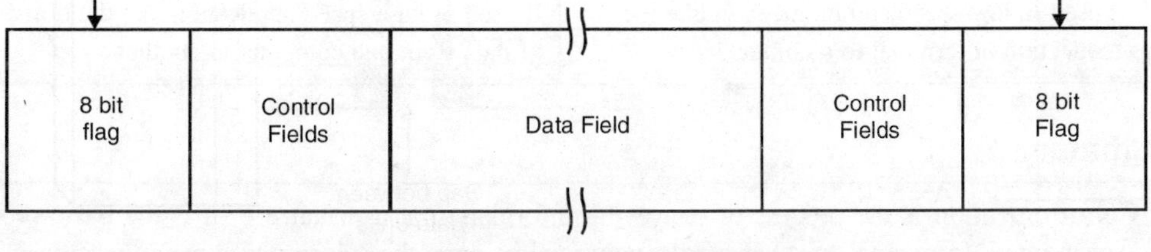

Fig. 5.66. Synchronous Frame Format.

In synchronous transmission, data are sent one bit after another without start/stop bits or gaps. The receiver should be capable enough to separate the bit stream into bytes for decoding purposes. Figure 6.12 shows the schematic of synchronous transmission.

In Fig. 5.67, the data stream is shown as 8 bit block for explanation purpose. But in reality, no such portions. When the data stream reaches the receiver, it counts the bits and groups them in eight bit units. If the sender wishes to send data in separate groups, the gap between groups must be filled with a special sequence of 0 and 1's that means idle.

As there is no start and stop bit to identify the blocks and also the grouping of bits should be performed by the receiver, it must be capable enough to keep an accurate count and error free reception.

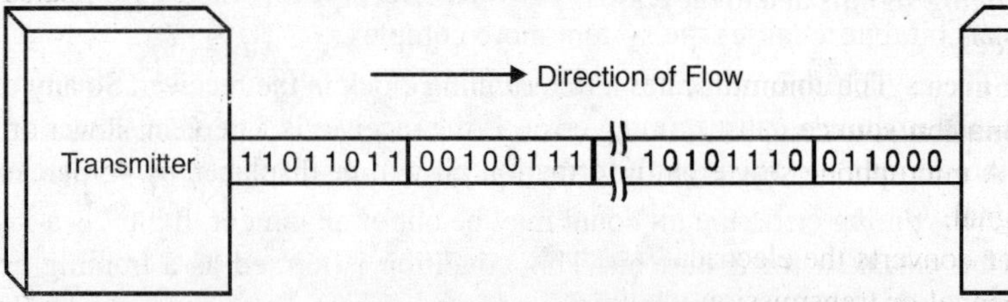

Fig. 5.67. Synchronous Transmission.

Comparisons between synchronous and asynchronous transmission techniques are as follows :

S. No.	Asynchronous Transmission	Synchronous Transmission
1.	In asynchronous transmission, information is transmitted character by character.	In synchronous transmission, blocks of about 100 characters are transmitted at high-speed on the transmission line.
2.	At the beginning of a character, a start bit is sent. The nature of the start signal is standardized and is 'understood' by the receiving end equipment which prepares to receive the coded character. The start signal is followed by the bits of coded character. If ASCII code is used, the seven coded bits and a parity bit are sent. Following the character code is stop signal.	A distinctive synchronization pattern is send at the beginning of the block. The synchronization pattern is followed by codes to identify sender and receiver and the message character. The message is terminated by an end of message character followed by check character to aid detection of any transmission error.
3.	It is cheap	It is expensive.
4.	It takes more time in data transmission.	It takes less time.
5.	It does not require any local storage at the terminal end.	It needs for local buffer storages at the two ends of the line to assemble blocks.
6.	It does not require any synchronized clocks.	It requires accurately synchronized clocks at both ends.
7.	The transmission line is idle during the time intervals between transmitting characters.	Synchronous transmission is efficient utilization of the transmission line.
8.	It used in low-speed communication like the connection of terminal to a computer.	It used in high speed application like the transmission of data from one computer to another.

Summary

- **Communication** is the process of conveying information at a distance or it is the basic process of exchanging information. In other words, communication is the science of transmitting, receiving and processing of information.
- The electronic equipments or devices which are used for the purpose of communication are known as communication equipments and integrating of different communication equipments for the purpose of communication form a **communication system**.

- Electrical communication system and it has three main basic components,
  1. Transmitter
  2. Communication channel or transmission media
  3. Receiver
- An electrical communication system in rather broad terms based on the implicit assumption that the message signal is a continuous time-varying waveform. We refer to such continuous-time signal waveforms as ***analog signals*** and to the corresponding information sources that produce such signals as ***analog sources***.
- **Information Source :** The communication system exists to convey a message. This message comes from the **information source**,
- **Transducer :** A microphone serves as the transducer that converts an acoustic speech signal into an electrical signal,
- The **transmitter** converts the electrical signal into a form that is suitable for transmission through the physical channel or transmission medium.
- The **communications channel** is the physical medium that is used to send the signal from the transmitter to the receiver. There are two types of channels, namely point to point channel and broadcast channels.
- **Noise** may be defined as unwanted signals which tend to interfere with the required signals.
- The function of the **receiver** is to recover the message signal contained in the received signal.
- **Destination** is the final stage which is used to convert an electrical signal in to its original form by using the output transducer.
- The communication system which uses the digital signals as input is known as **digital communication system**.
- The information source that generates the data is called a ***discrete (digital) source***.
- **Information Source :** The source output may be either an analog signal, such as audio or video signal, or a digital signal, such as the output of a computer which is discrete in time.
- The A/D produces analog-to-digital conversion (for analog source) including sampling and quantizing.
- **Sampling :** Pulse amplitude modulation (PAM) : Converting the continuous input to discrete sequence.
- **Quantization :** Pulse code modulation (PCM) : Encoding each quantized sample into a *digital word* including uniform and nonuniform quantization.
- The *source alphabet* will have letters, digits, and special characters available from the information source.
- The *symbol rate* is the rate at which the information sources the symbols/characters and its unit is symbol/sec.
- The process of efficiently converting the output of either an analog or a digital source into a sequence of binary digits is called ***source encoding* or *data compression.***
- There are few important parameters used for source encoder as follows :
- **Block size :** Block size gives the number of distinct code-words that the encoder can represent and depends upon the number of bits in the codeword. As an example, the block size of 2^5 *i.e.*, 32 code-words.
- **Code-word length :** It is number of bits used to represent the codeword. As example, if 5 bits are assigned to each code-word, then code-word length will be 5 bits.
- **Average data rate :** It is the output bits per second from the source encoder.
- The **source decoder** at the receiver converts the binary output of the channel decoder in to a symbol sequence.

- The purpose of the channel encoder is to introduce, in a controlled manner, some redundancy in the binary information sequence which can be used at the receiver to overcome the effects of noise and interference encountered in the transmission of the signal through the channel.
- Analog modulation methods are employed in analog communication system such as amplitude, frequency and phase modulation.
- In digital communication system, digital modulation techniques are used such as amplitude shift keying (ASK), frequency shift keying (FSK), phase shift keying (PSK) and minimum shift keying (MSK).
- Analog communication involves transmission of a continuous of values over a defined interval of time. Processing of analog information is generally not easy and it is more prominent to the noise. Further, there is no facility of error detection/correction in the analog system.
- The primary limitations of information transmission by electrical means are bandwidth and noise.
  1. Bandwidth Limitation
  2. Noise Limitation
- The **bandwidth** can be defined as the frequency range or the band of frequency needed for a particular transmission.
- Noise is another limitation of any communication system. It can be defined as an unwanted form of energy which tends to interfere with the transmission and reception of the desired signal in a communication system.
- Digital modulation is sometimes called **digital radio** because digitally modulated signals can be propagated through earth's atmosphere and used in wireless communication systems.
- **Data Communications** is the transfer of data or information between a source and a receiver. The source transmits the data and the receiver receives it.
- Data Communication is interested in the transfer of data, the method of transfer and the preservation of the data during the transfer process.
- In the data communication world, **data** means the information stored in digital form. A single unit of data is **datum** and the data is plural. *Data Communications* is the process of transferring the binary (digital) information between two or more points.
- Though data communication may seem a relatively new innovation, the history of data connections dates to the early 19th century. Through an array of technological developments that enjoyed commercial, government and military contributions, data communication evolved from simple wired connections to a robust global exchange of information.
- The importance of standards in the field of communication cannot be overstressed. Standards enable equipment from different manufacturer and with different operating characteristics to become components of the same network. Standards also enable different networks in different geographical locations (e.g., different countries and continents) to be interconnected.
- The basic purpose of a **data communications circuit** is to transfer digital information from one place to another using electronic circuit.
- The data communications needs to be translated into a form that the medium can understand. Data communication is further broken down into the ***Data Communication Model*** which consists of 5 parts
- Digital signals are non-continuous, they change in individual steps. They consist of pulses or digits with discrete levels or values.
- The **bit interval** is the time required to send one bit. The **bit rate** is the number of bit intervals per second.
- Nyquist proved that any arbitrary signal has been passed through a low pass filter of bandwidth B, the filtered signal can be reconstructed completely by making only 2B (exact) sample per second.
- According to **Nyquist's theorem**, if the signal consists of V discrete levels, the maximum data rate can be given as

$$\text{Maximum data rate} = 2B \log_2 V \text{ bits/sec}$$

- According to **Shannon's theorem**, that the maximum data rate of a noisy channel where bandwidth is B Hz, and has SNR is S/N, is given by.

$$\text{Maximum number of bits/sec} = B \log_2 (1 + S/N)$$

- Any communications channel has a direction associated with it, communication modes or connections between devices may be classified into three categories depending upon the direction of the flow of information:
  1. Simplex
  2. Half-duplex
  3. Full-duplex
  4. Full/Full Duplex
- **Simplex connections** are useful in situations where a device only receives or only sends data.
- **Half-Duplex :** This is a bidirectional connection, with the restriction that data can travel in one direction at a time.
- **Full-duplex** is a bidirectional connection in which data can travel in both directions at once.
- In the **full/ full duplex** mode transmission is possible in both the directions simultaneously but not between the same two stations.
- If more than 1 bit of information is transmitted over the data transmission medium at a time then it is considered a **Parallel Communication**. Byte-serial transmission conveys eight bits at a time through eight parallel channels.
- If only 1 bit of information can be transmitted over the data transmission medium at a time then it is considered a **Serial Communication**.
- Bit-serial transmission is normally just called **serial transmission** and is the chosen communications method in many computer peripherals.
- Serial data transfer is used for high transfer rates over long distances.
- At shorter distances, typically less than 15 feet, parallel data transfers are used.
- Serial data communication only transfers one bit at a time and does not have the timing problems (race conditions) of parallel data communications.
- When the timing information is known, the receiver is said to be **synchronized** with the transmitter, and accurate data transfer becomes possible.
- There are two basic methods of synchronization: synchronous transmission and asynchronous transmission.
- In **synchronous transmission**, a clock signal is used as a common source of reference by both the transmitter and the receiver.
- In **asynchronous transmission**, the beginning and end of each byte of data is marked by start and stop bits.
- In **asynchronous systems**, a separate timing channel is not used.
- Two distinct techniques are used in data communications to transfer data.
- Protocol may be mainly classified as either **connection-oriented** method or the **connectionless** method.
- With a **connection oriented protocol,** a logical connection is established between the end points (for example virtual circuit) before to the transmission of data. It requires a session connection be established before any data can be sent.
- Connection-oriented service involves three phases: connection establishment, data transfer, and connection termination.
- Transmission Control Protocol (TCP) is a connection-oriented protocol.

- **A connectionless service does** not require a session connection between sender and receiver.
- UDP (User Datagram Protocol) is a connectionless network protocol.
- **Line encoding** is the method used to represent the digital information on the media. A pattern, that uses either voltage or current, is used to represent the 1s and 0s of the digital signal on the transmission link.
- A signal element carries data element and it is the shortest unit (time wire) of a digital signal.
- Data element is the smallest entity that can represent a piece of information known a bit.
- "Data elements are what we need to send;
  Signal elements are what we can send.
  Data elements are being carried and signal
  Element are the carriers"
- The **data rate** is defined the number of data element (or bits) sent in one second. Its unit is bps (Bits per second).
- The **signal rate** is the number of signal element sent in one second. Its unit is baud.
- The relationship between data rate and signal rate is given by, $S = C \times N \times 1/r$.
- Baud rate determine the bandwidth. Although the actual bandwidth of a digital signal is infinite but the effective bandwidth is finite.
- The minimum bandwidth can be determined, $B_{min} = C \times N \times 1/r$.
- According to Nyquist theorem, maximum data of a channel is given as $N_{max} = 2 \times B \times \log_2 (L)$.
- There are some lines encoding schemes leaves residual **direct current (dc) components.** A dc signal has zero frequency so it cannot travel through a media and it is very difficult to handle dc components by channel.
- A long string of either logic 1's or logic 0's in NRZ produces a condition in which a receiver may loss its amplitude reference for optimum discrimination between received 1's and 0's. This state is known as **dc wandering or base line wandering.**
- **Unipolar encoding** technique is the simplest technique to represent the digital signal with only one level of amplitude.
- **Bipolar encoding** scheme utilizing the three voltage levels to represent digital data. These voltage levels are +V, 0 V and –V. in this scheme, there is no transition at the middle of the bit interval.
- Alternative Mark Inversion means alternately change in voltage level to represent mark (*i.e.*, binary bit '1' from telegraphy).
- **Pseudoternary** is a variation of the AMI, in this scheme binary bit '1' is represent with the zero level and binary bit '0' is represent with alternative +v and –v to voltage level.
- A twinned binary code increases the number of available symbols by increasing the number of signal levels.
- One way of ensuring that there are large numbers of 1's in any transmitted bit stream is to detect long strings of 0's and substitute for them code patterns containing mostly ones. At the receiver these code patterns are recognized and the inserted pattern is replaced with 0's. Such algorithms are called **binary N-zero substitution schemes** or BNZS schemes.
- When the digital encoding is symmetrical around 0 Volts, it is called a **Polar Code.**
- In **polar NRZ** encoding scheme, we use two levels of voltages and the level is always positive or negative.
- **Manchester encoding scheme** is the combination of RZ and NRZ-L encoding schemes. In this encoding scheme, the duration of the bit is divided into two halves. The voltage level remains at one level during the first half and moves to the opposite voltage level in the second half.

- Differential Manchester encoding continues the ideas of RZ and NRZ-I. We use the inversion at the middle of the each bit duration for synchronizing purpose but at the same time presence or absence of an additional transition at the beginning of the interval are used to determine the value of bit.

Review Questions

1. Explain different modes of transmissions.
2. What do you understand by term communication? Explain electronic communication systems in brief.
3. Draw and explain the basic block diagram of communication system.
4. Compare the digital communication system with the analog communication system.
5. Describe the term channel in communication system.
6. Describes the fundamental limitations of communication system.
7. Draw the digital communication block diagram and explain the working of each block.
8. Explain the difference between digital and analog signals.
9. Define signal and explain how they are classified.
10. Write the advantages and disadvantages of digital communication system.
11. The SNR for a channel having 1 MHz bandwidth is 63. Calculate the following for this channel:
    (*i*) Bit rate (*ii*) Signal level. [**Ans.** Bit rate = 6 Mbps and signal level = 35.99 dB]
12. What do you understand by data communication? Explain various communication modes.
13. Discuss the various standards in data communication.
14. Describe in detail about data communication circuits.
15. Given a channel with an intended capacity of 20 Mbps and bandwidth of this is 3 MHz. Determine the SNR required in order to achieve this capacity. [**Ans.** 100 or 20 dB]
16. Determine the maximum bit rate for a channel having bandwidth of 3100 Hz and SNR is 20 dB. And also calculate the number of levels required to transmit at the maximum bit rate.
    [**Ans.** Maximum bit rate=20.640 kbps, No. of levels = 16]
17. What does the Shannon capacity have to do with communication? Calculate the maximum bit rate for the noiseless channel with a bandwidth 3000 Hz transmitting a signal a signal with two signal levels.
    (**Hints :** Capacity = BW × $\log_2$ (1+ SNR), bit rate = 6kbps).
18. What is Nyquist theorem? What is the bit rate, if we want to digitize the human voice, assuming 8 bits per sample?
19. Write the short notes on following:
    (*i*) Digital Radio
    (*ii*) Shannon's limit for information capacity
    (*iii*) The advantages and disadvantages of digital communication system.
20. What is data encoding? How NRZ coding is implemented ?
21. How is differential Manchester encoding is different from others ?
22. Define and explain Shannon limit theorem.
23. Consider a long sequence of 1's followed by single 0 and then a long sequence of 1' s. sketch the waveform for this sequence using the following method of representing symbols 1 and 0. (Assume data stream—111101111).
    (*i*) On-off signaling
    (*ii*) RZ signaling
    (*iii*) NRZ signaling
    (*iv*) Manchester code
    (*v*) Differential Manchester encoding

24. Consider the binary data 01101001. Sketch the waveform for binary data, using following methods of representing symbols 1 and 0.

(*i*) On-off signaling

(*ii*) RZ signaling

(*iii*) NRZ signaling

(*iv*) Manchester code

(*v*) Differential Manchester encoding

25. Explain the Manchester and Differential Manchester encoding with examples.

26. Name the various data encoding techniques used in data transmission system and explain one of them in detail.

27. Write in detail about the Differential Manchester encoding.

28. What is data encoding? Explain Differential Manchester encoding with suitable examples.

29. What are the important factors to be considered before selecting a data encoding technique?

30. Explain unipolar encoding with an example. Give its disadvantages.

31. Explain the difference between Manchester and differential Manchester encoding with an example?

32. Define bipolar encoding?

33. Explain AMI encoding.

34. The following data stream is the unipolar encoded data. What is the data stream?

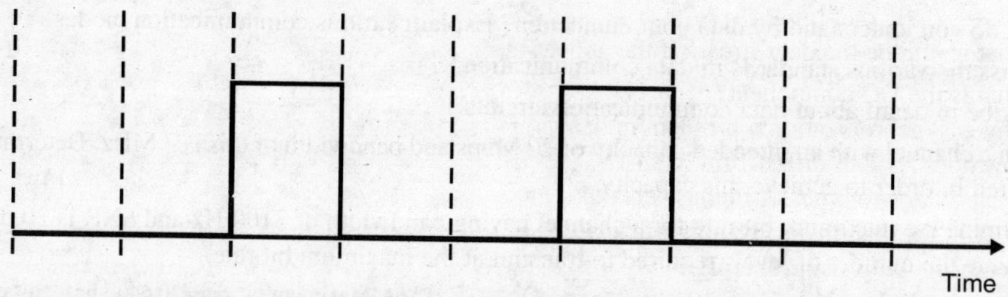

35. The following data stream is NRZ-L encoded. What is the data stream?

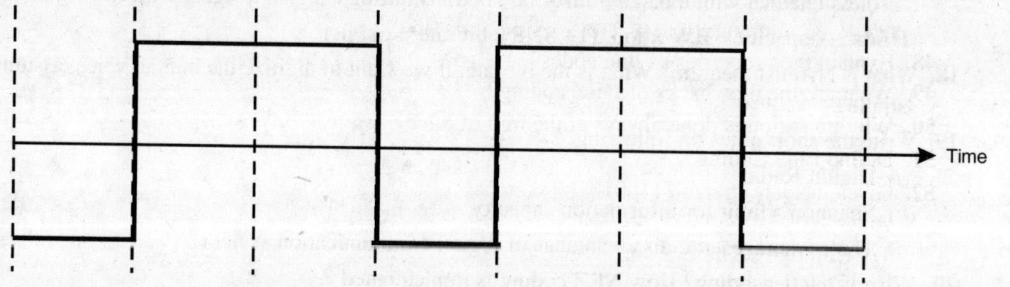

36. What would be the data stream if the above stream was NRZ-I encoded?

37. What is the data stream if the stream below is RZ encoded?

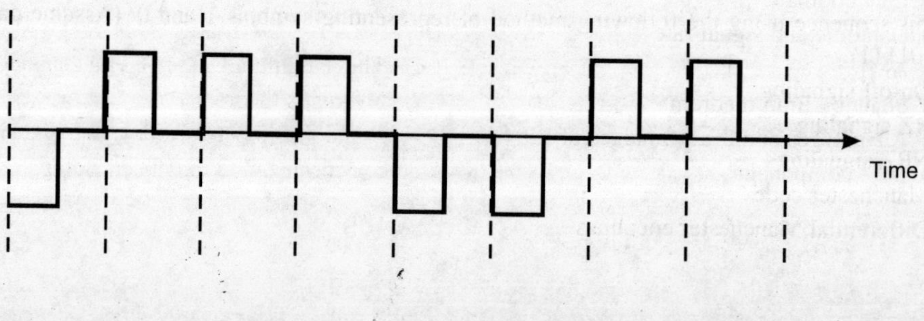

38. What is the data stream if the following figure represents Manchester encoding?

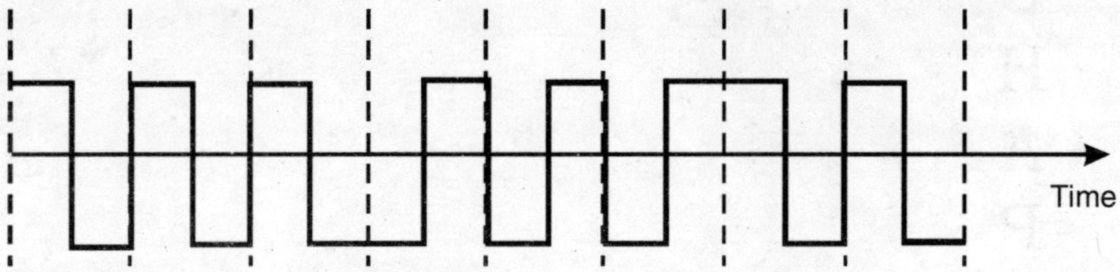

39. What would have been the data stream if the above figure represented differential Manchester encoding?

40. What is the data stream if the figure below represents AMI encoding?

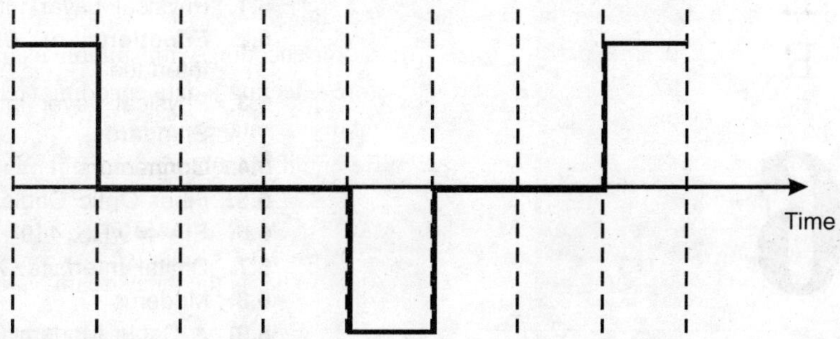

41. What would be the data stream if the above figure is pseudoternary encoding of the data stream?

42. What is the DC component?

43. Why is synchronization a problem in data communication?

44. How does NRZ-L differ from NRZ-I?

45. Discuss the two types of biphase encoding in use on networks.

46. What is the major disadvantage in using NRZ encoding? How do RZ encoding and biphase encoding attempt to solve the problem?

47. Given the data stream 1110010100. Sketch the transmitted sequence of pulses for each of the following line codes:

(*i*) Unipolar nonreturn – to – zero.
(*ii*) Polar nonreturn – to – zero.

48. Explain the properties of line codes?

49. What are the drawbacks of NRZ coding?

50. Why transition is done always at the mid of the bit interval?

51. Define Line Coding.

52. Given the data stream 1110010100 sketches the transmitted sequence of pulse for each of the following line codes:

(*i*) Unipolar nonreturn – to – zero.
(*ii*) Polar nonreturn – to – zero.
(*iii*) Unipolar return – to – zero.
(*iv*) Bipolar return – to – zero.
(*v*) Manchester code.

53. Write short notes on following:

(*i*) Line coding
(*ii*) Data element and data signal
(*iii*) Data rate and signal rate
(*iv*) NRZ-L and NRZ-I
(*v*) AMI
(*vi*) The maximum Data rate of a channel
(*vii*) Shannon limit theorem
(*viii*) Nyquist's theorem
(*ix*) Advantages and disadvantages of RZ
(*x*) Relationship between data rate and signal rate
(*xi*) D.C. component
(*xii*) Synchronization in data encoding

❑❑❑

CHAPTER 6

INSIDE THIS CHAPTER

INTERFACE STANDARDS

6.1. PHYSICAL LAYER INTERFACE

Why Interfaces are needed?

The physical layer defines the electrical, mechanical, procedural and functional specifications for activating, maintaining and deactivating the physical link between communicating network systems.

Before discussing the physical layer interfaces, it is better to know that why interfaces are required in the data communication. Often the source is not able to speak the same communication language as the medium and the data communications needs to be translated into a form that the medium can understand. Data communication is further broken down into the **Data Communication Model** which consists of data communication hardware as shown in Fig. 6.1.

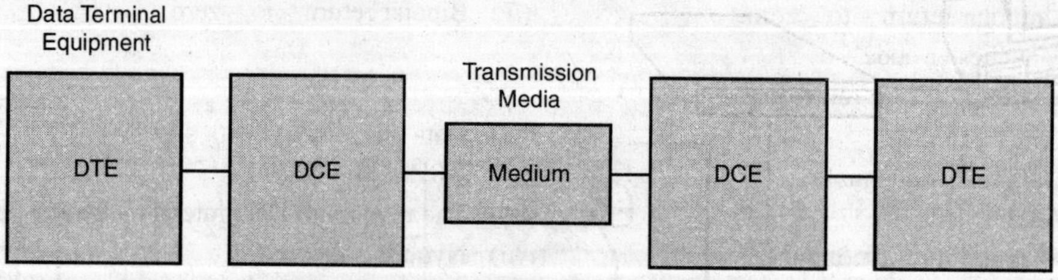

Fig. 6.1. Data communications model.

Data Terminal Equipment (DTE)

The **Data Terminal Equipment (DTE)** is the source of any binary digital device which is generating, transmitting, receiving or interpreting the data messages. Data terminal equipment includes the concept of terminals, clients, hosts and server. Terminals are devices used to input/output and display information, such as keyboards, printers and monitors.

Examples of DTE devices are: PCs, terminals, end user devices, Teletypewriter, video display terminal, Transactional terminal, intelligent terminal, smart terminal, workstation, line control unit, front end processor etc.

Data Communication Equipment (DCE)

The interface between the source and the medium, and the medium and the destination or receiver is called the **Data Communication Equipment (DCE)** and is a physical piece of equipment. Its job is to provide the interface from the DTE device to the medium. Examples of DCE devices are modems, channel banks, and routers.

An example of this would be your PC dialing into an Internet Service Provider (ISP) as shown in Fig. 6.2.

The output of the PC (DTE device) is an RS232 signal (we will discuss this later in this chapter) which is a digital communication consisting of logic ones and zeros. The medium i.e., the telephone line is made for carrying analog communications consisting of voice transmission. The output of the PC and the requirements of the phone line are mutually incompatible. The phone line does not understand digital signals and the PC does not understand analog transmission. The DCE device *i.e.*, the modem provides the conversion (the interface) so that digital data can communicate through the analog medium.

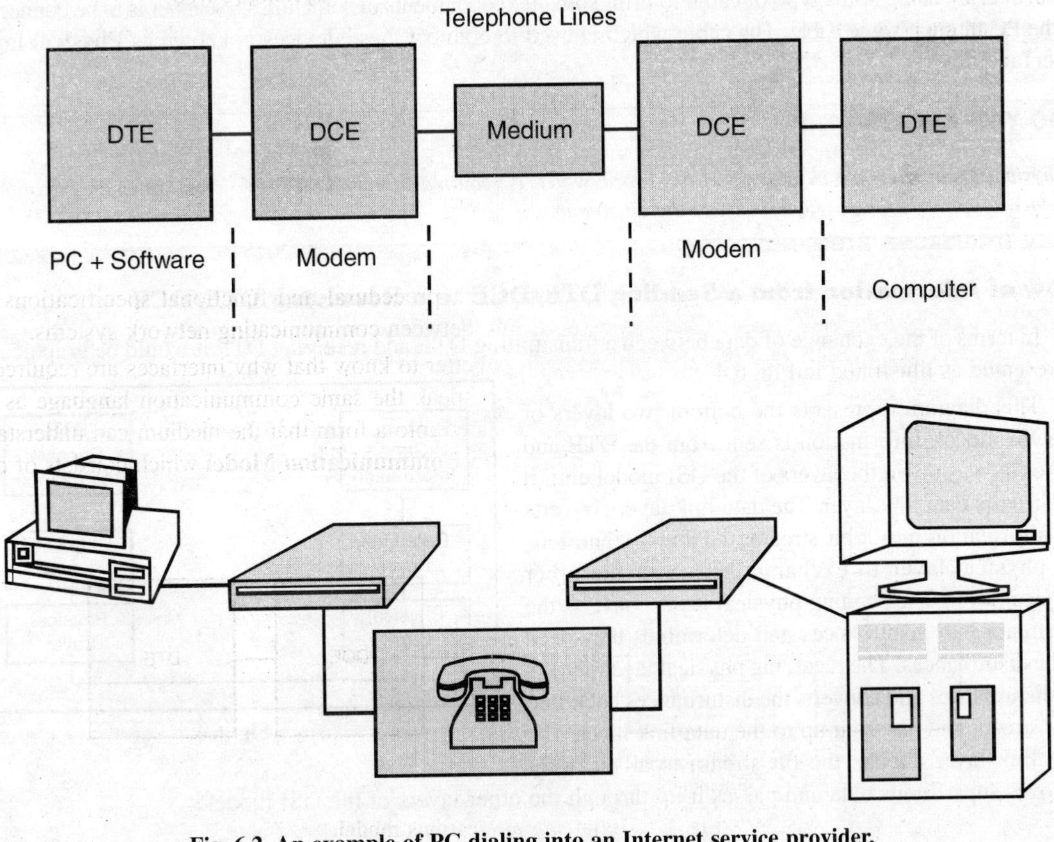

Fig. 6.2. An example of PC dialing into an Internet service provider.

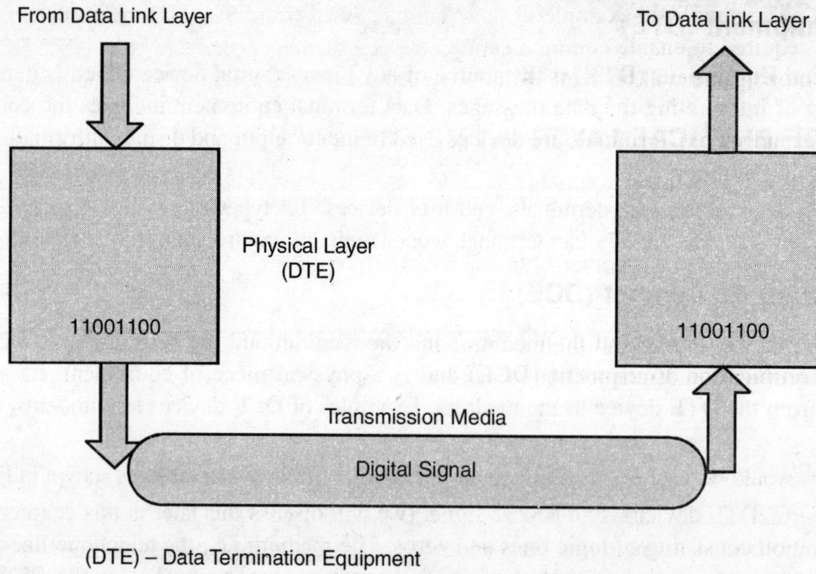

Fig. 6.3. Physical layer interface.

Physical layer interface is the type of hardware, which is required to connect the two or more communication devices to provide the physical media to the bit stream. For example, we connect our PC to the printer by using some type of cable to print some text documents or sometimes a scanner is to be connected to the PC through some cable. The cable which is used to connect these devices is known as **Physical layer interface device.**

DO YOU KNOW ▸

***Physical layer interface** is the type of hardware, which is required to connect the two or more communication devices to provide the physical media to the bit stream.*

Flow of Information from a Sending DTE/DCE to a Receiving DTE/DCE

In terms of the exchange of data between a transmitting DTE and receiving DTE it would be graphically represented as illustrated in Fig. 6.4.

Fig. 6.4.

This diagram represents the bottom two layers of the OSI model. Information is sent from the DTE and makes its way down the layers of the OSI model until it reaches the data link layer. The data link layer converts the information into a bit stream, 0's and 1's and tells the physical layer to exchange bits with the other physical layer. The sending physical layer converts the bit stream into disturbances and determines the speed of the disturbances. The receiving physical layer detects the disturbances and converts the disturbances back into a bit stream and passes it up to the data link layer. The data link layer checks the bit stream is all in order removes superfluous data and passes it up through the other layers of the OSI model.

The above description is a simplistic example of how data is passed from one DTE to another. It does not go into the protocols required to enable communication between the two systems to occur. The RS 232 standard defines these protocols and will be dealt with in detail throughout the rest of the assignment

6.2. FUNCTIONS OF PHYSICAL LAYER INTERFACE

1. Physical characteristics of interfaces and medium.
2. Representation of bits (Type of encoding scheme)
3. Data rate (transmission rate *i.e.* number of bits per second)
4. Synchronization of bits
5. Line configuration (Point to point or point to multipoint configuration)
6. Established and termination of connection to a communication medium
7. Physical topology

1. Physical Topology

The **physical topology** defines how devices are connected to make a network. For the sake of simplicity, the network wiring is symbolized as a single line between the two devices. In actuality, it may be a twisted pair of wires, a coaxial cable, and an optical fiber. Right now, we're merely focusing on the "shape" of the network, technically known as its ***topology***.

If we want to include more devices (sometimes called *nodes*) on this network, we have several options of network configuration to choose from:

Devices can be connected any or combinations of one or more configurations of the following:

(*a*) Star topology
(*b*) Ring topology
(*c*) Bus topology
(*d*) Mesh topology
(*e*) Hybrid topology
(*f*) Point to point topology

(a) Star Topology

The primary feature of the star topology is that each station is radially connected to a central node through a direct point to point connection. With devices known as "gateways" at branching

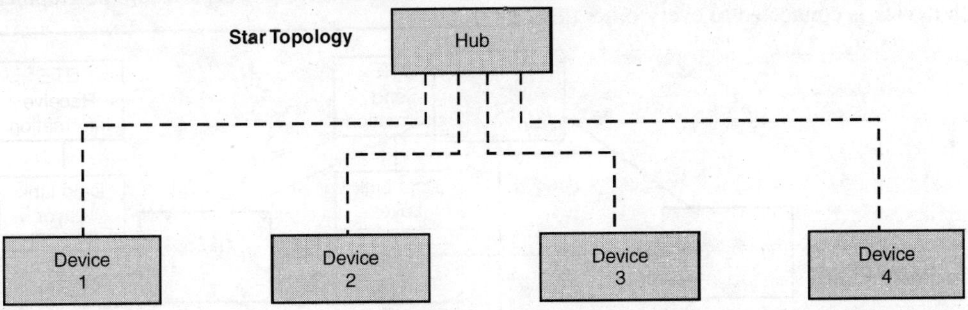

Fig. 6.5. Star topology.

points in the network, data flow can be restricted between nodes, allowing for private communication between specific groups of nodes. This addresses some of the speed and security issues of the simple bus topology. However, those branches could easily be cut off from the rest of the "star" network if one of the gateways were to fail. Figure 6.5 show the star topology.

(b) Ring Topology

With the ring topology, adjacent devices are interconnected in a closed loop configuration as shown in Fig. 6.6. This topology provides the best reliability with the least amount of wiring. Since each node has two connection points to the ring, a single break in any part of the ring doesn't affect the integrity of the network. The devices, however, must be designed with this topology in mind. Also, the network must be interrupted to install or remove nodes. As with bus topology, ring networks are *broadcast* by nature.

Ring Topology

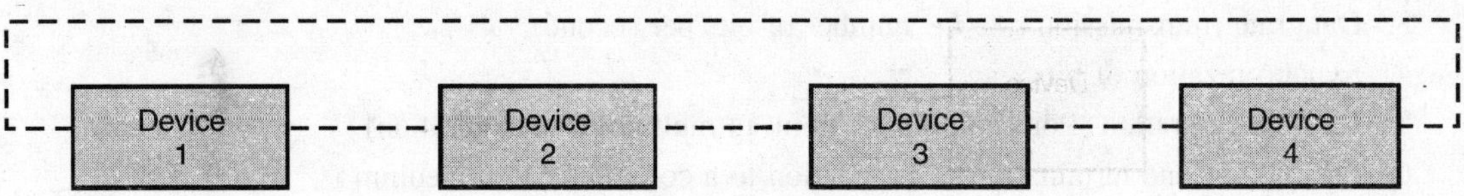

Fig. 6.6. Ring topology.

(c) Bus Topology

A linear bus topology is a multipoint circuit configuration where individual devices are connected by a common, shared communication channel (transmission media) as shown in Fig. 6.7. Very simple to install and maintain. Nodes can be easily added or removed with minimal wiring changes. On the other hand, the one bus network must handle *all* communication signals from *all* nodes. This is known as ***broadcast*** networking, and is analogous to a group of people talking to each other over a single telephone connection, where only one person can talk at a time (limiting data exchange rates), and everyone can hear everyone else when they talk (which can be a data security issue). Also, a break in the bus wiring can lead to nodes being isolated in groups.

Bus Topology

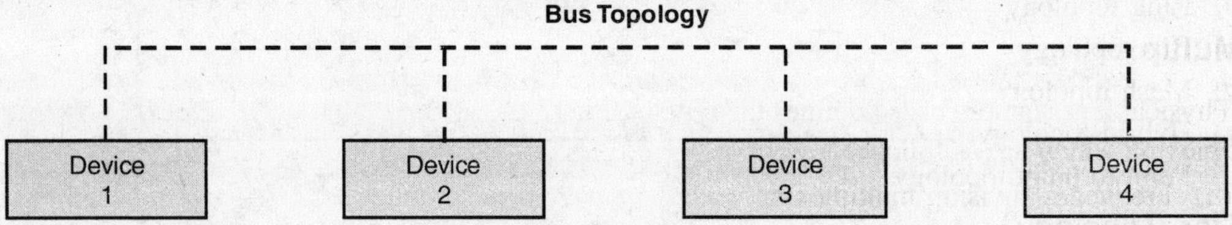

Fig. 6.7. Bus Topology.

(d) Mesh Topology

Each device is connected to every other device.

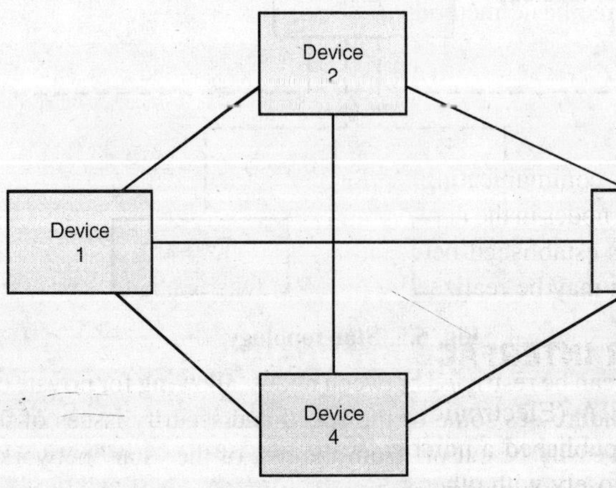

Fig. 6.8. Mesh topology.

(e) Hybrid Topology

Hybrid topology is that topology in which combination of two or more topology is used.

(f) Point to Point Topology

If we want to connect two digital devices with a network, we would have a kind of network known as "point-to-point:"

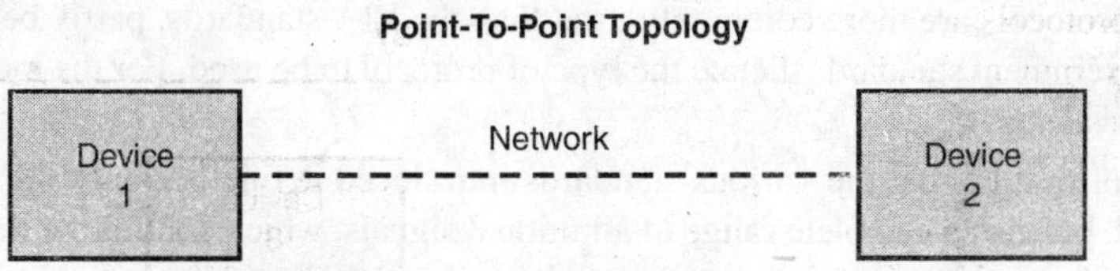

Fig. 6.9. Point to point topology.

Many network standards dictate the type of topology which is used, while others are more versatile. Ethernet, for example, is commonly implemented in a "bus" topology but can also be implemented in a "star" or "ring" topology with the appropriate interconnecting equipment. Other networks, such as RS-232C, are almost exclusively point-to-point; and token ring is implemented solely in a ring topology.

2. Communication Mode

It also defines the direction of transmission between the devices such as

(*a*) Simplex
(*b*) Half Duplex
(*c*) Full Duplex

3. Multiplexing

Physical layer interface also defines the type of multiplexing to be used. We have mainly three types of multiplexing which are as follows:

(*a*) Frequency division multiplexing,
(*b*) Time division multiplexing
(*c*) Wave division multiplexing

4. Circuit Switching

There are traditionally three basic methods of switching, namely:

1. Circuit switching
2. Packet switching.
3. Message switching

In circuit switching, two communicating stations are connected by a *dedicated* communication path which consists of intermediate nodes in the network and the links that connect these nodes. In circuit switching, an electrical dedicated path is established between the source and the destination before any data transfer take place. The electrical path may be realized by physical wires or radio or satellite links.

6.3. PHYSICAL LAYER INTERFACE STANDARDS

Founded in 1924, the EIA (Electronic Industries Alliance) is a U.S. organization of electronics manufacturers. The EIA has published a number of standards related to telecommunication and computer communication, and works closely with other associates such as ANSI and the ITU.

The primary EIA standards for telecommunication define the serial interface between modems and computers. The most popular are the RS-232-C, RS-449, RS-422, and RS-423 serial interfaces. The physical layer specifications define 37-pin (DB-37), 25-pin (DB-25), and 9-pin (DB-9) connectors and associated cable, as well as electrical characteristics such as the type of signal used on each pin and the timing of those signals.

Note that the EIA RS-232 standard is also the CCITT standard V.24. CCITT is part of the ITU. The CCITT V series protocols are more commonly used than the EIA standards, partly because they are used in Europe where government standards dictate the type of protocol to be used. For the most part, EIA standards have CCITT equivalents.

It is important to note that the various standards introduced define not only the form of the electrical signals to be used, but also a complete range of additional signals, which control the timing and order of data transfer across the appropriate interfaces. At large, these signals are said to form the physical layer interface standard. We have many types of physical layer interfaces such as, RS-232, X.21, RS-449, RS-550, V-35, 1.430, 1.431, T1, E1, etc. We shall discuss some of the important physical layer interfaces.

6.3.1. RS-232 Interface

RS232 interface is used to connect Data terminal equipment (DTE) and Data circuit terminating equipment (DCE). It is used for serial interface. In United States, a standard interconnection between business machine and modem is given by the RS-232 interface. In 1962 EIA has standardized some set of standard specifications to ensure compatibly between data sets (DCE) and terminal equipment (DTE). The interface uses a 25 pin connector, where each of the 25 pins has been assigned a specific function by Electronic Industries Association (EIA).

The RS-232 interface specifications limits the interconnecting cable to length of 15 meter with data rate of 20kbps.

The EIA RS-232 serial communication standard is a universal standard, originally used to connect teletype terminals to modem devices. In a modern computer the RS-232 interface is referred to as a COM Port. The COM Port uses a 9 pin D-Type connector to attach to the RS-232 cable. The RS232 standards define a 25 pin D-Type connector but IBM reduced this connector to a 9 pin device so as to reduce size and cost.

It also be defined by the EIA that the voltage levels with which data and control signals are exchanged between DCE and DTE. Each pin in the connector carries either binary 1 or 0 to indicate activation or deactivation of control functions or data values. The RS-232 interface can accommodate many different type of circuit operation, using different combination of circuit lines. For example, point to point dedicated system will require a minimum number of control lines in the interface.

Signaling Levels

As stated earlier the physical layer converts the bit stream 0's and 1's into disturbances. Within the RS 232 standards the disturbances are represented as voltages along the transmission medium, usually copper wire between the sending and receiving DTE/DCE. The strength of the voltage travelling along the copper wire determines whether it recognized as a 0 or 1 by the receiving physical layer. This is graphically represented as shown in Fig. 6.10.

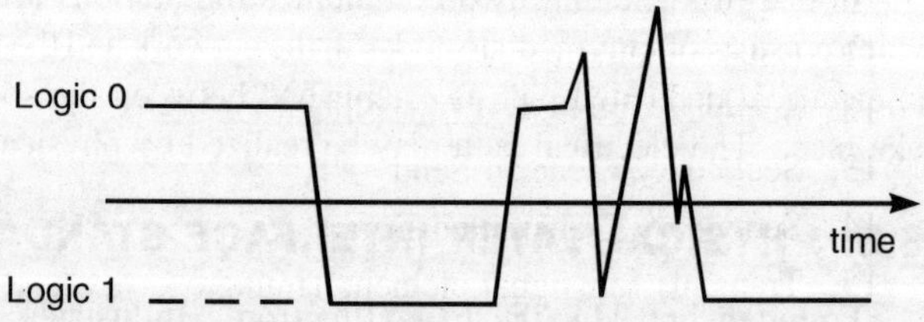

Fig. 6.10. Representations of 0 and 1 in RS 232.

The diagram shows that a 1 is represented by the electrical signal which is a negative voltage and a 0 is a positive voltage. The receiving physical layer is able to recognize these positive and negative voltages and convert them into a recognizable bit stream of which the data link layer can interpret. The passing of disturbances by the physical layer is graphically represented as follows.

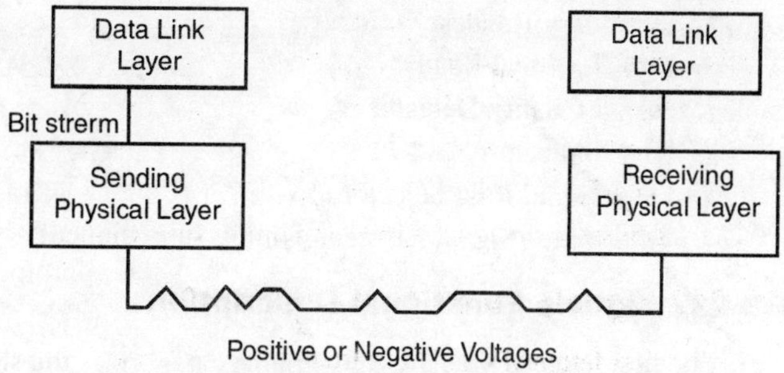

Fig. 6.11. The passing of disturbances by the physical layer.

As can be seen in this illustration, the sending physical layer converts the bit stream into voltages, transmits them along the transmission medium, the receiving physical layer recognizes the voltages as either positive or negative and converts them back into 0's and 1's so they are recognizable by the data link layer.

The differences in the voltage become important when cable maximum lengths are being reached between DTE and DCE. The larger range of voltage at the receiver means it may be able to pick up a weaker signal which may have been caused by a long cable.

The specific voltage levels for the RS 232 standard are :

TABLE 6.1. Voltage Levels for the RS 232 Standard

State	Voltage Level TX	Voltage Level RX
Logic 0	+5v to +15v	+3v to +25v
Logic 1	–5v to –15v	–3v to –25v

RS-232 Signals

The number preceding each signal name correspond to the pin number defined in the standard

1. Protective Ground
2. Transmitted Data
3. Received Data
4. Request to Send
5. Clear To Send
6. Data Set Ready
7. Signal Ground
8. Received Line Signal Detect (Carrier Detect)
9. +P (for testing only)
10. –P (for testing only)
11. (unassigned)
12. Secondary Received Line Signal Detect
13. Secondary Clear To Send
14. Secondary Transmitted Data
15. Transmission signal element Timing
16. Secondary Received Data
17. Receiver Signal Element Timing
18. (unassigned)

19. Secondary Request To Send
20. Data Terminal Ready
21. Signal Quality Detector
22. Ring Indicator
23. Data Signal Rate Selector
24. Transmitter Signal Element Timing (unassigned)

RS-232 Signals Functional Description

The first letter of the EIA signal name categorizes the signal into one of five groups, each representing a different "circuit":

1. A - Ground
2. B - Data
3. C - Control
4. D - Timing
5. S - Secondary channel

Pin-1 : Protective Ground

This pin is usually connected to the frame of one of the devices, either the DCE or the DTE, which is properly grounded. The sole purpose of this connection is to protect against accidental electric shock and usually this pin should not be tied to Signal Ground.

This pin should connect the chassis (shields) of the two devices, but this connection is made only when connection of chassis grounds is safe and it is considered optional.

Pin-2 : Transmit Data

Serial data (primary) is sent on this line from the DTE to the DCE. The DTE holds this line at logic 1 when no data are being transmitted. A "On" (logic 0) condition must be present on the following signals, where implemented, before data can be transmitted on this line: CA, CB,CC and CD (Request To Send, Clear To Send, Data Set Ready, Data Terminal Ready).

Pin-3 : Receive Data

Serial data (primary) is sent on this line from the DCE to the DTE. This pin is held at logic 1 (Mark) when no data are being transmitted, and is held "Off" for a brief interval after an "On" to "Off" transition on the Request To Send line, in order to allow the transmission to complete.

Pin-4 : Request To Send

This pin enables transmission circuits. The DTE uses this signal when it wants to transmit to the DCE. This signal, in combination with the Clear To Send signal, coordinates data transmission between the DTE and the DCE.

A logic 0 on this line keeps the DCE in transmit mode. The DCE will receive data from the DTE and transmit it on to the communication link.

The Request To Send and ClearTo Send signals relate to a half- duplex telephone line. A half duplex line is capable of carrying signals on both directions but only one at a time. When the DTE has data to send, it raises Request To Send, and then waits until the DCE changes from receive to transmit mode. This "On" to "Off" transition instructs the DCE to move to "transmit" mode, and when a transmission is possible, the DCE sets Clear To Send and transmission can begin.

On a full duplex line, like a hard-wired connection, where transmission and reception can occur simultaneously, the Clear To Send and Request To Send signals are held to a constant "On" level.

An "On" to "Off" transition on this line instructs the DCE to complete the transmission of data that is in progress, and to move to a "receive" (or "no transmission") mode.

Pin-5 : Clear To Send

Clear To Send signal is an answer signal to the DTE. When this signal is active, it tells the DTE that it can now start transmitting (on Transmitted Data line). When this signal is "On" and the Request To Send, Data Set Ready, and Data Terminal Ready are all "On", the DTE is assured that its data will be sent to the communications link. When "Off", it is an indication to the DTE that the DCE is not ready, and therefore data should not be sent.

When the Data Set Ready and Data Terminal Ready signals are not implemented, in a local connection which does not involve the telephone network, the Clear To Send and Request To Send signals are sufficient to control data transmission.

Pin-6 : Data Set Ready

On this line the DCE tells the DTE that the communication channel is available (i.e., in an automatic calling system, the DCE (modem) is not in the dial, test or talk modes and therefore is available for transmission and reception). It reflects the status of the local data set, and does not indicate that an actual link has been established with any remote data equipment.

Pin-7 : Signal Ground

This pin is the reference ground for all the other signals, data and control.

Pin-8 : Receive Line Signal Detect or Data Carrier Detect

The DCE uses this line to signal the DTE that a good signal is being received (a "good signal" means a good analog carrier that can ensure demodulation of received data).

Pin-9 : +P

This pin is held at +12 volts DC for test purposes.

Pin-10 : –P

This pin is held at -12 volts DC for test purposes.

Pin-12 : Secondary Receive Line Signal Detect

This signal is active when the secondary communication channel is receiving a good analog carrier (same function as the Receive Line Signal Detect signal).

Pin-13 : Secondary Clear To Send

Secondary Clear To Send signal is an answer signal to the DTE. When this signal is active, it tells the DTE that it can now start transmitting on the secondary channel (on the Secondary Transmitted Data line).

Pin-14 : Secondary Transmitted Data

Serial data (secondary channel) is sent on this line from the DTE to the DCE. This signal is equivalent to the Transmitted Data line except that it is used to transmit data on the secondary channel.

Pin-15 : Transmission Signal Element Timing

The DCE sends the DTE a clock signal on this line. This enables the DTE to clock its output circuitry which transmits serial data on the Transmitted Data line.

The clock signal frequency is the same as the bit rate of the Transmitted Data line. A "On" to "Off" transition should mark the center of each signal element (bit) on the Transmitted Data line.

Pin-16 : Secondary Receive Data

Serial data (secondary channel) is received on this line from the DCE to the DTE. When the secondary channel is being used only for diagnostic purposes or to interrupt the flow of data in the primary channel, this signal is normally not provided.

Pin-17 : Receiver Signal Element Timing

The DCE sends the DTE a clock signal on this line. This clocks the reception circuitry of the DTE which receives serial data on the Received Data line.

The clock signal frequency is the same as the bit rate of the Received Data line (BB). The "On" to "Off" transition should indicate the center of each signal element (bit) on the Received Data line.

Pin-19 : Secondary Request To Send

The DTE uses this signal to request transmission from the DCE on the secondary channel. It is equivalent to the Request To Send signal.

When the secondary channel is only used for diagnostic purposes or to interrupt the flow of data in the primary channel, this signal should turn "On" the secondary channel unmodulated carrier.

Pin-20 : Data Terminal Ready

When on, tells the DCE that the DTE is available for receiving. This signal must be "On" before the DCE can turn Data Set Ready "On", thereby indicating that it is connected to the communications link.

The Data Terminal Ready and Data Set Ready signals deal with the readiness of the equipment, as opposed to the Clear To Send and Request To Send signals that deal with the readiness of the communication channel.

When this signal is "Off", it causes the DCE to finish any transmission in progress and to be removed from the communication channel.

Pin-21 : Signal Quality Detector

This line is used by the DCE to indicate whether or not there is a high probability of an error in the received data. When there is a high probability of an error, it is set to "Off", and is "On" at all other times.

Pin-22 : Ring Indicator

On this line the DCE signals the DTE that there is an incoming call. This signal is maintained "Off" at all times except when the DCE receives a ringing signal.

Pin-23 : Data Signal Rate Selector

The DTE uses this line to select the transmission bit rate of the DCE. The selection is between two rates in the case of a dual rate synchronous connection, or between two ranges of data rates in the case of an asynchronous connection.

Typically, when this signal is "On", it tells the DCE (modem) that the receive speed is greater than 600 baud.

Pin-24 : Transmitter Signal Element Timing

The DTE sends the DCE a transmit clock on this line. This is only when the master clock is in the DTE.

An "On" to "Off" transition should indicate the center of each signal element (bit) on the Transmitted Data line.

It is to be noted that the pin names are the same for the DCE and DTE. The Transmit Data (pin number 2) is a transmit line on the DTE and a receive line on the DCE, Data Set Ready (pin number 6) is a receive line on the DTE and a transmit line on the DCE, and so forth.

Electrical Signal Characteristics

Voltage levels defined in the standard

Data signals	"0", "Space"	"1", "Mark"
Driver (Required)	5 to 15	–5 to –15 Volts
Terminator (expected)	3 to 25	–3 to –25 Volts
Control signals	"Off"	"On"
Driver (Required)	–5 to –15	5 to 15 Volts
Terminator (expected)	–3 to –25	3 to 25 Volts

Physical Connectivity

The RS 232 standard defines the requirements of the physical connectors which interconnect the DTE to the DCE and to the transmission medium. Having the correct connectors allows the equipment to be interconnected but does not necessarily make it work.

Common connectors in use are as follows:

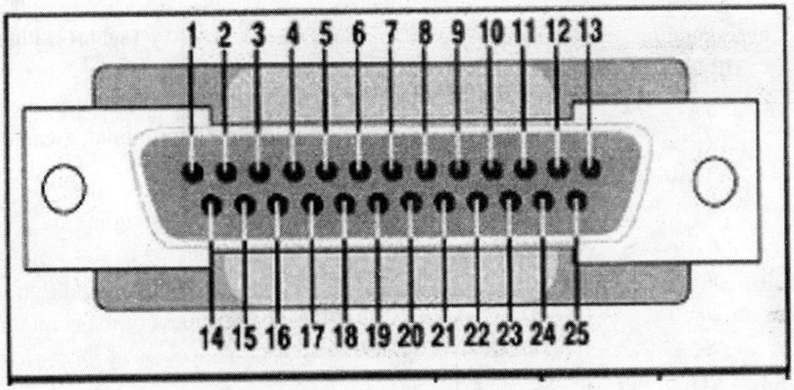

Fig. 6.12. Pin 25 DB-25 connector

Pin description		EIA Circuit	From DCE	To DCE
1	Frame Ground	AA		
2	Transmitted Data	BA		D (Data)
3	Received Data	BB	D	
4	Request to Send	CA		C (Control)
5	Clear to Send	CB	C	
6	Data Set Ready	CC	C	
7	Signal Gnd/Common Return	AB		
8	Received Line Signal Detector	CF	C	
11	Undefined			
12	Secondary Received Line Signal Detector	SCF	C	
13	Secondary Clear to Send	SCB	C	
14	Secondry Transmitted Data	SBA	C	
15	Transmitter Signal Element Timing	DB	T (Timing)	
16	Secondary Received Data	SBB	D	
17	Receiver Signal Element TIming	DD	T	
18	Undefined			
19	Secondary Request to Send	SCA		C
20	Data Terminal Ready	CD		C
21	Signal Quality Detector	CG		C
22	Ring Indicator	CE	C	
23	Data Signal Rate Selector (DCE)	CI	C	
24	Transmitter Signal Element Timing	DA		T
25	Undefined			

Fig. 6.13. The 25 pin RS 232 connector defined in the standard

The 25 pin RS 232 connector is the one defined in the standard shown in Fig. 6.14.

This connector is nine pin RS 232 connector referred as DB-9 and developed by IBM. On a PC, this is commonly referred to as the serial port.

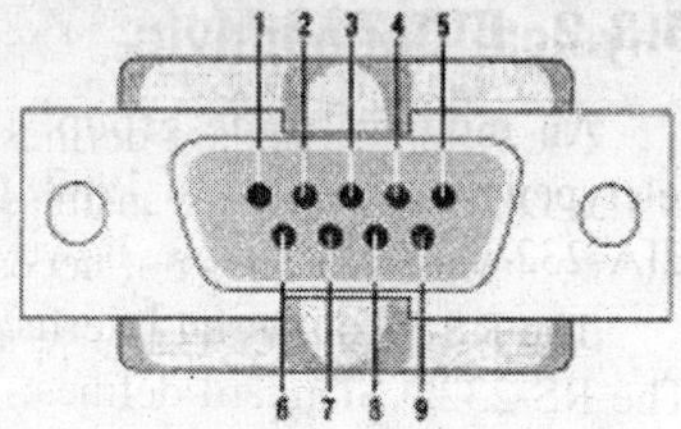

Fig. 6.14. Pin 9 DB-9 Connector

TABLE 6.2. DB 9 Pin Assignments

DB-9 Pin	Corresponding DB-25	Signal	Function
2	3	RxD	Received Data
3	2	TxD	Transmited Data
8	5	CTS	Clear to send
7	4	RTS	Request to Send
6	6	DSR	Data set ready
5	7	SG	Signal Ground
4	20	DTR	Data Terminal Ready
1	8	DCD	Data Carrier Detect

The connectors allow the physical components to interconnect; however, how the cabling is wired to the various pins determines whether the sending and receiving physical layers will be able to communicate.

A 9 pin connector will be used to explain how the physical layers communicate.

DTE

- At the physical layer it can be a terminal, computer, printer, fax etc
- Sends data on transmitted data (TXD) line
- Receives on received data (RXD) line
- Asserts request to send (RTS) when ready to receive
- Waits for clear to send (CTS) to be asserted before sending
- Asserts data terminal ready (DTR) when ready and waiting (to receive)
- Expects data set ready (DSR) to be asserted before performing (sending)
- Expects Data Carrier Detect (DCD) to be asserted before receiving

DCE

- At the physical layer the DCE takes the bit stream converts them into a signal (voltages) and then introduces the signal onto the telecommunications medium. Normally the DCE is known as a modem.
- The wiring is the inverse as that of the DTE.
- The most important lines are RxD, TxD, and SG; the others are used with modems, printers and plotters to indicate internal states.

DO YOU KNOW►

RS232 interface is used to connect Data terminal equipment (DTE) and Data circuit terminating equipment (DCE).

6.3.2. EIA-232-D

An industry trade group, the Electronic Industries Association (EIA), defined it originally for teletypewriter devices. In 1987, the EIA released a new version of the standard and changed the name to EIA-232-D. Many people, however, still refer to the standard as RS-232C, or just RS-232.

The RS-232D Serial Interface Standard added the mechanical characteristics to the RS-232C Standard. The RS-232D Standard defines:

- The mechanical characteristics of the interface
- The electrical characteristics of the interface
- The function of each signal
- Subsets of the signals for certain applications

The European version of RS-232D is defined in:

- V.24 - mechanical standard
- V.28 - electrical standard

Mechanical Characteristics of the RS-232D

(*i*) The connector is a DB25 connector. DB9 is not universally accepted.
(*ii*) The connector gender is male at the DTE and female at the DCE.
(*iii*) The assignments of signals to pins.
(*iv*) The maximum cable length is 50 ft.
(*v*) The maximum cable capacitance = 2500 pF. Typical cable has 50 pF/foot capacitance.

Electrical Characteristics of the RS-232D

The transmitter side generates a voltage between +5 and +25 volts for a space (digital 0 or low) and generates a voltage between –5 and –25 volts for a mark (digital 1 or high). (Fig. 6.15)

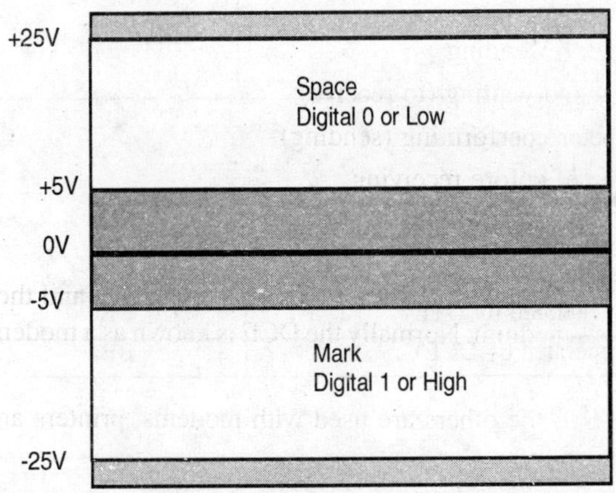

Fig. 6.15

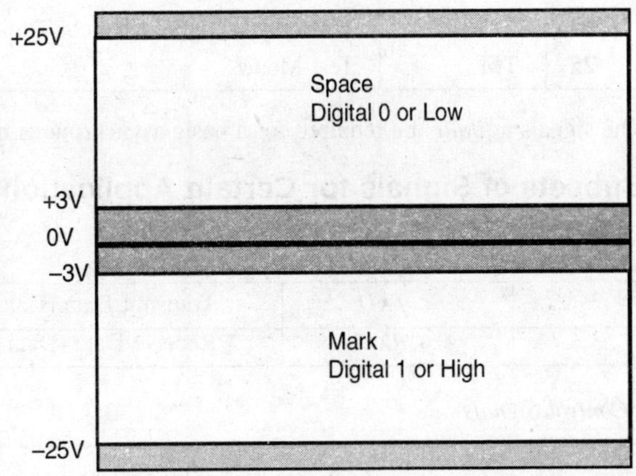

Fig. 6.16

The receiving side recognizes a space as any voltage between +3 and +25V and a mark as any voltage between –3 and –25 V. The standard allows for a voltage loss through the cable and noise immunity by reducing the receive requirements to +/-3 volts. (Fig. 6.16)

TABLE 6.3. Function of Each Signal

Pin	Name	Description	EIA Circuit
1	*GND*	Chassis Ground	AA
2	*TXD*	Transmit Data (TXD)	BA
3	*RXD*	Receive Data (RXD)	BB
4	*RTS*	Ready to Send	CA
5	*CTS*	Clear to Send	CB
6	*DSR*	Data Set Ready (DCE Ready)	CC
7	*SGND*	Signal ground	AB
8	*DCD*	Carrier Detect (CD or RLSD)(RLSD - Received Line Signal Detector)	CF
9	not used		
10	not used		
11	not used		
12	DCD2	Secondary Carrier Detect (SRLSD)	SCF
13	CTS2	Secondary Clear to Send	SCB
14	TXD2	Secondary Transmit Data	SBA
15	TxSigC	Transmitter Signal Element Timing - DCE	DB
16	RXD2	Secondary Receive Data	SBB
17	RxSig	Receive Signal Element Timing - DCE	DD
18	LL	Local Loopback	
19	RTS2	Secondary Ready to Send	SCA
20	*DTR*	Data Terminal Ready (DTE Ready)	CD
21	SQ/RL	Signal Quality/Remote Loopback	CG
22	*RI*	Ring Indicator	CE
23	DSRS	Data Signal Rate Selector	CH/CI
24	TxSigT	Transmitter Signal Element Timing - DTE	DA
25	TM	Test Mode	

The signals in *italic* are required for a basic asynchronous modem connection.

Subsets of Signals for Certain Applications

Data Signals

2	*TXD*	Transmit Data (Data generated by DTE)	BA
3	*RXD*	Receive Data (Data generated by DCE)	BB

Control Signals

4	*RTS*	Ready to Send	DTE wishes to transmit
5	*CTS*	Clear to Send	DCE ready to receive
6	*DSR*	Data Set Ready (DCE Ready)	DCE powered on & ready to go
20	*DTR*	Data Terminal Ready (DTE Ready)	DTE powered on & ready to go
22	*RI*	Ring Indicator	Phones ringing

Test Modes

18	LL	Local Loopback	Initiate Local Loopback Self-Test
21	SQ/RL	Signal Quality/Remote Loopback	Initiate Remote Loopback Self-Test
25	TM	Test Mode	Initiate Test Mode

Synchronous Control Signals

21	SQ/RL	Signal Quality/Remote Loopback	Error in received data
23	DSRS	Data Signal Rate Selector	DTE can dynamically select 1 of 2 data rates

Synchronous Timing Signals

15	TxSigC	Transmitter Signal Element Timing - DCE	DCE generated
17	RxSig	Receive Signal Element Timing - DCE	DCE generated
24	TxSigT	Transmitter Signal Element Timing - DTE	DTE generated

Ground/Shield

1	*GND*	Chassis ground	Shield DTE side only for noise protection. Do NOT connect to signal ground
7	*SGND*	Signal ground	Signal return path

6.3.3. Limitations of RS-232

RS-232 has some serious shortcomings as an electrical interface.

Firstly, the interface presupposes a common ground between the DTE and DCE. This is a reasonable assumption where a short cable connects a DTE and DCE in the same room, but with longer lines and connections between devices that may be on different electrical busses, this may not be true. We have seen some spectacular electrical events causes by "uncommon grounds".

Secondly, a signal on a single line is impossible to screen effectively for noise. By screening the entire cable one can reduce the influence of outside noise, but internally generated noise remains a problem. As the baud rate and line length increase, the effect of capacitance between the cables introduces serious crosstalk until a point is reached where the data itself is unreadable.

Using low capacitance cable can reduce crosstalk. Also, as it is the higher frequencies that are the problem, control of slew rate in the signal (*i.e.*, making the signal more rounded, rather than square) also decreases the crosstalk. The original specifications for RS-232 had no specification for maximum slew rate.

Voltage levels with respect to ground represent the RS 232 signals. There is a wire for each signal, together with the ground signal (reference for voltage levels). This interface is useful for point-to-point communication at slow speeds. For example, port COM1 in a PC can be used for a mouse, port COM2 for a modem, etc. This is an example of point-to-point communication: one port, one device. Due to the way the signals are connected, a common ground is required.

This implies limited cable length-about 30 to 60 meters maximum. (Main problems are interference and resistance of the cable.) Shortly, RS 232 was designed for communication of local devices, and supports one transmitter and one receiver.

6.4. CONNECTORS

Connectors attach components together. Several types of connectors are available, serving various purposes. For example, connectors are used to:

(a) Connect network interface cards, such as an Ethernet card, to a cable;

(b) Connect cable segments (e.g., thin coax to thin coax); and

(c) Terminate a segment.

In this last category, connectors actually connect the cable to a terminating resistor or an array of resistors and are consequently known as *terminators*. The type of connectors used is usually a function of cable type.

Connectors are also classified on their genders, and they do indeed "mate". You can use your imagination about how a connector's gender is derived, but a "plug" is usually label a clear specification of exactly which of a mating pair of connectors should be used in the particular application.

Connectors are also frequently labeled by their type.

Three common types are

1. DB type,
2. Centronics, and
3. DIN.

DB (data bus) connector serves as an interface between a computer and a peripheral device such a printer or external modem; they are distinguished as a rectangular row of "pins" (for male connectors) or "holes" (for female connectors). Several types of DB connectors exit and are distinguished by the number of pins they contain. Common types include DB-9 (a 9-pin serial or video interface), DB-15 (A 15-pin video interface), DB-25 (a 25-pin serial interface – RS-232 – or parallel printer interface), and DB-37 (a 37-pin serial interface based on RS-422).

DIN (Deutsche Industrie Norm, a German industrial standard) and centronics connectors are similar to DB-type connectors except DIN connectors are circular instead of rectangular, and centronics connectors contain "teeth" instead of pins. DIN connectors are typically used to connect a keyboard to computers.

Other connector types include coax, v-type, fiber, video, SCSI, and modular. A brief destination of these connectors based on cable type follows.

6.4.1. Registered Jack (RJ)

A **registered jack** (**RJ**) is a standardized physical network interface — both jack construction and wiring pattern — for connecting telecommunications or data equipment to a service provided by a local exchange carrier or long distance carrier. The standard designs for these connectors and their wiring are named **RJ11**, **RJ14**, **RJ21**, **RJ48**, etc. Many of these interface standards are commonly used in North America, though some interfaces are used world-wide.

The physical connectors that registered jacks use are mainly of the modular connector and 50-pin miniature ribbon connector types. For example, RJ11 uses a 6 position 2 conductor (6P2C) modular plug and jack, while RJ21 uses a 50-pin miniature ribbon connector.

There is much confusion over these connection standards. The six-position plug and jack commonly used for telephone line connections may be used for RJ11, RJ14 or even RJ25, all of which are actually names of interface standards that use this physical connector. The RJ11 standard dictates a 2-wire connection, while RJ14 uses a 4-wire configuration, and RJ25 uses all six wires. The RJ abbreviations, though, only pertain to the wiring of the jack (hence the name *registered jack*); it is commonplace but not strictly correct to refer to an unwired plug connector by any of these names.

The most familiar registered jack is probably the RJ11. This is a modular connector wired for one plain old telephone service line (using two wires out of six available positions), and is found in most homes and offices in most countries of the world for single-line telephones. Essentially all one, two, and three line analog telephones made today (2009) are meant to plug into RJ11, RJ14, or RJ25 jacks, respectively.

RJ14 is similar, but for two lines, and *RJ25* is for three lines. RJ61 is a similar registered jack for four lines. The telephone line cord and its plug are more often a true RJ11 with only two conductors.

6.4.2. RJ-45

Ethernet cables are terminated with RJ45 connectors. Often termed an *8P8C connector*, this means there are eight positions which are occupied by eight contacts and, it fits into a socket called a *module*.

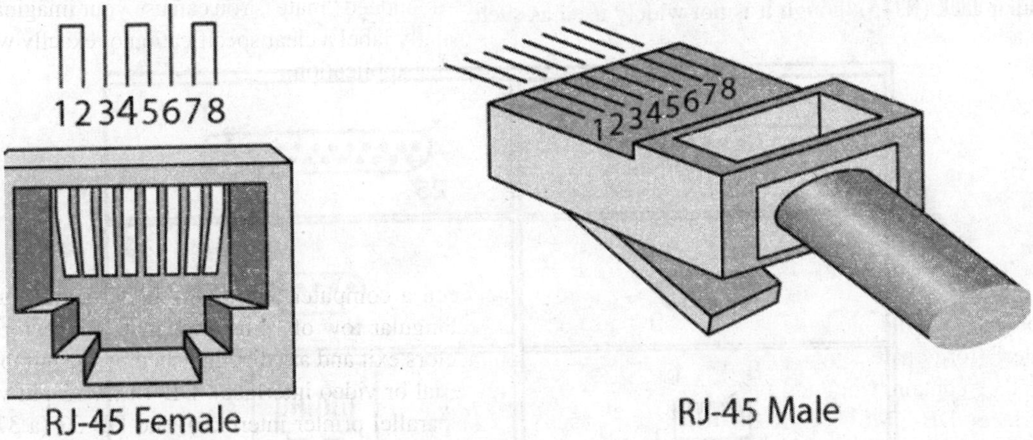

Fig. 6.17. RJ-45 connector.

When viewed end-on, RJ45 connectors are *oblong*. Do not confuse this connector with, the *RJ11 connector*. The RJ11 standard, also called 6P4C, is widely used to connect DSL routers to telephone lines. An RJ11 plug when viewed end-on is *square-ish* not oblong. The smaller 4P4C connector is used for American telephone cables.

RJ45 is a keyed connector, meaning the connector can be inserted in only one way.

Refer the Fig. 6.18, RJ45 connector wiring plan. There is no need to strip the eight cables bare before they are located inside the connector. The plug contacts have teeth that bite through the insulation in a technique called *insulation displacement*. The contacts must be punched down firmly in order to make a reliable contact.

Fig. 6.18. Connectors RJ-45 and RJ-11.

An RJ45 crimping tool is a worthwhile investment as it crimps all eight contracts at once. An automated cable tester is a nice-to-have accessory as it tests each conductive line at once. Should for some reason, the connector fail to make, it is often best to cut off the failed end and start again. A small bag of 8P8C connectors is a wise purchase. *Never* push an uncrimped connector into a module. The still proud contacts will foul the contact tangs, and prove a real nuisance to remove. That's experience talking.

Fig. 6.19. A correctly assembled RJ45 connector.

Although unnecessary for most inside wiring jobs, RJ45 connectors can be covered using protective *boots*. Available in a range of colours, boots are suitable for colour coding the otherwise grey and buff cables. When assembling ad-hoc cables, snag-*less* boots are preferable.

6.4.3. RS232 on RJ-45

RJ-45 (Registered Jack-45) is an eight-wire connector used commonly to connect computers onto local-area networks (LAN), especially Ethernets. In other words, RJ-45 is a single-line jack for digital transmission over ordinary phone wire, either untwisted or twisted. The interface has eight pins or positions. For faster transmissions in which you're connecting to an Ethernet 10BASET network, you need to use twisted pair wire. RS232D, EIA/TIA - 561 standard is applied when connecting to or from a serial port with a 8 position Modular Jack (RJ45) though it is not widely used as such.

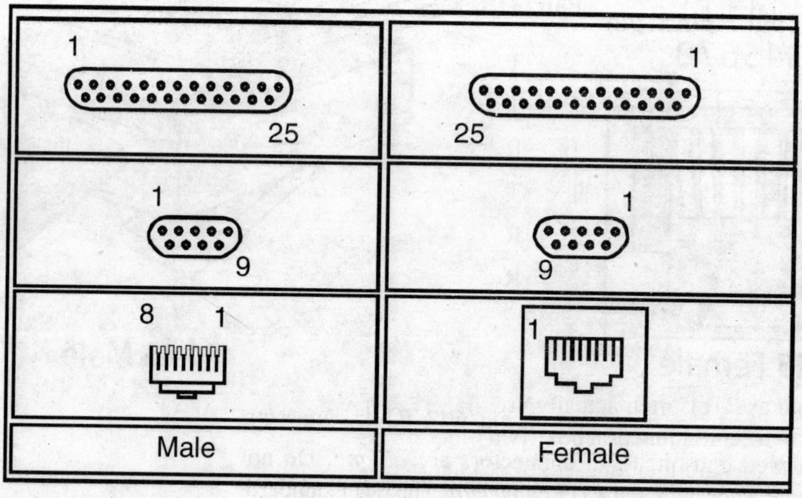

Fig. 6.20.

Figure 6.21 shows the details of Pin connections to connect DB-9 to RJ-45.

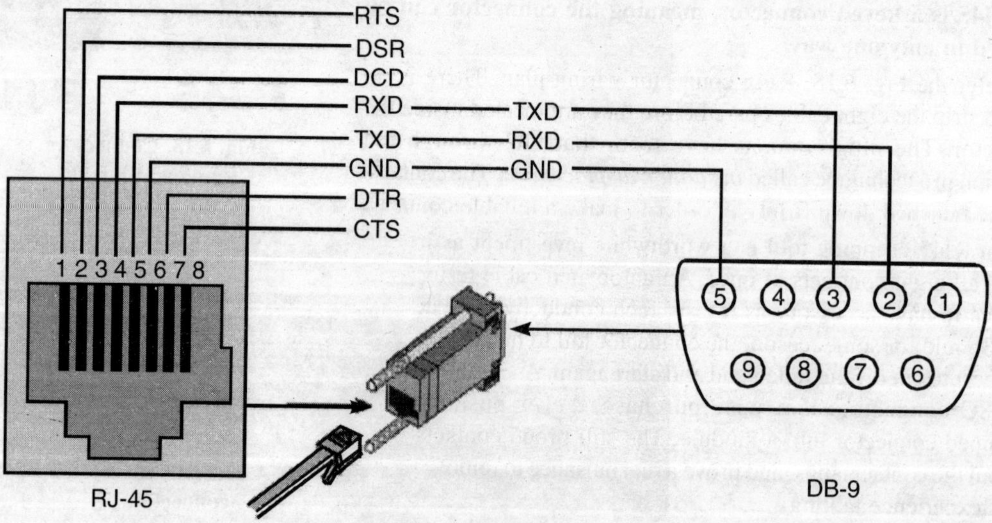

Fig. 6.21. Interconnections of DB-9 and RJ-45.74.

6.4.4. RJ-11

The six-position plug and jack commonly used for telephone line connections may be used for RJ11, RJ14 or even RJ25, all of which are actually names of interface standards that use this physical connector. The RJ11 standard dictates a 2-wire connection, while RJ14 uses a 4-wire configuration, and RJ25 uses all six wires. The RJ abbreviations, though, only pertain to the wiring of the jack (hence the name "registered jack"); it is commonplace but not strictly correct to refer to an unwired plug connector by any of these names. Typically you'll find the RJ-11 and RJ-14 on your Plain Old Telephone Service (POTS) though your RJ25 typically is found in an Integrated Services Digital Network (ISDN) Line. It is the communication standards that lets telephone lines to carry digital, voice, video, and network services.

position	RJ25 pin	RJ14 pin	RJ11 pin	Pair	T/R	±	Cat 5e/6 colors	Colors	Old colors
1	1			3	T	+	white/green	white/green	orange
2	2	1		2	T	+	white/orange	white/orange	black
3	3	2	1	1	R	–	blue	blue/white	red
4	4	3	2	1	T	+	white/blue	white/blue	green
5	5	4		2	R	–	orange	orange/white	yellow
6	6			3	R	–	green	green/white	blue

RJ11 Wiring

Cables sold as RJ11 often actually use 6P4C RJ14 connectors (six position, four conductor), with four wires running to a central junction box. Two of its six possible contact positions connect tip and ring, and the other two conductors are then unused. 6P2C and 6P6C can also be found in stores.

The conductors other than the two central tip and ring conductors are in practice used for various things such as a ground for selective ringers, low voltage power for a dial light, or for 'anti-tinkle' circuitry to prevent pulse dialing phones from ringing the bell on other extensions. With tone dialing, anti-tinkle measures are not required.

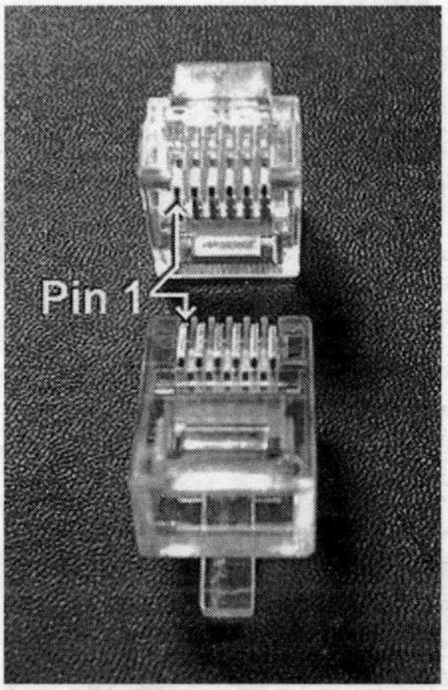

Fig. 6.22. 6P6C connector showing the location of pin 1.

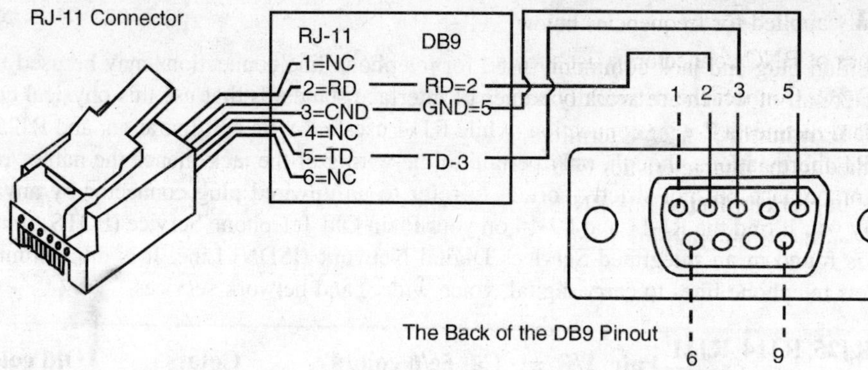

Fig. 6.23. Illustrates the interconnection of RJ-11 and DB-9.

6.4.5. BNC Connector

The **BNC connector (Bayonet Neill-Concelman connector)** is a common type of RF connector used for the coaxial cable which connects to various types of radio, television, and other radio-frequency electronic equipment.

The connector was named after its bayonet mount locking mechanism and its inventors, Paul Neill and Carl Concelman. Neill worked at Bell Labs and also invented the N connector; Concelman worked at Amphenol and also invented the C connector.

What does the BNC Connector do?

The BNC connector is configured to process both analog and serial digital interface video transmissions. The connector is also capable of handling transmissions for audio interface as well.

Sizes Available with BNC Connectors

Typically, BNC connectors are manufactured with 50 and 75 ohm impedance capability, with the 50 ohm impedance good for frequencies of up to 4 GHz and the 75 ohm impedance for up to 2 GHz.

How can a BNC Connector be used?

Many types of electronic test equipment components are configured for the BNC connector. Aviation equipment also frequently employs the use of BNC connectors as well. A third application is with amateur radio antenna connections, making it a popular option with ham radio operators.

Early in the development of consumer-based access to the Internet, the BNC connector was used with Ethernet networks. This application has mainly fallen out of favor, since modern Ethernet network components tend to use a construction that does not include coaxial cables.

Other Names for the BNC Connector

The BNC connector is often called other names. Among them are the Baby Neill-Concelman connector, the Baby N connector, the British Naval connector, and the Bayonet Nut connector.

Fig. 6.24.

It is usually applied for frequencies below 3 GHz and voltages below 500 Volts.

Three types of BNC connectors exist such as

- **BNC-T :** Connect the network boards in the PC to the network.
- **BNC Terminator :** A BNC terminator is a special connector that includes a resistor that is carefully matched to the characteristics of the cable system.
- **BNC barrel connector :** A BNC barrel connector connects to Thin net cables.
- **BNC cable connector :** It attach cable segment to the T-connectors.

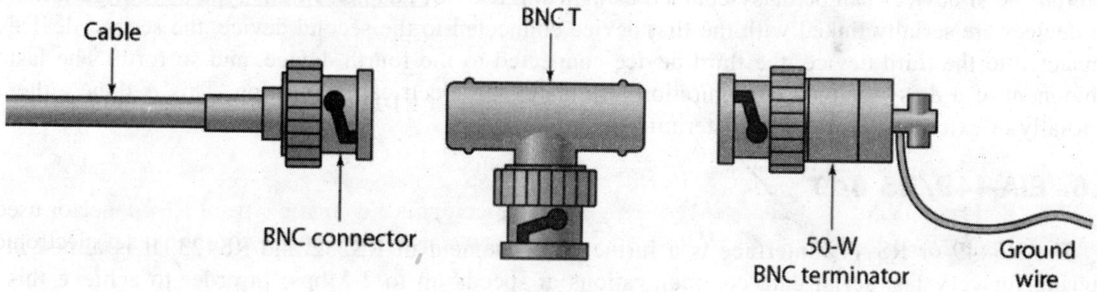

Fig. 6.25. BNC connectors.

6.5. FIBER OPTIC CABLES CONNECTORS

Several different types of fiber-optic connectors exist. Before the issuance of the EIA/TIA-568A standard, the most popular were ST connectors, which are similar to BNC connectors are those connectors in that you push and turn the connectors to mate them to a device. SC connectors are those approved by the EIA/TIA-568 A standard, by virtue of which they are destined to eclipse the popularity of the ST type. SC connectors, which are also called 568SC, are available in single and dual varieties, the latter designed to aid differentiation between transmit and receive fibers. SMA connectors use a threaded coupling mechanism – you attach them by screwing one end onto another. SMA connectors are designed to meet stringent military specifications. Another type of fiber-optic connector is the MIC (medium interface connector) used in FDDI networks (Figure 6.26).

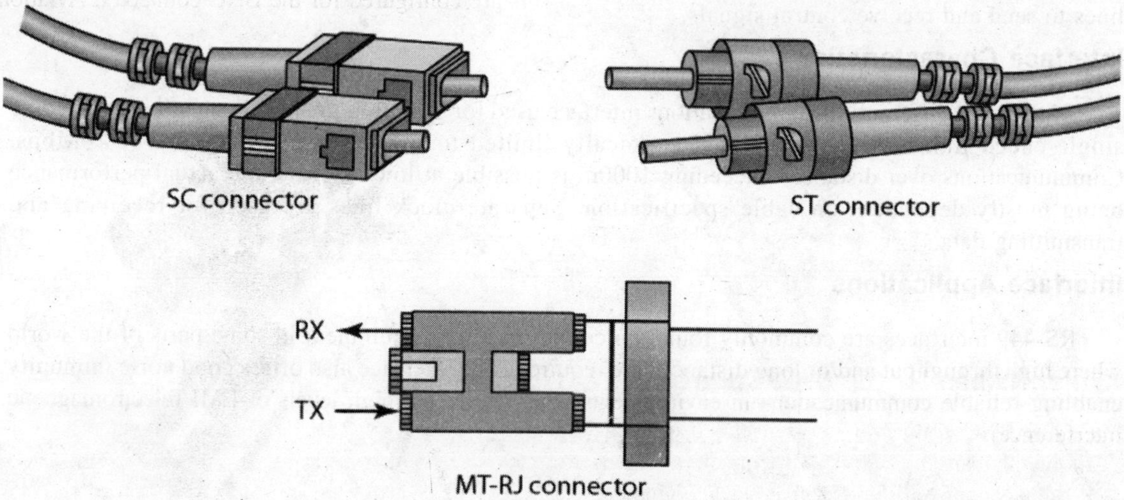

Fig. 6.26.

Other Connectors

Two other examples of connectors found in computer communications and networking environments include v.35 and SCSI connectors. V.35 connectors are high-speed serial interfaces typically found on routes, switches, high-speed modems, and CSU/DSUs. One common application is connecting routers to CSU/DSUs. SCSI (small computer systems interface; pronounced "scuzzy") connectors are high-speed parallel interfaces that enable computer peripheral devices such as scanners, CD-ROM drives, tape drives, and hard disk drives to connect directly to a computer. In most applications, up to seven different SCSI devices can be daisy-chained using a single SCSI adapter. In this type of configuration, the devices are serially linked with the first device connected to the second device, the second device connected to the third device, the third device connected to the fourth device, and so forth. The last component of a daisy-chained configuration terminates the electrical connection. This is done either internally or externally using a SCSI terminator.

6.6. EIA-449/RS 449

The RS449 or RS-449 interface is a further enhancement of RS232 and RS423. It is aimed at catering for very fast serial data communications at speeds up to 2 Mbps. In order to achieve this RS449 makes some changes when compared to RS232 to the way in which the signals are referenced, while still being able to retain some compatibility with RS232.

The **RS-449** specification, also known as **EIA-449** or **TIA-449**, defines the functional and mechanical characteristics of the interface between data terminal equipment and data communications equipment.

RS 449 is General purpose 37-Position Interface for Data terminal Equipment and Data Circuit-Terminating Equipment employing serial binary data interchange.

A serial mechanical interface standard for transmission of balanced and unbalanced signals between a variety of higher-end computer, media, and multimedia peripherals.

The standard specifies DC-37 and DE-9 for the primary and secondary data circuits.

EIA-449 allows a maximum data rate of 10 Mbps and uses a 37- or 9-pin connector. EIA-449 cabling uses EIA-422 differential pairs to send and receive data, and EIA-423 single-ended lines to send and receive control signals.

Interface Characteristics

RS-449 is a differential communications interface used for synchronous communications with some single-ended link management signals, typically limited to a maximum throughput of 10Mbps. Communications over distances exceeding 1000m is possible at low bit rates, the actual performance being mostly dependent on cable specification. Separate clock lines are used for receiving and transmitting data.

Interface Applications

RS-449 interfaces are commonly found on communications equipment in some parts of the world where high throughput and/or long distances are required. The interface also offers good noise immunity enabling reliable communications in environments where there are high levels of EMI (electromagnetic interference).

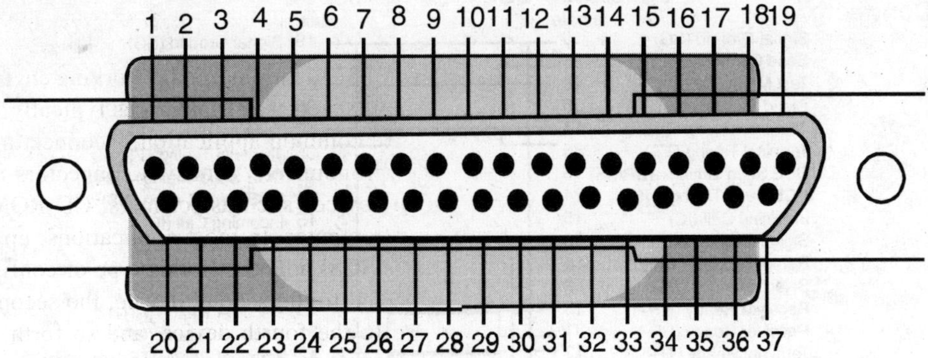

Fig. 6.27.

TABLE 6.4. Pin Configuration of EIA-449

Pin EIA A	B	Description Circuit	From	To DCE	DCE
1			Shield		
2		SI	Signalling Rate Indicator	*C	
4	22	SD	Send Data		*D
5	23	ST	Send Timing	*T	
6	24	RD	Received Data	*D	
7	25	RS	Request to Send		*C
8	26	RT	Request Timing	*T	
9	27	CS	Clear to Send	*C	
10		LL	Local Loopback		*C
11	29	DM	Data Mode	*C	
12	30	TR	Terminal Ready		*C
13	31	RR	Receiver Ready	*C	
14		RL	Remote Loopback		*C
15		IC	Incoming Call	*C	
16		SR	Signaling Rate Selector		*C
17	35	TT	Terminal Timing		*T
18		TM	Test Mode	*C	
19		SG	Signal Ground		
20		RC	Receive Common		
28		IS	Terminal in Service		*C
32		SS	Select Standby		*C
33		SQ	Signal Quality	*C	
34		NS	New Signal	*C	
36		SB	Standby Indicator		*C
37		SC	Send Common		

Signal Type : D = Data, C = Control, Timing

Note : On the DB37 connector that is commonly used for RS 449; PINs 3 and 21 are undefined. B = Return.

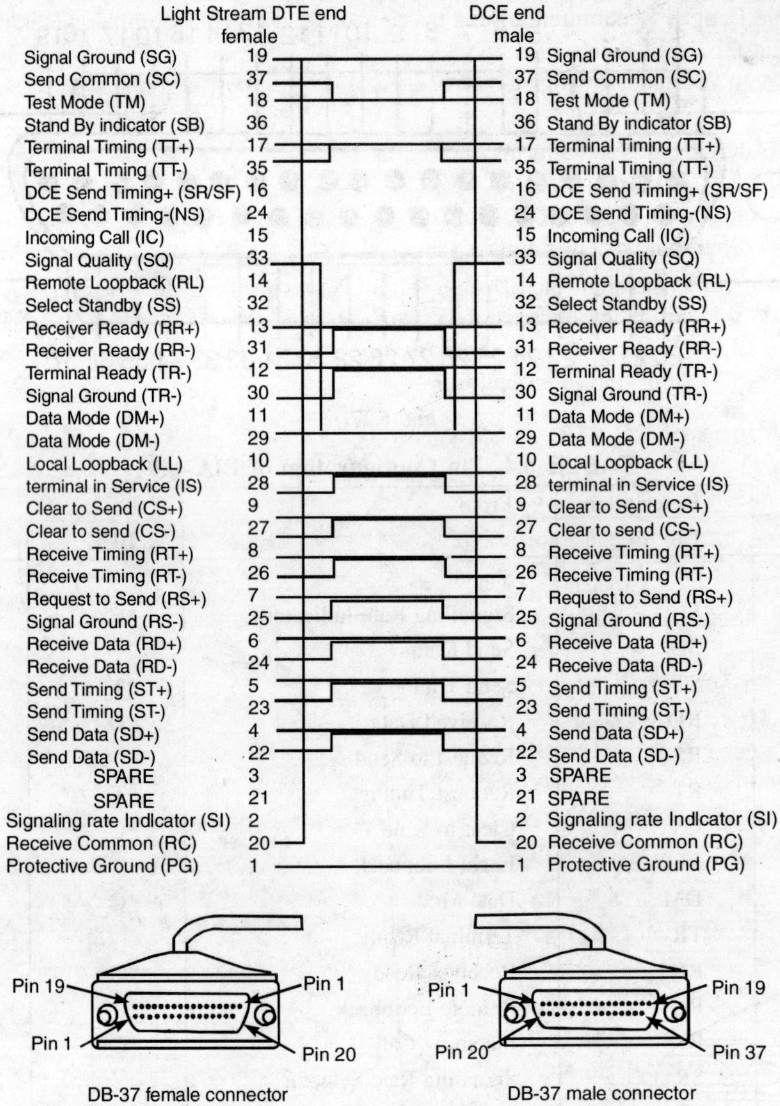

Fig. 6.28. Interfacing of Rs 449 DB-37 Connector

TABLE 6.5. DB 9 Pin Assignments

Pin	Signal
1	Shield
2	Secondary Receiver Ready
3	Secondary Send Ready
4	Secondary Receiver Data
5	Signal Ground
6	Received Common
7	Secondary Request to Send
8	Secondary Clear to Send
9	Send Common

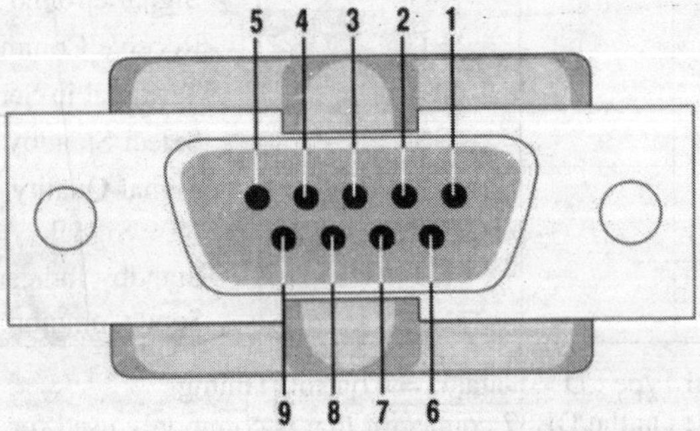

Fig. 6.29. EIA 449 secondary interface.

RS449 Cable Length Recommendations : The cable length will determine maximum bus speed, or the maximum transfer rate is determined by the cable length (examples): 2Mbps out to 50 feet, 1Mbps to 100 feet, 512kbps to 200 feet, 256kbps to 400 feet, 128kbps to 800 feet, 56kbps to 1600 feet, 1.2kbps to 3000 feet.

Normally EIA422 and 423 systems may not be connected together. EIA-449 cabling of 422 sends and receives data as differential pairs and control signal as single-ended, but for 423 cabling it sends and receives single-ended data and control signals. Receiving the single ended signal in 423 is accomplished by grounding the 'B' side of the differential receiver at the connector.

So if the system does not follow the EIA-449 (cabling) specification than one (B) side of the differential receiver of the 423 side will be grounded at the connector forcing the differential driver on the 422 side to drive ground. The EIA422 side also uses a termination resistor between the ends of its differential receiver, providing a serious 120 ohm short to ground for the 423 driver.

Pin Configurations of RS 449

TABLE 6.6. RS-449- 9 Pin Connector Pin Out

Pin	Signal Name	Signal Function	Pin	Signal Name	Signal Function
1	Shield	Ground	-	-	-
2	Sec Receive Ready	Control from DCE	6	Receive Common	Ground
3	Sec Send Data	Data to DCE	7	Sec Request to Send	Control from DCE
4	Sec Receive Data	Data from DCE	8	Sec Clear to Send	Control from DCE
5	Signal Ground	Ground	9	Send Common	Ground

EIA-449-1 was rescinded in January, 1986. Superseded by EIA/TIA-530-A, the final version EIA-449-1 was withdrawn in September, 2002.

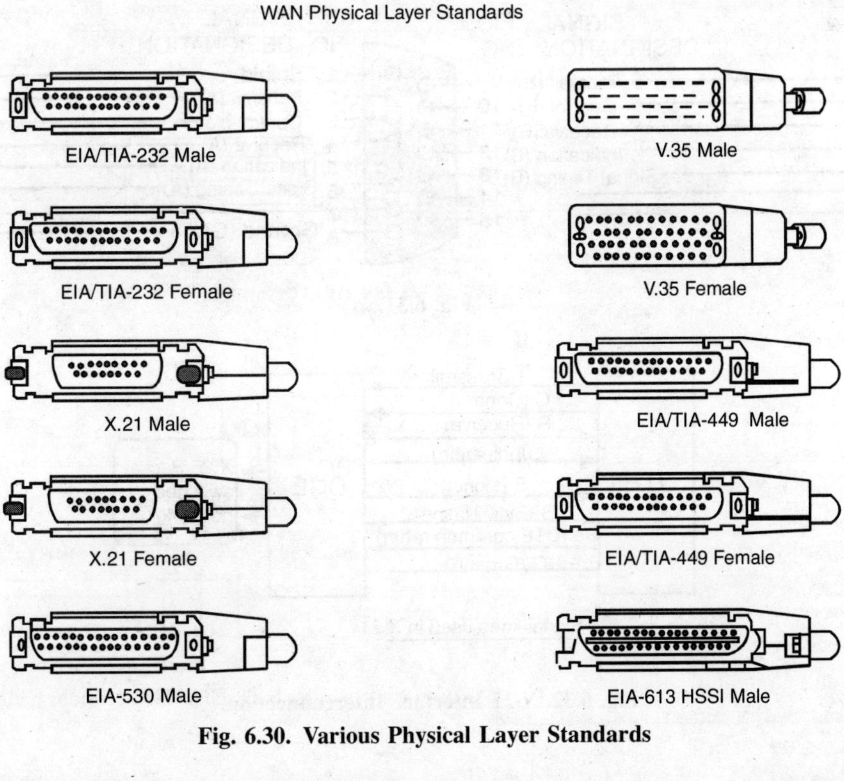

Fig. 6.30. Various Physical Layer Standards

6.7. DIGITAL INTERFACE: X.21

X.21 Physical Layer Interface

The X.21 interface was recommended by the CCITT in 1976. It is defined as a digital signalling interface between customers (DTE) equipment and carrier's equipment (DCE). And thus primarally used for telecom equipment.

All signals are balanced. Meaning there is always a pair (+/-) for each signal, like used in RS422. The recommendation specifies how the DTE, sets up and clears calls by exchanging signals with the carrier's equipment, the DCE.

Voltages:	+/- 0.3Vdc
Speeds:	Max. 100Kbps (X.26)
	Max. 10Mbps (X.27)

Figure 6.31 shows the connections between the DTE and DCE.

X.21 interface is also used as the terminations interface for digital leased circuits of n x 64kbps. The connector, which is known as DB-15 is shown in the Fig. 6.31. The names and functions of the eight wires defined by X.21 are given in the following Fig. 6.32. The physical connector has 15 pins, but not all of them are used. The DTE uses the T and C lines to transmit data and control information, respectively. The DCE uses the R and I lines for data and control. The S line contains a signal stream emitted by the DCE to provide timing information, so the DTE knows when each bit interval starts and stops. At the carrier's option, a B line may also be provided to group the bits into 8-bit frames. If this option is provided, the DTE must begin each character on a frame boundary. If the option is not provided, both DTE and DCE must begin every control sequence with at least two SYN characters, to enable the other one to deduce the implied frame boundaries. All the signal lines use the balanced drivers and receivers.

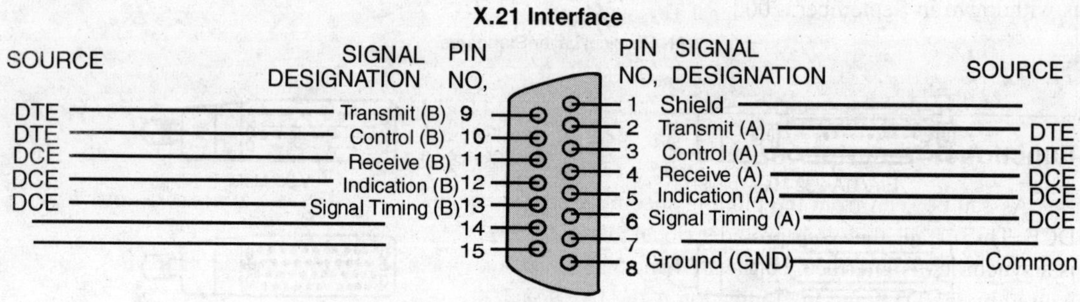

Fig. 6.31.

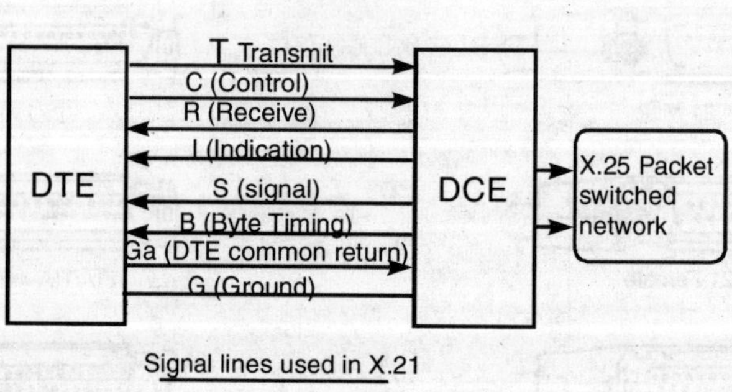

Fig. 6.32. X.21 Interface Interconnection.

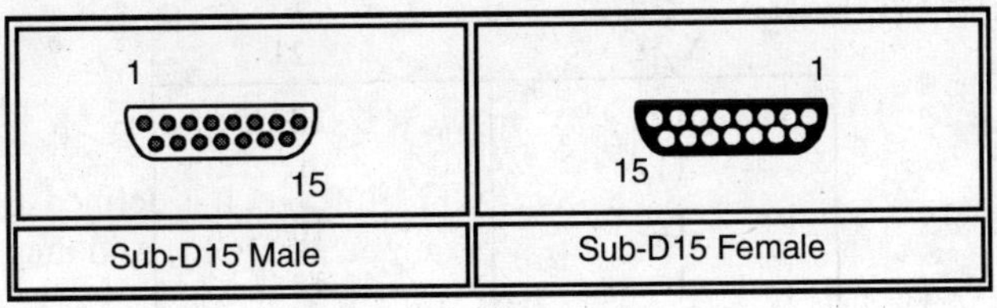

Fig. 6.33. D-15 X.21.

TABLE 6.7. Pinning according to ISO 4903.

Pin number	Signal	DTE	DCE
1	Shield	-	
2	Transmit (A)	Out	In
3	Control (A)	Out	In
4	Receive (A)	In	Out
5	Indication (A)	In	Out
6	Signal Timing (A)	In	Out
7	Unassigned		
8	Ground	-	-
9	Transmit (B)	Out	In
10	Control (B)	Out	In
11	Receive (B)	In	Out
12	Indication (B)	In	Out
13	Signal Timing (B)	In	Out
14	Unassigned		
15	Unassigned		

Functional Description

As can be seen from the pinning specifications, the Signal Element Timing (clock) is provided by the DCE. This means that your provider (local telco office) is responsible for the correct clocking and that X.21 is a synchronous interface. Hardware handshaking is done by the Control and Indication lines. The Control is used by the DTE and the Indication is the DCE one.

Cross-Cable pinning

TABLE 6.8. X.21 Cross Cable

X.21	X.21
1	1
2	4
3	5
4	2
5	3
6	7
7	6
8	8
9	11

TABLE 6.9. X.21 Cross Cable

X.21	X.21
10	12
11	9
12	10
13	14
14	13
15	

A variation of X.21 known as X.21 bis is a CCIR recommendation that defines the analog interface to allow access to the digital circuit switched networks using an analog circuit. X.21 bis provides procedures for sending and receiving addressing information which enable a DTE to establish switched circuits with other DTEs.

DO YOU KNOW▶

The CCIR recommendation X.21 defines the physical layer interfaces between a node (DTE) and the network connection (DCE).

6.8. MODEMS

*A modem is a **mo**dulator/**dem**odulator*. The analog signals encode the digital information at the time of modulation and decode it back during demodulation to transmit the data.

Modems allow digital transmissions over analog telephone lines. They allow people to connect their computers with other computers, corporate office LANs, and the Internet. Modems are used to establish connections of the telephone company's *local loop,* which is the analog copper cable that runs between homes and businesses to the telephone company central office (CO). The phone company then switches the call like any other voice call to create a point-to-point link to the destination.

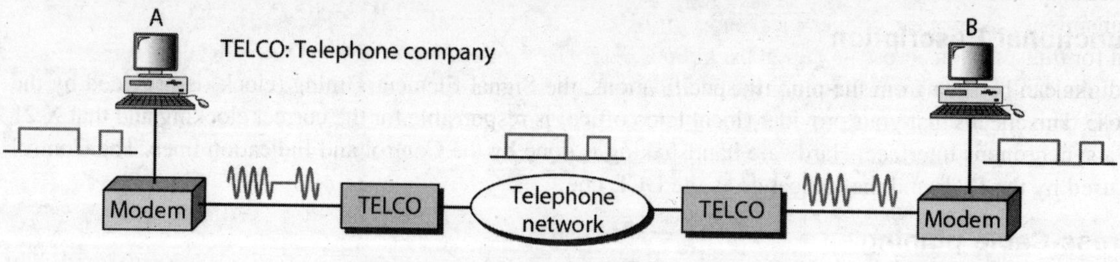

Fig. 6.34.

It converts digital signals from a computer to analog signals so they can be sent over a telephone line. The receiving computer will normally have a modem which will convert the analog signals back to digital signals. A modem may be an internal modem which is a card inside your computer or it may be external with a connection to a serial RS-232 line on your computer.

The telephone jack that will plug into the modem is called an RJ-11 jack. Speed of the modem is measured in bits per second (bps). There are a set of V-series standards developed by the International Telecommunications Union which indicate the speed of the modem.

DO YOU KNOW ▶

A device which carries out the two important tasks of modulating and demodulating the analog signals is called a ***modem****.*

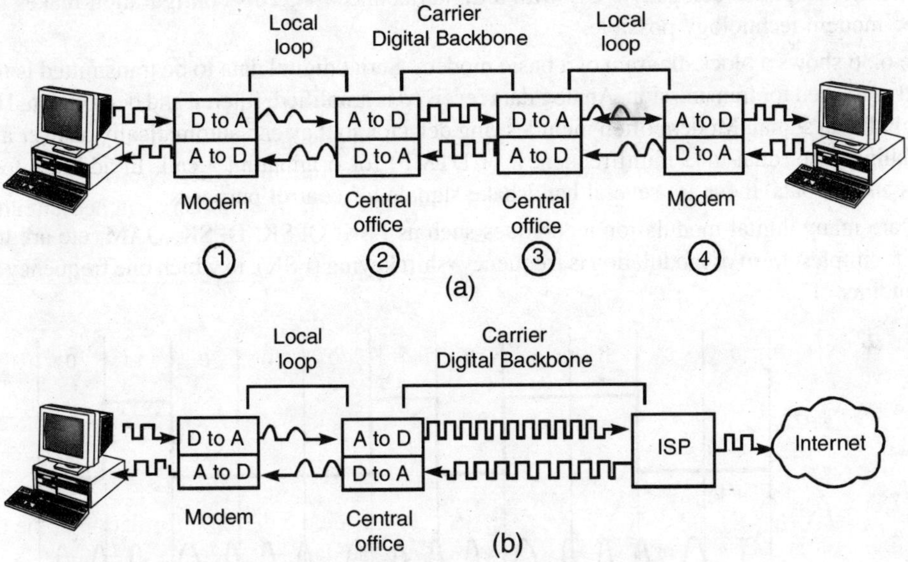

Fig. 6.35. Role of MODEM

Before the Internet, the usual connection was for one user to dial another user, or for a user to dial into a modem at a bulletin board, place of business, or a school (rather than connect through the Internet, as discussed next). In these cases, the connection is modem to modem across the PSTN, as shown at the top in Fig. 6.35(*a*). Note that in the figure, the computer's outgoing digital signal is converted to an analog signal for transmission across the copper local loop to the CO. At the CO, the analog signal is converted back to digital for transmission across its digital backbones. At the destination CO, the signal is once again converted from digital to analog and transmitted to the end user's modem, where it is converted back to digital. Remember all those conversions.

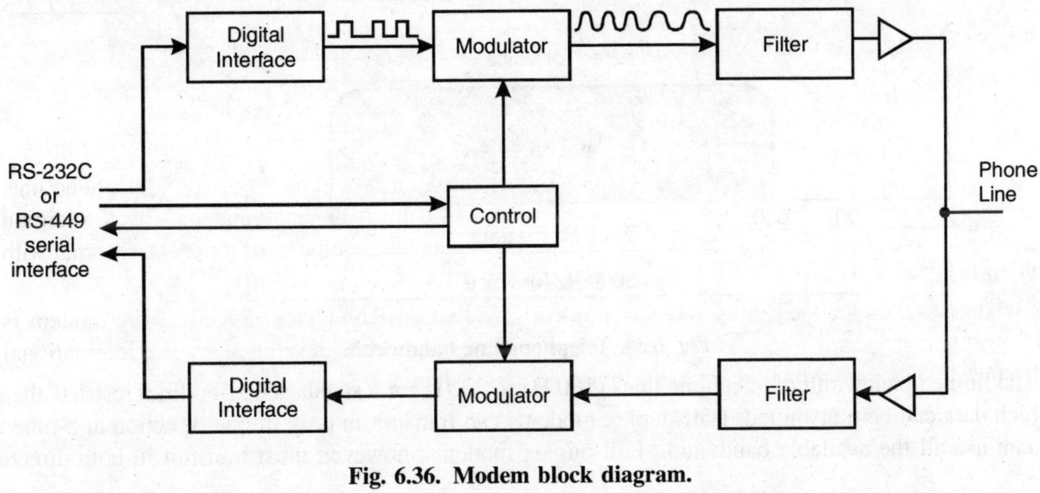

Fig. 6.36. Modem block diagram.

Today, most users dial into the Internet, where they connect with other users via e-mail, collaborative applications, instant messaging, and so on. Users also connect with their home networks over the Internet by using secure VPN technology. This allows users to dial into a local number and connect to the corporate office across the Internet. The bottom half of Fig. 6.35 (*b*) illustrates how most modem connections are made in today's Internet-connected environment. The important point of this configuration is that the ISP is connected to the telephone company's CO with a digital connection. This configuration makes high-speed 56-Kbit/sec modem technology possible.

Figure 6.36 shows a block diagram of a basic modem. Serial digital data to be transmitted is modulated, filtered and amplified for transmission. Analog data received is amplified, filtered and demodulated to produce a serial digital data signal. Modem often includes ring detector so they can automatically answer a telephone call and dialing tone (dual-tone multifrequency, or DTMF) for originating a call. In addition to the serial data inputs and outputs, there are several handshake signals for control purposes.

There are many digital modulation techniques such as FSK, QPSK, DPSK, QAM, etc are used in the Modem. The simplest form of modulation is frequency–shift keying (FSK), in which one frequency represents a '0' and another '1'.

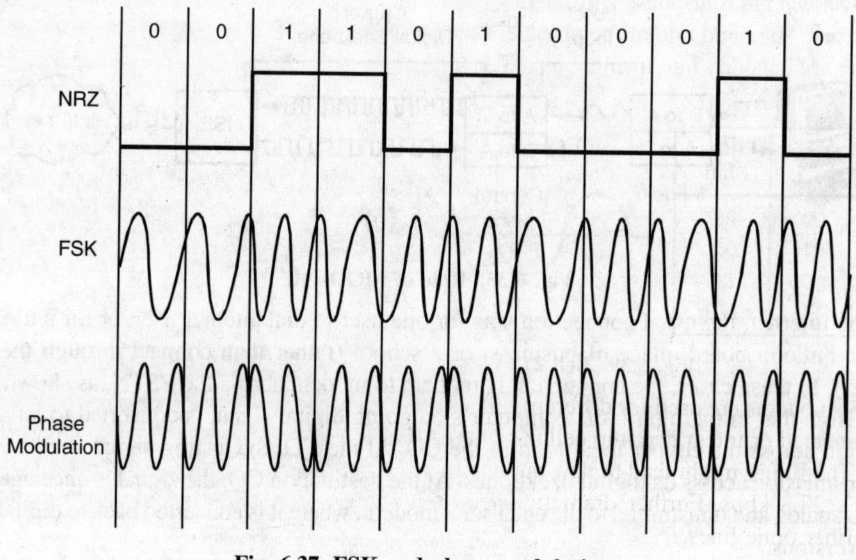

Fig. 6.37. FSK and phase modulation.

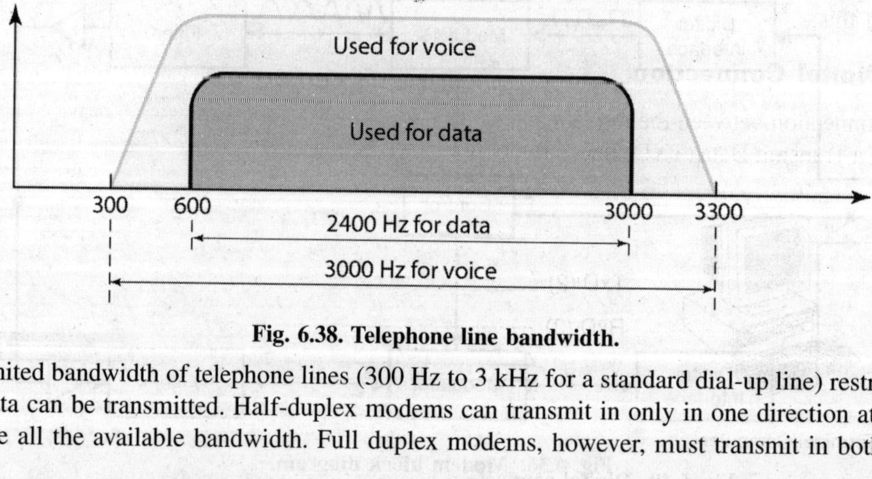

Fig. 6.38. Telephone line bandwidth.

The limited bandwidth of telephone lines (300 Hz to 3 kHz for a standard dial-up line) restricts the rate at which data can be transmitted. Half-duplex modems can transmit in only in one direction at a time and thus can use all the available bandwidth. Full duplex modems, however, must transmit in both directions simultaneously. Unless two separate lines are used, the bandwidth must be shared. Each end transmits using different frequencies. To identify which end should transmit using which frequencies, a distinction is made between the originating modem and the answering modem.

The older modem standard in common use is the Bell 103, which uses FSK modulation to transmit up to 300 bps, full duplex. Each modem contains receiver filters that allow only the signal from the other modem to be received, so the modem is not confused by its own transmissions.

It uses the following frequency assignments:

Originate Modem: 0 = 1070 Hz
1 = 1270 Hz

Answer Modem: 0 = 2025 Hz
1 = 2225 Hz

Another common standard is Bell 212A, which provides 1200 bps full duplex operation. It uses the dibit phase modulation. The transmit carrier frequency is 1200 Hz for the originate modem and 2400 Hz for the answering modem. The term dibit refers to the fact that the two bits are encoded in each phase shift. Figure 6.39 shows the four phase shift angles used in 212 A type modem and an example waveform with a 2400 Hz carrier. The baud rate of the phase encoded signal is the rate of the rate of phase changes. Since the two data bits are encoded into a single phase change, the 1200 bps data stream produces a 600 baud analog signal.

Dibit	Phase Shift
00	90°
01	0°
10	180°
11	270°

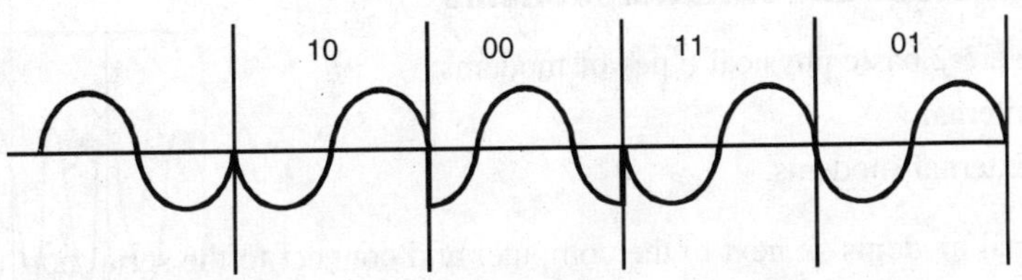

Fig. 6.39. Bell 212A Phase Modulation Dibit waveforms.

Today's modems are used for different functions. They act as textual and voice mail systems, facsimiles, and are connected or integrated into cellular phones and in notebook computers enabling sending data from anywhere. The future might lead to new applications. Modem speeds are not expected to be increased much over today's 28.8 kbps. Further dramatic speed increases will require digital phone technology such as ISDN and fiber optic lines.

New applications might be implemented such as simultaneous voice and data. Videophones are an example of this.

6.8.1. Digital Connection

The connection between the modem and terminal/computer is a digital connection. A basic connection consists of a Transmit Data (TXD) line, a Receive Data (RXD) line and many hardware hand-shaking control lines.

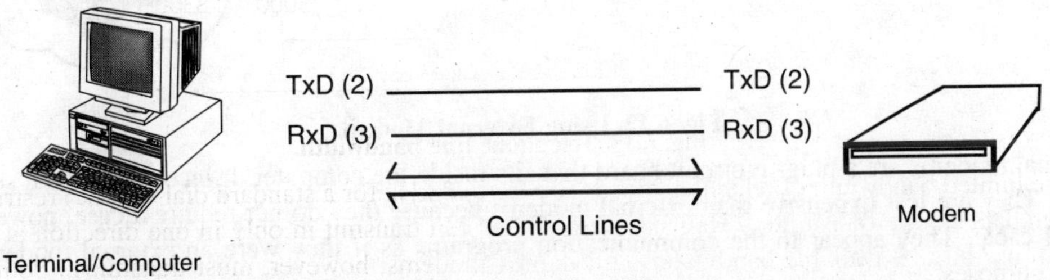

Fig. 6.40. Digital connection between the modem and terminal.

The control lines determine whose turn it is to talk (modem or terminal), if the terminal/computer is turned on, if the modem is turned on, if there is a connection to another modem, etc.

6.8.2. Analog Connection

The connection between the modem and outside world (phone line) is an analog connection. The Voice Channel has a bandwidth of 0-4 kHz but only 300 - 3400 Hz is usable for data communications.

The modem converts the digital information into tones (frequencies) for transmitting through the phone lines. All of the tones are restricted to the narrow 300-3400 Hz Voice Band.

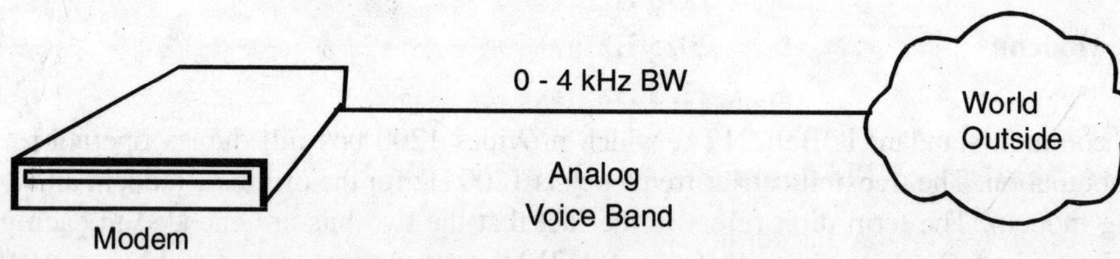

Fig. 6.41. Analog connection. The connection between the modem and outside world (phone line).

6.8.3. External and Internal Modems

There are 2 basic physical types of modems:

1. Internal
2. External modems.

External modems sit next to the computer and connect to the serial port using a straight through serial cable.

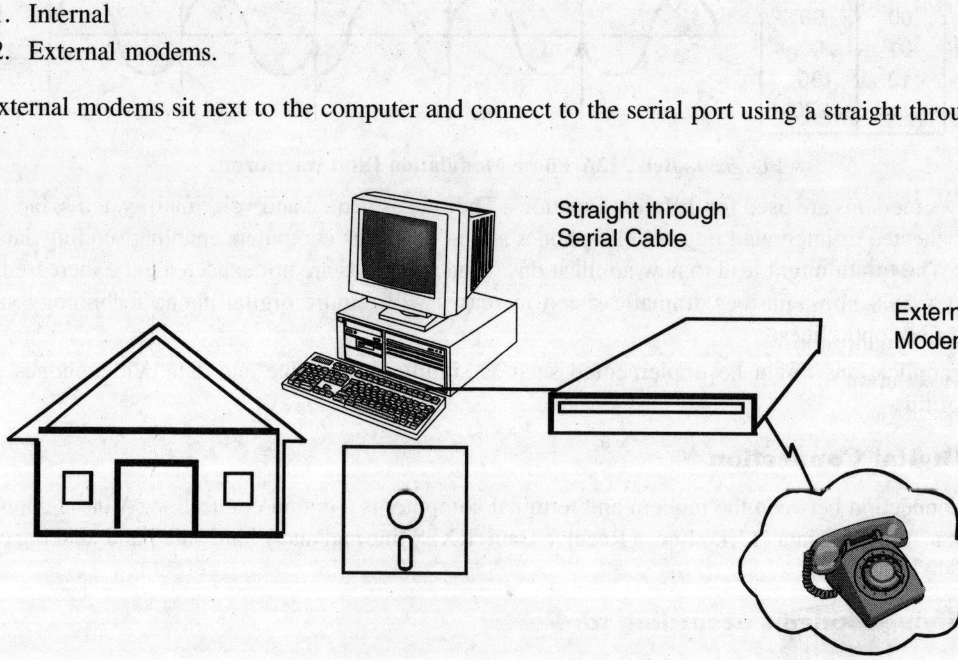

Fig. 6.42. Using External Modem.

Internal modems are a plug-in circuit board that sits inside the computer. It incorporates the serial port on-board. They are less expensive than external modems because they do not require a case, power supply and serial cable. They appear to the communication programs as if they were an external modem for all intensive purposes.

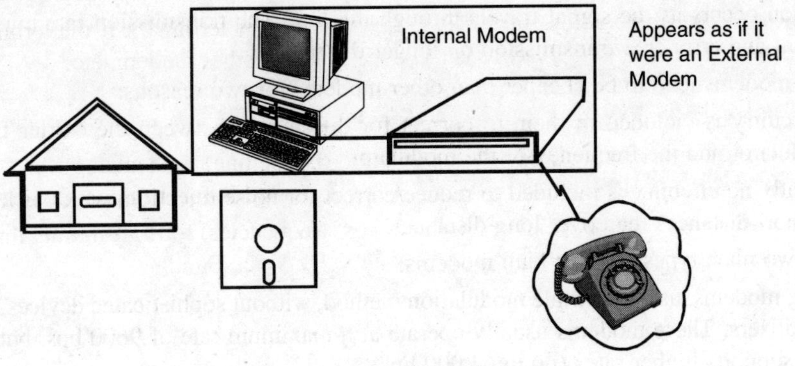

Fig. 6.43. Using Internal Modem.

6.8.4. Classification of Modems

The modems can be classified according to their characteristics:

1. Range
   (*i*) Short Haul
   (*ii*) Voice Grade (VG)
   (*iii*) Wideband
2. Line Type
   (*i*) Dial-up
   (*ii*) Leased
   (*iii*) Private
3. Operation Mode
   (*i*) Half Duplex
   (*ii*) Full Duplex
   (*iii*) Simplex
4. Synchronization
   (*i*) Asynchronous
   (*ii*) Synchronous
5. Modulation
   (*i*) AM
   (*ii*) FM/FSK
   (*iii*) PM
   (*iv*) TCM
6. Data Rate

1. Classifying Modems according to: Range

(i) Short Haul

Short haul modems are cheap solutions to systems of short ranges (up to 15 km), which use private lines and are not part of a public system. Short haul modems can also be used, even if the end-to-end length of the direct connection is longer than 15 km, when both ends of the line are served by the same central office in the telephone system. These lines are called "local loops". Short haul modems are distance-sensitive, because signal attenuation occurs as the signal travels through the line. The transmission rate must be lowered to ensure consistent and error-free transmission on longer distances.

Short haul modems tend to be cheaper than other modems for two reasons:

(*a*) No circuitry is included in them to correct for differences between the carrier frequency of the demodulator and the frequency of the modulator.

(*b*) Generally no circuitry is included to reduce/correct for noise rejection, which is less of a problem over short distances than over long distances.

There are two main types of short haul modems:

1. Analog modems, using a simple modulation method, without sophisticated devices for error control or equalizers. These modems usually operate at a maximum rate of 9600 bps, but there are some which supports higher rates (up to 64,000 bps).
2. Line drivers increase the digital signal, which transmit to the communication channel and do not transmit the carrier signal, as conventional modems. Line drivers are very cheap and tiny and connect to the RS232 connector of the terminal (since they lack a power supply, they use the signal voltage of the DTE-DCE interface for DC power supply).

(ii) Voice Grade (VG)

Voice-grade modems are used for unlimited destination, using a moderate to high data rate. These modems are expensive and their maintenance and tuning are sophisticated. Communication channels are leased lines and dial-up.

Voice-band telephone network is used for data transmission. A user-to-user connection may be either dedicated or dialed. The links in the connection are the same in the two cases, and the only difference for the user is that for some impairment (particularly attenuation and delay distortion), a dedicated (*private* or leased) line is guaranteed to meet certain specifications, whereas a dialed connection can only be described statistically.

(iii) Wideband

Wideband modems are used in large-volume telephone-line multiplexing, dedicated computer-to-computer links. These modems exceed high data rates.

2. Classifying Modems according to: Line Type

(i) Leased, Private

Leased, private or dedicated lines (usually 4-wire) are for the exclusive use of "leased-line" modems - either pair (in a simple point-to-point connection) or several (on a multidrop network for polling or a contention system). If the medium is the telephone network, their transmission characteristics are usually guaranteed to meet certain specifications, but if the link includes any radio transmission, the quality of it may be as variable as that of a switched (*i.e.* nondedicated) line.

(ii) Dial-Up

Dial-up modems can establish point-to-point connections on the PSTN by any combination of manual or automatic dialing or answering. The quality of the circuit is not guaranteed, but all phone companies establish objectives. The links established are almost always 2-wire because 4-wire dialing is tedious and expensive.

(iii) Two and Four-Wires Lines

A four-wire (4W) line is a pair of two-wire (2W) lines, one for transmitting and one for receiving, in which the signals in the two directions are to be kept totally separate. Perfect separation can be maintained only if the four-wire configuration is sustained from transmitter to receiver. The lines may be combined in a

4W/2W network (often called a *hybrid* or a *hybrid transformer*) at any point in the signal path. In this case impedance mismatches will cause reflections and interference between the two signals.

3. Classifying Modems According to : Operation Mode

(i) Half Duplex

Half duplex means that signals can be passed in either direction, but not in both simultaneously. A telephone channel often includes an echo-suppressor, allowing transmission in only one direction; this renders the channel half-duplex. Echo suppressors are slowly being replaced by echo cancellers, which are theoretically full-duplex devices.

When a modem is connected to a two-wire line, its output impedance cannot be matched exactly to the input impedance of the line, and some part of its transmitted signal (usually badly distorted) will always be reflected back. For this reason half- duplex receivers are disabled (received data is clamped) when their local transmitter is operative.

Half-duplex modems can work in full-duplex mode.

(ii) Full Duplex

Full duplex means that signals can be passed in either direction, simultaneously. Full duplex operation on a two-wire line requires the ability to separate a receive signal from the reflection of the transmitted signal. This is accomplished by either FDM (frequency division multiplexing) in which the signals in the two directions occupy different frequency bands and are separated by filtering, or by Echo Canceling (EC).

The implication of the term *full-duplex* is usually that the modem can transmit and receive simultaneously at *full* speed. Modems that provide a low-speed reverse channel are sometimes called *split-speed* or *asymmetric* modems.

Full duplex modems will not work on half-duplex channels.

(iii) Simplex

Simplex means that signals can be passed in one direction only. A remote modem for a tele-metering system might be simplex and a 2-wire line with a common unidirectional amplifier is simplex.

(iv) Echo Suppressor and Echo Canceller

At the junction between the local loop, which is usually a 2-wire circuit, and the trunk, which is a 4-wire circuit, echoes can occur. The effect of the echo is that a person speaking on the telephone hears his own words after a short delay. Psychological studies have shown that this is annoying to many people, often making them stutter or become confused. To eliminate the problem of echoes, echo suppressors are installed on lines longer than 2000 km. (On short lines the echoes come back so fast that people cannot detect them). An *echo suppressor* is a device that detects human speech coming from one end of the connection and suppresses all signals going the other way. The device compares the levels at its two input ports, and if it decides, for example that the other end is talking, it inserts an attenuator in the return (echo) path, and vice versa.

Echo suppressors have several properties that are undesirable for data communication. First, they prevent full- duplex data transmission, which would otherwise be possible, even over the 2-wire local loop (by allocating part of the bandwidth to the forward channel and part to the reserve channel). Even if half-duplex transmission is adequate, they are a nuisance because the time required to switch directions can be substantial. Double-talking totally confuses them, and the attenuation may be switched in and out repeatedly. Furthermore, they are designed to reverse upon detecting human speech, not digital data.

To reduce these problems, when echo suppressors detect a specific tone they shut down, and remain shut down as long as the carrier is present (this is an example of *inband signaling*, where control signals that activate and deactivate internal control functions lie within the band accessible to the user). This disabling is usually done during initial handshaking by one modem transmitting an *answer* tone in either 2100 Hz (CCITT standard) or 2225 Hz (modems following the old Bell 103 standard).

Echo suppressor are slowly being replaced by ECs, which allow a certain amount of double-talking and do not require "capture" time for any one talker to assume control of the connection.

4. Classifying Modems according to : Synchronization

(i) Asynchronous Modems

The common modem used today is asynchronous modem. In **asynchronous** transmission, data is coded into a series of pulses, including a start bit and a stop bit. A start bit is sent by the sending modem to inform the receiving modem that a character is to be sent. The character is then sent, followed by a stop bit designating that the transfer of that bit is complete. Each byte is placed between a stop and a start bit. Each modem must operate with the same start and stop bit sequence. Each modem must operate at the same baud rate. Each modem must have the same parity settings for the data checking in order to communicate correctly. In asynchronous transmission parity checking is defined.

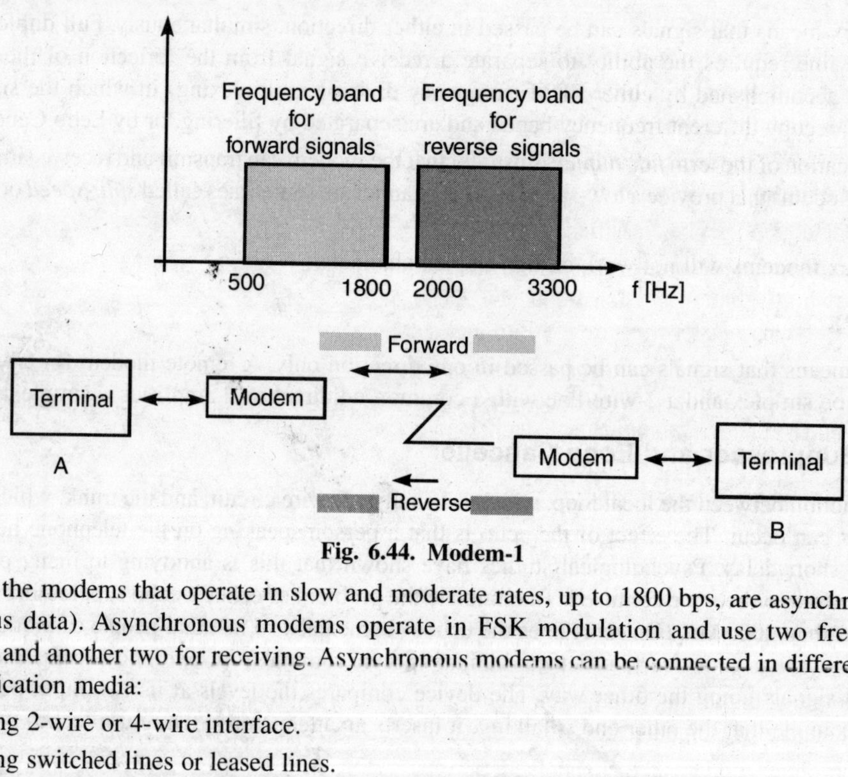

Fig. 6.44. Modem-1

Most of the modems that operate in slow and moderate rates, up to 1800 bps, are asynchronous (using asynchronous data). Asynchronous modems operate in FSK modulation and use two frequencies for transmission and another two for receiving. Asynchronous modems can be connected in different options to the communication media:

- Using 2-wire or 4-wire interface.
- Using switched lines or leased lines.
- Using interface to call unit/automatic answer, when dialing-up.

In a 2-wire line, full duplex operation can be achieved by splitting the channel into two sub-channels.

Asynchronous Data

Asynchronous data is not accompanied by any clock, and the transmitting and receiving modems know only the nominal data rate. To prevent slipping of the data relative to the modems' clocks, this data is always grouped in very short blocks (characters) with framing bits (start and stop bits). The most common code used for this is the seven-bit ASCII code with even parity.

(ii) Synchronous Modems

Synchronous modems can be faster than asynchronous. They depend on timing to communicate. In **synchronous** data transmission, data is sent via a bit-stream, which sends a group of characters in a single stream. Data is transmitted in frames with synchronization bits which are used to be sure the timing of the transmission and reception of data is accurate. In order to do this, modems gather groups of characters into a buffer, where they are prepared to be sent as such a stream. In order for the stream to be sent, synchronous modems must be in perfect synchronization with each other. They accomplish this by sending special characters, called synchronization, or syn, characters. When the clocks of each modem are in synchronization, the data stream is sent. Synchronous modems are normally used on dedicated leased lines.

Synchronous modems are one of binary synchronous communications protocol (bisync), high level data link control (HDLC), or synchronous data link control (SLDC). Three methods can be used to control synchronization:

(*a*) Additional clock signal

(*b*) Guaranteed state change - Clocking is part of the data signal.

(*c*) Over sampling - The receiver samples the signal much faster than the data is sent. The extra samples can be used to be sure the clock is synchronized

Synchronous modems operate in the audio domain, at rates up to 28800 bps in audio lines, used in telephones systems (using synchronous data). The usual modulation methods are the phase modulation and integrated phase and amplitude (at higher rates than 4800 bps).

In synchronous modems, equalizers are used, in order to offset the misfit of the telephone lines. These equalizers are inserted in addition to the equalizers that sometimes already exist in the telephone lines.

These equalizers can be classified into three main groups:

(*a*) ***Fixed/statistical equalizer*** **:** These equalizers offset the signal according to the average of the known attenuation in each frequency. Tuning the equalizer is sometimes done in the factory and stays fixed, usually they are used to operate at low rates in a dial up line.

(*b*) ***Manually adjusted equalizer*** **:** These equalizers can be tuned to optimal performance to a given line. These equalizers should be re-tuned when the line is replaced and periodically. Specially, it should be tuned frequently when the line is of a low quality and it's parameters are changed frequently. Tuning is done using a button inside the modem (or on the external board).

(*c*) ***Automatic equalizer*** **:** These equalizers are tuned automatically when the connection is established. Depending on the line quality in a specific moment, in a process of about 15ms to 25ms, after the first tuning, the equalizer samples the line continually and adjusts itself to the changed conditions, so the modem operates at each moment under optimal conditions. The fitness process operates, in some modems, at rates of 2400 times in a second.

Synchronous modems operate in the same manner asynchronous modems. However, synchronous modems operate at higher rates and since the requirements to transmit at these rates are increasing, most of the innovations are implemented for synchronous modems.

In synchronous modems the channel can be split for several consumers at various speeds. Modems that have this ability are called **SSM - Split System Modem**. These modems can use a simple split or a split using multipoint connection.

Synchronous data is accompanied by a clock signal. Synchronous data is almost always grouped in blocks, and it is the responsibility of the data source to assemble those blocks with framing codes and any extra bits needed for error detecting and/or correcting according to one of many different protocols (BISYNC, SDLC, HDLC, etc.). The data source and destination expect the modem to be transparent to this type of data, conversely, the modem can ignore the blocking of the data.

5. Classifying Modems according to : MODULATION

Communication channels like telephone lines are usually analog media. Analog media is a bandwidth limited channel. In the case of telephone lines the usable bandwidth frequencies is in the range of 300 Hz to 3300 Hz.

Data communication means moving digital information from one place to another through communication channels. These digital information signals have the shape of square waves and the meaning of "0" and "1".

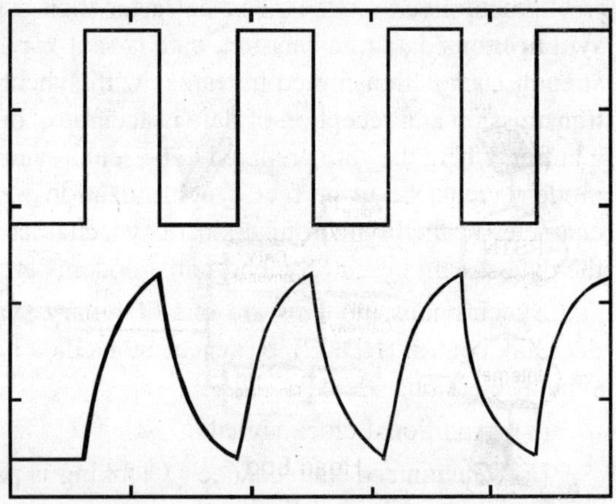

Fig. 6.45. Modem–Modulation.

If such digital signals were transmitted on analog media the square waves of the digital signals would be distorted by the analog media as shown in Fig. 6.45. The receiver which receives these distorted signals will be unable to interpret accurately the incoming signals. These digital signals must be converted into analog signals so that the communication channels can carry the information from one place to another. The technique which enables this conversion is called ***modulation*.**

6.8.5. Limited Distance Modems

Limited distance Modems are designed for the complete frequency band of the non-loaded copper transmission line and their application is limited to short distances only as the transmission medium distortions and attenuation increases with distance. The distance limitation is a function of bit-rate and cable characteristics. As the distance increasing, the transmission speeds decreasing. The transmission rate at 20 km distance is 1200bps and at 8 km it is 19200 bps on 26 gauge cable. It usually requires four-wire unloaded connection between modems.

6.9. A CABLE MODEM (CM)

Figure 6.46 shows that the two-way cable television system adds a cable modem transmission system (CMTS) at the head-end and a cable modem (CM) at the customer's location. The CMTS also provides an interface to other networks such as the Internet. The regional cable head (RCH) normally serves up to 400,000 subscribers. The RCHs feed the distribution hubs, each of which serves up to 40,000 subscribers. The distribution hub plays an important role in the new infrastructure. Modulation and distribution of signals are done here; the signals are then fed to the fiber nodes through fiber-optic cables. The fiber node splits the analog signals so that the same signal is sent to each coaxial cable. Each coaxial cable serves up to 1000 subscribers. The use of fiber-optic cable reduces the need for amplifiers down to eight or less. One reason for moving from traditional to hybrid infrastructure is to make the cable network bidirectional (two-way).

A **cable modem** is a device that MOdulates/DEModulates data signals on a coaxial cable and divides the high data rate signals into digital signals designated for a specific user. Cable modems are often asymmetrical modems as the data transfer rate in the downstream direction is typically much higher than the data transfer in the upstream direction. The typical gross (system) downstream data rates range between 30-40 Mbps and gross upstream data rates typically range up to 2 Mbps.

Usually 500 to 2000 users share the gross data transfer rate on a cable system.

Cable modems also have the requirement to divide the high-speed digital signals into low-speed connections for each user. In 2001, the average data rates for a cable modem users was approximately 720 kbps.

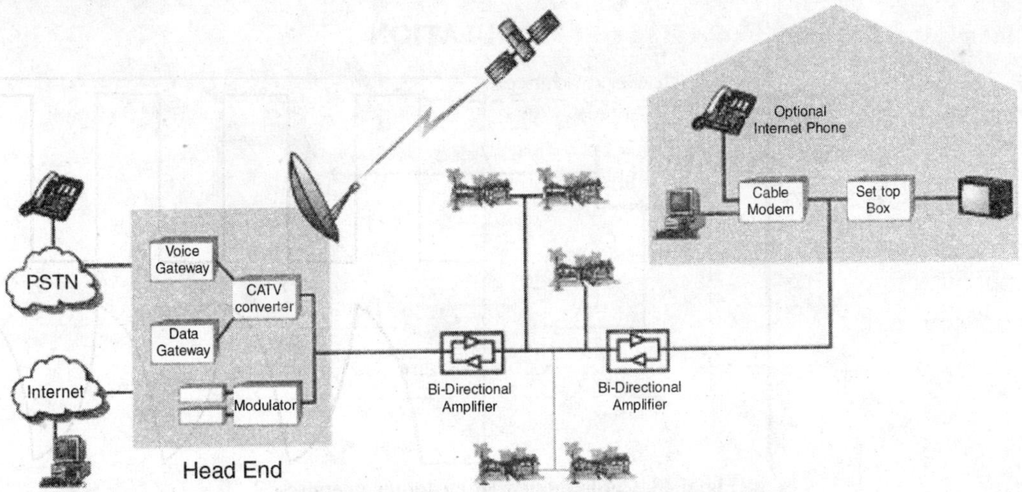

Fig. 6.46. shows a two-way cable television system.

> **DO YOU KNOW▸**
>
> *Downstream data are modulated using the 64-QAM modulation technique.*

Figure 6.47 shows a block diagram of a cable modem. This diagram shows that a cable modem has a tuner to convert an incoming 6 MHz RF channel to a low frequency baseband signal. This signal is demodulated to a digital format, demultiplexed (separated) from other digital channels, and is decompressed to a single data signal. This data signal is connected to a computer typically in Ethernet (e.g. 10 Base T Ethernet) or universal serial bus (USB) data format. Data that is sent to the modem is converted to either audio signals for transfer via a telephone line (hybrid system) or converted to an RF signal for transmission back through the cable network.

> **DO YOU KNOW▸**
>
> *Upstream data are modulated using the QPSK modulation technique.*

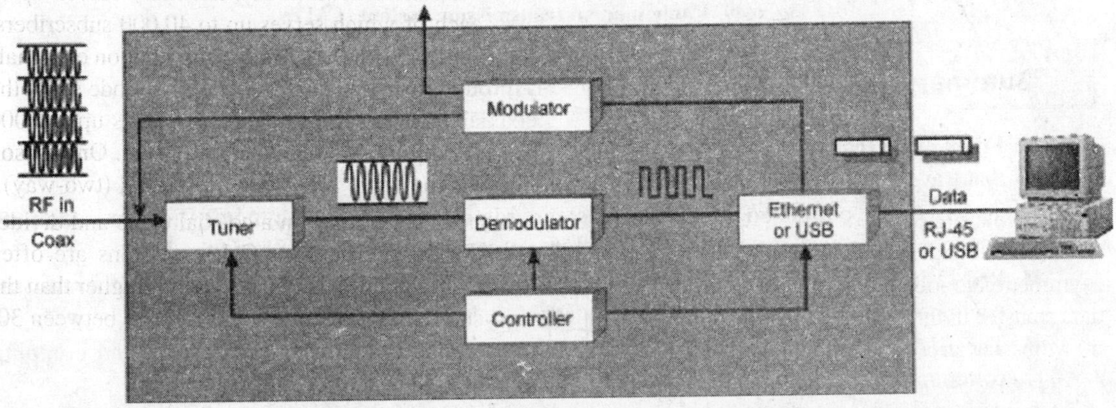

Fig. 6.47. A simple block diagram of Cable modem

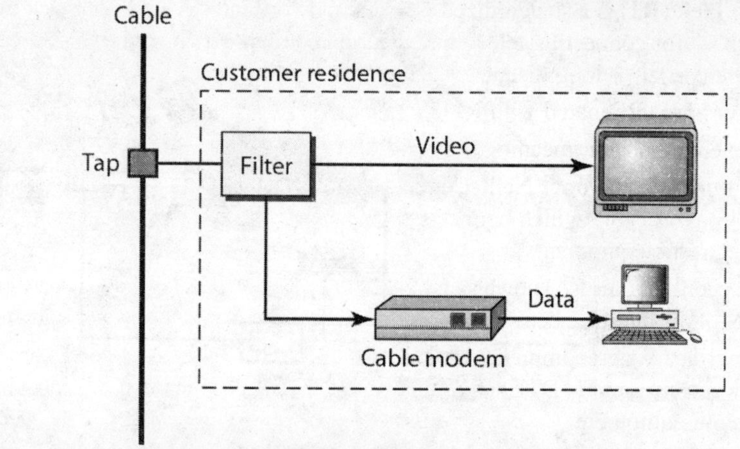

Fig. 6.48. Cable modem at customer premises

6.10. CABLE MODEM TRANSMISSION SYSTEM (CMTS)

The cable modem transmission system (CMTS) is installed inside the distribution hub by the cable company. It receives data from the Internet and passes them to the combiner, which sends them to the subscriber. The CMTS also receives data from the subscriber and passes them to the Internet. Figure 6.49 shows the location of the CMTS.

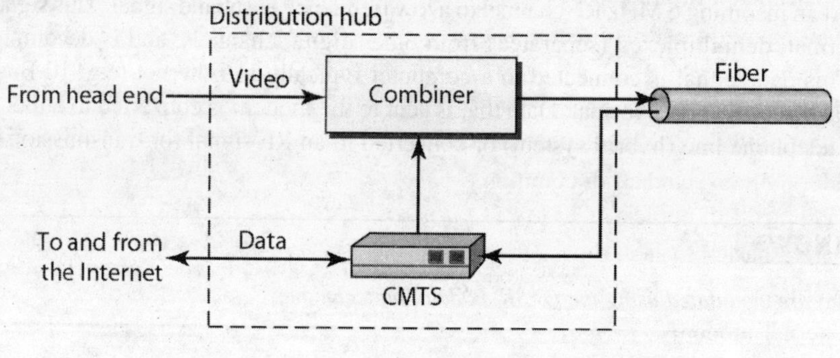

Fig. 6.49. Cable modem transmission system (CMTS)

Summary

- **Fiber optic transmission** is the transmission of information using light waves as a carrier signal that travel through an optical waveguide (optical fiber).
- The EIA has published a number of standards related to telecommunication and computer communication, and works closely with other associates such as ANSI and the ITU.
- RS232 interface is used to connect Data terminal equipment (DTE) and Data circuit terminating equipment (DCE). It is used for serial interface.
- The EIA has published a number of standards related to telecommunication and computer communication, and works closely with other associates such as ANSI and the ITU.
- RS232 interface is used to connect Data terminal equipment (DTE) and Data circuit terminating equipment (DCE). It is used for serial interface.

- A **registered jack** (**RJ**) is a standardized physical network interface — both jack construction and wiring pattern — for connecting telecommunications or data equipment to a service provided by a local exchange carrier or long distance carrier.
- Ethernet cables are terminated with RJ45 connectors.
- RJ45 is a keyed connector, meaning the connector can be inserted in only one way.
- The **BNC connector** (**Bayonet Neill-Concelman connector**) is a common type of RF connector used for the coaxial cable which connects to various types of radio, television, and other radio-frequency electronic equipment.
- The **RS-449** specification, also known as **EIA-449** or **TIA-449**, defines the functional and mechanical characteristics of the interface between data terminal equipment and data communications equipment.
- The X.21 interface was recommended by the CCITT in 1976. It is defined as a digital signalling interface between customers (DTE) equipment and carrier's equipment (DCE). And thus primarally used for telecom equipment.
- **EIA/TIA-530**; High Speed 25-Position Interface for Data Terminal Equipment and Data Circuit-Terminating Equipment, Including Alternative 26-Position Connector EIA530 Defines a cabling interface between a DTE and DCE device, with pin outs.
- **Category 1, which have the balanced electrical characteristics of EIA-422-A, and Category 2, which have the unbalanced electrical characteristics of EIA-423**.
- *A modem is a* ***modulator/demodulator****.* The analog signals encode the digital information at the time of modulation and decode it back during demodulation to transmit the data.
- A device which carries out the two important tasks of modulating and demodulating the analog signals is called a **modem**.
- There are many digital modulation techniques such as FSK, QPSK, DPSK, QAM, etc are used in the Modem.
- Half-duplex modems can transmit in only in one direction at a time and thus can use all the available bandwidth. Full duplex modems, however, must transmit in both directions simultaneously.
- The older modem standard in common use is the Bell 103, which uses FSK modulation to transmit up to 300 bps, full duplex.
- There are 2 basic physical types of modems:
  1. Internal
  2. External modems.
- Short haul modems are cheap solutions to systems of short ranges (up to 15 km), which use private lines and are not part of a public system.
- Voice-grade modems are used for unlimited destination, using a moderate to high data rate.
- Leased, private or dedicated lines (usually 4-wire) are for the exclusive use of "leased-line" modems - either pair (in a simple point-to-point connection) or several (on a multidrop network for polling or a contention system).
- Dial-up modems can establish point-to-point connections on the PSTN by any combination of manual or automatic dialing or answering.
- An *echo suppressor* is a device that detects human speech coming from one end of the connection and suppresses all signals going the other way.
- Echo suppressor are slowly being replaced by ECs, which allow a certain amount of double-talking and do not require "capture" time for any one talker to assume control of the connection.
- In **asynchronous** transmission, data is coded into a series of pulses, including a start bit and a stop bit.

- Synchronous modems can be faster than asynchronous. They depend on timing to communicate. In **synchronous** data transmission, data is sent via a bit-stream, which sends a group of characters in a single stream.
- Three methods can be used to control synchronization:
  (*a*) Additional clock signal
  (*b*) Guaranteed state change - Clocking is part of the data signal.
  (*c*) Over sampling - The receiver samples the signal much faster than the data is sent. The extra samples can be used to be sure the clock is synchronized
- Communication between two devices might work only when the interface is defined and agreed. For modems, the standards define techniques used for modulation, for error correction for data compression, and other attributes.
- The **X-modem protocol** divides the data into blocks. Each block contains a sequence number of the block, 128 Bytes of the data and 4 Bytes of checksum.
- In the Y-modem protocol, each block contains 1024 Bytes. The checksum size is 4 Bytes.
- **Z-modem** is a public domain program written by Chuck Forsberg at Omen Technology.

Review Questions

1. Explain the following :
   [1] RJ-45
   [2] ETA-449 digital interface
2. Discuss the v-series modem standards.
3. Define and explain Shannon limit theorem.
4. What is the maximum data rate of channel? Explain the theorems related to it.
5. What do you understand by physical layer interface and explain the functions of it.
6. Describe physical layer interface device RS-232.
7. What is the effect of limited bandwidth on digital signal?
8. Write about the various transmission media and derive the expression for Shannon limit on data rate.
9. What does the Shannon capacity have to do with communication? Calculate the maximum bit rate for the noiseless channel with a bandwidth 3000 Hz transmitting a signal a signal with two signal levels.
   (Hints : Capacity = BW $\times \log_2$ (1+ SNR), bit rate = 6kbps)
10. What is Nyquist theorem? What is the bit rate, if we want to digitize the human voice, assuming 8 bits per sample?
11. What does Shannon capacity have to do with communications?
12. Explain what crosstalk is and what is needed to reduce it?
13. What is the significance of Shannon capacity?
14. What are the functions of a DTE? What are the functions of a DCE? Give an example of each.
15. What standard organizations are involved in DTE-DCE interface standards? Name some popular DTE-DCE standards.
16. What implementations of EIA-232 are available? How are they different?
17. What is the purpose of null modem? Describe the data pins of null modem.
18. How is X.21 able to eliminate most of the control circuits of EIA standards?
19. What factors affect the data rate of a link?
20. Define the bandwidth of a line. What is the bandwidth of a traditional telephone line?

21. What is the difference between a balanced circuit and an unbalanced circuit?

22. What is the relationship between the data rate and the distance that the data can reliably travel on an EIA interface?

23. What does the mechanical specifications, electrical specifications and functional specifications of EIA-232 describe?

24. Explain the working of MODEM by using a simple block diagram.

25. What do you understand by synchronous and asynchronous MODEM ?

26. Compare synchronous and asynchronous MODEM.

27. Explain cable MODEM in brief.

28. Write short notes on following:

(*i*) Physical layer interface

(*ii*) RS-232

(*iii*) X.21

(*iv*) Data rate of channel

(*v*) RJ-45,

(*vi*) RJ-11,

(*vii*) BNC connector

(*viii*) EIA-449

(*ix*) Digital Interface

(*x*) Modem

□□□

CHAPTER 7

TRANSMISSION MEDIA

INSIDE THIS CHAPTER

7.1. INTRODUCTION

On any network, the various entities must communicate through some form of media. Human can communicate through telephone lines or sound wave in the air; computer can communicate through cables, light and radio waves. So the media used to transmit and receive the information from one point to another point, is called the **Transmission Media**. A transmission medium may be broadly defined as anything can carry information from source to destination. In data communication, the information is usually a signal that is the result of a conversion of data from another form. Figure 7.1 shows the simple block diagram of sender–receiver using transmission media.

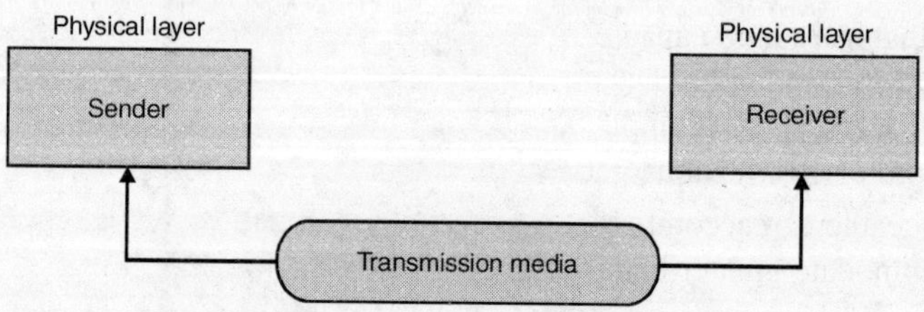

Fig. 7.1

We know that, computer and other telecommunications devices use signals to represent the data. These signals are transmitted in the form of electromagnetic energy from one device to another device, which is propagated through transmission media. The characteristics and quality of a data transmission are defined by the properties of medium and character of the signal. The Transmission media used are mainly classified according to the frequency used as following (Refer Fig. 7.2) :

DO YOU KNOW ▸

*A **transmission medium** may be broadly defined as anything can carry information from source to destination.*

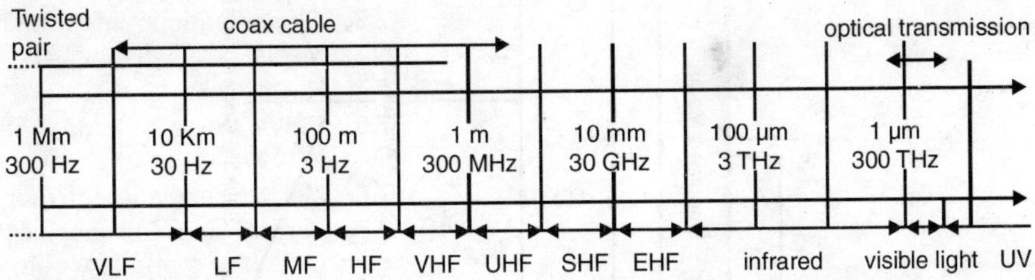

VLF = Very Low Frequency
LF = Low Frequency
MF = Medium Frequency
HF = High Frequency
VHF = Very High frequency
UHF = Ultra High Frequency
SHF = Super High Frequency
EHF = Extra High Frequency
UV = Ultraviolet Light

Frequency and wave length
$\lambda = c/f$
wave length λ, speed of light $c \cong 3 \times 10^8$ m/s, and frequency f.

Fig. 7.2. Frequency band used in communication.

7.2. CHARACTERISTICS OF TRANSMISSION MEDIA

Each type of transmission media has special characteristics that make it suitable for a specified type of service. These characteristics are as follows :

1. Bandwidth
2. Band usage (Baseband or Broadband)
3. Attenuation
4. Cost
5. Installation requirements
6. Immunity from EM interference

In computer networking, the term bandwidth refers to the measure of the capacity of a medium to transmit data. A medium that has a high capacity, for example, has a high bandwidth, whereas a medium that has limited capacity has a low bandwidth. Data transmission rates frequently are stated in terms of the bits that can be transmitted per second. For example, an Ethernet LAN theoretically transmits 10 million bits per second and has a bandwidth of 10 Megabits per second (Mbps).

The bandwidth that a cable can accommodate is determined in part by the length of the cable. A short cable generally can accommodate greater bandwidth than a long cable.

7.3. CLASSIFICATIONS OF TRANSMISSION MEDIA

Generally, transmission media are classified (as shown in Fig. 7.3) in two categories such as :

1. Guided transmission media
2. Unguided transmission media

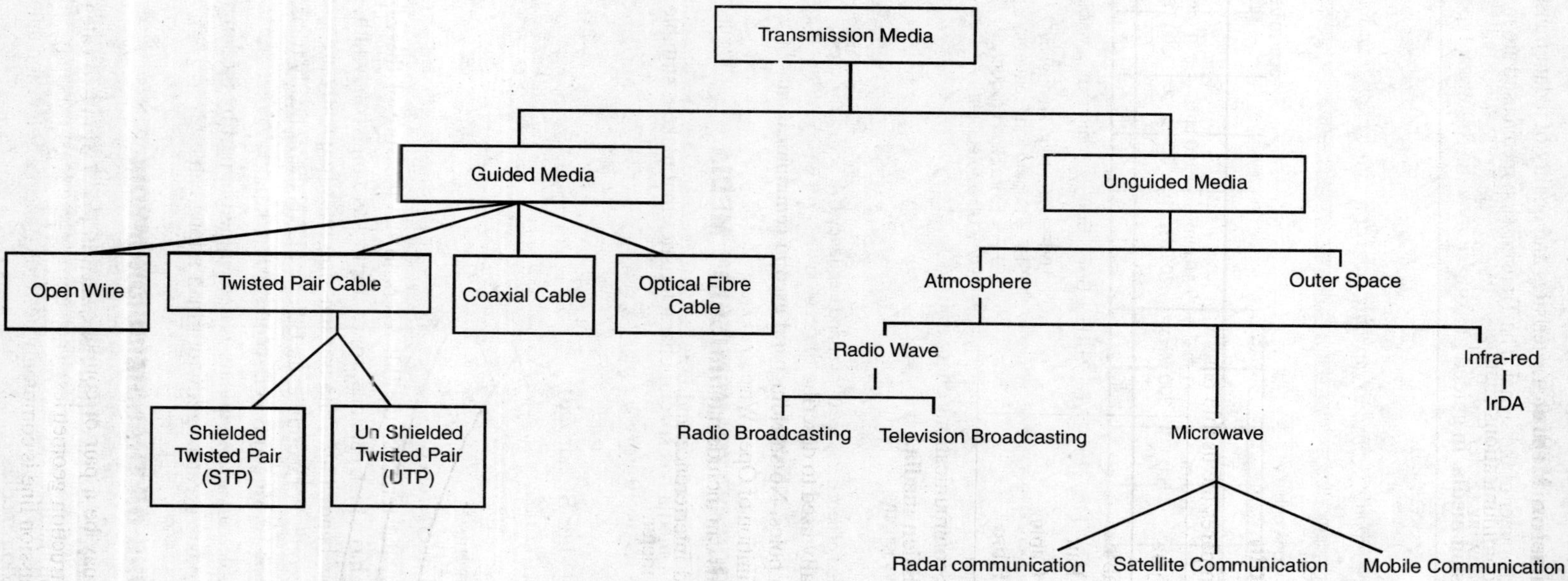

Fig. 7.3. Classification of transmission media.

7.3.1. Guided Transmission Media

Transmission Media or communications channels can be a physical connection between correspondents, known as **Guided transmission media.** In guided transmission media following types of medium are there :

1. Open Wire
2. Twisted pair cables
3. Coaxial cables and
4. Optical fibers cables

Media versus Bandwidth

The following table compares the usable bandwidth between the different Guided Transmission Media

Cable Type	Bandwidth
Open Cable	0 - 5 MHz
Twisted Pair	0 - 100 MHz
Coaxial Cable	0 - 600 MHz
Optical Fibre	0 - 10 GHz

Transmission Media or communications channels can be a physical connection between correspondents, known as **Guided transmission media.**

7.3.1.1. Open Wire

Open Wire is traditionally used to describe the electrical wire strung along a telephone poles. There is a single wire strung between poles. No shielding or protection from noise interference is used. We are going to extend the traditional definition of Open Wire to include any data signal path without shielding or protection from noise interference. This can include multiconductor cables or single wires. This media is susceptible to a large degree of noise and interference and consequently not acceptable for data transmission except for short distances less than 8 meter.

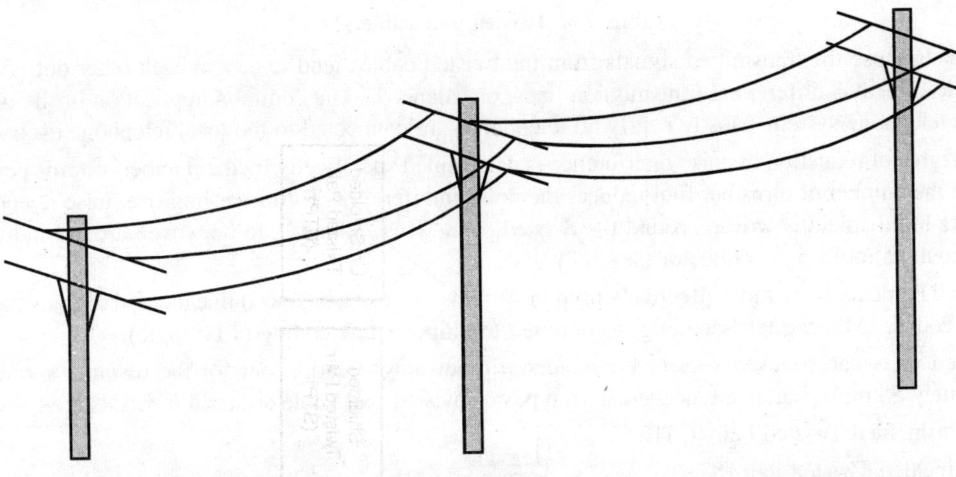

Fig. 7.4. Open wire.

A transmission line may be a pair of conducting wires held apart by an insulator or dielectric. They come in a variety of construction geometries. The simplest and least expensive form is **two-wire (ribbon) cable**. This type of transmission line is commonly used to connect a Television antenna to television receiver.

This line is essentially the same as the two wire open line except that uniform spacing is assured by embedding the two wires in a low loss dielectric, usually polyethylene. Since the wires are embedded in the thin ribbon of polyethylene space is partly air and partly polythene. Figure 7.4 shows the open wire cable.

Fig. 7.5.

7.3.1.2. Twisted Pair Cables

Twisted pair cable consists of two insulated copper wires, typically 1mm thick, twisted together. The wires are twisted in a helical form, just like DNA molecule as shown in Fig. 7.6. This twisting reduces the sensitivity of the cable to EMI (Electromagnetic interference) and also reduces the tendency of the cable to radiate radio frequency noise that interferes with nearby cables and electronic components. Each pair would consist of a wire used for the positive data signal and a wire used for the negative data signal. Any noise that appears on one wire of the pair would occur on the other wire. Because the wires are opposite polarities, they are 180 degrees out of phase (180 degrees - phasor definition of opposite polarity). When the noise appears on both wires, it cancels or nulls itself out at the receiving end.

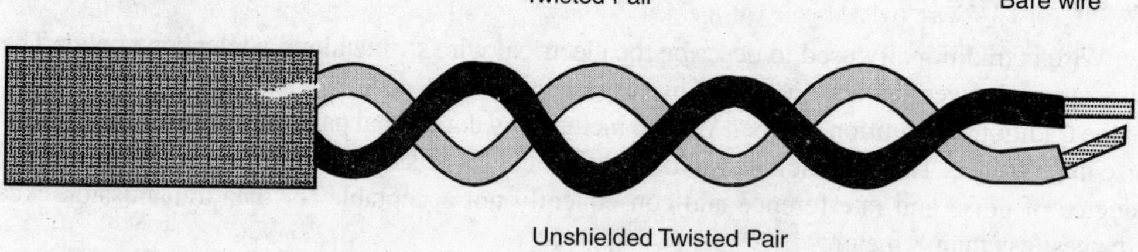

Fig. 7.6. Twisted pair cable.

This is because the transmitted signals from the twisted cables tend to cancel each other out. Twisted pairs are described as differential transmission lines, or balanced. The common application of the twisted pair is the telephone system. Mostly nearly all telephones are connected to the local telephone exchange.

The degree of reduction in noise interference is determined specifically by the number of turns per foot. Increasing the number of turns per foot reduces the noise interference. To further improve noise rejection, a foil or wire braid shield is woven around the twisted pairs. This "shield" can be woven around individual pairs or around a multi-pair conductor (several pairs).

Twisted Pair cables are most effectively used in systems that use a balanced line method of transmission: polar line coding (Manchester Encoding) as opposed to unipolar line coding (TTL logic).

Twisted pairs can be used several kilometers without amplification but for the distance more than approximately 5 km, repeaters are needed. Two types of twisted pair cable are used in LANs are :

(*i*) Unshielded Twisted Pair (UTP)

(*ii*) Shielded Twisted Pair (STP)

Unshielded Twisted Pair

Cables without a shield are called **Unshielded Twisted Pair** or UTP. Twisting the wires together results change of characteristic impedance of the cable. Typical impedance for UTP is 100 ohm for Ethernet 10BaseT cable.

UTP or Unshielded Twisted Pair cable is used on Ethernet 10BaseT and can also be used with Token Ring. It uses the RJ line of connectors (RJ45, RJ11, etc.)

Unshielded Twisted Pair Cable (UTP) consists of two conductors and each surrounded by some insulating material. This is the most common type of cable used in house wiring and in telecommunication.

Shielded Twisted Pair

Cables with a shield are called **Shielded Twisted Pair** and commonly abbreviated STP. Shielded pair cable contains two wires surrounded and separated by a solid dielectric. The dielectric is contained within a copper braid that shields the conductors from external noise sources. The entire construction is housed in a flexible, waterproof cover. Twisted pairs may be used for transmitting either digital or analog signals. The bandwidth depends on the thickness of the wire and the distance traveled.

STP or Shielded Twisted Pair is used with the traditional Token Ring cabling or ICS - IBM Cabling System. It requires a custom connector. IBM STP (Shielded Twisted Pair) has a characteristic impedance of 150 ohms.

Figure 7.7 (*a*), (*b*) and (*c*) shows the (*a*) Two-wire twisted-pair Unshielded cable (*b*) Two-wire twisted-pair shielded cable and (*c*) Four pair Unshielded twisted cable CAT-5 type.

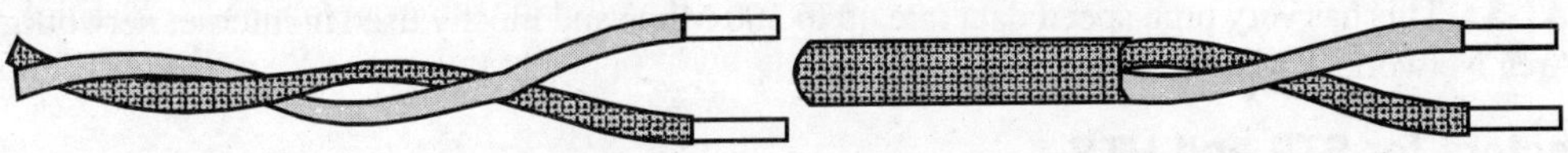

(a) Two-wire twisted-pair Unshielded cable (b) Two-wire twisted-pair shielded cable

(c) Four pair Unshielded twisted cable CAT-5 type

Fig. 7.7. (*a*), (*b*) and (*c*). Types of Twisted-pair cables

The use of this kind of cable is limited by two factors :

(*a*) Attenuation (*b*) Cross-talk

There are three principle sources of attenuation.

Resistance (or impedance) losses are simply the loss resulting from the resistance of the wires. This loss is minimized by the choice of a metal with low resistivity. Copper is chosen for this reason. (Gold is even better, and is in fact used on satellites to reduce losses.)

Dielectric losses are caused by the heating effects when a varying electric field passes through a dielectric (insulator).

Radiation losses occur because the cable acts as an antenna. All these losses increase with frequency. When a transmission line can act as an antenna, it can also act as a receiver. Lines prone to radiation loss are also susceptible to **pick-up**, or **cross-talk**. The shielded pair is designed to reduce this pick-up.

All these lines have strong attenuation at frequencies above 1MHz. They are generally used for low bit-rate communication. Two-wire ribbon cable is standard for the connection of individual telephone receivers. Twisted pair(s) is the normal method of connection for computer terminals and short high bit-rate connections.

Attenuation increases with both frequency and length. It is usually specified in dB/m at a particular frequency. Because of this fact, it is not possible to give hard-and-fast rules concerning the bandwidth availability of transmission lines. A twisted pair can support rates of several Mbps over short distances (meters), but over long distances (kilometers) will be completely unsuitable at these data-rates.

Categories of Unshielded Twisted–Pair (UTP) Cable

The Electronic Industries Association (EIA) has developed standards to classify UTP cable in to five categories. Categories are specified by cable quality, with 5 as the highest and 1 as the lowest.

CAT-1 : This cable is used for voice communication in telephone system.

CAT-2 : This cable is suitable for voice as well as for data transmission with the data transfer rate is up to 5 Mbps.

CAT-3 : This type of cable requires minimum 3 to 4 twists per feet of length and capable of data transferring rate up to 10 Mbps.

CAT-4 : Capable of transferring all types of data with the speed of 66 Mbps.

CAT-5 : This has very high speed data rate up to 100 Mbps and mostly used in internet networking and Local Area Network (LAN).

Connectors for STP and UTP

The common connector mostly used is RJ-45 (Registered Jack) for STP and UTP cables. It has 8 conductors and this connector is a keyed connector, means connector can be inserted in only in one direction.

Applications of STP and UTP

(*a*) These cables are used for Local area network.

(*b*) Used for local loop in telephone lines.

(*c*) Used to provide high data rate connection used in DSL lines.

(*d*) Used in telephone lines to provide voice and data channel.

7.3.1.3. Coaxial cable

Coaxial cable gets its name because two conductors share a common axis; Coax is shorthand for Co-axial. For long distances, or data-rates in excess of several Mbps, coaxial cable is used. Generally the coaxial cable is used for higher frequency ranging from 10 kHz to 500 MHz.

Components of **coaxial cables** are as follows (refer Fig. 7.8(*a*)):

(*i*) *A center (inner) conductor :* Usually solid copper wire and sometimes is made of stranded wire.

(*ii*) *An outer conductor :* Forms a tube surrounding the center conductor. This conductor can consist of braided wires, metallic foils or both. The outer conductor, oftenly called shield, serves as a ground and also protects the inner conductor from electromagnetic interference.

(*iii*) *An insulation layer :* This layer keeps the outer conductor spaced evenly from the inner conductor.

(*iv*) A plastic encasement (jacket) protects the cable from damage.

Coaxial cable has a central wire, surrounded by a dielectric, in turn concentrically sheathed in a braided conductor. The cable is finally surrounded in a water-proof, flexible sheath. Coaxial cable is familiar is the cable used to connect our television to the Ariel or to the cable operator. The main advantage of this method of construction is its resistance to radiation losses. The outer conductor acts to shield out any external fields, whist preventing any internal fields escaping. The outer shield protects the inner conductor from outside electrical signals.

The essential function of coaxial cables is to transmit high frequency energy and signals with low loss, with reflection less matching and without Phase and Amplitude distortion. For such transmission, Characteristic Impedance and Attenuation are important concepts. Characteristic Impedance (Z_0) is determined by the ratio of dielectric OD to conductor diameter; hence a large cable and a very small cable can have the same Z_0 but the smaller cable will have higher losses.

The problem in high energy transmission is the losses because of (*i*) cable dielectric, (*ii*) cable construction, (*iii*) radiation, and (*iv*) interference with adjoining circuits.

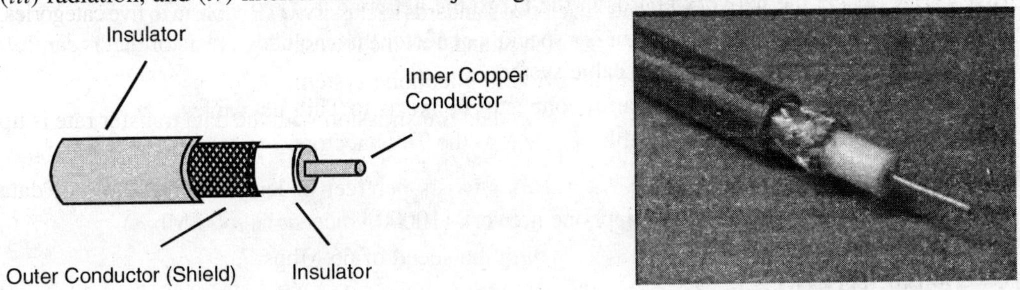

(a) Coaxial cable (Braid Shield) (b) RG-59 coaxial cable

Fig. 7.8 (a) and (b). Coaxial cables

The coaxial cable construction has been invented to overcome these problems, in which signals and energy are contained within an enclosed space. It consists of a core conductor, surrounded by pure PTFE dielectric (solid or foamed or air-spaced), covered with a metal shield, and finally encased in an overall jacket.

The shield provides the return path as well as the confinement for RF energy, and is generally grounded. The shield is typically tied to system ground, and the signal is carried as a voltage on the center wire. In the video world, Coax is usually 75 ohms impedance.

Transmission through coaxial cables is called an **unbalanced transmission** because the centre conductor and the shield are not reversible (as opposed to the balanced construction of a twisted pair). Shield can be SPC (silver plated copper) round wire single or double braid, round wire served (helically wrapped), flat foil served (helically wrapped with drain wires), foil and braid combination with drain wires, and other materials.

Jackets provide outer cover to keep the cable clean, smooth and to prevent ingress of moisture into shield, as well as for isolation of shield. Materials used are PTFE (our specialty), VFG (Fibre Glass braid covered with lacquer Varnish, over sintered PTFE tape moisture seal), PVC or PU. High temperature melt extrudible jackets such as FEP (Fluorinated Ethylene Propylene) and PFA (Per Fluoro Alkoxy) are also likely to be available in future.

The distance between the outer conductor (shield) and inner conductor plus the type of material used for insulating the inner conductor determine the cable properties or impedance. Typical impedances for coaxial cables are 75 ohms for Cable TV, 50 ohms for Ethernet Thinnet and Thicknet. The excellent control of the impedance characteristics of the cable allow higher data rates to be transferred than Twisted Pair cable.

Until the advent of optical fibre, coaxial lines were the standard method of long-distance, high bit-rate communication. They are expensive (but getting cheaper, especially as demand rises), and only used where necessary. Typical attenuation values for coaxial cable are 10 dB/Km at 10 KHz, 50 dB/Km at 500MHz. For very long-haul routes, repeaters and equalizers are necessary.

Category of Coaxial Cable

Coaxial cables are categorized by their Radio Government (RG) rating. Each RG number denotes a unique set of physical specifications. These different standards are categorized on the basis of physical specification including the dimensions of inner conductor, type of insulator and its thickness, construction and type of shielding.

(*a*) RG-59: Impedance is 75ohm and used for cable television.

(*b*) RG-58: Impedance is 50ohm and used for thin Ethernet.

(*c*) RG-11: Impedance is 50ohm and used for thick Ethernet.

Coaxial Connectors

There are numbers of coaxial connectors are available but mostly used is BNC (Bayonet-Neill-concelman). Three types of BNC connectors exist such as :

(*i*) **BNC-T :** Connect the network boards in the PC to the network.

(*ii*) **BNC Terminator :** A BNC terminator is a special connector that includes a resistor that is carefully matched to the characteristics of the cable system.

(*iii*) **BNC barrel connector :** A BNC barrel connector connects to Thin net cables.

(*iv*) **BNC cable connector :** It attach cable segment to the T-connectors.

Applications of Coaxial Cable

(*a*) Mostly used in analog and digital telephone network (10000 voice signal/600 Mbps).

(*b*) Used for Cable TV.

(*c*) Used in Ethernet LAN

Noise Immunity : Unbalanced Line versus Balanced Line

It is very interesting to analyzing the transmission of the signal in balanced and unbalanced transmission lines. When sending a signal down a guided media, two wires are required. One wire carries the information and the other carries the reference. There are two methods used to transmit the signal on the wire:

(*a*) Unbalanced lines (*b*) Balanced lines

Unbalanced Lines

Unbalanced lines consist of two wires. One wire carries the signal and the other is the reference line called the common. The common wire is usually at ground potential. Often the common wire will also be used as a shield for noise immunity.

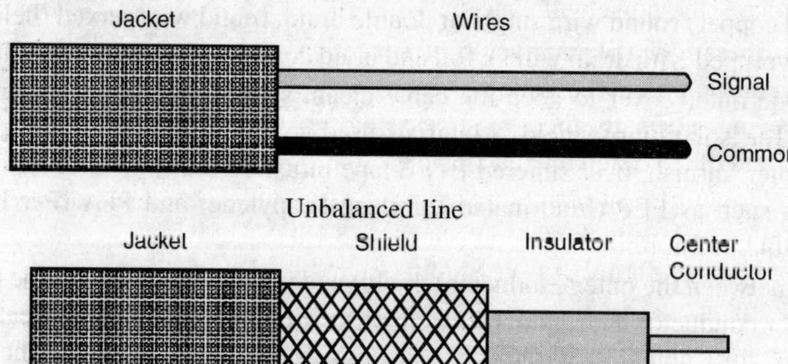

Fig. 7.9. Unbalanced line using common as shield.

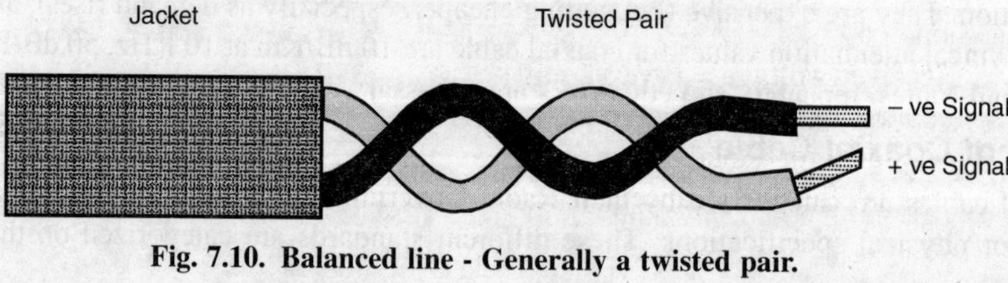

Fig. 7.10. Balanced line - Generally a twisted pair.

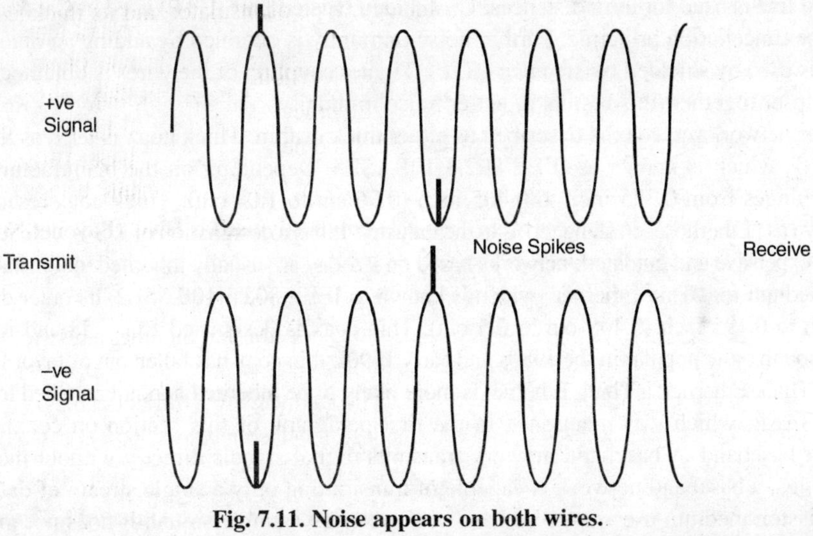

Fig. 7.11. Noise appears on both wires.

Unbalanced lines have difficulty with noise immunity as any EMI (Electromagnetic Interference) noise will appear on the signal lead. Unbalanced lines are used for short distances because of the inherent problem with noise immunity.

Balanced Lines

Balanced lines do not have a common. The signal information is carried on both wires. One wire carries the signal called the positive (+ve) signal and the other carries a signal 180 degrees out of phase called the negative (–ve) signal.

Often the wires are twisted together in order to tightly couple the wires electrically. The aim is to have any noise that appears on one wire to appear on the other wire. Because the signals are 180 degrees out of phase, the noise will cancel.

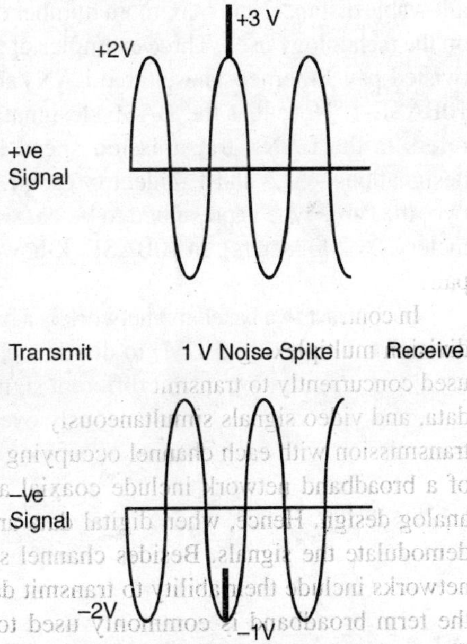

Fig. 7.12. 1 volt noise spike appears on both wires.

During normal operation, the receiver subtracts the two transmitted signals. For the point where the noise spikes appear in the preceding drawing, normally the +ve signal would have an amplitude of +2V and the –ve signal would have an amplitude of –2V. The receiver would subtract these two voltages electronically to obtain a +4V signal.

Received voltage = (+ve Signal) – (–ve Signal) = +2V – (–2 V) = +4V

But there is a voltage spike that appears on both wires with an amplitude of +1V.

This would be disastrous for an unbalanced line as the information is only carried on one wire but let's see what happens to the unbalanced line:

Received voltage = (+ve Signal) – (–ve Signal) = +3V – (–1 V) = +4V

The noise spike adds to the +ve Signal for a total instantaneous voltage of +3 V. Similarly the noise spike adds to the –ve Signal for a total instantaneous voltage of –1 V. When subtracted, the resulting voltage is still +4 V. The noise is effectively cancelled.

Unbalanced line is used for long distances. Unshielded twisted pair (UTP) used in Cat 5 and Cat 6 cable utilizes the noise cancellation principle. Further noise immunity is obtained by adding an outer shield that is grounded; this is used by shielded twisted pair (STP). Tighter coupling of the wires is obtained by having the twists wound closer together; this results in better noise immunity.

In computer networking, coax is described as either thick or thin. Thick coax is used as the medium for "Thick Ethernet", which is known as IEEE 802.3 10BASE5. Depending on the manufacturer, the cable's outer diameter ranges from 0.375 inch to 0.405 inch (0.96 cm to 1.04 cm). Thick coax resembles a garden hose and is known as "Etherhose" in slang terms in the industry. It has a designation of RG-8 with 50 Ω impedance. This medium is expensive and outdated; networks based on it today are usually inherited, not installed. Thin coax is used as the medium for "Thin Ethernet" which is known as IEEE 802.3 10BASE2. Its outer diameter ranges from 0.175 inch to 0.195 inch (0.448 cm to 0.5 cm). Thin coax is designated RG – 58 and it too has 50 Ω impedance. Although quite popular in the 1980s and early 1990s, thin coax has fallen out of favor for UTP, and as is the case with Thick Ethernet, a Thick Ethernet is more likely to be inherited and not installed today.

The term BASE, which was mentioned in the first paragraph of this section on coaxial cable, is an abbreviation for baseband. A baseband network transmits digital signals directly without modulating their transmission. Thus, a baseband network is capable of transmitting only a single stream of data. This means that the transmission medium uses its entire bandwidth to carry a single signal. It does not mean, however, that the channel cannot be shared. Using multiplexing techniques such as time division multiplexing (TDM), nodes connected to a baseband network can share the medium, but they can only transmit twisted-pair cable, coaxial cable, and fiber-optic cable. Various topologies are also available, including star, ring, and bus. Baseband networks traditionally have been characterized by their limitations on transmission speed, maximum allowable distance, and maximum number of nodes. These limitations vary considerably, however, depending on the technology used. Three examples of baseband networks include thick coaxial Ethernet, and unshielded twisted-pair Ethernet. These three LANs are respectively designated by IEEE as 10BASE5, 10BASE2, and 10BASE-T. Note that the BASE designation implies that these are baseband LANs. The 10 designation refers to the LANs' transmission speed limitation, which in each of these cases is 10 Mbps. The last designations – 5 ,2, and T – identify each LAN's maximum cable length or cable type. In the case of the first two, the cable type is presumed to be coaxial and the number represents the maximum cable length (5 = 500 meters; 2= 200 meters). In 10BASE-T, however, the T represents cable type, specifically, unshielded twisted-pair.

In contrast to a baseband network is a broadband network. A broadband network typically uses frequency division multiplexing (FDM) to divide a channel's bandwidth into smaller, distinct channels, which can be used concurrently to transmit different signals. Thus, a broadband network is capable of transmitting voice, data, and video signals simultaneously over the same cable. Cable television, for example, uses broadband transmission with each channel occupying a bandwidth of approximately 6 MHz. The transmission media of a broadband network include coaxial and fiber-optic. Broadband networks are generally based on an analog design. Hence, when digital data are transmitted, modems are needed at each end to modulate and demodulate the signals. Besides channel separation, two additional defining characteristics of broadband networks include their ability to transmit data over long distances and their high bandwidth. Consequently, the term broadband is commonly used today in networking circles to refer to high-bandwidth access. Comparing the two, baseband is relatively simple and inexpensive to install, requires inexpensive interfaces, and is ideally suited for digital transmission. Broadband equipment is much more expensive, is based on analog signaling, and requires expensive amplifiers to strengthen its signal and personnel to maintain it.

Comparison between Baseband and Broadband Cable

Baseband is simple and inexpensive to install and require inexpensive interfaces. It offers a single digital channel with a data rate of about 10 Mbps over a distance of 1 km.

Using off-the-shelf co-axial cable. For most data communication application baseband is perfectly adequate.

Broadband, on the other hand, requires experienced radio frequency engineers to plan the cable and amplifier layout and to install the system. Skilled professionals are also required to maintain the system and periodically tune the amplifiers during its use. The head end also requires careful servicing because a failure of the head end brings the network down. The broadband interfaces are also more expensive than the baseband one. However, broadband offers multiple channels and can transmit data, voice and television on the same cable for tens of kilometer if need be. For most applications, the additional bandwidth of broadband does not justify its complexity and expense so baseband is more widely used.

Advantages

1. Low maintenance cost.
2. Simple to install and tap.
3. Better resistance to signal noise over longer distance.

Disadvantages

1. Limited distance and topology.
2. Low security, easily tapped.
3. Difficult to make major changes to the cabling topology.
4. Cable cost is higher than that of twisted pair. Although electronic, support components are less expensive.

Twin Axial Cable

We'll mention one more type of copper cable for completeness, because it exists in some computer networks, albeit proprietary ones those are a vestige of the mainframe days. It's called *twin axial cable* (twinax for short). It is much like coaxial cable except that there are two inner conductors instead of one. Because it is not used in the networks we discuss in this book, we'll stop right here.

7.3.1.4. Optical Fiber

Fiber optic transmission is the transmission of information using light waves as a carrier signal that travel through an optical waveguide (optical fiber). For many years it has been appreciated that the use of optical (light) waves as a carrier wave provides an enormous potential bandwidth. Optical carriers are in the region of 10^{13} Hz to 10^{16} Hz, *i.e.* three to six orders of magnitude higher than microwave frequencies. However, the atmosphere is a poor transmission medium for light waves. Optical communication only became a widespread option with the development of low-loss dielectric waveguide.

Principle of Fiber Optics

Optical Fibers work on the principle that the core refracts the light and the cladding reflects the light. The core refracts the light and guides the light along its path. The cladding reflects any light back into the core and stops light from escaping through it.

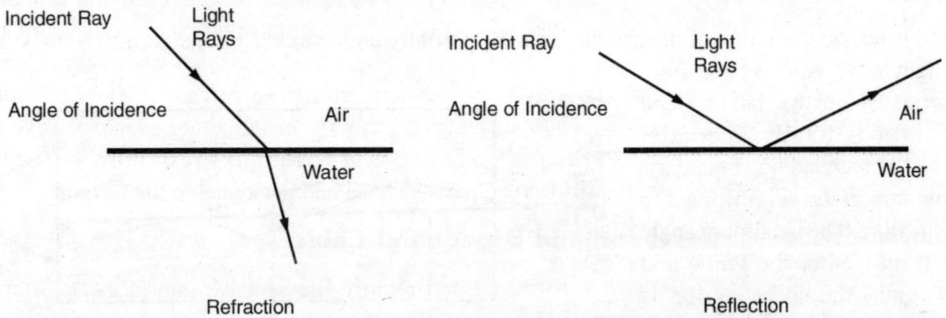

Fig. 7.13. Principle of optical fiber.

An important characteristic of Fiber Optics is Refraction. Refraction is the characteristic of a material to either pass or reflect light. When light passes through a medium, it "bends" as it passes from one medium to the other. If the angle of incidence is small, the light rays are reflected and do not pass into the water. If the angle of incident is great, light passes through the media but is bent or refracted. (refer Fig. 7.13)

A fiber is a relatively long, thin strand of transparent substance, usually glass, capable of conducting an electromagnetic wave of optical wavelength (at the visible or near-visible region of the frequency spectrum). A fiber optic waveguide offers some capability to confine the transmission along its center axis or near its center axis confining the lightwaves to its interior by means of internal reflection.

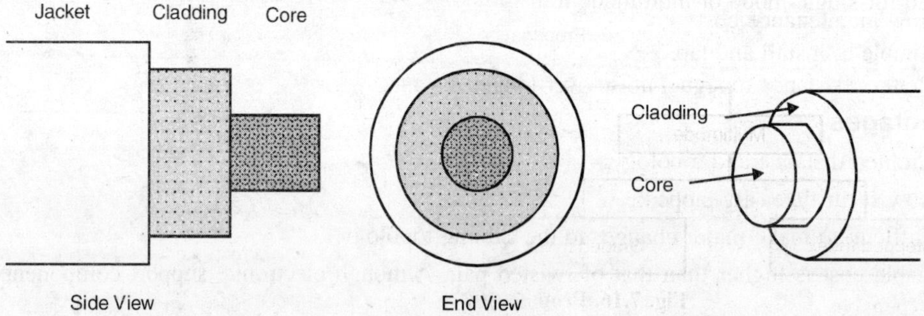

Fig. 7.14. The basic structure of a fiber optic waveguide.

The size of most fibers is from 10 to 200 microns (1/100th to 1/5th of a mm). Optical fibers are typically used in a unidirectional mode (i.e., data moves in only one direction). Because of this, every transmission system requires at least two fibers one for transmission and one for reception.

Each fiber consists of a **core**, thinner than a human hair, surrounded by a **cladding** sheath with a much lower refractive index refer Fig. 7.14. The purpose of the cladding of the fiber is redirecting the light back into the center (core) of the fiber as it hits the cladding material. Light signals introduced at one end of the cable are conducted along the cable as the signals are reflected from the sheath. When all the light is contained in an optical waveguide such as an optical fiber it is called total **internal reflection**.

Optical fiber has a core that typically has an index of refraction that is higher than the index of refraction of the surrounding cladding. The index of each part is chosen so that virtually all the light that reaches the interface between the core and the cladding is reflected back into the core. If the difference in index is not sufficient, light will be refracted into the cladding, resulting in a loss of signal.

For the light to enter into the fiber and travel through the core, the light rays must be within an acceptance angle. Acceptance angle is the angle at which light can strike the core of a fiber and still propagate down its length. A large acceptance angle makes alignment of a light source to a fiber less difficult.

Optical Transmission

In simple terms, the action of a waveguide can be partially understood by considering the rays down the fibre. A light-wave entering the fibre is either refracted into the cladding, and attenuated, or is totally internally reflected at the core/cladding boundary. In this manner it travels along the length of the fibre. The maximum angle at which it may enter the guide and travel by total internal reflection is termed the **acceptance angle.**

Figure 7.15 shows the basic process for fiber optic transmission.

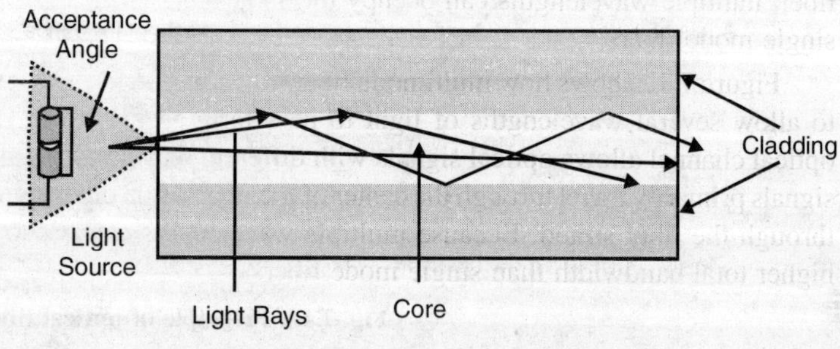

Fig. 7.15. Fiber optic transmission.

This diagram shows a light source (a LED) that produces light signals that enter the end of the fiber. Light that reaches the fiber within the acceptance angle can travel through the fiber core. This example shows that light rays travel down the core of the fiber. When light rays reach the edge of the fiber, they are reflected back into the center of the fiber. This zigzag process continues for the light rays until they emerge at the end of the fiber.

Propagation Modes of Optical Fiber

Fiber transmission lines can have one or more modes of operation. The modes of fiber transmission are the paths (modes) that can be taken by light rays to travel through the fiber. Fiber transmission lines can be constructed for single mode or multimode transmission shown in Fig. 7.16.

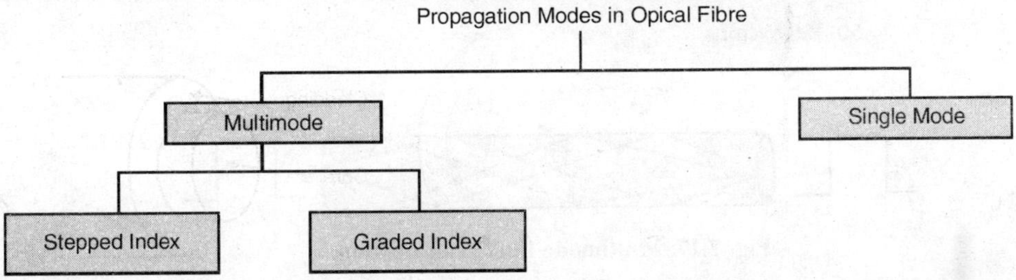

Fig. 7.16. Propagation modes in optical fiber.

Propagation Modes

1. **Multimode**: A **multi-mode** fiber has a number of possible rays that light of a particular frequency may take. There are two types of multimode which are as follows:
   (*a*) Stepped index,
   (*b*) Graded index
2. **A single mode** : A single-mode fiber is a fiber that only has one acceptable ray-path per frequency.

1. Multimode

A **multi-mode** fibre has a number of possible rays that light of a particular frequency may take. Multimode fiber transmission allows optical signals to propagate in more than one transmission mode. Multimode fiber means that when photons are injected into the fiber they have more than one way of traveling in the core of the fiber.

Multimode fibers are made by creating the diameter of the fiber core relatively large compared to single mode fibers, typically 100 to 200 micrometers (40 to 80 thousandths of an inch). Because multimode fibers allow light rays to travel down the fiber core on different paths, this creates distortion of optical pulses. The distortion relative to single mode fiber limits multimode fiber to support higher data rates at shorter distances.

In general, multiple modes are undesirable because they are susceptible to dispersion effects. However, multimode fiber is less expensive to manufacture and less fragile than single mode fiber, so it is often used in less demanding applications such as short distance local area network (LAN) systems. In multimode fiber, multiple wavelengths can occupy the same fiber strand; this allows in a higher total bandwidth than single mode fibers.

Figure 7.17 shows how multimode fiber transmission uses a relatively wide (50-125 micron) fiber strand to allow several wavelengths of light to pass through the fiber. This example shows that relatively wide optical channel allows optical signals with different wavelengths to make it through the fiber. Some of these signals primarily travel through the center of the fiber while other signals travel from side to side (by refraction) through the fiber strand. Because multiple wavelengths can occupy the same fiber strand, this allows in a higher total bandwidth than single mode fibers.

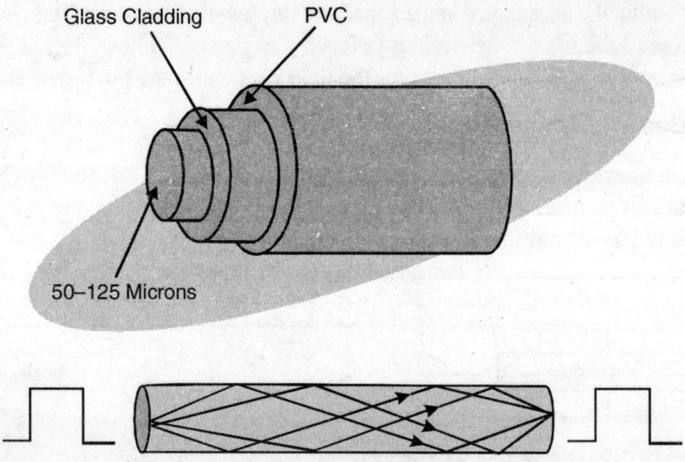

Fig. 7.17. Multimode fiber transmission.

(a) Stepped-index fiber : In this type of fiber, the core has a uniform refractive index throughout. This generally has a core diameter of 100 mm to 500 mm. Figure 7.18 shows a stepped index fibre.

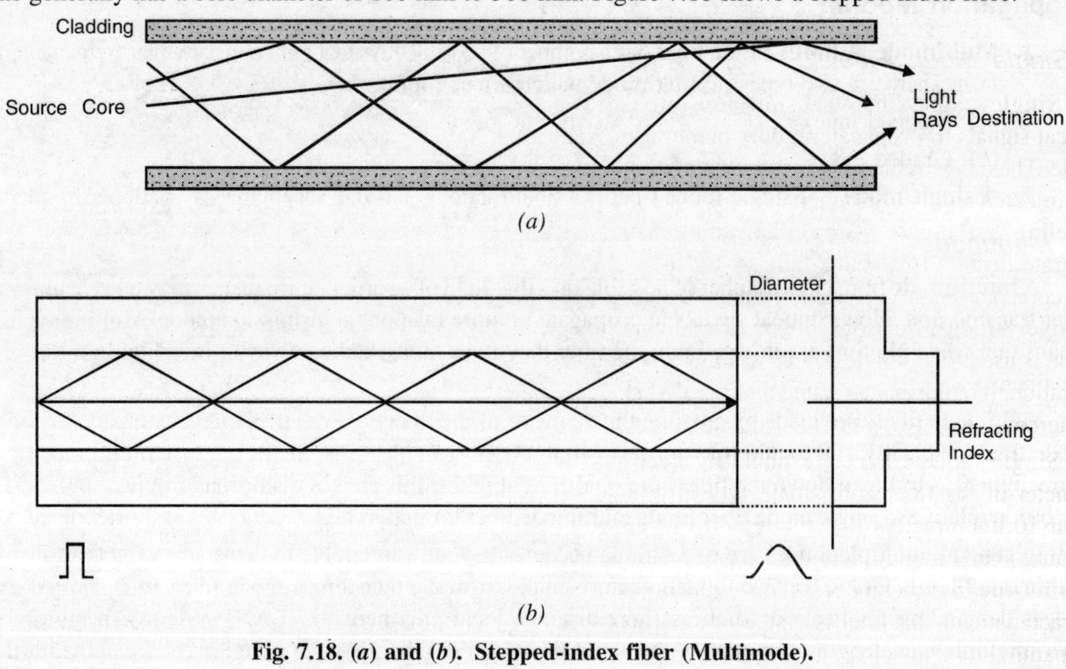

Fig. 7.18. ***(a)* and *(b)*. Stepped-index fiber (Multimode).**

Step Index has a large core the light rays tend to bounce around, reflecting off the cladding, inside the core. This causes some rays to take a longer or shorter path through the core. Some take the direct path with hardly any reflections while others bounce back and forth taking a longer path. The result is that the light rays arrive at the receiver at different times. The signal becomes longer than the original signal. LED light sources are used. Typical Core: 62.5 microns.

(b) Graded-index fiber : Grade Index has a gradual change in the Core's Refractive Index as shown in Fig. 7.19. This causes the light rays to be gradually bent back into the core path. In this type of fiber, the core has a refractive index that gradually decreases as the distance from the centre of the fiber increases. The result is a better receive signal than Step Index. LED light sources are used. Typical Core: 62.5 microns.

Both Step Index and Graded Index allow more than one light source to be used.

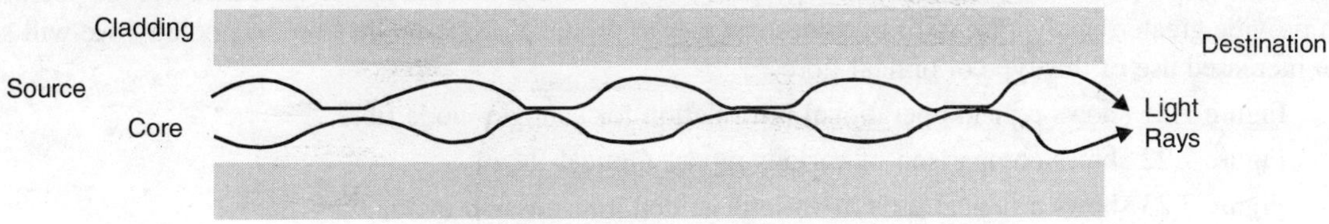

(a)

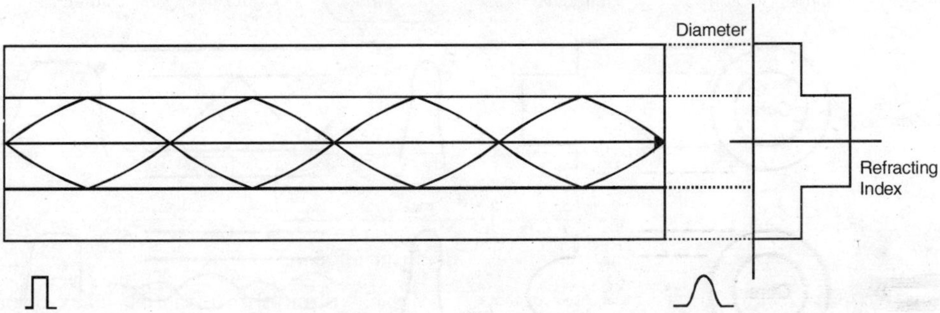

(b)

Fig. 7.19 (*a*) and (*b*). Graded-index fiber (Multimode).

2. Single Mode Fiber

Single mode fiber transmission only allows optical signals to propagate in only one transmission mode. The mode of the fiber means that when photons are injected into the fiber, it has only one way of traveling in the core of the fiber. Single Mode has separate distinct Refractive Indexes for the cladding and core. The light ray passes through the core with relatively few reflections off the cladding. Single Mode is used for a single source of light (one colour) operation. It requires a laser and the core is very small: 9 microns.

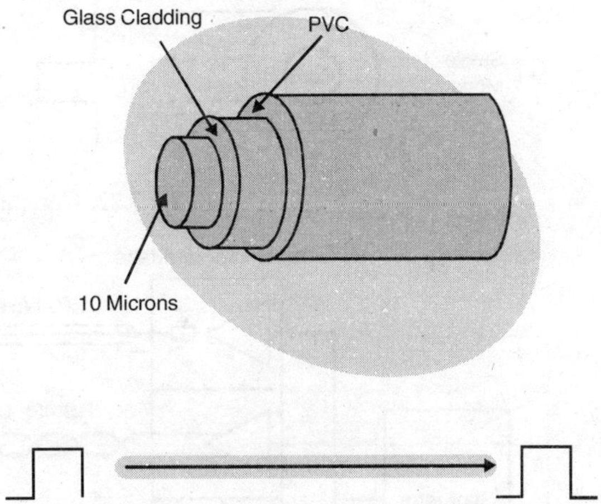

Fig. 7.20. Single-mode fiber.

Single mode fibers are made by keeping the diameter of the fiber core small, typically less than about 10 micrometers (0.4 thousandths of an inch). Because single mode fibers limit the number of light rays as the name suggests, the distinguishing characteristic of this fiber is that allows only a single ray path. The radius of the core of this type of fiber is much less than that of the other two, however it does have a uniform refractive index, refer Fig. 7.20.

Figure 7.20 shows how single mode fiber transmission uses a very small (10 micron) fiber strand to only allow a specific narrow wavelength of light to pass through the fiber. Because the optical channel is so small, the signals that make it through the fiber primarily travel directly through the center of the fiber.

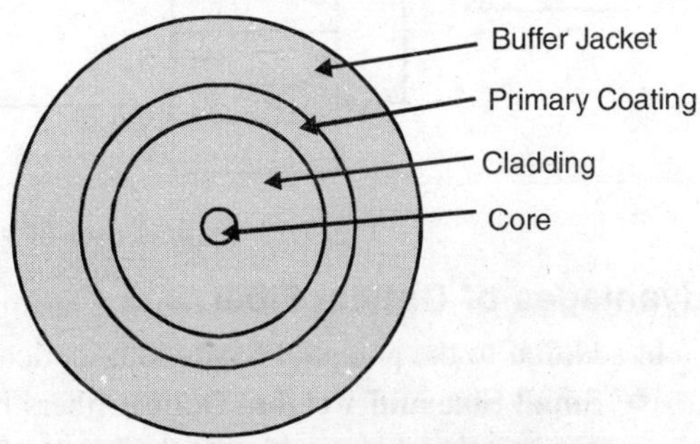

Fig. 7.21. Constructional information for a Single-mode fiber.

We find that the cost of production increases, the complexity of transmitter and receiver increases, while the dispersion decreases. This latter property change means that the single-fibre also has the potential to provide greater bandwidth. As it becomes cheaper to produce single-mode fibre technology, we will see an increased use of this type of optical fibre.

Figure 7.21 shows typical operational information for a single-mode fibre.

Figure 7.22 shows comparison of various modes Optical Fibers.

Figure 7.23 shows a typical uses of various guided transmission media.

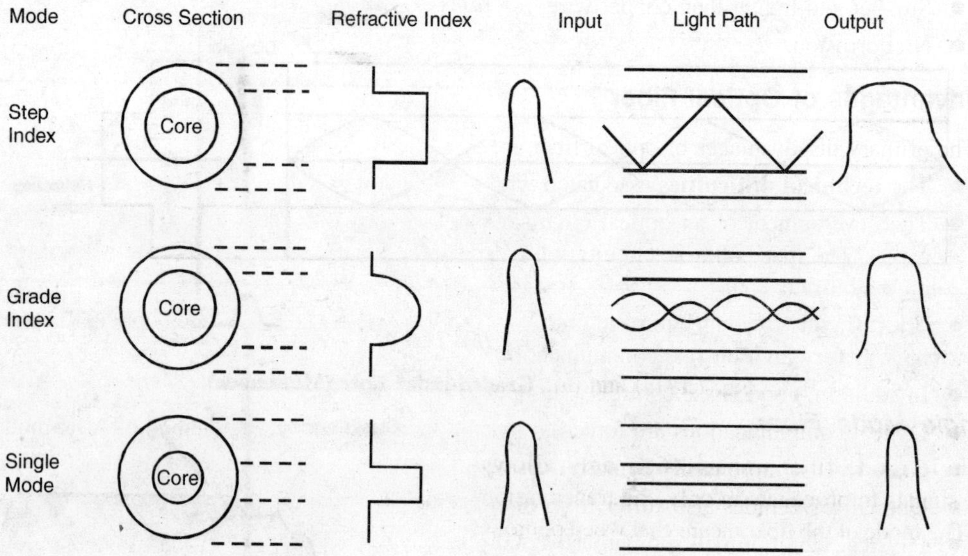

Fig. 7.22. Comparison of optical fibers.

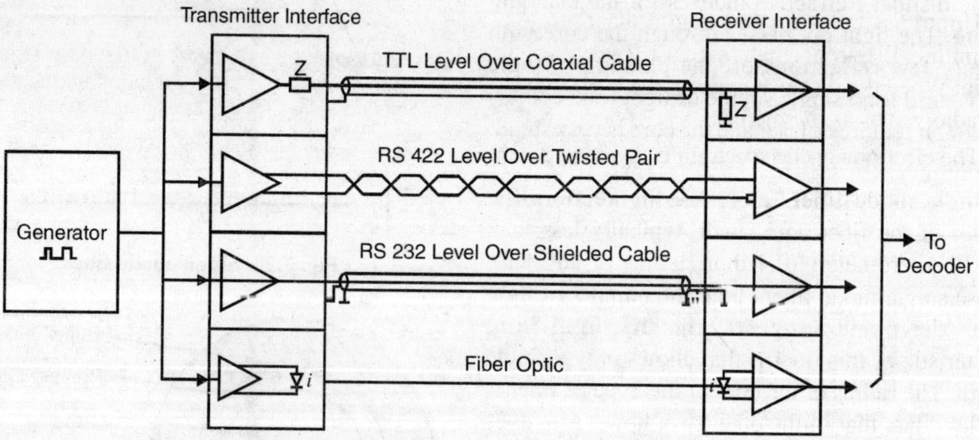

Fig. 7.23. Typical uses of various guided transmission media.

Advantages of Optical Fiber

In addition to the potential bandwidth, optical fiber communication offers a number of benefits :

- **Small Size and weight.** Optical fibers have very small diameters. A very large number of fibers can be carried in a cable the thickness of a coaxial cable.
- **Electrical isolation.** Optical fibers are almost completely immune from external fields. They do not suffer from cross-talk, radio interference, etc.

- **Extremely high immunity to EM interference**
- **Large bandwidth**
- **Security.** It is difficult to tap into an optical line. It is extremely difficult to tap into an optical line unnoticed.
- **Low transmission loss.** Modern optical fiber now has better loss characteristics than coaxial cable. Fibers have been fabricated with losses as low as 0.2 dB/km.
- Longer distances than copper wire.
- Smaller and lighter than copper wire.
- No corrosion.

Disadvantages of Optical Fiber

The primary disadvantages of optical fiber are :

- The **technical difficulties** associated with reliable and cheap connections,
- The development of an optical circuit technology that can match the potential data-rates of the cables. The speed of these circuits, which are electronically controlled, is usually the limiting factor on the bit-rate.
- The difficulty of connection and **high-cost** of associated circuitry result in optical fibers being used only in very high bit-rate communication.
- In addition, good phase control of an optical signal is extremely difficult.
- Optical communications are forced to use the comparatively crude method of ASK modulation.
- Glass fiber is more **easily broken** than wire.
- Optical waveguides also suffer from **dispersion**
- Physical vibration will show up as signal noise.

7.3.2. Unguided Transmission Media

Unguided transmission media make use of atmosphere and outer space which provides a means of transmitting Electromagnetic signals. This type of transmission is commonly referred as **wireless transmission media.** When the distance is large, the cost of cable lying become very expensive, it can be advantageous to use radio waves as the media.

The electromagnetic spectrum is divided up into a number of bands, as shown in Table 7.1.

TABLE 7.1. The higher frequencies of the electro-magnetic spectrum

Description	Frequency	Wavelength
High frequency	3-30 MHz	100-10 m
VHF	50-100 MHz	6-3 m
UHF	400-1000 MHz	75-30 cm
Microwaves	$3 \times 10^9 - 10^{11}$ Hz	10 cm - 3 mm
Millimetre waves	$10^{11} - 10^{12}$ Hz	3 mm - 0.3 mm
Infrared	$10^{12} - 6 \times 10^{14}$ Hz	0.3 mm – 0.5 μm
Light	$6 \times 10^{14} - 8 \times 10^{14}$ Hz	0.5 μm – 0.4 μm
Ultra-violet	$8 \times 10^{14} - 8 \times 10^{17}$ Hz	0.4 μm – 10^{-9} m
X-rays	$10^{17} - 10^{19}$ Hz	10^{-9} m – 10^{-13} m
Gamma rays	$> 10^{19}$ Hz	$< 10^{-13}$ m

DO YOU KNOW ▶

Unguided transmission media *make use of atmosphere and outer space which provides a means of transmitting Electromagnetic signals.*

We have already studied the propagation of radio waves directed by any man-made structure such as transmission lines, coaxial lines wave guides, and strip or single-wire lines. In this chapter we shall discuss the unguided propagation of radio waves, that is, in terrestrial atmosphere, over and through the earth and in outer space. Unguided wave propagation is also used in geophysics, in the study of the upper atmosphere, radio astronomy, the study of the sun, stars and nebular inside and outside our Galaxy.

In all the above applications (despite the differences) of unguided radio wave propagation, the common aspect is that all use a radio link, consisting of transmitter, a receiver and a propagation medium. In radio transmission, the atmosphere around the earth or outer space is the natural propagation medium.

7.4. RADIO LINKS

There are three commonly used radio links. Figure 7.24 shows the simplest one in which transmitter and receiver located at its terminal points. In Fig. (7.24), the transmitted waves reach the receiver after reflection or scattering in the ionosphere. in other cases radio waves can reach the receiver through diffraction round the globe (Fig. 7.24), scattering the troposphere (Fig. 7.25) or some other way (space wave Fig. 7.24), Figure 7.25 shows a line-or-sight radio relay system (a microwave link). In which there are number of repeaters. At each repeater, the R.F. signal is converted back to the I.F. band, amplified, translated to a microwave frequency, and restrained in the direction of the next repeater at a frequency different from the received one. Figure (7.24) shows a scatter radio link. In which the transmitted signal dies not reach the receiver R_X directly. Instead transmitted signal from T_X illuminated a man-made or natural body, *C*, which differs from the surrounding medium in electrical characteristics.

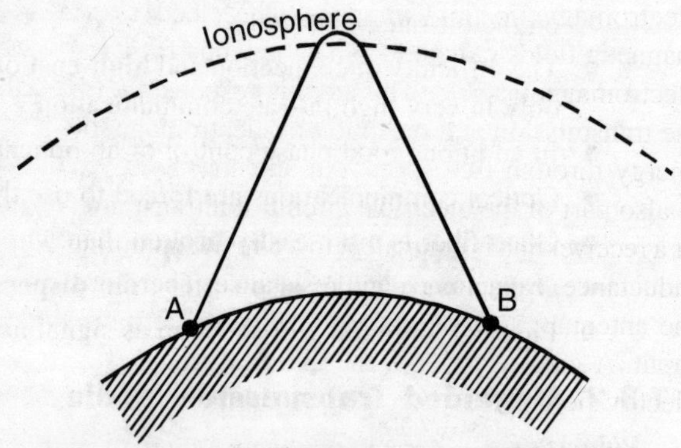

Fig. 7.24. A simple radio link (*T*- transmitter, *R*- receiver)

As a result, the body *C* reflects or scatters the incident radiation, and the reflected and forward-scattered signal is picked up at R_X (tropospheric scatter, ionospheric scatter, reflection from moon or artificial earth satellite used as passive reflectors). In case of Radar transmitted and receiver system is located together, *C* in the target or targets.

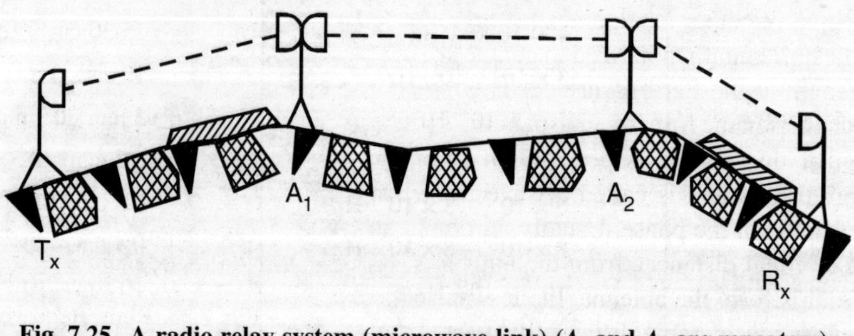

Fig. 7.25. A radio relay system (microwave link) (A_1 and A_2 are repeaters).

7.5. EM WAVE PROPAGATION : INTRODUCTION

An electromagnetic wave consists of two primary components – an electric field and a magnetic field. The electric field results from the force of voltage and the magnetic fields results from the flow of current. Although electromagnetic fields that are radiated are commonly considered to be waves, under certain circumstances their behavior makes them appear to have some of the properties of particles. In general, however, it is easier to picture electromagnetic radiation in space as horizontal and vertical lines of force oriented at right angles to each other. These lines of forces are made up of an electric field (*E*) and a magnetic field (*H*), which together makes up the electromagnetic field in space. The electric and magnetic fields radiated from an antenna from the electromagnetic field. This field is responsible for the transmission and reception or electromagnetic energy through free space. An antenna, however, is also part of the electrical circuit of a transmitter or a receiver and is equivalent to a circuit containing inductance, capacitance and resistance. Therefore, the antenna can be expected to display definite voltage and current relationships with respect to a given input. A current through the antenna produces a magnetic field and a charge on the antenna produces an electric field. These two fields combine to form the induction field.

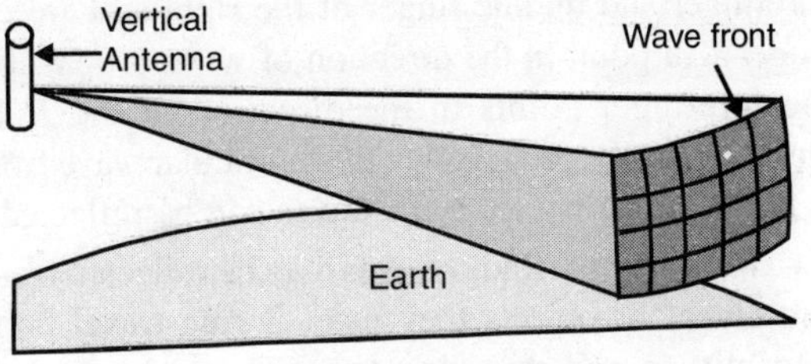

Fig. 7.26. Vertical polarization

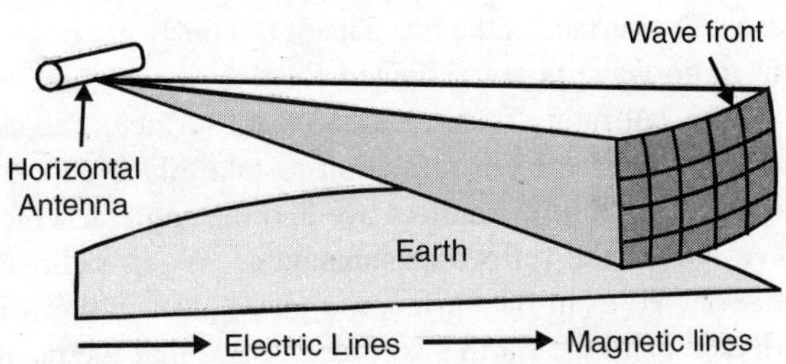

Fig. 7.27. Horizontal polarization

Polarization : For maximum absorption of energy form the electromagnetic fields, the receiving antenna must be located in the plane of polarization. This places the conductor of the antenna at right angles to the magnetic lines of force moving through the antenna and parallel to the electric lines, causing maximum induction. Normally, the plane of polarization of a radio wave is the plane in which the *E* field propagates with respect to the earth. If the *E* field component of the radiated wave travels in a plane perpendicular to the Earth's surface (vertical), the radiation is said to be vertically polarized, as shown in Fig. 7.26. If the *E* field propagates in a plane parallel to the Earth's surface the position of the antenna in space is important because it affects the polarization of the electromagnetic wave.

When the transmitting antenna is close to the ground, vertically polarized waves cause greater signal strength along the Earth's surface. On the other hand, antennas high above the ground should be horizontally polarized to get the greatest possible signal strength to the Earth's surface. The radiated energy from an antenna is in the form of an expanding sphere. Any small section of this sphere is perpendicular to the direction the energy travels and is called a wave front. All energy on a wave front is in the phase. Usually all points on the wave front are at equal distances from the antenna. The farther the wave front is form the antenna, the less spherical the wave appears. At a considerable distance the wave front is from the antenna, the less spherical the wave appears. At a

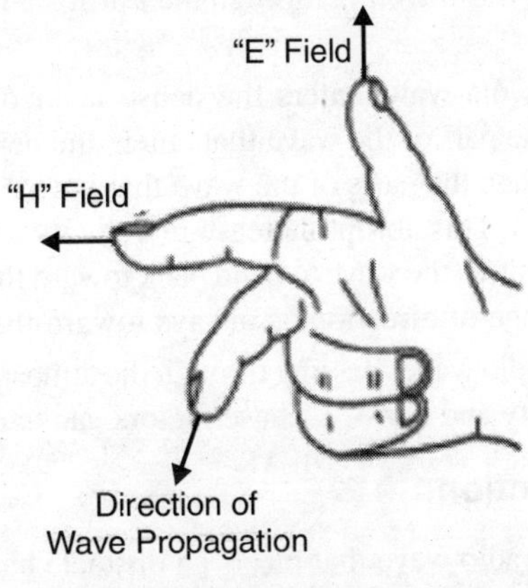

Fig. 7.28. Right-hand rule for propagation

considerable distance the wave front can be considered as a plane surface at a right angle to the direction of propagation. If you know the directions of the *E* and *H* components, you can use the "right-hand rule" as shown in Fig. (7.28) to determine the direction of wave propagation. This rule states that if the thumb, forefinger and middle finger of the right and are extended so they are mutually perpendicular. The middle finger will point in the direction of wave propagation if the thumb points in the direction of the *E* field and the forefinger points in the direction of the *H* field. Since both the *E* and *H* fields reverse direction simultaneously, propagation of a particular wave front is always in the same direction (away from the antenna). With the atmosphere, radio waves can be reflected, refracted and diffracted like light and heat waves.

Reflection : Radio waves may be reflected from various substances or objects they meet during travel between the transmitting and receiving sites. The amount of reflection depends on the reflecting material. Smooth metal surfaces of good electrical conductivity are efficient reflectors of radio waves. The surface of the Earth itself is a fairly good reflector. The radio wave is not reflected form a single point on the reflector, but rather from an area on its surface. The size of the area required form reflection to take place depends on the wavelength of the radio wave and the angle at which the wave strikes the reflecting substances. When radio waves are reflected from flat surfaces, a phase shift in the wave occurs. Figure 7.28 shows two radio waves being reflected from the Earth's surface. Notice that the positive and negative alternations of radio waves (*A*) and (*B*) are in phase with each other in their paths towards the Earth's surface. After reflection takes place, however, the waves are approximately 180° out of phase from their initial relationship. The amount of phase shift that occurs is not constant. It depends on the polarization of the wave and the angle at which the wave strikes the reflecting surface. Radio waves that keep their phase relationships after reflection normally produce a stronger signal at the receiving site. Those that are received out of the phase produce a weak or fading signal. The shifting in the phase relationships of reflected radio waves is one of the major reasons for fading.

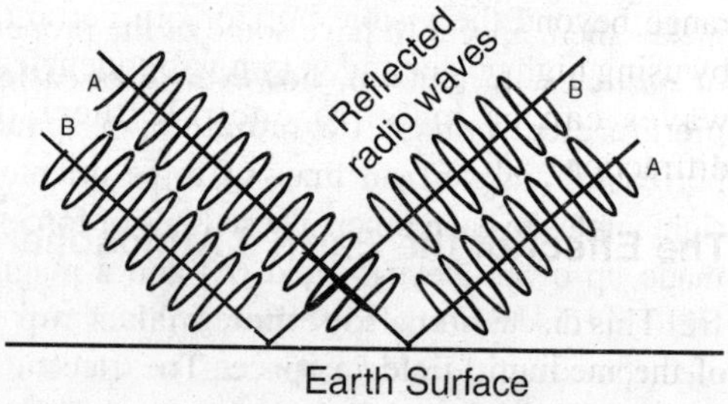

Fig. 7.29. Reflection.

Refraction : Another phenomenon common to most radio waves is the bending of the waves as they move from one medium into another in which the velocity of propagation is different. This bending of the waves is called refraction. This simple principle applies to radio waves as changes occur in the medium through which they are passing. As an example, the radio wave shown in Fig. 7.30 is traveling through the Earth's atmosphere at a constant speed.

As the wave enters the dense layer of electrically charged ions, the part of the wave that enters the new medium first travels faster than the parts of the wave that have not yet entered the new medium. This abrupt increase in velocity of the upper part of the wave causes the wave to bend back toward the Earth. This bending, or change of direction, is always toward the medium that has the lower velocity of propagation.

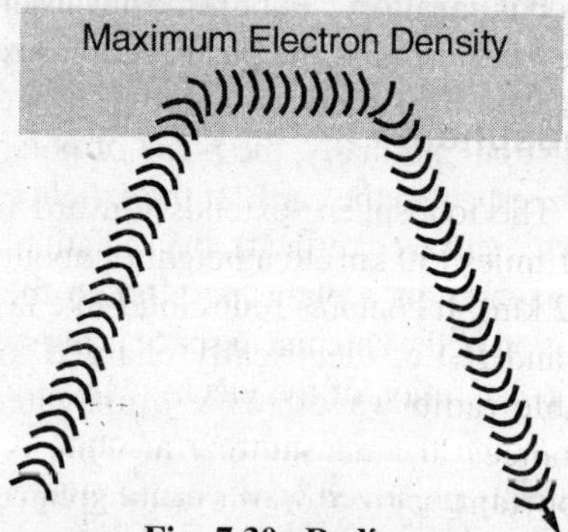

Fig. 7.30. Radiowave refraction.

Radio waves passing through the atmosphere are affected by certain factors, such as temperature, pressure, humidity and density. These factors can cause the radio waves to be refracted.

Diffraction

A radio wave that meets an obstacle has a natural tendency to bend around the obstacle as illustrated in Fig. 7.31. The bending, called **diffraction**, results in a change of direction of part of the wave energy form the normal line-of-sight path. This change makes it possible to receive energy around the edges of an obstacle as shown in figure (view *A*) or at some distance below the highest point of an obstruction, as shown in figure (view *B*). Although diffracted R.F. energy usually is weak, it can still be detected by a suitable receiver.

The principal effect of diffraction extends the radio range beyond the visible horizon. In certain cases, by using high power and very low frequencies, radio waves can be made to encircle the Earth by diffraction.

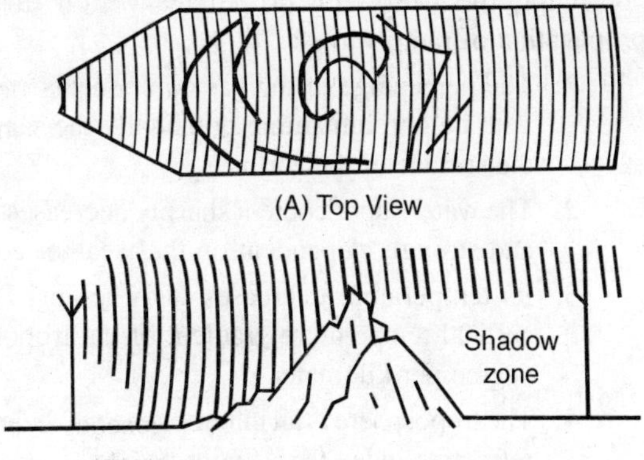

Fig. 7.31. Diffraction around an object.

The Effect of the Earth's Atmosphere on Radio Waves

This discussion of electromagnetic wave propagation is concerned mainly with the properties and effects of the medium located between the transmitting antenna and the receiving antenna. While radio waves traveling in free space have little outside influence affecting them, radio waves traveling within the Earth's atmosphere are affected by varying conditions. The influence exerted on radio waves by the Earth's atmosphere adds many new factors to complicate what at first seems to be a relatively simple problem. These complications are because of a lack of uniformity within the Earth's atmosphere. Atmospheric conditions vary with changes in height, geographical location, and even with changes in time (day, night, season, and year). Knowledge of the composition of the Earth's atmosphere is extremely important for understanding wave propagation. The Earth's atmosphere is divided into three separate regions, or layers.

Stratosphere

The stratosphere is located between the troposphere and the ionosphere. The temperature through this region is considered to be almost constant there are little water vapors present. One stratosphere has relatively little effect on radio waves because it is a relatively calm region with little or no temperature changes.

Ionosphere

The ionosphere extends upward from about 31.1 miles (50 km) to a height of about 250 miles (402 km). It contains four cloud-like layers (*D, E, F1* and F_2) of electrically charged ions, which enable radio waves to be propagated to great distances around the Earth. This is the most important region of the atmosphere for long distance point-to-point communication.

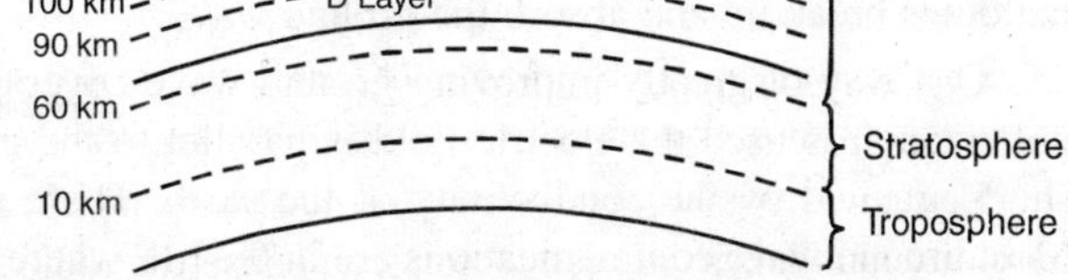

Fig. 7.32. Layers of the earth's atmosphere.

Troposphere : The troposphere is the portion of the earth's atmosphere that extends from the surface of the earth to a height of about 3.7 miles (6 km) at the north pole or the south pole and 11.2 miles 918 km) at the equator. Virtually all weather phenomena take place in the troposphere. The temperature in this region decreases rapidly with altitude, clouds from and there may be much turbulence because of variations in temperature, density and pressure. These conditions have a great effect on the propagation of radio waves.

It has the following properties which influence the propagation of radio waves:

1. The percentage of the gas components does not vary with height, remaining practically the same at it is at the surface of the earth.
2. The water vapor content sharply decreases with height and strongly dependent on the weather conditions.
3. Its temperature decreases with height. The average vertical temperature gradient of the troposphere in 6 degree per kilometer.
4. The troposphere is an inhomogeneous dielectric whose refractive index varies with height.

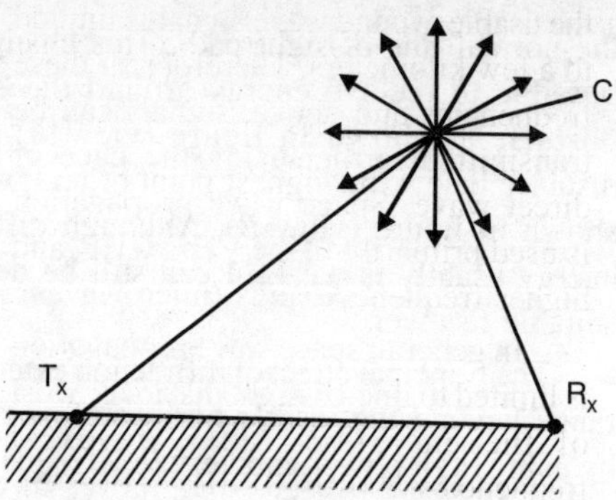

Fig. 7.33. Scatter radio link (T_X-transmitter, R_X-receiver).

7.6. MECHANISM/MODES OF PROPAGATION

When an electromagnetic wave is radiated form a transmitting antenna, part of the radiated energy travels along or near the earth's surface, and another part travels from the antenna upward into space. The energy that stays close to the surface of the earth is called the **ground wave**. The energy that travels upward into space is called the **sky wave.** Energy that travels directly from the transmitting antenna to the receiving antenna, without following the surface of the earth or radiating toward the sky, is called **space wave** or **direct wave.**

7.6.1. Ground Waves

As just stated, the ground wave is a radio wave that travels along the surface of the earth. Ground waves are the primary mode of propagation in the LF and MF bands. The longer wavelengths in these bands tend to follow the curvature of the earth due to diffraction and actually travel beyond the horizon, as shown in Fig. 7.34. The radio wave propagation due to diffraction cover a distance not exceeding 3000 km to 4000 km. However, as the frequency increases, the ground wave is more effectively absorbed by the irregularities on the earth's surface, because hills, mountains, trees and buildings become significant relative to the transmitted wavelength. For example, at 30 kHz, the wavelength is 10 km and even mountains are relatively insignificant compared to this value. But, at 3 MHz, the wavelength is only about 100m, short enough that hills, trees, and large buildings break up and absorb the ground wave.

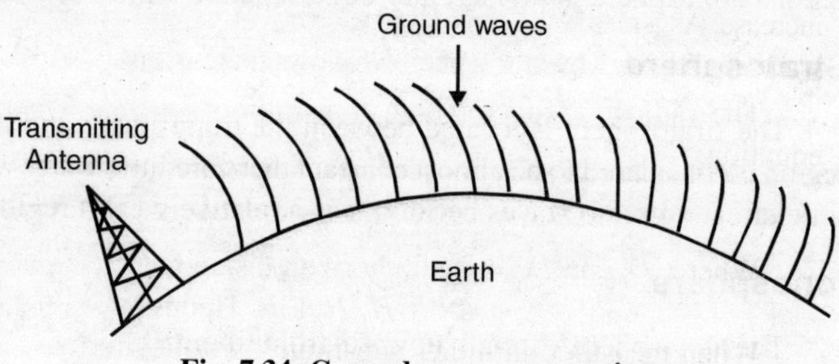

Fig. 7.34. Ground wave propagation.

One way of greatly improving ground wave coverage is to use vertical polarization. When horizontal polarization is used the electric field is parallel to the earth's surface, and any ground wave is effectively short-circuited by the conductivity of the earth. There are few communications services in the LF band. Most ground wave communications are in the MF where antenna size is more practical. In LF band full-size quarter wave antenna does present problem.

7.6.2. Space Waves

Figure 7.35 illustrates the concept of space wave in which energy radiated form the transmitter travels directly in a straight line to receiving antenna. When the transmitted signal is increased above 4 or 5 MHz,

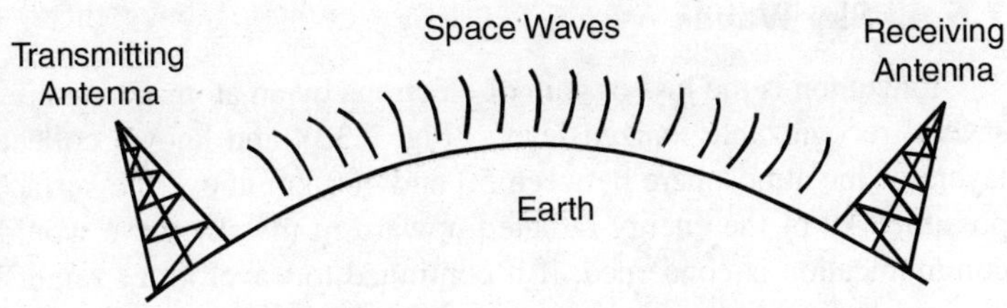

Fig. 7.35. Space wave propagation.

the usable ground wave signal is limited to a few kilometers. Therefore, at these frequencies and above, signals can be transmitted farther using the pace or direct wave. Space wave propagation is used primarily in the VHF, UHF, and higher frequency bands (microwaves).

In general, space wave propagation is limited to line-of-sight distance. Most of the energy in radio waves at frequencies above 30 MHz moves through space in straight lines like light waves, although some of the energy at these higher frequencies follows the curvature of the earth, and good VHF reception is often obtained considerably beyond line-of-sight distances. At still higher frequencies, such as those in the UHF and SHF ranges, normal transmissions from conventional antenna are limited to line-of-sight.

When the transmitting and receiving antennas are within sight of each other, the signal is considered to be a space wave. It follows, then, that to increase the range of space wave propagation, you need only to increase the height of either or both the transmitting and receiving antennas. Example of increased antenna height is satellites and aircraft antennas. In these circumstances, propagation ranges are greatly extended.

7.6.3. Radio Horizon

In space wave transmission, there is a phenomenon called the radio horizon, the distance of which is about one third greater than that of the optical horizon. This phenomenon is called by refraction in the earth's lower atmosphere. The refraction occurs because the density of the earth's atmosphere decreases linearly as height increase. As a result, the top of the wave travels slightly faster than the bottom of the wave, effectively bending the wave slightly downward, where is follows the curvature of the earth beyond the optical horizon.

The radio horizon for both transmitting and receiving antennas can be calculated by the following equation:

$$D_T = 4.12\sqrt{H_T} \quad or \quad D_R = 4.12\sqrt{H_R} \qquad ...(7.1)$$

Where, D_I and D_R = Radio horizon distance in kilometer

$H_I(H_R)$ = Height of transmitting (receiving) antenna in meter

When measured in miles, substitute the following:

$$D = \sqrt{2H} \qquad ...(7.2)$$

Where, D = Horizon distance in mile

H = Height of antenna in feet

The maximum space wave communications distance is the sum of the numbers obtained by using Eq. (7.1) for both antennas.

$$D_{max} = 4.12\sqrt{H_T} + 4.12\sqrt{H_R} \qquad ...(7.3)$$

$$D_{max} = D_T + D_R \qquad ...(7.4)$$

Example 7.1. A receiver is located 76 km from the transmitter. The transmitting antenna is 81 km high. Find the height of the receiving antenna.

Solution. Algebraic manipulation of Eq. (7.2) gives

$$H_R = \left(\frac{D_{max} - 4.12\sqrt{H_T}}{4}\right)^2 = \left(\frac{76 - 4.12\sqrt{81}}{4}\right)^2$$

$$H_R \approx 100 \text{ m}$$

7.6.4. Sky Waves

Ionization is the loss or gain of electrons by an atom. There are different degrees of ionization, forming several recognizable ionized layers (Fig. 7.32), and known collectively as the ionosphere. These ionized layers of the atmosphere between 50 and 400 km above the surface of the earth make sky wave reception possible. All of the energy radiated upward in the sky wave would be wasted, so far as earthbound radio communication is concerned, if it continued to travel in a straight line off into space. However, at certain frequencies and radiation angle, the ionosphere reflects radio waves much the same way a mirror reflects light. Also, radio waves at other frequencies and angles are refracted (bend) in such a manner that they return to earth because of decreasing refractive index. These type of wave propagation is known as **sky wave** or ionospheric wave, as shown if Fig. 7.36).

The amount of refraction depends on three factors:

(1) The frequency of the wave *f*,

The density of the ionized layer (refractive index)

$$n = \sqrt{1 - \frac{81N}{f^2}}$$

(2) The angle at which the wave enters the ionosphere.

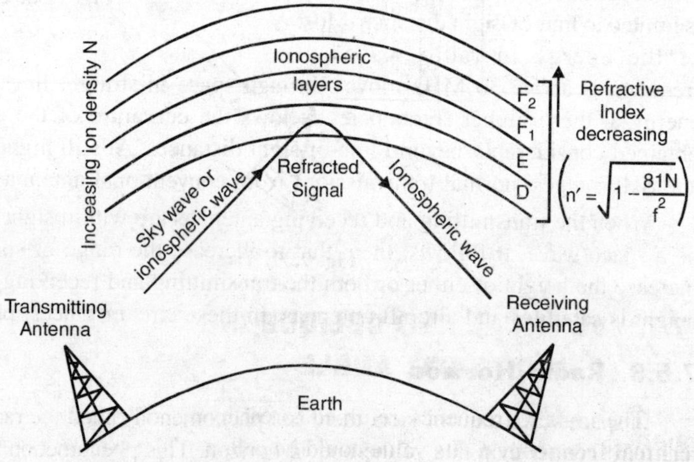

Fig. 7.36. Sky wave ionospheric wave propagation.

When conditions are just right, the wave will be refracted enough to return to earth (Fig. 7.36). The refracted radio waves may return to earth very far from the point when they originated.

Long distance communications often makes use of this characteristic by sing carrier frequencies in the MF and HF bands, because waves radiated at these frequencies can be refracted back to earth with comparative ease. Almost all HF propagation, and night-time long distance MF propagation, is by ionospheric wave. However, waves at frequencies above 30 MHz are much more likely to penetrate the ionosphere and continue moving out into space. In the UHF and SHF bands, a very small percentage to the wave's energy is refracted back to earth.

7.6.5. Tropospheric Scatter

Tropospheric scatter, commonly called troposcatter, is a special case of sky wave propagation used for frequencies higher than those in standard sky wave propagation techniques. These higher frequencies are directed to the troposphere instead of to the ionosphere.

The troposphere is the upper region of the earth's atmosphere, ranging from about 10 km to 16 km above the surface of the earth. Propagation by tropospheric scatter uses the properties of the troposphere as a reflector of UHF signals in much the same manner as the ionosphere is used to reflect lower-frequency signals. However, the troposphere only partially reflects the signal, scattering the energy, some extremely small portion of which is scattered in the forward direction or in the direction of the receiver. Local in homogeneities in the troposphere cause scattering of radio wave, and scattered waves can cover ranges up to 1000 km away from the transmitter. It is a very reliable technique for extending UHF signal communications paths beyond the horizon; however, it is not cost effective.

The scattering process, illustrated in Figure 7.37 shows directional transmitting and receiving antennas aimed so that their beams intercept in the troposphere. Note that much of the energy is scattered in undesired directions and is lost.

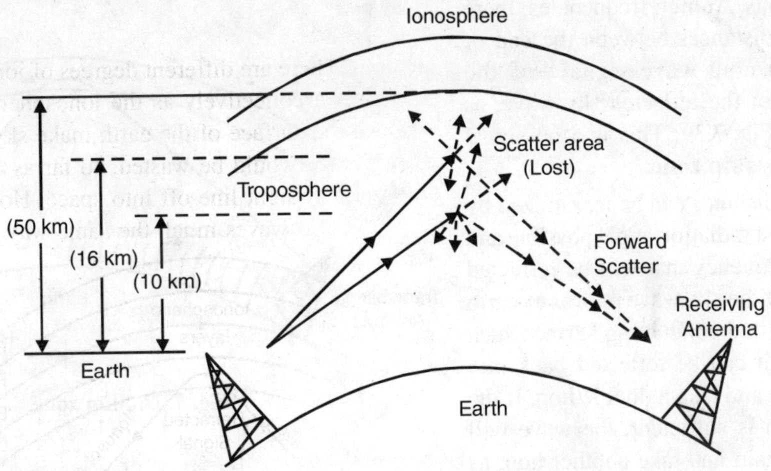

Fig. 7.37. Tropospheric scatter propagation.

7.7. VERTICAL AND OBLIQUE INCIDENCE—CRITICAL FREQUENCY AND CRITICAL ANGLE

The highest frequency returned to earth when radiated upward in the vertical direction is called the **critical frequency** f_c. Its value could be calculated by the relation

$$f_c = 9\sqrt{N_{max}}$$

where f_c is expressed in MHz and N_{max} in the ionic density in m^3. The value of the critical frequency depends on the condition of the ionosphere. Since ionospheric conditions change form hour to hour, day to day, month to month, season to season, and year to year, the critical frequency also changes constantly.

Lowering the radiation angle from the exact vertical direction allows the wave to travel longer through the ionized layer, causing a greater degree of refraction. In practical terms, this means that, with one limitation, signal frequencies higher than the critical frequency can be refracted back to earth. The limitation is that, for any given frequency, there is a **critical angle** above which the signal will not be refracted enough to return to earth. Figure 7.38 shows how signals above the critical angle penetrate the ionospheric layer while signals below the critical angle return to earth.

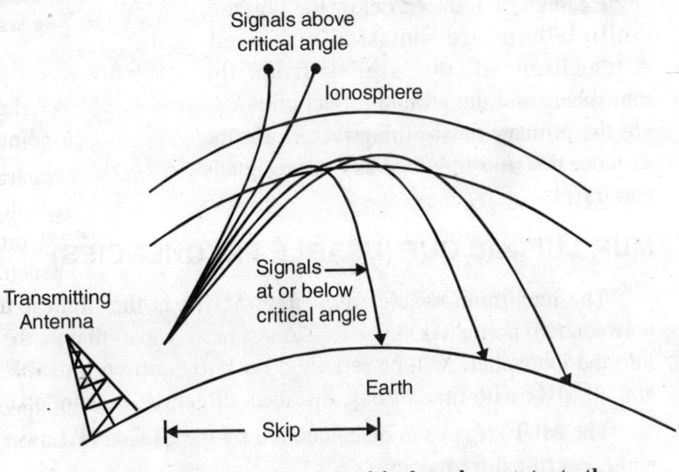

Fig. 7.38. Signals above the critical angle penetrate the ionosphere.

7.8. SKIP DISTANCE, SKIP ZONE AND MULTIPLE HOP TRANSMISSION

In Fig. 7.38 note that as the radiation angle decreases the distance signal wave travels over the earth increase. This distance is known as the **skip distance** and is an important consideration in long distance communications. At high frequencies, there may be long distances between the end of the usable ground wave signal and the reappearance of the reflected sky wave, as illustrated in Fig. 7.39. This is a no-signal zone known as **skip zone.**

The skip distance can be maximized by using the lowest radiation angle possible and the highest frequency that will be refracted at that angle. If the sky wave returns to earth and strikes a good conducting surface such as salt water, it can be reflected back into the ionosphere and take a double hop. If the signal strength is sufficient, the wave will be reflected again and take another hop, as shown in Figure 7.40. This concept is known as **multiple-hop transmission** and is used extensively for long-distance communications.

Under the best conditions, the maximum distance of a single hop is about 3200 km. Limiting factors are the frequency used and the radiation angle, and since the radiation angle cannot be reduced below the horizon, multiple-hop transmission is used. Attenuation of the signal from the ionosphere and the ground reflection points are the primary determining factors for the distance that, multiple-hop sky wave signals can travel.

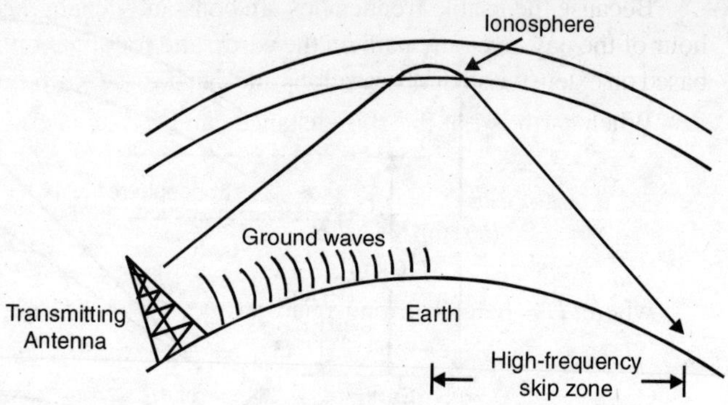

Fig. 7.39. Skip zone.

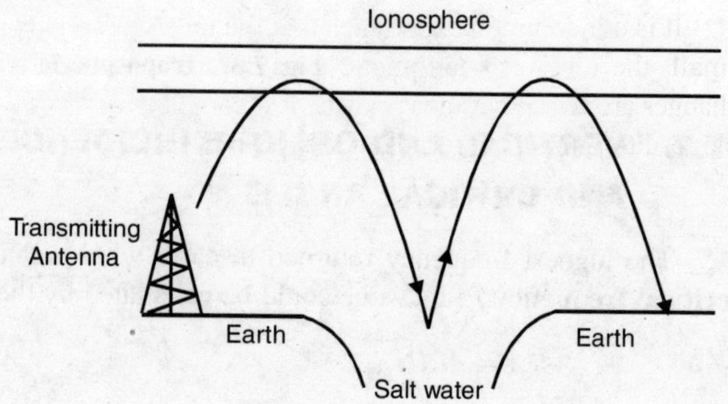

Fig. 7.40. Multiple-hop transmission.

MUF, LUF and OUF (USABLE FREQUENCIES)

The maximum usable frequency (MUF) is the highest frequency that can be used for transmission between two points via sky wave (ionospheric wave) that is, the highest frequency at which a signal radiated into the ionosphere will be refracted back to earth with **usable strength.** The MUF varies between 8 MHz and 30 MHz with time of day, distance, direction, season, and solar activity.

The MUF (f_{muf}) can be calculated by the SECANT Law ($f_{muf} = f_c \sec i$) where, i is angle of incidence and f_c is critical frequency

A frequency lower than the MUF can actually only be used, because it will also be refracted. However, as the signal frequency decreases, the signal energy absorbed by the ionosphere increases, drastically reducing the usable signal strength. Below a certain frequency, sometimes called the **lowest usable frequency** (LUF), the RF signal will be totally absorbed by the ionosphere. Therefore, operating at the MUF will produce the maximum received signal strength.

The MUF changes constantly with atmospheric conditions, and the LUF is border-line for refraction. Therefore, to obtain the most reliable ionospheric propagation, the **optimum usable frequency** (OUF) is used. The OUF is well above the LUF but far enough below the MUF that the signal is not appreciably affected by the constant atmospheric changes, thus providing relatively reliable ionospheric wave communications.

Because the usable frequencies are constantly changing, charts are available that predict them for every hour of the day, over any path on the earth, and for any month of the year. The predictions on these charts are based on extensive solar observations and can be used to optimize sky wave/ionospheric wave communications.

When earth is flat the skip distance can be calculated by the following equation

$$D_{skip} = 2H \sqrt{\frac{f_{muf}^2}{f_c^2} - 1}$$

where, H = height of ionosphere layer,

f_{muf} = = maximum usable frequency and

f_c = critical frequency for the layer under consideration.

7.9. FADING

It is our common experience that the received signal strength varies. If the changes in signal strength are small, the receiver's automatic gain control (AGC on AVC) may compensate. On the other hand, if the changes are large, the signal may be lost completely. This variation in signal strength is called **fading** and may be the result of multipath reception as shown in Fig. 7.41 and Fig. 7.42.

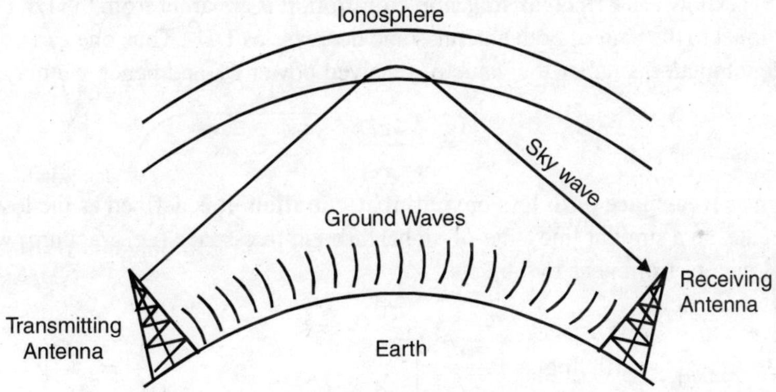

Fig. 7.41. Received signal arrives via ground and sky waves paths.

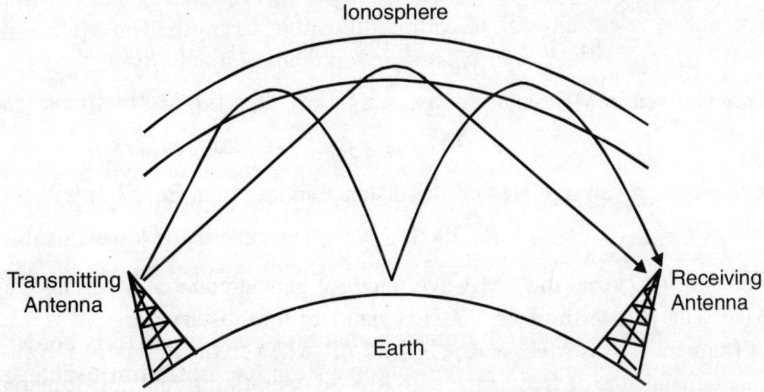

Fig. 7.42. Received signal arrives via both single and double hope sky waves paths.

Multipath fading exists in more or less in all type of communication systems. Several methods can be used to reduce fading such as installing highly directive receiving antenna, stagger stacking etc.

7.10. FRISS POWER TRANSMISSION EQUATION AND FREE SPACE PATH LOSS

Consider an isotropic antenna (radiator) which is radiating average power P_T (watt) in all directions, is placed in free-space (*i.e.*, homogeneous and loss-free medium of dielectric constant unity). A receiving antenna may be positioned so that it collects maximum power from the incoming wave. Let P_R be the received power at the receiver in Watts, one can derive (using the concept of power flux density or Poynting vector).

$$\frac{P_R}{P_T} = G_T \ G_R \left(\frac{\lambda}{4\pi D}\right)^2$$

or

$$P_R = \frac{P_T G_T G_R}{\left(\frac{4\pi D}{\lambda}\right)^2} \qquad ...(7.5)$$

where G_T = maximum directivity gain of transmitting, antenna

G_R = maximum directivity gain of the receiving, antenna

λ = wave length of the signal radiated in meters = c/f and

D = distance between the antennas.

The Eq. (7.5) is known as **Friis transmission equation**. It is apparent from the Eq. (7.5) that the received power is proportional to the gain of both antennas and decrease as $1/D^2$. Thus one can note from the Eq. (7.5) that larger the denominator, smaller the value of received power P_R and hence we may write

$$L_p = \left(\frac{4\pi D}{\lambda}\right)^2 \qquad ...(7.6)$$

and is known as **free-space path loss** or **spatial attenuation**. It is defined as the loss incurred by an EM wave as it propagates in a straight line (line-of-sight) through free-space (i.e., vacuum) with no absorption or reflection of EM energy from near by objects.

Converting to *dB* yields

$$L_{p\,(dB)} = 10 \ \log_{10} \left(\frac{4\pi f D}{c}\right)^2 ; \qquad \left(\frac{1}{\lambda} = \frac{f}{c}\right)$$

$$= 20 \ \log_{10} \left(\frac{4\pi f D}{c}\right)$$

$$= 20 \ \log_{10} \frac{4\pi}{10C} + 20 \ \log_{10} f + 20 \ \log_{10} D \qquad ...(7.7)$$

when the frequency is given in MHz and the distance in km, then Eq. (7.7) will take the form

$$L_{p\,(dB)} = 32.4 + 20 \ \log_{10} f_{(MHz)} + 20 \ \log_{10} D \qquad ...(7.8)$$

and when the frequency is given in GHz and the distance in km, then Eq. (7.7) will take the form

$$L_{p\,(bB)} = 92.4 + 20 \ \log_{10} f_{GHz} + 20 \ lof_{10} D_{(un)} \qquad ...(7.9)$$

It is clear from Eq. (7.7) that the EM wave (energy) gets attenuated in strength even while travelling through ideal space. The first term of Eq. (7.7) is standard unit distance loss (it is 32.5 dB for 1 km from transmitter when frequency is in MHz and it is 92.4 dB when frequency is in GHz). The second term of Eq. (7.7) shows the dependence of free-space path loss on operating frequency. It increases with increase in frequency. One can see from Eq. (7.6) that for the same distance an EM wave at 100 MHz will be attenuated by 40 dB more than an EM wave at 1 MHz, The third term of Eq. (7.7) shows the dependence of free-space path loss on distance from the transmitter antenna. An EM wave propagating from a transmitting antenna the free-space path loss at 10 km distance will be more by 6 dB as compared to the attenuation at 5 km from the transmitter. If the distance is increased 10 times from the reference level, the free-space path loss will increase by 20 dB.

7.11. HIGH FREQUENCY (HF) PROPAGATION

The propagation of HF consists of two waves.

1. The sky-wave
2. The ground-wave

The **sky-wave**, that bounces from the ionosphere, and the **ground-wave**, that propagates along the ground. Because of the long wavelengths, HF communications are not affected by the troposphere, and can usually bend around even large objects such as hills. The ionosphere permits communication over great distances: these are the frequencies of the *radio broadcasting*.

7.11.1. Very High Frequency (VHF) and Ultra High Frequency (UHF) Propagation

These frequencies are not affected by the troposphere, but are too high to exploit the ionosphere. In addition their decreased wavelength makes them increasingly less able to diffract around obstacles. At UHF in particular, areas in deep radio shadow can experience very poor reception. In urban areas, most reception is due to scattered arrivals. Medium wavelength transmission is usually better than VHF. VHF is used for FM radio stations and for TV transmissions VHF and UHF are used.

When the received signal consists of a number of scattered arrivals, an effect called **fading** occurs. According to the location, the many arrivals may interfere constructively or destructively. The phase relationship between the arrivals changes as the location of receiver changes. This causes a moving receiver to experience fluctuations between strong and weak signals. This effect is known as fading.

Radio and television **broadcasting** is the natural method for communicating unselec-tively with large numbers of correspondents, or when one or both of the correspondents are moving. Mobile communications employ networks of radio transmitter/receiver stations.

7.12. MICROWAVE AND MILLIMETER WAVE PROPAGATION

Microwave frequencies are usually completely blocked by obstructions. They need **line-of-sight** geometries. In addition, the wavelengths are small enough to become effected by water vapor and rain. Above 10 GHz these effects become important. Above 25 GHz, the effects of individual molecules become important. Water and oxygen are the most important gases. These have resonant absorption lines at 23, 69 and 120 GHz shown in Fig. 7.43. Above these frequencies; absorption is increasingly affected by the Earth's *black-body* spectrum. These peaks leave "windows" that may be used for communication, notably at 38 GHz and 98 GHz.

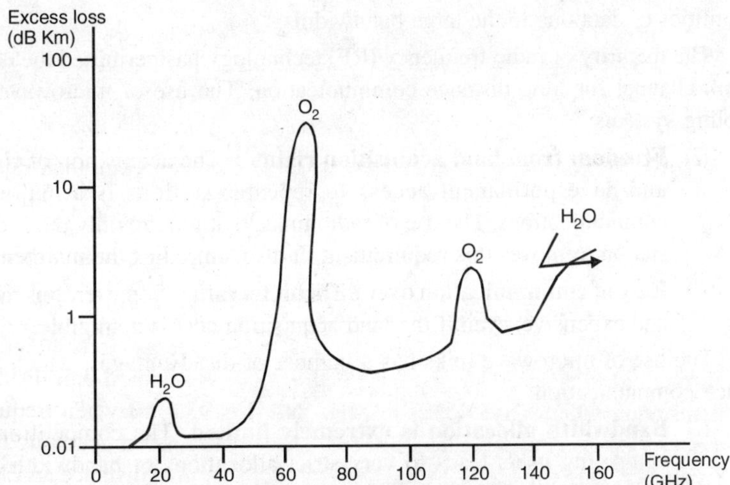

Fig. 7.43. Graph of attenuation due to water and oxygen for microwave and millimeter wave.

The use of the atmosphere as a transmission medium means that control over the noise environment is not possible. There are a number of sources of noise. Natural noise is classified by its source. **Atmospheric noise** dominates the low frequencies, up to 2MHz. The primary cause of this noise is electrical discharges in the atmosphere: lightning. So-called **galactic noise** is radio noise from outside the solar system. It is important up to 200 GHz. The sun also generates a small noise contribution at these frequencies.

Microwave links are high bandwidth line-of-sight links. Satellite links use geostationary satellites to provide inter-continental high-bandwidth communications. Optical links (i.e. the transmission through free-space of light waves) can be used for short, high bandwidth transmission. Microwave frequencies are normally used by Radar, satellite, and mobile communication.

7.13. APPLICATIONS OF WIRELESS MEDIA

1. Microwave link communication
2. Satellite communication
3. Mobile communications
4. Infrared

7.13.1. Microwave Link Communication

Microwave transmission is line of sight transmission. The Transmit station must be in visible contact with the receive station (refer Fig. 7.44). This sets a limit on the distance between stations depending on the local geography. Typically the line of sight due to the Earth's curvature is only 50 km to the horizon. Repeater stations must be placed so the data signal can hop, skip and jump across the country.

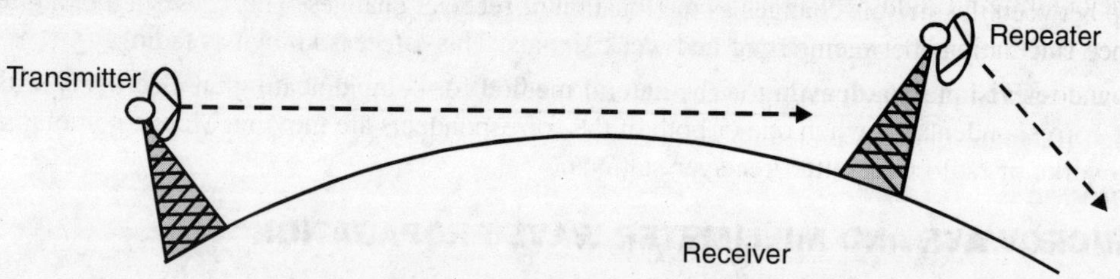

Fig. 7.44. Microwave link communication.

Microwaves operate at high operating frequencies of 3 to 10 GHz. This allows them to carry large quantities of data due to the large bandwidth.

The maturity of radio frequency (RF) technology has permitted the use of microwave links as the major trunk channel for long distance communication. The use of microwave links has major advantages over cabling systems:

(*i*) **Freedom from land acquisition rights :** The acquisition of rights to lay cabling, repair cabling, and have permanent access to repeater stations is a major cost in the provision of cable communications. The use of radio links, that require only the acquisition of the transmitter/receiver station, removes this requirement. It also simplifies the maintenance and repair of the link.

(*ii*) **Ease of communication over difficult terrain :** Some terrains make cable laying extremely difficult and expensive, even if the land acquisition cost is negligible.

The use of microwave links has a number of disadvantages, which mainly arise from the use of free-space communication:

(*i*) **Bandwidth allocation is extremely limited.** The competition for RF bandwidth from various competing users leads to very strict allocations of bandwidth. Unlike cabling systems, that can increase bandwidth by laying more cables, the radio frequency (RF) bandwidth allocation is finite and limited. In practice, bandwidth allocations of 50MHz in the carrier range 300MHz to 1GHz are typical.

(*ii*) **Atmospheric effects.** The use of free-space communication results in susceptibility to weather effects, particularly rain.

(*iii*) **Transmission path needs to be clear.** Microwave communication requires line-of-sight, point-to-point communication. The frequency of repeater stations is determined by the terrain. Care must be taken in the system design to ensure freedom from obstacles. In addition, links must be kept free of future constructions that could obstruct the link.

(*iv*) **Interference.** The microwave system is open to RF interference.

(*v*) **Costs.** The cost of design, implementation and maintenance of microwave links is high.

(*vi*) The modern urban environment presents a particular challenge, in that bandwidth allocation, RF interference, link obstruction and atmospheric pollution place maximum constraints on the system simultaneously. However, urban environments also have the highest land acquisition values too. Many modern cities have found it cost effective to build a single, very high tower to house an entire city's trunk communication microwave dishes. These towers are now a common feature of the modern urban landscape.

As the demand for bandwidth increases, microwave links will become increasingly unable to deliver. The use of increased carrier frequencies in the millimeter wave region would be advantageous. However, for technical reasons, no efficient methods of producing large quantities of millimetre power have been found. This is a necessity, given the increase in atmospheric attenuation at millimeter wave frequencies.

Advantages

1. They require no right of way acquisition between towers.
2. They can carry high quantities of information due to their high operating frequencies.
3. Low cost land purchase: each tower occupies small area.
4. High frequency/short wavelength signals require small antenna.

Disadvantages

1. Attenuation by solid objects: birds, rain, snow and fog.
2. Reflected from flat surfaces like water and metal.
3. Diffracted (split) around solid objects
4. Refracted by atmosphere, thus causing beam to be projected away from receiver.

7.13.2. Satellite Communication

In the last two decades satellites have become the most dominant vehicles of long distant communications. Modern satellite communication has been made possible by combining the skills and knowledge of space technology with those of microelectronics industry. This has been allowed increasingly heavier, more sophisticated satellites to be launched into the orbit for ever-lower cost. The evolution of satellite is a perfect example of how the boundary between the computer and communication industries is quickly dissolving. Satellite communication is one of the most rapidly growing and evolving technologies bringing with it multitude of business opportunities in the decades to come.

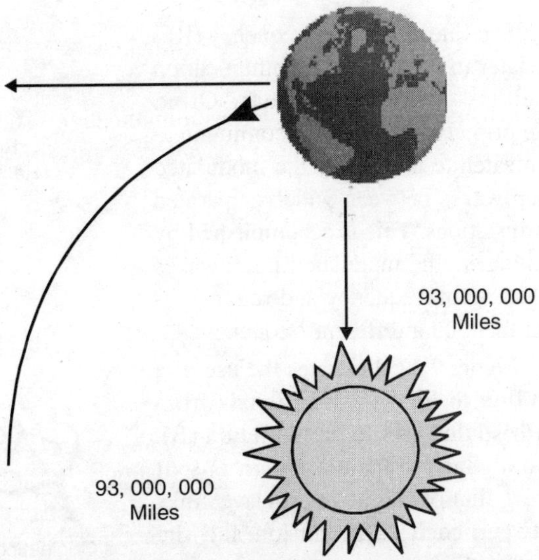

Fig. 7.45. The Earth orbits the sun with angular velocity

The concept of satellite communication was first conceived by Arthur C. Clarke groundbreaking article in October 1945. Signals are beamed into the space by an uplink antenna, received by an orbiting satellite electronically processed, broadcast back to the earth by a downlink antenna and received by an earth station located anywhere in the satellites footprint.

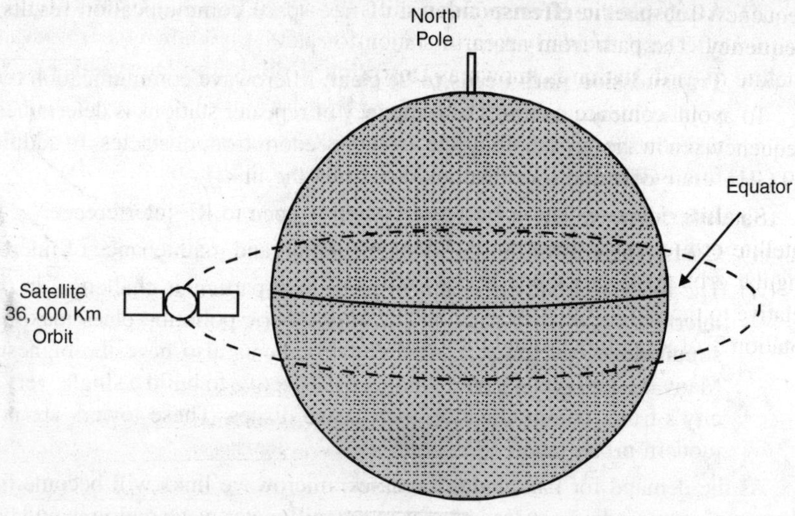

Fig. 7.46.

Most of the communication satellites are parked into the "Clarke's Belt" or the "geosynchronous "arc at 22,247 miles directly above the equator. The circle above the earth is unique because in this orbit the velocity of satellite matches that of the surface of the earth below. Each satellite thus appears to be in a fixed orbital position in the sky. This allows the stationary antenna to be permanently aimed towards any chosen geo synchronous satellite.

A **satellite** is an object that orbits another object. Gravity pulls the satellite closer to the primary object it orbits, but the satellite is moving perpendicular to that pull so quickly that the satellite continually avoids colliding with the primary object. All masses that are part of the solar system, including the Earth, are satellites of the Sun, or satellites of those objects, such as the Moon.

In common usage, the term is usually used to describe an **artificial satellite.** A gravitational force serves as the centripetal force needed to make the object circle the primary object. The motion of the satellite around its primary gravitational source is known as freefall.

Communication Satellite System

Satellite is a radio frequency (RF) repeater in orbit. Most communication satellites are in Geostationary Orbit. The primary purpose of a communication satellite is to relay the modulated microwaves between widely separated earth stations. This is accomplished by picking up the modulated microwaves on a certain frequency and retransmitting them on a different frequency.

Figure 7.47 illustrates the use of a satellite to transmit the signals from earth station (A) to earth station (B). The signal propagates by means of direct line-of- sight microwave links between each earth station and the satellite. Signals from a ground station (A) are sent to the satellite on one

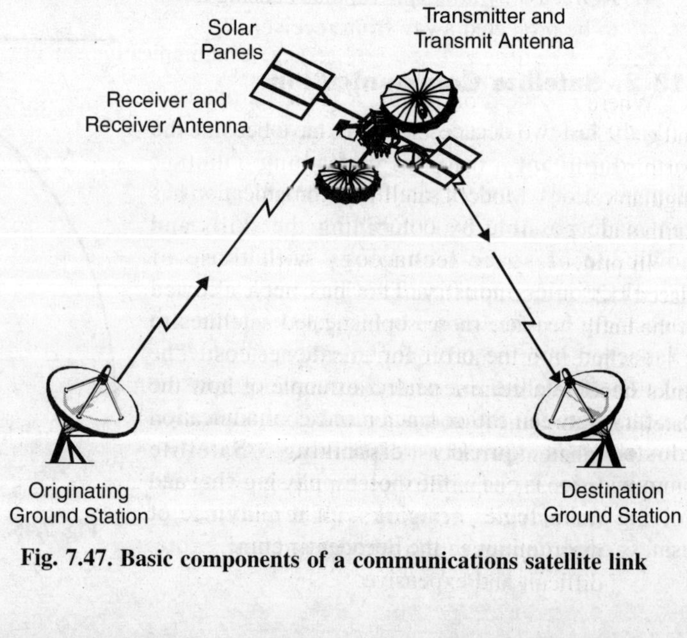

Fig. 7.47. Basic components of a communications satellite link

frequency. The satellite **Transponder** shifts the signal from the **Uplink frequency** to a different **Downlink frequency**. The path from an earth station to satellite is known as **Uplink** and the returning path from a satellite to earth station is known as **Downlink**. **Uplink and Downlink Frequencies**

To avoid influence and any filtering of signal we use in which all the frequencies are in GHz. Uplink frequency is always greater than the Downlink Frequency. The frequency range for Uplink frequency is 5.9 GHz – 6.4 GHz and for downlink is 3.7 GHz –7.2 GHz.

Satellite communication became a possibility when it was realized (by Sir Arthur C. Clarke) that a satellite orbiting at a distance of 36000 Km from the Earth would be **Geostationary**, *i.e.* would have an angular orbital velocity equal to the Earth's own orbital velocity. It would thus appear to remain stationary relative to the Earth if placed in an equatorial orbit. This is a consequence of Kepler's law that the period of rotation T of a satellite around the Earth was given by:

$$T = \frac{2\pi r^{3/2}}{\sqrt{(g.R^2)}}$$

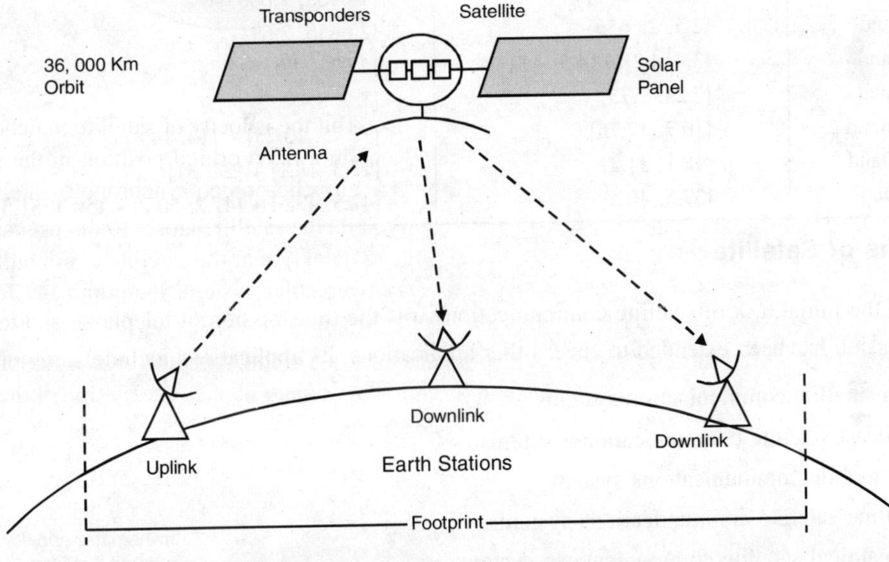

Fig. 7.48. Communication satellite system

Where r is the orbit radius, R is the Earth's radius and $g = 9.8\ ms^{-2}$ is the acceleration due to gravity at the Earth's surface. As the orbit increases in radius, the angular velocity reduces, until it is coincident with the Earth's at a radius of 36000Km.

In principle, three geostationary satellites correctly placed (120 degree apart) can provide complete coverage of the Earth's surface shown in Fig. 7.49.

For intercontinental communication, satellite radio links become a commercially attractive proposition. Satellite communication has a number of advantages:

- Inherent broadcast capability
- Inherent capability of by passing the whole terrestrial networks. The laying and maintenance of intercontinental cable is difficult and expensive.

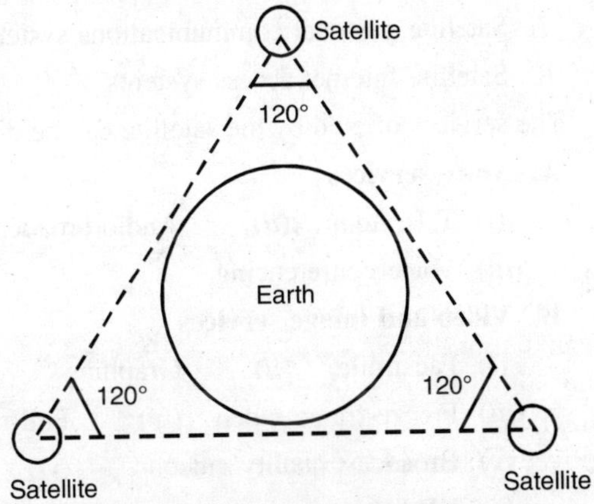

Fig. 7.49. Geostationary satellites providing global coverage

- The heavy usage of intercontinental traffic makes the satellite commercially attractive.
- Satellites can cover large areas of the Earth.
- High speed network access.
- Easy deployment and configuration.

Satellite communication is limited by four factors :

- Technological limitations preventing the deployment of large, high gain antennas on the satellite platform.
- Over-crowding of available bandwidths due to low antenna gains.
- The high investment and maintenance cost.
- High atmospheric losses above 30GHz limit carrier frequencies.

Frequency bands reserved for satellite communications are as follows:

Band	Downlink (GHz)	Uplink (GHz)
L-Band	[1.53, 1.559]	[1.6265, 1.6605]
S-Band	[2.5, 2.6655]	[2.6655, 2.69]
C-Band	[3.4, 4.2] + [4.5, 4.8]	[5.725, 7.075]
X-Band	[7.25, 7.75]	[7.9, 8.4]
Ku-Band	[10.7, 12.75]	[12.75, 13.25] + [14, 14.8]
Ka-Band	[18.1, 21.2]	[17.3, 18.1] + [27, 31]
V-Band	[37.5, 40.5]	[42.5, 43.5] + [47.2, 50.2] + [50.4, 51.5]

Applications of Satellite

Although the initial task of satellite communications was the transmission of telephone and television signals, its mission has been extended to cover other applications. Its applications include:

1. Fixed satellite communication systems
2. Broadcast satellite communications systems
3. Land mobile communications system
4. Maritime satellite communications systems
5. Aeronautical satellite communications systems
6. Radio determination satellite systems
7. Satellite personal communications systems
8. Satellite Internet access systems

The services offered by the satellite can be divided into the following categories :

A. Voice services

(*i*) Telephony (*ii*) Audio broadcasting – DAB

(*iii*) Voice conferencing

B. Video and image services

(*i*) Facsimile (*ii*) Graphics

(*iii*) Freeze-frame video (*iv*) Full-motion video

(*v*) Broadcast quality video (*vi*) Teletext/videotext

(*vii*) HDTV

C. Data services

(*i*) Electronic mail (*ii*) Database access

(*iii*) File transfer (*iv*) Remote data monitoring

(*v*) Short message transmission (*vi*) Paging

(*vii*) Electronic funds transfer

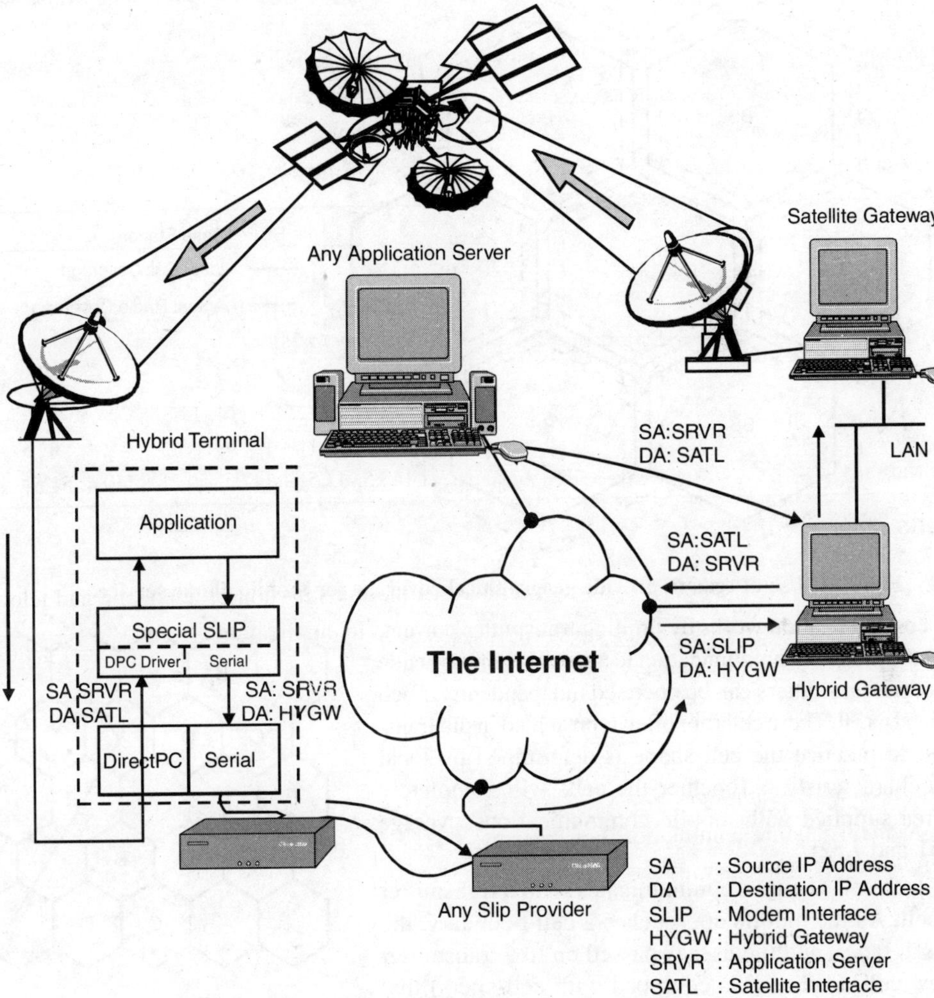

Fig. 7.50. Various Satellite Applications

7.13.3. Mobile Communications

The use of mobile radio-telephones has seen an enormous boost in the 1980s and 1990s.

Mobile communications are usually allocated bands in the 50MHz to 1GHz band. At these frequencies the effects of scattering and shadowing are significant. Lower frequencies would improve this performance, but HF bandwidth is not available for this purpose. The primary problems associated with mobile communication at these frequencies are:

(*i*) Maintaining transmission in the fading circumstances of mobile communication.

(*ii*) The extensive investigation of propagation characteristics required prior to installation.

The use of digital mobile telephones has a number of advantages:

(*i*) Access to national and international telephone system.

(*ii*) Privacy of communication.

(*iii*) Data independent transmission.

(*iv*) An infinitely extendable number of channels.

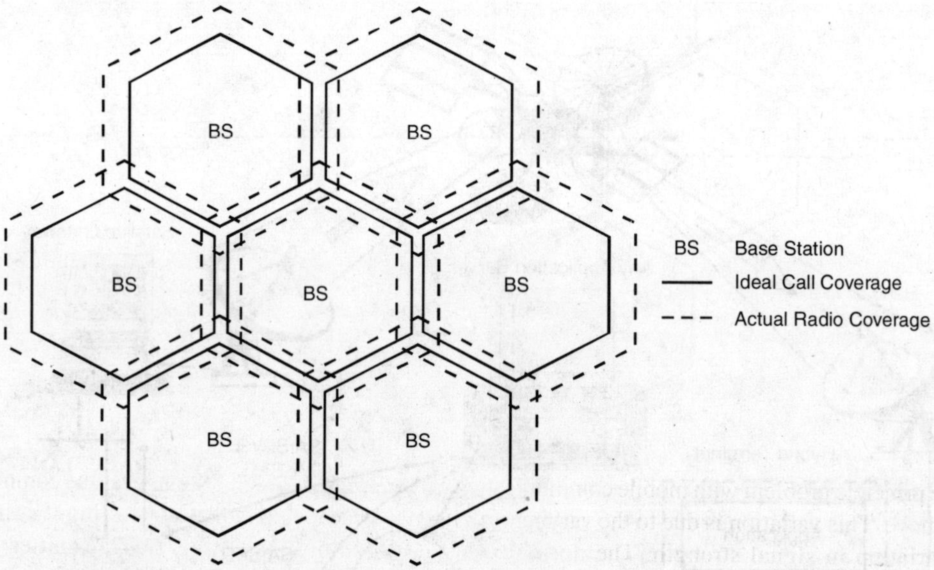

Fig. 7.51. Use of cells to provide geographical coverage for mobile phone service.

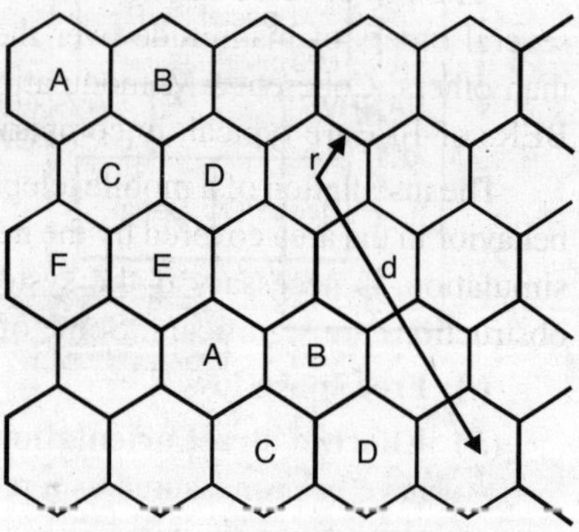

Fig. 7.52. Frequency re-use in cells.

Mobile communication works by limiting transmitter powers. This restricts the range of communication to a small region. Outside this region, other transmitters can be operated independently. Each region is termed a **cell**. These cells are often represented in diagrams as hexagons. In practice the cell shape is determined by local propagation characteristics. Together the cells will completely cover the area supplied with mobile communication coverage (Figures 7.51 and 7.52).

Within each cell, the user communicates with a transmitter within the cell. As the mobile approaches a cell boundary, the signal strength fades, and the user is passed on to a transmitter from the new cell. Each cell is equipped with cell-site(s) that transmit/receive to/from the mobile within the cell. Within a single cell, a number of channels are available. These channels are (usually) separated by frequency. Then a mobile initiates a call, it is assigned an idle channel within the current cell by the **mobile-services switching centre (MSC)**. User uses the channel within the cell until he reaches the boundary. He is then allocated a new idle channel within the next cell.

The heart of the mobile telephone network is the MSC. Its task is to acknowledge the paging of the user, assign user a channel, broadcast user dialed request, return the call. In addition it automatically monitors the signal strength of both transmitter and receiver, and allocates new channels as required. This latter process, known as **hand-off**, is completely hidden to the user, although is a major technical problem. In addition, the MSC is responsible for charging the call, shown in Fig. 7.53. The decision making ability of the MSC relies to a great extent on modern digital technology. It is the maturity of this technology that has permitted the rapid growth of mobile communications.

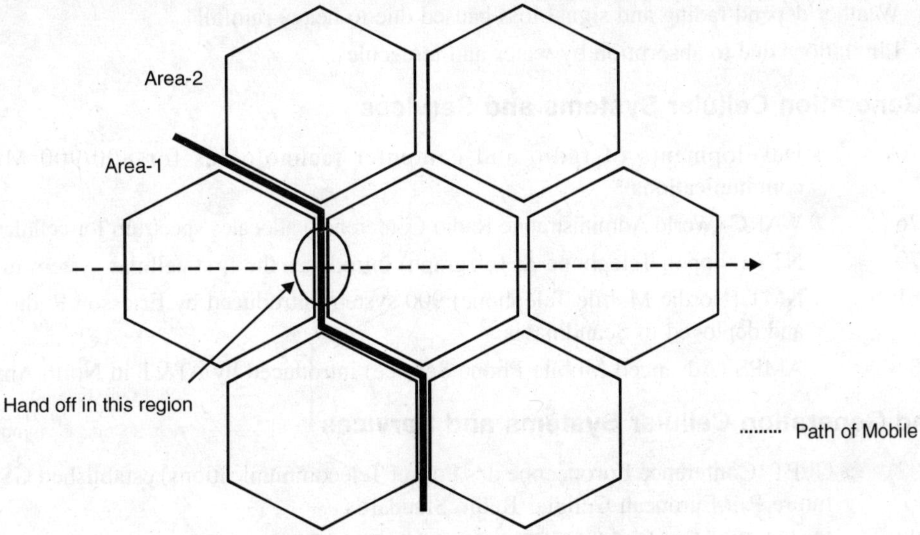

Fig. 7.53. Hand-off between cells

The principle problem with mobile communication is the variation in signal strength as the communicating parties move. This variation is due to the varying interference of scattered radiation—**fading**. Fading causes rapid variation in signal strength. The normal solution to fading, increasing the transmitter power, is not available in mobile communication where transmitter power is limited.

The presence of fading has a severe effect on the BER (Bit-Error Rate). It can reduce the BER by several orders of magnitude over the non-fading case. In addition, some modulation schemes cope better than others. Coherent AM modulation is better than coherent FM in the fading environment. In any event BERs of 10^{-3} are typical, in comparison with BERs of 10^{-6} for non-fading channels.

The installation of a mobile telephone system requires a large initial effort in determining the propagation behavior in the area covered by the network. Propagation planning, by a mixture of observation and computer simulation, is necessary if the system is to work properly. At UHF and VHF frequencies, the effects of obstructions are significant. Some of the effects that need to be considered are :

(*i*) **Free space loss.**

(*ii*) **Effect of street orientation.** Streets have a significant waveguide effect. Variations of up to 20dB have been measured as a result of street direction.

(*iii*) **Effects of foliage.** Propagation in rural areas is significantly affected by the presence of leaves. Variations of **18dB** between summer and winter have been observed in forested areas.

(*iv*) **Effect of tunnels.** Tunnels can introduce signal attenuation of up to 30dB according to the tunnel length and frequency.

Frequencies for Mobile Communication

VHF/UHF ranges for mobile radio :

(*i*) Simple and small antenna for car.

(*ii*) Deterministic propagation characteristics and reliable connection.

SHF and higher for directed radio links :

(*i*) Simple and small antenna

(*ii*) Large bandwidth

Wireless LANs uses frequencies in UHF to SHF range :

(*i*) Some system planned to used up to EHF

(*ii*) Weather depend fading and signal loss caused due to heavy rainfall

(*iii*) Limitations due to absorption by water and molecules.

First Generation Cellular Systems and Services

1970s	Developments of radio and computer technologies for 800/900 MHz mobile communications
1976	WARC (World Administrative Radio Conference) allocates spectrum for cellular radio
1979	NTT (Nippon Telephone & Telegraph) introduces the first cellular system in Japan
1981	NMT (Nordic Mobile Telephone) 900 system introduced by Ericsson Radio System AB and deployed in Scandinavia
1984	AMPS (Advanced Mobile Phone Service) introduced by AT&T in North America

Second Generation Cellular Systems and Services

1982	CEPT (Conference Europeenne des Post et Telecommunications) established GSM to define future Pan-European Cellular Radio Standards
1990	Interim Standard IS-54 (USDC) adopted by TIA (Telecommunications Industry Association)
1990	Interim Standard IS-19B (NAMPS) adopted by TIA
1991	Japanese PDC (Personal Digital Cellular) system standardized by the MPT (Ministry of Posts and Telecommunications)
1992	Phase I GSM system is operational
1993	Interim Standard IS-95 (CDMA) adopted by TIA
1994	Interim Standard IS-136 adopted by TIA
1995	PCS Licenses issued in North America
1996	Phase II GSM operational
1997	North American PCS deploys GSM, IS-54, IS-95
1999	IS-54: North America
	IS-95: North America, Hong Kong, Israel, Japan, China, etc
	GSM: 110 countries

The aims of the GSM system were :

(*i*) Good speech quality

(*ii*) Low terminal cost

(*iii*) Low service cost

(*iv*) International roaming

(*v*) Ability to support hand-held portables

(*vi*) A range of new services and facilities (ISDN)

Third Generation Cellular Systems and Services

IMT-2000 (International Mobile Telecommunications-2000):

Fulfill one's dream of anywhere, anytime communications a reality.

Key Features of IMT-2000 include:

(*i*) High degree of commonality of design worldwide;

(*ii*) Compatibility of services within IMT-2000 and with the fixed networks;

(*iii*) High quality;

(*iv*) Small terminal for worldwide use;

(*v*) Worldwide roaming capability;

(*vi*) Capability for multimedia applications, and a wide range of services and terminals.

7.14. FREE SPACE OPTICAL TRANSMISSION

Let us discuss free space communication using optical. Free space optical communication systems use an optical carrier signal to transfer information through the air (free space) between two or more optical receivers or transceivers. Figure 7.54 shows free space optical transmission. This figure shows that free

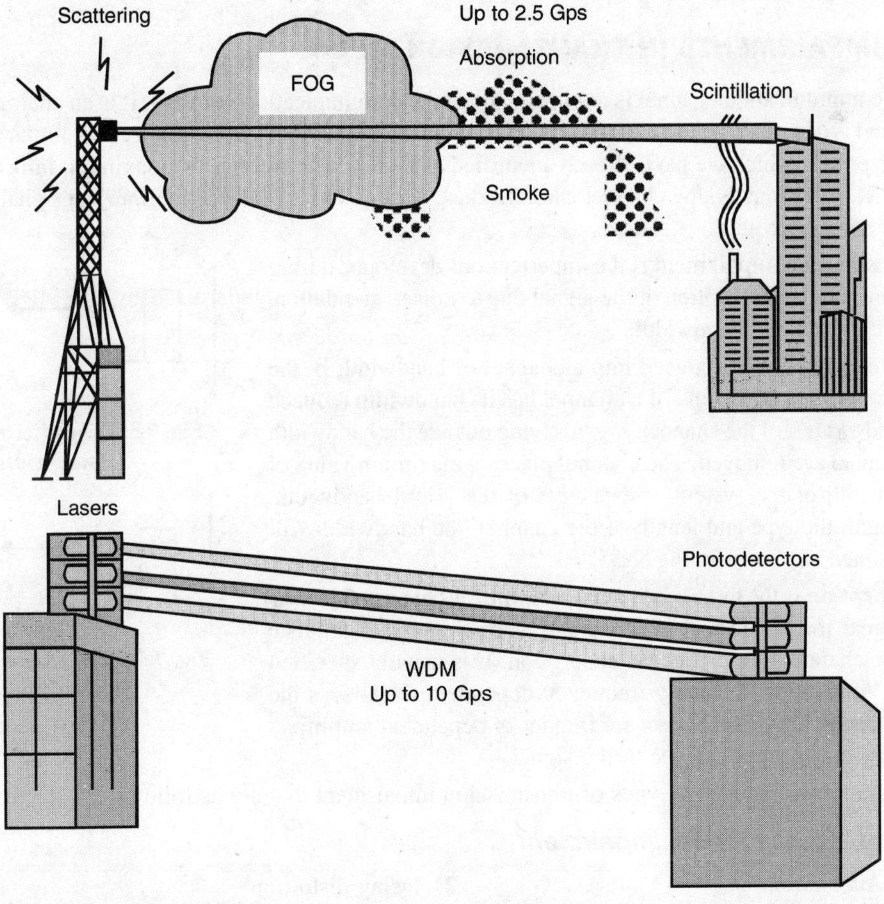

Fig. 7.54. Free space communications.

space optical transmission systems lose some of their energy from signal scattering, absorption and scintillation. Optical signal scattering occurs when light signals are redirected as they pass through water particles. Optical signal absorption occurs as some optical energy is converted to heat as it strikes particles (such as smog). Scintillation occurs when heated such as from smokestacks air cause a bending of the optical beam. This example shows that it is possible to transmit multiple light wave signals on different wavelengths (WDM) to increase the overall data transmission rate.

Infrared

Signal having the frequency range from 300 GHz to 400 THz (400×10^{12} Hz) are used for short range communication, are known as **Infrared signal**. These signals have very high frequency and cannot penetrate walls. That property makes these signals to have communication within the Line-of-Sight (LOS) only. Because of this property these signals cannot interfere with another system in the next room. But due to this property, Infrared signals cannot be used for longer distance and even we cannot use these signals in the open sun light because Sun rays contain Infrared wave and that can interfere with the communication.

Applications

1. Having the frequencies almost 400 THz, it has very high bandwidth for transfer of data with the very high data rate. This property is used to transfer the data in mobile handsets and wireless mouse, keyboard, etc. These port is called IrDA port.
2. Infrared signals are used in heat seeking guidance system, electronic photography and astronomy.

7.15. IMPAIRMENTS IN TRANSMISSION MEDIA

The communication channel is a crucial part of the communication network. It is an analogue part of the system, and is described in terms of the analogue; quantities bandwidth, absorption etc. It limits the bandwidth and noise power, which we have already identified as factors determining the maximum information rate of the channel. Additionally, the channel can introduce various kinds of distortion into the signal, and also has a delay associated with it.

Transmission impairment is the imperfections developed during the transmission and reception of the signal due to noise, attenuation, distortion and limited bandwidth.

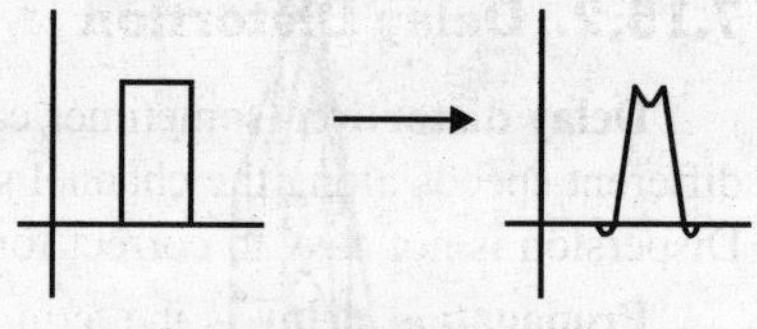

Fig. 7.55. The effects of limited bandwidth

When a signal is introduced into a channel of bandwidth **B**, the signal emerging from the end of a channel has its bandwidth reduced to the bandwidth B of the channel. Signals lying outside the bandwidth of the channel are removed. The channel places a maximum value of the bandwidth of the system, irrespective of the signal bandwidth. According to the type and length of the channel, the bandwidth will be determined as shown in Fig. 7.55.

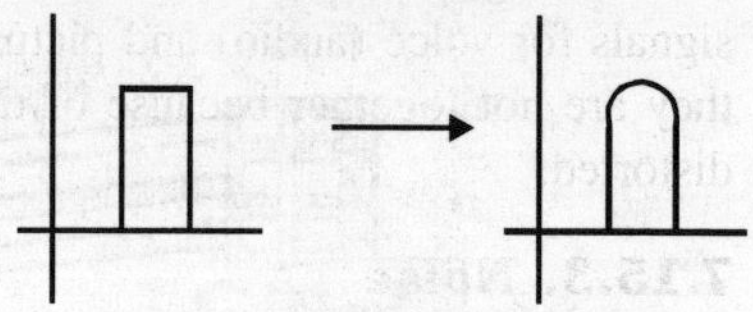

Fig. 7.56. The effects of channel absorption

Absorption is the term used to describe **the loss of signal power** as the signal moves through the channel, shown in Fig. 7.56. The longer the channel, the higher the absorption. It is usually specified in dB/m. Absorption is usually frequency dependent. It reduces the available bandwidth. **Equalizers** are frequency dependent amplifiers that restore the spectral balance of the signal.

Generally, we have three types of transmission impairment that are as follows:

Types of Transmission impairment

1. Attenuation
2. Delay distortion
3. Noise
4. Distortion

7.15.1. Attenuation

Attenuation is the decrease of signal strength as it propagates along a channel or loss of energy through the channel (guided or unguided), refer Fig. 7.57. **Repeaters** are amplifiers placed along the channel that restore the signal power.

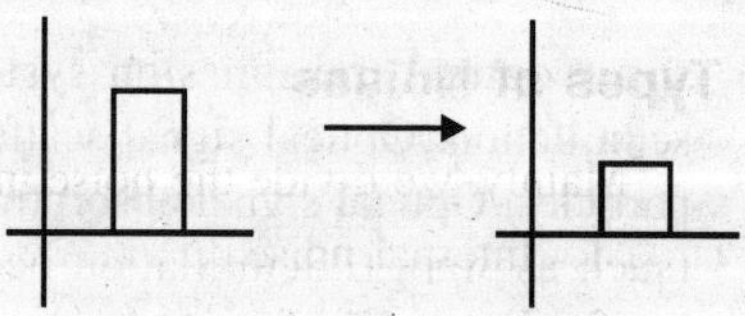

Fig. 7.57. The effects of signal attenuation.

In the guided channel, the loss may be in the form of power loss, heat loss, dielectric loss, leakage loss etc. And if the channel is unguided then due to interference of signal energy loss occur. The distance between the repeaters is determined so as to minimize any errors in the signal due to attenuation effects, for instance to maintain a minimum signal to noise ratio (SNR) along the channel. Some attenuation affects are due to absorption. The amount of attenuation is measured in dB/m. Some typical values are shown in Table 7.2.

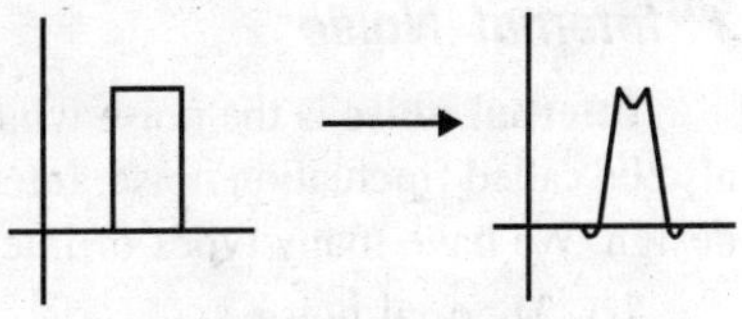

Fig. 7.58. The effects of dispersion (delay distortion).

TABLE 7.2. Some Typical Media Loss Values.

Transmission medium	Frequency	Loss (dB/Km)
Copper wire pair (0.3 cm diameter)	1 kHz	0.05
Twisted pair (16 gauge)	10 kHz	2
	100 kHz	3
	300 kHz	6
Coaxial cable	100 kHz	1
	1 MHz	2
	3 MHz	4
Fibre-optic cable	4×10^{14} Hz	5

7.15.2. Delay Distortion

Delay distortion (sometimes called **Dispersion**) describes the effect of different frequencies moving at different speeds along the channel shown in the Fig. 7.58. This introduces **changes in shape of the signal**. Dispersion is not easy to correct for, and channels are sometimes limited by its effect.

Propagation delay is the term used to describe the time taken for the signal to pass along the channel. Electromagnetic signals travel at velocities near that of light in vacuum of 3×10^8 m/s. For example, in Television broadcasting, we transmit audio and video signals, known as composite video signal, with the different carrier signals for voice (audio) and picture (video) signals. Now, when both the signals arrived at the receivers, they are not together because of different speed and time delay, with the result the composite signal get distorted.

7.15.3. Noise

Noise is the term used to describe all signals present at the receiver that are not part of the message signal or it may be defined as the unwanted electrical signals present at the receiver and lies between the passband of the signals. Noise is an unwanted EM energy, which interfere with the ability of receiver to detect the wanted signal. The effect of noise on a signal is shown in Fig. 7.59. The source of these noise signals vary widely.

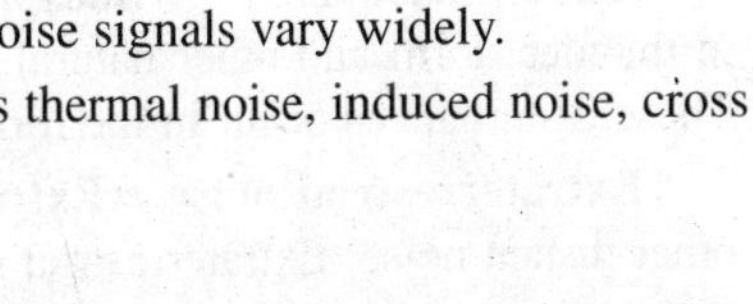

Fig. 7.59. The effects of noise on a signal

Depending upon the medium we have many types of noise such as thermal noise, induced noise, cross talk and impulse noise.

Types of Noises

Mainly, we divide the noise in two categories such as;

1. Internal noise
2. External noise

1. Internal Noise

Internal noise is the noise which is generated within the system or device itself. This type of noise can also be called fluctuation noise. Internal noise can be treated quantitatively may be reduced by proper system design. We have many types of internal noises; such as

(*i*) Thermal noise,
(*ii*) Transit time and
(*iii*) Shot noise
(*iv*) Partition noise
(*v*) Crystal noise
(*vi*) Flicker noise

We will be discussing some of the internal noise such as Thermal noise, Transit time noise and shot noise.

Thermal Noise : Thermal noise is always present. It is due to the random motion of electric charges present in all media. The noise generated in a resistance or the resistive component of any impedance is random. It may be called thermal-agitation noise. It can be generated externally or internally at the receiver. A common property of all thermal noise is that it may be described by the relation:

$$N = kTB \text{ watts}$$

where, $k = 1.38 \times 10^{-23}$ J/ k, (known as *Boltzman's Constant*), **T** is the absolute temperature in degrees Kelvin (273 + °C), and **B** is the receiver bandwidth.

Transit-Time Noise : It is normally observed in semiconductor devices. If the time taken by an electron to travel from the emitter to the collector of a transistor is comparable with the time period of the signal, some charge carries diffuse back to the emitter or source. This process gives rise to an input admittance in which the conductance components increases with the frequency. The small currents induced in the input of the device by random fluctuations in the output current become of great importance at such frequencies and produced random noise.

Shot Noise : Shot noise is present in active devices due to random behavior of charge carries. It arises in electronic devices because of discrete nature of current flow in the device. In solid state semiconductor devices it arises from the random diffusion of minority carriers and the random generation and recombination electron hole pairs. When this noise is amplified and fed to loudspeaker it sounds like a shower of lead shots striking a metal plate and hence the name **shot noise**.

2. External Noise

External noise may be defined as that type of noise which is created outside the receiver. There are various forms of external noise, are classified as

1. Atmospheric noise,
2. Industrial noise
3. Extraterrestrial noise.

Atmospheric noise : Atmospheric noise resulting from natural disturbance such as lighting discharge in thunder storm and other natural electrical disturbances which occur in the atmospheric is called static. These disturbances occur in the form of impulses and are random in nature.

Extraterrestrial noise : Extraterrestrial noise arises from extraterrestrial bodies such as Sun and other distant noise. Extraterrestrial noise may be divided into following groups:

(*i*) Solar noise
(*ii*) Cosmic noise

Industrial noise : It is also called man-made noise. This noise is produced by such sources as electrical motors, automobiles and aircraft ignition; switch gears and leakage from high voltage transmission lines. Manmade noise is impulsive in nature.

Impulse noise is common in low-frequency circuits and arises from electric fields generated by electrical switching. It appears as bursts at the receiver, and when present can have a catastrophic effect due to its large power. Other people's signals can generate noise: **cross-talk** is the term given to the pick-up of radiated signals from adjacent cabling. When radio links are used, **interference** from other transmitters can be problematic.

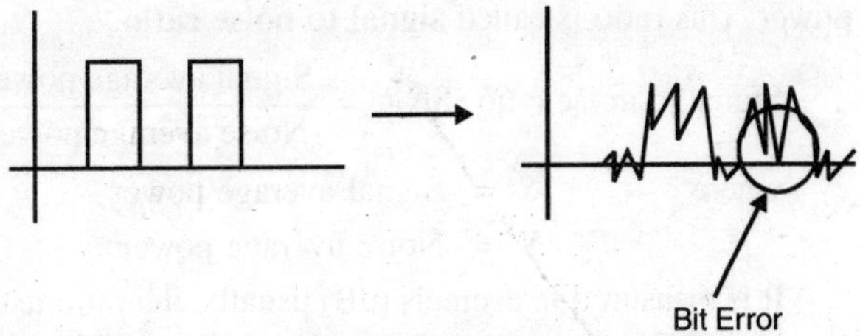

Fig. 7.60. A bit error in a transmitted signal due to channel effects

The combined effects of attenuation, absorption, dispersion, and noise may result in **bit errors** occurring in the transmitted signal as shown in Figure 7.60.

DO YOU KNOW▶

Transmission impairment *is the imperfections developed during the transmission and reception of the signal due to noise, attenuation, distortion and limited bandwidth.*

7.16. SHANNON CAPACITY

In 1924, Henry Nyquist, has realized that even a perfect channel (i.e. noiseless channel) has a finite transmission capacity. He derived as expression for the maximum data rate for a finite bandwidth noiseless channel.

Sampling is a process by which an analog signal is converted in to discrete time signal. The number of samples to be taken depends upon maximum signal frequency present; sampling theorem gives the complete idea about the sampling of the signal.

Sampling theorem states that :

(*i*) A bandlimited signal of finite energy, which has no frequency components higher that B Hz, is completely recovered from the knowledge of its samples taken at the rate of 2B samples per second, where B is the bandwidth of a signal. 2B is normally referred as sampling rate or Nyquist data rate.

(*ii*) A bandlimited signal of finite energy, which has no frequency components higher that B Hz, is completely described by specifying the values of the signal at instants of time separated by 1/2B seconds.

Nyquist proved that any arbitrary signal has been passed through a low pass filter of bandwidth B, the filtered signal can be reconstructed completely by making only 2B (exact) sample per second. Sampling the line faster than 2B times per second is of no use because the higher frequency components such sampling could recover have already been filtered out.

According to **Nyquist's theorem,** if the signal consists of V discrete levels, the maximum data rate can be given as

$$\text{Maximum data rate} = 2B \log_2 V \text{ bits/sec} \qquad ...(7.10)$$

Let us consider an example; a noiseless channel of kHz cannot transmit binary (i.e. two levels) signals at a rate encoding 6000 bits per sec.

In 1948, Claude Shannon carried Nyquist's work further and extended it to the practical care of a channel subject to random noise. All practical channels are noisy channels and random noise deteriorates the situation rapidly. It is always that random or thermal noise present due to the motion of the molecules in the system. The amount of thermal noise present is determined by the ratio of the signal power to the noise power. This ratio is called **signal to noise ratio**.

$$\text{Signal to noise ratio (SNR)} = \frac{\text{Signal average power}}{\text{Noise average power}}$$

where S = Signal average power

N = Noise average power

It is measured in decibels (dB) usually, the ratio itself is not quoted, and instead, the quantity $10\log_{10}$ S/N is given. For example an S/N ratio of 10 is 10dB, a ratio of 100 is 20dB, and a ratio of 1000 is 30dB and so on.

Shannon Limit for Information Capacity

According to **Shannon's theorem**, that the maximum data rate of a noisy channel where bandwidth is B Hz, and has SNR is S/N, is given by.

$$\boxed{\text{Maximum number of bits/sec} = B \log_2 (1 + S/N)} \quad ...(7.11)$$

Shannon's result was derived from information-theory arguments and applies to any channel subject to thermal noise. Now, for examples, a channel of 3 kHz bandwidth with a signal to noise ratio of 30 dB can never transmit much more than 30,000 bps. This is this maximum data rate in noisy channel.

This theorem is concerned with the rate of transmission of information over a transmission media or communication channel. The term communication channel includes all the features and components parts of the transmission system which introduce noise, or limit the bandwidth.

Shannon's theorem defines that it is possible, in principle, to device a means whereby a communication system will transmit information with an arbitrarily small probability of error provided that the information rate 'R' is less than or equal to a 'C' the channel capacity.

$$R = C \quad ...(7.12)$$

where R = Information rate

C = Channel capacity

It is commonly known as Shannon-Hartley capacity theorem. According to this the system capacity (C) is a function of the average received signal power 'S', the average noise power 'N', and bandwidth 'B'. They are related as:

$$\boxed{C = B \log_2 (1 + S/N)} \quad ...(7.13)$$

If bandwidth (B) is in hertz (Hz) and base is 2 of logarithmic then unit of channel capacity, is bits/sec.

Theoretically as we have discussed in Eq. (7.11), it is possible to transmit information over such a communication channel at any rate R, where $R = C$, with an arbitrarily small error.

The important feature of the theorem is that it indicates that for $R = C$, error free transmission is possible in the presence of noise.

There is another statement associated with this theorem, that is given a source of M equally likely messages, with $M >> 1$, which is generating information at a rate R and given a channel with a channel capacity C. if, now $R > C$, the probability of error is close to unity for every possible set of M transmitter signals.

If the information rate R exceeds a specific value of channel capacity C, the error probability approaches unity, as M increases.

The channel capacity 'C' decides the maximum permissible rate at which an error free transmission of information is possible through it.

Effect of Noise on Digital Signals

Noisy Channel : So far this information seems to imply there is no upper bound for the data rate given the maximum frequency. Unfortunately, this is not true for two reasons.

First, more signal components means subtler changes among them. For Example, suppose a signals amplitude must be less than or equal to 5 voltas and that each components is determined by amplitude only. If we used two components defined by 2.5 v and 5v, the signal differ by 2.5 v. However, using 16 signal components requires a difference of about 1/3 v between adjacent amplitudes. The receiver must be more sophisticated to be able to detect smaller differences. If the difference becomes too small, we eventually exceed the ability of a device to even detect then.

The Second reason occurs because many channels are subject to noise, which means a transmitted signal can be distorted. If the distortion is too large the receiver cannot reconstruct the signal. For Example, consider the digital signal in Fig. 7.61 (*a*). We use a digital signal simply because it is easier to illustrate.

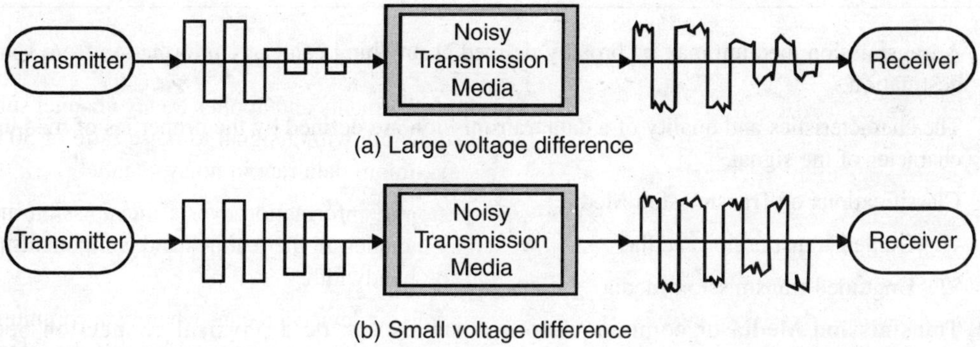

Fig. 7.61.

The transmitter sends two signals, each of which oscillates between two voltage levels. However, the transmitted signal differs from it. The distortion is not too great, so the received signal still pretty clearly defines two voltage levels. Thus, it would not be too difficult to reconstruct them.

Figure 1.10(*b*) shows a similar situation , except in this case the original two voltage levels differ by less. Now when noise occurs the two distorted signals overlap voltage levels and it is difficult, if not impossible, to reconstruct the original signal from the received one.

Shannos Result : We have learned that noise can alter and possibly destroy information. Whether the information can be reconstructed depends on how powerful the noise.

Example 7.2. Consider a noiseless channel with a bandwidth of 3000 Hz transmitting a signal with two signal levels.

Solution. The maximum bit rate can be calculated as

$$\text{Bit Rate} = 2 \times 3000 \times \log_2 2 = 6000 \text{ bps}$$

Example 7.3. Consider the same noiseless channel, transmitting a signal with four signal levels (for each level, we send two bits).

Solution. The maximum bit rate can be calculated as:

$$\text{Bit Rate} = 2 \times 3000 \times \log_2 4 = 12{,}000 \text{ bps}$$

Example 7.4. Consider an extremely noisy channel in which the value of the signal-to-noise ratio is almost zero. In other words, the noise is so strong that the signal is faint. For this channel the capacity is calculated as

$$C = B \log_2 (1 + SNR) = B \log_2 (1 + 0)$$

$$C = B \log_2 (1) = B \times 0 = 0$$

Example 7.5. We have a channel with a 1 MHz bandwidth. The SNR for this channel is 63; what is the appropriate bit rate and signal level?

Solution. First, we use the Shannon formula to find our upper limit.

$$C = B \log_2 (1 + SNR)$$

$$= 10^6 \log_2 (1 + 63) = 10^6 \log_2 (64) = 6 \text{ Mbps}$$

Then we use the Nyquist formula to find the number of signal levels.

$$\mathbf{4\ Mbps = 2 \times 1\ MHz \times \log_2 V \rightarrow V = 4}$$

Summary

- A transmission medium may be broadly defined as anything can carry information from source to destination.
- The characteristics and quality of a data transmission are defined by the properties of medium and character of the signal.
- Classifications of Transmission Media :
  1. Guided transmission media
  2. Unguided transmission media
- Transmission Media or communications channels can be a physical connection between correspondents, known as **Guided transmission media.**
  1. Open Wire
  2. Twisted pair cables
  3. Coaxial cables and
  4. Optical fibers cables
- The simplest and least expensive form is **two-wire (ribbon) cable.**
- **Twisted pair** cable consists of two insulated copper wires, typically 1 mm thick, twisted together.
- Twisted Pair cables are most effectively used in systems that use a balanced line method of transmission: polar line coding (Manchester Encoding) as opposed to unipolar line coding (TTL logic).
- Two types of twisted pair cable are used in LANs are :
  (*i*) Unshielded twisted pair (UTP)
  (*ii*) Shielded twisted pair (STP)
- Coaxial cable gets its name because two conductors share a common axis; Coax is shorthand for Co-axial. The coaxial cable is used for higher frequency ranging from 10 kHz to 500 MHz.
- **Fiber optic transmission** is the transmission of information using light waves as a carrier signal that travel through an optical waveguide (optical fiber).
- Optical Fibers work on the principle that the core refracts the light and the cladding reflects the light.

- Each fiber consists of a **core**, thinner than a human hair, surrounded by a **cladding** sheath with a much lower refractive index.
- An electromagnetic wave consists of two primary components – an electric field and a magnetic field.
- Radio waves may be reflected from various substances or objects they meet during travel between the transmitting and receiving sites. The amount of reflection depends on the reflecting material.
- A radio wave that meets an obstacle has a natural tendency to bend around the obstacle.
- Ionization is the loss or gain of electrons by an atom. There are different degrees of ionization, forming several recognizable ionized layers.
- The highest frequency returned to earth when radiated upward in the vertical direction is called the **critical frequency** f_c. Its value could be calculated by the relation.
- The **sky-wave**, that bounces from the ionosphere, and the **ground-wave**, that propagates along the ground.
- **Propagation Modes of Optical Fiber :**
  (*i*) **Multimode**: A **multi-mode** fiber has a number of possible rays that light of a particular frequency may take. There are two types of multimode which are as follows:
    (*a*) Stepped index,
    (*b*) Graded index
  (*ii*) **A single mode** : A single-mode fiber is a fiber that only has one acceptable ray-path per frequency.
- **Advantages of Optical Fiber :**
  (*a*) Small Size and weight
  (*b*) Electrical isolation
  (*c*) Extremely high immunity to EM interference
  (*d*) Large bandwidth
  (*e*) Security
  (*f*) Low transmission loss
  (*g*) Longer distances than copper wire
  (*h*) No corrosion.
- **Disadvantages of Optical Fiber :**
  (*a*) Technical difficulties
  (*b*) High cost
  (*c*) Glass fiber is more easily broken than wire
  (*d*) Optical waveguides also suffer from dispersion
- **Unguided transmission media** make use of atmosphere and outer space which provides a means of transmitting Electromagnetic signals.
- There are three types of RF (Radio Frequency) Propagation :
  (*i*) Ground Wave
  (*ii*) Ionospheric wave
  (*iii*) Line of Sight (LOS) Propagation.

- **Applications of wireless media:**
  - (i) Microwave link communication
  - (ii) Satellite communication
  - (iii) Mobile communications
  - (iv) Infrared
- **Transmission impairment** is the imperfections developed during the transmission and reception of the signal due to noise, attenuation, distortion and limited bandwidth.
- **Types of Transmission impairment:**
  - (i) Attenuation
  - (ii) Delay distortion
  - (iii) Noise
  - (iv) Distortion

Review Questions

1. Briefly discuss about the light sources used in optical communication.
2. Discuss about the losses in optical communication.
3. What do you understand by term transmission media? Name various types of transmission media used in communication.
4. Define guided transmission media and explain advantages of twisted pair cable over open wire cable.
5. Explain the working principle of twisted pair cable and describe the various types of twisted pair cable.
6. Write the advantages and disadvantages of twisted pair cable.
7. Explain the working principle of coaxial cable. Write the advantages and disadvantages of coaxial cable over twisted pair cable.
8. Explain the working principle of optical fibre cable. Write the advantages of optical fibre cable over coaxial cable and twisted pair cable.
9. What is the purpose of cladding in an optical fibre. Discuss its density relative to core.
10. Explain the various types of propagation modes in optical fibre cable.
11. Write the advantages and disadvantages of optical fibre cable.
12. Compare the various types of guided transmission media.
13. Define the term unguided transmission media and explain various types of it.
14. Compare the various types of unguided transmission media.
15. What do you understand by term transmission impairment in transmission media?
16. Explain various types of transmission impairments.
17. What is delay distortion? How it affects the data rate of channel?
18. Define and explain Shannon limit theorem.
19. What is the maximum data rate of channel? Explain the theorems related to it.
20. Describe the characteristics of the following transmission media. Give their limitations also:
    - (i) Twisted pair
    - (ii) Wireless

21. Write about the various transmission media and derive the expression for Shannon limit on data rate.
22. Name the advantages of optical fibre over twisted pair and coaxial cable.
23. What does the Shannon capacity have to do with communication? Calculate the maximum bit rate for the noiseless channel with a bandwidth 3000 Hz transmitting a signal a signal with two signal levels. **(Hints:** capacity = BW × $\log_2$ (1+ SNR), bit rate= 6kbps)
24. Explain the three ways for wireless data to be propagated.
25. What is Nyquist theorem? What is the bit rate, if we want to digitize the human voice, assuming 8 bits per sample?
26. What is basic difference between guided and unguided media?
27. Which parts of electromagnetic spectrum are used for communication?
28. Explain the three major classes of guided media.
29. What is the major advantage of shielded twisted pair over unshielded twisted pair?
30. Why is coaxial pair superior to twisted pair cable?
31. Discuss the modes for propagating light along optical channels.
32. Why are communication satellites in geosynchronous orbit?
33. What is a handoff in cellular telephony?
34. What does Shannon capacity have to do with communications?
35. Explain what crosstalk is and what is needed to reduce it?
36. Describe the components of fiber-optic cable. Draw its picture.
37. Describe the layers of the atmosphere. What types of radio communication utilize each?
38. How does ionosphere propagation work? What are the uses for this type of propagation?
39. Draw and explain unshielded twisted pair cable. Give its advantages and disadvantages.
40. Draw and explain shielded twisted pair cable. Give its advantages and disadvantages.
41. Discuss coaxial cable with a neat diagram.
42. On what principle does an optical fiber operate? Explain.
43. Draw a well labeled diagram of a fiber optic cable and discuss its working principle. Give its advantages and disadvantages.
44. Explain the various type of propagation of radio waves.
45. How can communication take place through unguided media? Explain in reference to terrestrial microwave and satellite communication.
46. Explain the basic principle of cellular telephony.
47. How can be the performance of a transmission media is measured?
48. What is the significance of Shannon capacity?
49. What are the factors which should be considered while evaluating the suitability of a particular medium to a specific application?
50. The attenuation of a signal is –10 dB. What is the final signal power if it was originally 5 watts?
51. Given that the speed of light is 186,000miles/second and a satellite is at geosynchronous orbit. How long would it take for a signal to go from earth station to the satellite (minimum time).
52. If the throughput at the connection between a device and the transmission media is 5 kbps, how long does it take to send 100,000 bits out of this device?
53. A line has a signal-to-noise ratio of 1000 and a bandwidth of 4000 kHz. What is the maximum data rate supported by this line?

54. We measure the performance of a telephone line (4 kHz ok bandwidth). When the signal is 10 volts, the noise is 5 millivolts. What is the maximum data rate supported by this telephone line?

55. What is the relationship between the data rate and the distance that the data can reliably travel on an EIA interface?

56. Write short notes on following:

(*i*) Classification of transmission media

(*ii*) Transmission impairments

(*iii*) Wireless media

(*iv*) Infrared transmission media

(*v*) Data rate of channel

(*vi*) skip distance

(*vii*) free-space path loss

□□□

V

UNIT

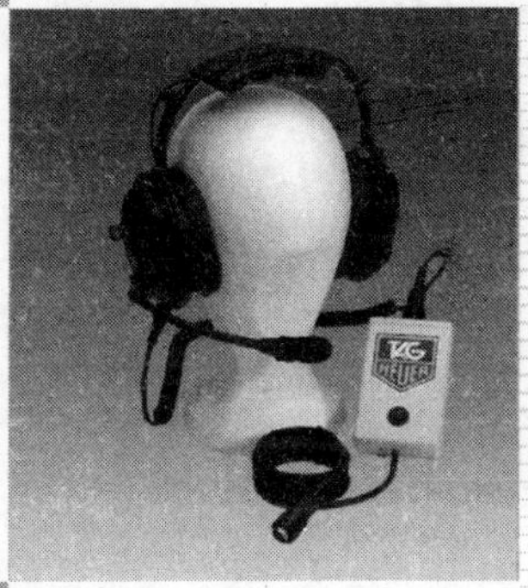

➤ Error Detection and Correction

➤ Switching

SYLLABUS

Error Detection and Correction : Types of Error, Error Detection—Redundancy Check (Longitudinal, Vertical and Cyclic), Checksum, Error Correction-Hamming Code.

Switching : Circuit Switching (Space-Division, Time Division and Space-Time Division), Packet Switching (Virtual Circuit and Datagram Approach), Message Switching.

CHAPTER

8

Error Detection and Correction

INSIDE THIS CHAPTER

8.1. INTRODUCTION

No physical medium can guarantee error-free transmission of data. Errors occur because of interference from the environmen t surrounding the transmission medium or because of artifacts introduced by the equipment or the medium itself. In case of copper wires, for example, thermal noise, electrical noise, and impulse noise are some of the most common causes. The networks must be able to transfer data from one device to another device with the complete accuracy. Many factors can alter or wipe out one or more bits of given data.

Transmission errors are caused by:

1. Thermal noise (Shannon)
2. Impulse noise (e.g., arcing relays)
3. Signal distortion during transmission (attenuation)
4. Crosstalk
5. Voice amplitude signal compression (companding)
6. Quantization noise (PCM)
7. Jitter (variations in signal timings)
8. Receiver and transmitter out of synch

Reliable systems must have a mechanism for detecting and correcting such errors. *The most important role of the data link layer is to provide error-free transmission.* To do this, it uses an error checking scheme to detect errors, and overcomes them by retransmitting corrupted frames. Many different error detection and correction methods have been devised.

DO YOU KNOW▸

Errors occur because of interference from the environment surrounding the transmission medium.

8.2. CATEGORIES OF TRANSMISSION ERRORS

Errors are introduced in the data bits during their transmission across a data network. These errors can be categorized as :

(*i*) Content errors, and
(*ii*) Flow integrity errors.

8.2.1. Content Errors

Content *errors* are the errors in the content of a message, e.g., a binary 1 may be received as a binary 0. A block of data may have single bit error or multiple errors. Multiple errors in a data block are referred to as *burst error.* Length of burst error is defined from the first corrupted bit to the last corrupted bit. For example, burst error in Fig. 8.1 is six bits long. An error bust may contain uncorrupted bits as well.

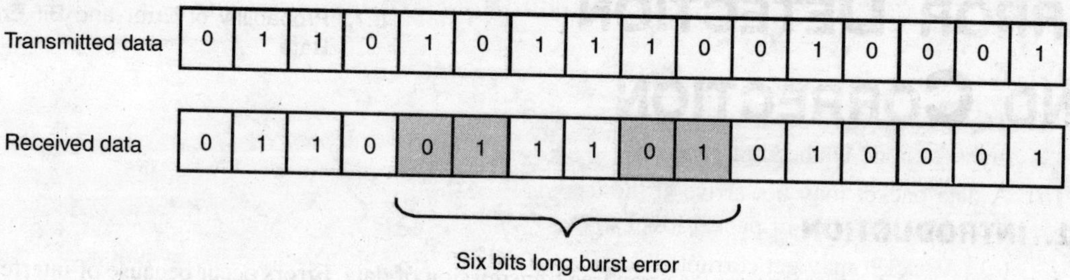

Fig. 8.1. Burst error.

There can be several causes of content errors.

Signal Impairment : Distortion of electrical signal and noise are the main causes of content errors. Thermal noise causes sporadic bit errors. An impulse can last long enough to corrupt several bits and therefore leads to burst errors.

Loss of synchronization : The receiver samples the received signal to extract the digital information by using a clock that is synbchronized to the clock loop. Too few transitions in extracted from the received signal using a filter and a phase lock loop. Too few transitions in the line signal may cause the phase lock loop to drift leading to loss of synchronization, which in turn results in sampling at wrong instants. Errors introduced are in form of burst till the synchronization is reestbalished between transmitter and receiver clocks.

Scramblers : Scramblers is used to ensure synchronization of transmitter and receiver clocks. It randomizes the serial data before its transmission. Strings of zerors and ones are thus avoided in the transmitted signal. Scrambling is done using a shift register and exclusive OR gates.

Scramblers have unfortunate property of multiplying errors. If an error is received, it gets multiplied as it shifts through the shift register. The errors multiplied depend on the number of taps taken from the shift register.

Transmission channel switching : Transmission channels of telecommunication network are usually protected, i.e. if the main path fails due to any reason (cut in fiber, fading of radio signal), the signals are switched over to standby path (alternate fiber route, alternate radio frequency). The switchover may take about 10 to 50 ms, and it causes momentary loss of digital signal. These momentary disturbances are unnoticeable for voice channels but lead to burst errors in data signals.

8.2.2. Flow Integrity Errors

Data is sent as data packets across a data network. It is not necessary that data packets sent by a source to the destination take the same path. In Fig. 8.2, packet 1 goes via nodes A, B and D. Packet 2 goes directly from node A to node D. Packet 3 starts its journey from nodes A to D via nodes B and C. Flow of these data packets from the source to the destination may be affected in several ways.

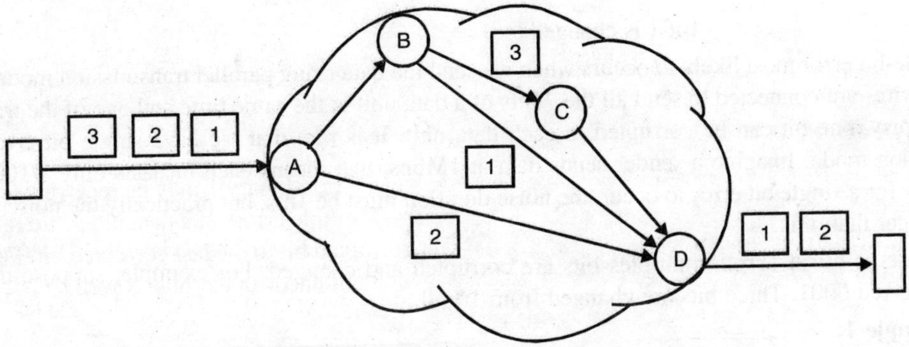

Fig. 8.2. Flow integrity errors.

(*i*) The packets may be delivered out-of-sequence. Packet 2 takes shorter path and therefore it may arrive at node D ahead of packet 1.

(*ii*) A data packet may not arrive at the destination at all. Packet 3 reaches up to node C, but is lost thereafter. Reasons of packet loss can be several :

- (*a*) A packet may get corrupted due to content errors and may be rejected by a node.
- (*b*) It may be delivered to wrong address due to errors that could not be detected.
- (*c*) A node may not forward the packet due to congestion in the networl. When a node experiences congestion (*i.e.*, it has long queue of packets), it starts discarding data packets.
- (*d*) Each data packet has limited lifetime in the network. This is called *time to live* (TTL). TTL can be defined in terms of number of hops. This is done to avoid circulating packets (packets that lose the way). A network node dumps all the TTL expired packets.

(*iii*) The sender at times expects from the receiver an acknowledgement of having received a packet. If the acknowledge is not received within a defined time interval, the sender resends the packet. It may happen the receiver sends the acknowledgement, but the acknowledgement from the receiver is lost in the network. In such situation, the receiver will receive duplicate data packets eventually.

Thus a packed may be delivered to a wrong destination, or may be delivered out of seuqnece, or may not be delivered, or delivered in duplicate. All these are instances of flow integrity errors.

Systematic Code

Error detection and correction involves addition of check bits. A *systematic code* is a code in which each code word includes the unaltered data word followed or preceded by a separate group of check bits.

8.3. TYPES OF ERRORS

There are two types of errors occur in data communication, namely:

1. Single-Bit Error
2. Burst Error

In case of **single-bit error** only one bit of the data is corrupted and rest all the bits are received correctly. That means only one bit from the data unit is changed either from 1 to 0 or 0 to 1.

For example, suppose the data frame 1111 is sent over the transmission line and at the receiver end the data frame is received as 1**0**11. The second bit is changed from 1 to 0.

Data sent	1111
Data received	1**0**11
	↑

Bit 1 is changed to 0

Single-bit error most likely to occurs when we send the data using parallel transmission mode. Suppose, if seven wires are connected to send all the 7 bits of a data unit at the same time and one of the transmission wire is noisy, one bit can be corrupted in each data unit. It is rare that we have single bit error in serial transmission mode. Imagine a sender sends data at 1Mbps, that means each bit lasts only 1/1000000s i.e. 1ms. Now for a single bit error to occur, the noise duration must be 1ms, but practically the noise duration is much longer than this.

In case of **burst error**, multiples bits are corrupted and changed. For example, suppose data sent is 1111, received 0001. Three bits are changed from 1 to 0.

Example 1:

Data sent	1111
Data received	**000**1
	↑

3 bits are changed from 1 to 0

Let us discuss one more example of burst error, in this case data is sent 0100010001, but received as 0101110101. Three bits are changed from 0 to 1. From the example, a burst length also can be seen, *i.e.* 5 bits in this case.

Example 2:

The **burst length** is measured from the first corrupted bit to the last corrupted bit. Some bits in between may be received correctly.

As we know that noise duration is more than the duration of one bit that means when noise affects data, it affects a set of bits. Serial transmission is most likely to be affected by the noise. It is obvious that the number of bits affected depends on the duration of noise and the data rate. Suppose we have data rate 1 kbps and a noise duration is 1/100s, than 10 bits are affected.

8.4. ERROR DETECTION

Error detection is the first stage of error correction. Error detection is simpler and easy than error correction. There are some mechanism used to detect the error in data communication based a concept of **redundancy**. In this concept a shorter group of bits may be appended to the end of each data unit before the transmission and they are discarded as soon as accuracy of the transmission has been confirmed.

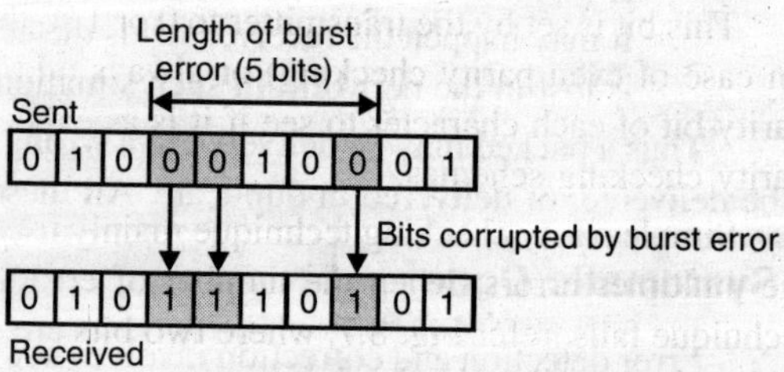

Fig. 8.3.

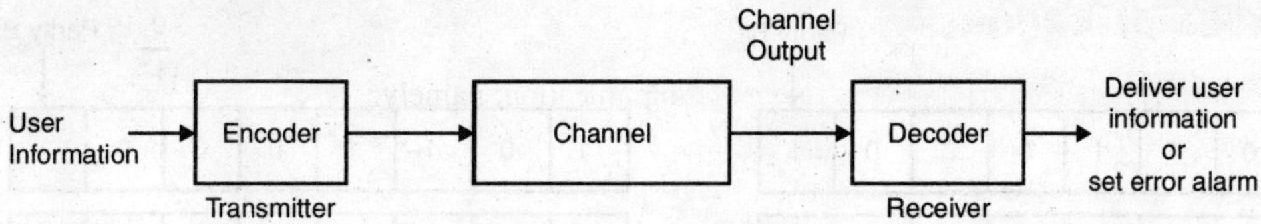

Fig. 8.4. General error detection system.

Figure 8.4 shows a general error detection system. In this figure, it is shown that user information is fed to encoder which will add some extra bits to the information and at the receiving end decoder, check the pattern.

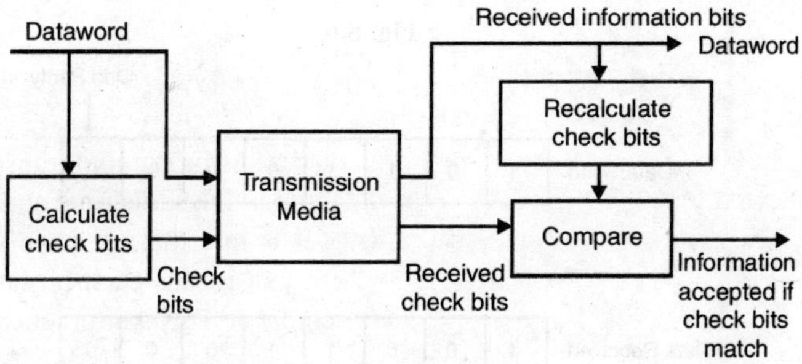

Fig. 8.5. Detection system using check bits.

Figure 8.4 shows the detection system using check bits. In this system check bits are calculated and appended to the dataword before transmitting to the receiver. At the receiver side, check bits are recalculated and with the received check bits. If check bits are same information is accepted otherwise it is discarded.

8.5. TYPES OF ERROR DETECTION TECHNIQUES

Types of error detection techniques using Redundancy concept are as follows:

1. Parity check
   (*i*) Simple parity check
   (*ii*) Two dimensional parity check
2. Cyclic Redundancy Check
3. Checksum

8.5.1. Parity Checking

Parity checking is a simple error detection method used with character oriented protocols. There are two types of parity checking techniques, such as **Even parity** and **Odd parity**. One bit in every character bit sequence is reserved for parity.

This bit is set by the transmitter to 0 or 1 so that the total number of 1 bits in the character is always even (in case of even parity checking) or always odd (in case of odd parity checking). The receiver checks the parity bit of each character to see if it is as expected. Figure 8.6 illustrates the method for an even and odd parity checking schemes.

Simple parity checking technique mainly used to detect the single error only. However, it can also detect the multiples errors, when the number of errors is odd. In case when the number of errors is even, this technique fails, refer Fig. 8.7, where two bits are changed. The receiver takes that there is no errors because odd parity is still maintained.

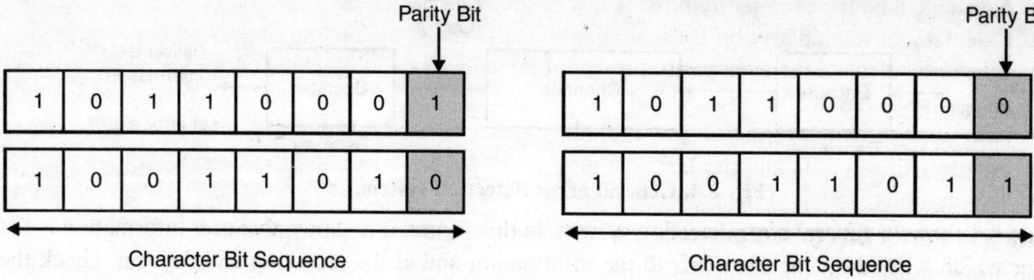

(*a*) Even parity error checking (*b*) Odd parity error checking

Fig. 8.6

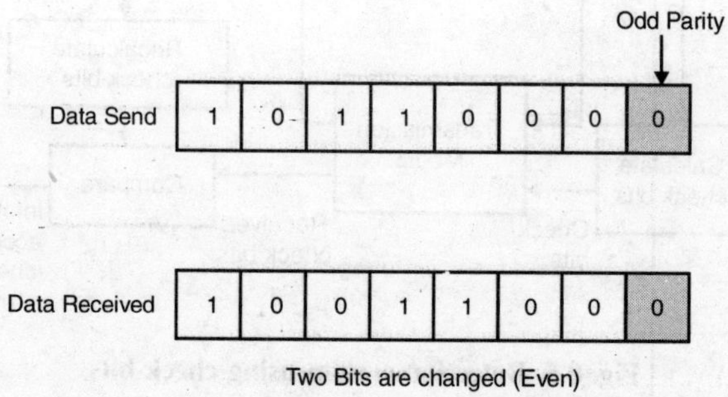

Fig. 8.7. Case of multiple errors such as 2 errors

DO YOU KNOW▸

There are two types of parity checking techniques, such as ***Even parity*** *and* ***Odd parity****.*

Two Dimensional Parity Check

As we have seen that simple parity check technique is not enough when numbers of errors are even. Another approach is the **Two-Dimensional Parity Check**; in this we have many data units and they are organized in rows and columns. In this technique we calculate the parity bit for each data unit. For example, we have four data units shown in four rows and five (suppose 5 bits in a data unit) columns, in Fig. 8.8. We then calculate the even parity bit for each column and write a new row of five bits, they are the parity bits for the whole block. It is to be noted that first parity bit in the fifth row is determined based on all first bits and second parity bit is based on all second bits and so on. Now we append the parity bits to the original data and then send to the receiver through channel. After appending the parity bits the data word is referred as codeword.

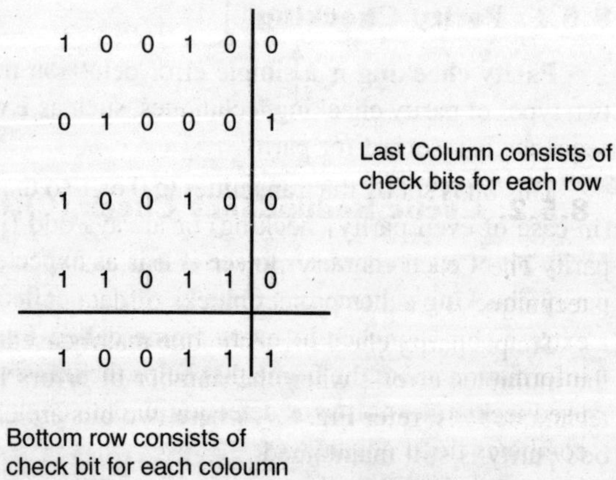

Fig. 8.8. Two dimensional parity check

Example: Suppose we have 4 data units; 10010, 01000, 10010, 11011.

Steps:

- Arrange the data units into row and columns as shown below.
- Calculate the even parity bit for first row and write at the extreme right.

- Now calculate the even parity bits for rest of the rows.
- Calculate the even parity bit for first column and write at the bottom of that column.
- Likewise calculate the even parity for rest of the columns and write the parity bits at the bottom of respective columns.
- Now write the bits of the first row including parity bit, than write the second row bits, third row bits, fourth row bits and the last row bits (i.e. parity bits for all the columns).
- This complete block of data units is sent to the receiver.

Data and Parity bits to sent —10010 **0** 01000 **1** 10010 **0** 11011 **0** **100111**

Two dimensional parity check increases the possibility of detecting burst errors. Still *there is one problem while detecting the burst errors, when two bits in one data unit are corrupted and two bits in exactly the same positions in another data unit are damaged, the checker will not detect an error.*

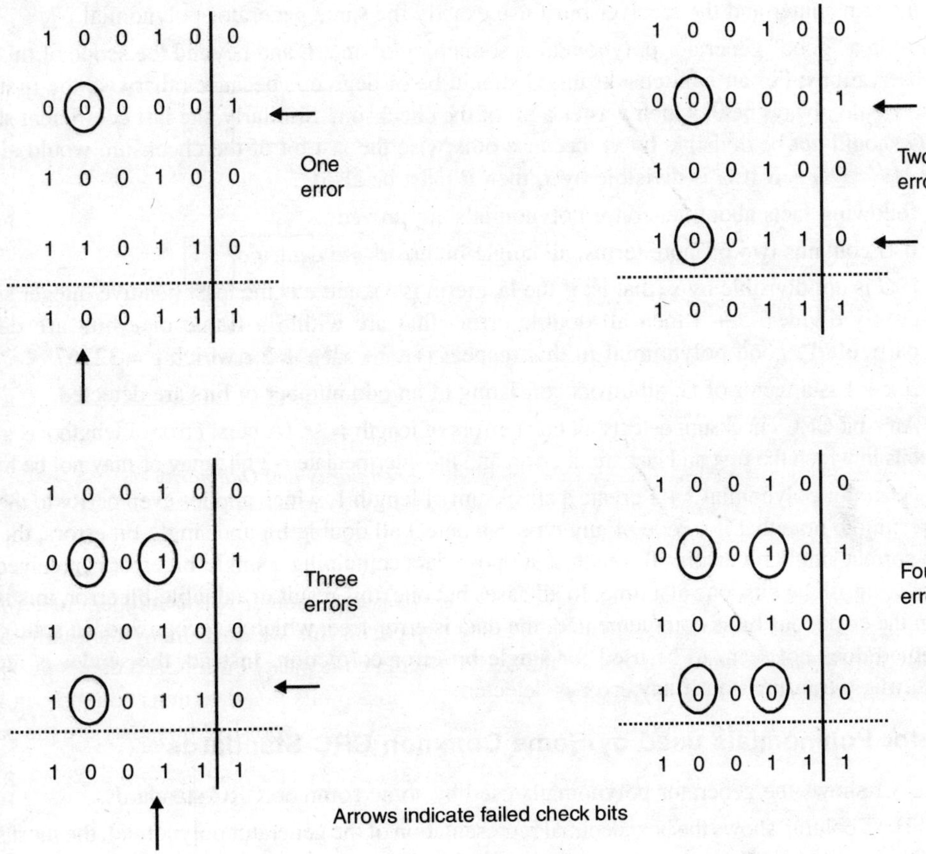

Fig. 8.9.

8.5.2. Cyclic Redundancy Check (CRC)

The CRC is a very powerful but easily implemented technique to obtain data reliability. The **CRC** technique is used to protect blocks of data called Frames. Using this technique, the transmitter appends an extra n- bit sequence to every frame called **Frame Check Sequence (FCS)**. The FCS holds redundant information about the frame that helps the transmitter detect errors in the frame. The CRC is one of the most used techniques for error detection in data communications. The technique gained its popularity because it combines three advantages:

- Extreme error detection capabilities.

- Little overhead.
- Ease of implementation.

Cyclic Redundancy Check (CRC) is a more sophisticated method used with bit-oriented protocols. It is also called the **polynomial code** — so named because it treats the single bits in a bit sequence as the coefficients of an imaginary polynomial.

For example, the bit sequence 101001 represents the polynomial:

$$1x^5 + 0x^4 + 1x^3 + 0x^2 + 0x^1 + 1x^0 = x^5 + x^3 + x^0$$

This polynomial is said to be of *order* 5, because its highest-order term is x^5.

CRC relies on the use of a **generator polynomial**, the choice of which is fairly arbitrary, except that the high and low order bits of its bit sequence must be 1 (e.g., 10111 is acceptable, but 0101 and 10110 are not) and that the transmitter and the receiver must use exactly the same generator polynomial.

Choosing a "good" generator polynomial is something of an art, and beyond the scope of this text. Two simple observations: For an r-bit checksum, G should be of degree r, because otherwise the first bit of the checksum would always be 0, which wastes a bit of the checksum. Similarly, the last coefficient should be 1 (that is, G should not be divisible by x), because otherwise the last bit of the checksum would always be 0 (because $Mx^r = QG + R$ if G is divisible by x, then R must be also).

The following facts about generator polynomials are proved :

- If G contains two or more terms, all single-bit errors are detected.
- If G is not divisible by x (that is, if the last term is 1), and e is the least positive integer such that G evenly divides $x^e + 1$,then all double errors that are within a frame of e bits are detected. A particularly good polynomial in this respect is $x^{15} + x^{14} + 1$ for which $e = 32767$.
- If $x + 1$ is a factor of G; all errors consisting of an odd number of bits are detected.
- An r-bit CRC checksum detects all burst errors of length = $\leq r$. (A burst error of length r is a string of r bits in which the first and last are in error, and the intermediate r-2 bits may or may not be in error.)

The generator polynomial $x + 1$ create a checksum of length 1, which applies even parity to the message. It is interesting to note that if a code of any type can detect all double-bit and single-bit errors, then it can in principle correct single-bit errors. To see this, suppose data containing a single-bit error is received. Imagine complementing all the bits, one at a time. In all cases but one, this result in a double-bit error, this is detected. But when the erroneous bit is complemented, the data is error-free, which is recognized. In spite of this, the CRC method does not seem to be used for single-bit error correction. Instead, the sender is requested to repeat the whole transmission if any error is detected.

Generator Polynomials used by Some Common CRC Standards

Table 8.1 shows the generator polynomials used by some common CRC standards.

The "Hex" column shows the hexadecimal representation of the generator polynomial; the most significant bit is omitted, as it is always 1.

TABLE 8.1. Generator polynomials used by some common CRC standards

Common Name	r	Generator	
		Polynomial	Hex
CRC-12	12	$x^{12} + x^{11} + x^3 + x^2 + x + 1$	80F
CRC-16	16	$x^{16} + x^{15} + x^2 + 1$	8005
CRC-CCITT	16	$x^{16} + x^{12} + x^5 + 1$	1021
CRC-32	32	$x^{32} + x^{26} + x^{23} + x^{22} + x^{16} + x^{12} + x^{11} + x^{10} + x^8 + x^7 + x^5 + x^4 + x^2 + x + 1$	04C11D87

Operation of CRC

Given a message (bit sequence) *m* represented by the polynomial $m(x)$, and a generator polynomial $g(x)$ of order *k*, *m* is appended with exactly *k* zero bits. The result is equivalent to the polynomial $p(x) = xkm(x)$. $p(x)$ is divided by $g(x)$ and the remainder is added to $p(x)$ to produce the final polynomial $q(x)$:

$$q(x) \ = \ p(x) + (p(x) \ \textbf{rem} \ g(x))$$

It follows that $q(x)$ is divisible by $g(x)$. The transmitter calculates *q* using *m* and *g* and then transmits *q* instead of *m*. If any of the bits in *q* are changed due to transmission errors, $q(x)$ has a very high probability of no longer being divisible by $g(x)$. This is checked by the receiver to detect transmission errors. If there are no errors, the receiver extracts *m* from *q* (by discarding the *k* low-order bits) and this will be identical to the original message.

The polynomial calculation procedures described above serve only as a conceptual model. In practice, equivalent binary operations are used instead. The binary operations are surprisingly simple, to the point that the whole process is usually implemented in hardware. Calculating *q* involves the following steps:

1. Given a bit sequence *m* and a generator polynomial bit sequence *g* of order *k*, append *m* with exactly *k* zero bits to produce bit sequence *p*. For example:

$$m = 100101110101$$

$$g = 10111 \ (k = 4)$$

$$p = 1001011101010000$$

2. Divide *p* by *g*, using modulo-2 arithmetic, to produce a remainder *r*. Modulo-2 arithmetic operations are similar to binary operations, except that there are no borrows or carries. Addition and subtraction in modulo 2 are identical to the binary *exclusive or* operation. Continuing with our example:

$$r = 1110$$

3. Add *r* to *p* to produce *q*:

$$q = 1001011101010000 + 1110$$

$$= 1001011101011110$$

The effectiveness of CRC depends to a large extent on the choice of the generator polynomial. In practice, polynomials of order 16 are widely used. Hence 16 bits (two octets3) are needed for storing the remainder, which appears as the checksum field in the frame. Such polynomials are more than 99% effective in catching a variety of errors.

In brief CRC can detect following types of errors:

- CRC can detect all burst error that affects an odd numbers of bits.
- CRC can detect, all burst error of length less than or equal to the degree of the polynomial.
- CRC can detect all single bit error and double bits errors.

Important Steps:

1. Always divisor will be given in the examination problem.
2. A string of zeros to be added to the data before staring the division. The number of zeros must be one less than divisor. Refer example-1, where number of bits in the divisor is five and number of zeros added in the data are 4 (5 – 1 = 4).
3. While dividing, divisor is subtracted from corresponding bits of dividend bits.
4. Left most bit is always to be discarded after the subtraction from the divisor.
5. If next to the left most bit is zero than instead of subtracting divisor subtract a string of zeros and the value of quotient is zero.
6. Divisor should always begin with the left most bit 1.

7. If the left most bit is zero in the remainder is zero in intermediate steps of division, it is discarded. But if the again next bit is also zero then it is retained and subtract from the string of zeros by pulling a higher bit from dividend to remainder.
8. The remainder, which we get is the CRC parity bits and to appended at the end of data. For the valid remainder CRC, the number of bits in CRC should be one less than divisor.

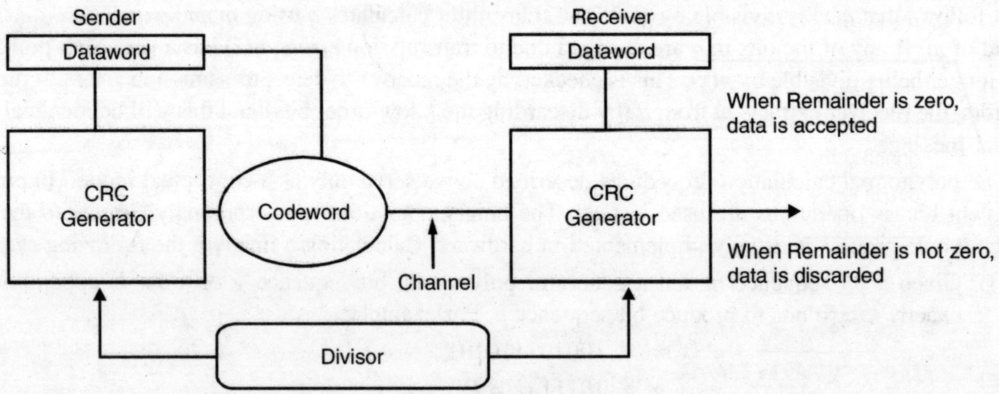

Fig. 8.10. A simple block diagram of CRC

9. The data is to be transmitted is the data strings followed by the CRC parity bits (remainder).
10. Now at the receiver end, data string followed by CRC parity is to be divided by the same divisor. If the remainder comes all zeros than data is correct and accepted as shown in example-1. In case remainder is not zeros, that means data received is not correct and discarded.
11. Figure 8.8, shows a simple block diagram of CRC error detection method. In this at the sender end, CRC generator will do the modulo-2 division and remainder is appended to the dataword that is then known as codeword. Codeword is send through the channel to the receiver. At the receiver end, CRC checker will perform the operation of division. Codeword is divided by the same divisor and if the remainder is zero, data is accepted and if the remainder is not zero than data is incorrect and discarded.

Example 8.1. Let a data - 11000110 is to send and the divisor- 11101, find the CRC (redundancy bits).

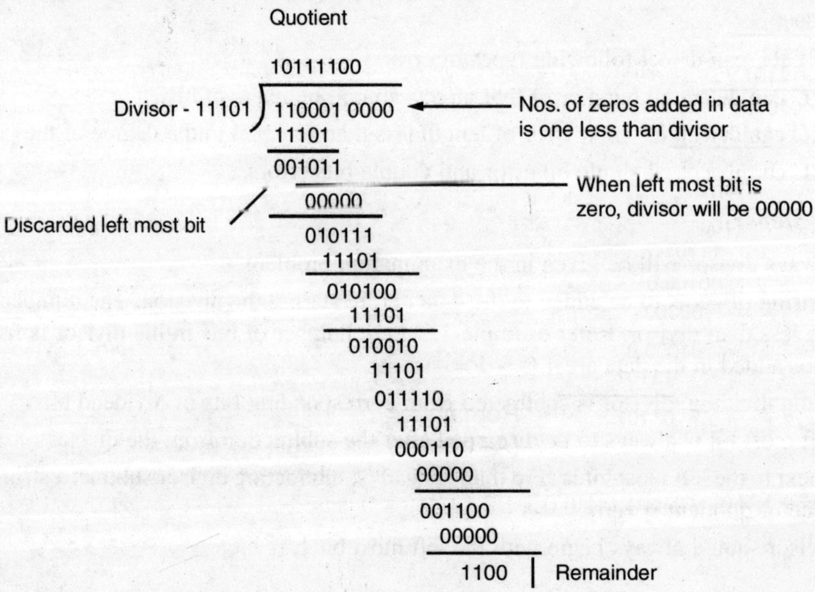

Now, CRC parity bits, i.e. remainder is 1100. Before sending these bits are to be appended to the data. The codeword (data along with the redundancy bits) will become 11000110**1100** and to be transmitted.

At the receiver end, the received data is to be divided by the same divisor and verify the remainder.

Case-1: Received correct data

Suppose data received is correct then remainder will be all zeros as shown below.

```
          10111100
        ____________
11101  / 110001101100
         11101
         -----
         001011
           00000
           ------
           010111
             11101
             ------
             010100
               11101
               ------
               010011
                 11101
                 ------
                 011101
                   11101
                   ------
                   00000 0
                     00000
                     ------
                     00000 0
                       00000
                       -----
                        0000   Remainder
```

(All bits are zeros, i.e. means data received is correct)

Case-2: Received incorrect data

Suppose data received is incorrect then remainder will not be all zeros as shown below.

```
          10111101
        ____________
11101  / 110001111100
         11101
         -----
         001011
           00000
           ------
           010111
             11101
             ------
             010101
               11101
               ------
               010001
                 11101
                 ------
                 011001
                   11101
                   ------
                   00100 0
                     00000
                     ------
                     01000 0
                       11101
                       -----
                       01101   Remainder
```

(All bits are not zeros, i.e. means data received is incorrect)

Example 8.2. Let a data 1 0 1 0 0 1 1 0 1 0 0 0 0 0 is to send and the divisor 1 0 0 1 1 1, find the CRC (redundancy bits).

```
                     101111110        Discard quotient
100111 | 10100110100000
         100111
         ------
           111010
           100111
           ------
            111011
            100111
            ------
             111000
             100111
             ------
              111110
              100111
              ------
               110010
               100111
               ------
                101010
                100111
                ------
                 11010   <==  Remainder
```

The codeword that is to be transmitted will became 1 0 1 0 0 1 1 0 1 **1 1 0 1 0.**

At the receiver end, the codeword is again divided by the divisor 1 0 0 1 1 1.

```
                     101111110   <==  Discard qoutient
100111 | 10100110111010
         100111
         ------
           111010
           100111
           ------
            111011
            100111
            ------
             111001
             100111
             ------
              111101
              100111
              ------
               110100
               100111
               ------
                100111
                100111
                ------
                     0   <==  Remainder equals zero (no errors)
```

8.5.3. Checksum Codes

One popular method for error detection is the use of checksums. **Checksums** have the advantage of operating on a long sequence of data. Typically the data is 8 bit character data (or ASCII data) in a file or character frame. Each character byte is summed into a checksum variable. As the data is summed into the checksum variable, the carries are ignored. The checksum variable is usually appended to the end of the data field and is called the **checksum field**. The checksum field can be one or more bytes. The ability to detect errors in a multi byte sequence is affected by the number of bytes in the sequence and the size of the checksum field.

In ones complement arithmetic, a negative integer *-A* is represented as the complement of *A;* that is each bit of *A* is inverted. When adding numbers in ones complement arithmetic, a carryout from the MSB needs to be added to the result.

Checksum Generator

At the transmitter end, the checksum generator divides the data unit in to equal segments of bits (usually 16). These segments are added using ones complement arithmetic in such a way so that the

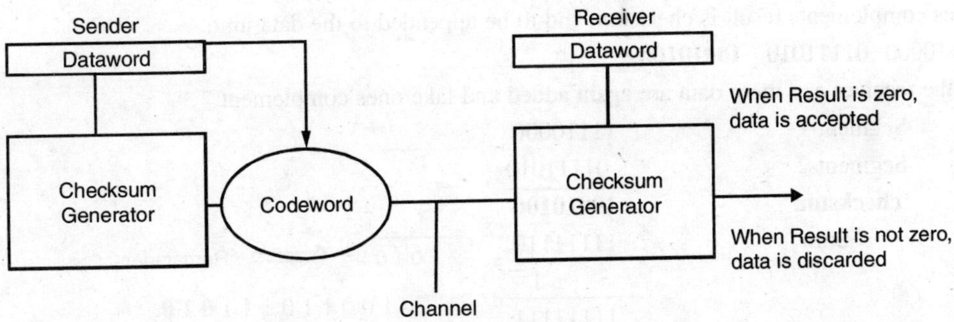

Fig. 8.11. A simple block diagram of checksum error detection method

total will be having the same number of bits. The total sum is complemented and appended to the end of the data unit as redundancy bits, called the **checksum field**. The codeword is now transmitted across the network. **(Refer to Fig. 8.9)**

Checksum Checker

At the receiver end, the codeword is again subdivided in to equal segments and added all segments, as it was done at the transmitter end. Than the sum is complemented and gets the result. If the data unit received at the receiver is correct than the checksum field should be zero and if the result is not zero, the data unit contains an error and receiver discard it.

Steps

At the transmitter and the receiver ends following steps are taken:

1. The data unit is divided into *m* sections. The number of bits in each segment is *n*.
2. All the segments are added using ones complement to get the sum.
3. The total is complemented and becomes the checksum. The checksum is send with the data.
4. The data unit is divided into *m* sections. The number of bits in each segment is *n*.
5. All the segments are added using ones complement to get the sum.
6. The total is complemented and if the result is zero, the data is accepted otherwise they are rejected.

There is one limitation of this technique is as follows:

If one or more bits of a section are corrupted and the corresponding bit or bits of opposite value in a second section is also damaged. In this case the sum of those columns will remained same and error is not detected by the receiver.

Example 8.3. Let us take the data unit to be transmitted is

1111000001111010.

Data unit has 16 bits, divides in two segments of 8 bits and add them.

Segment-1	11110000
Segment-2	01111010
	01101010
	1
Sum	01101011

Now complement the sum,

	01101011
Ones complement	10010100

Ones complements result is checksum and to be appended to the data unit,

11110000 01111010 **10010100**

At the receiver end these data are again added and take ones complement.

Segment-1	11110000
Segment-2	01111010
checksum	**10010100**
Sum	11111110
	1
	11111111
Ones complement	00000000 Result is zero that means there is no error.

Let us consider that one of the bits of the data unit is changed.

Bit changed from 1 to 0

↓

11010000

01111010

10010100

11011110

1

11011111

Ones complement 00100000; Result is not zero that means there is some error.

8.6. ERROR CORRECTION

There are two approaches for the error corrections, namely:

1. Feedback Error Correction
2. Forward Error Correction

Feedback Error Correction

In **feedback error correction**, only error detection technique is used and whenever errors occur, sender will resend that frame or packet to the receiver.

- **Error detection:** Adding enough "extra" bits to deduce that there is an error but not enough bits to correct the error.

- If only error detection is employed in a network transmission than retransmission is necessary to recover the frame (data link layer) or the packet (network layer). At the data link layer, this is referred to as **ARQ (Automatic Repeat request).**

DO YOU KNOW▶

*In **feedback error correction**, only error detection technique is used and whenever errors occur, sender will resend that frame or packet to the receiver.*

Forward Error Correction

In **forward error correction**, not only error detection technique is used but automatic error correction also takes place.

Error correction: It requires enough additional bits (redundant bits) to deduce what the correct bits must have been.

Examples, Hamming Codes

FEC = Forward Error Correction found in MPEG-4.

The purpose of **forward error correction (FEC)** is to improve the capacity of a channel by adding some carefully designed redundant information to the data being transmitted through the channel. The process of adding this redundant information is known as channel coding. Convolutional coding and block coding are the two major forms of channel coding. Convolutional codes operate on serial data, one or a few bits at a time. Block codes operate on relatively large (typically, up to a couple of hundred bytes) message blocks. There are a variety of useful convolutional and block codes, and a variety of algorithms for decoding the received coded information sequences to recover the original data.

DO YOU KNOW▶

*In **forward error correction**, not only error detection technique is used but automatic error correction also takes place.*

8.6.1. Hamming Code

Hamming coders are preferred over the Single-Parity Check coders when detection as well as correction is desired. *Hamming codes are capable of detecting up to two bits in error and have the ability to correct single bit errors.*

If more error-correcting bits are included with a message, and if those bits can be arranged such that different incorrect bits produce different error results, then bad bits could be identified. In a 7-bit message, there are seven possible single bit errors, so three error control bits could potentially specify not only that an error occurred but also which bit caused the error.

A mathematician named Richard W. Hamming developed the Hamming code at working at Bell Telephone Laboratories in 1950. It is an error-correcting code used for correcting transmission errors in synchronous data streams. However, the hamming code is used to detect only single bit errors but there is process that can be used to detect burst errors.

Now, to understand how errors can be handled, it is necessary to look closely at what an error really is. The Hamming code requires the addition of overhead to message, thus increasing the length of a transmission. The concept of underlying error correction can be easily understood by examine the simplest that is single bit errors. Single bit errors can be detected by adding a redundant or parity bit.

DO YOU KNOW▸

***Hamming codes** are capable of detecting up to two bits in error and have the ability to correct single bit errors.*

A single redundant bit can detect single bit errors in any sequence bits because it must distinguish between only two states either error or no error. The two state of bit, *i.e.* 0 and 1, are sufficient for this level of detection. These two states are enough only to identify that either error is there or not but not sufficient to correct it. The secret of error correction therefore is to locate that invalid bit or bits. An error occurs when the receiver reads a '0' bit as '1' or '1' bit as '0' bit. To correct the error, the receiver simply reverses the value of the altered bits.

Hamming bits are inserted into a character at random locations. Altogether, the combination of the data bits and hamming bit is known as **hamming code**.

To calculate the number of redundent Hamming bits needed for a given character length, relationship between the character bits and the Hamming bits must be established. Normally a frame consists of '*m*' data (message) bits and 'r' redundant or Hamming or check bits. Let the total lengths be '*n*' *i.e.* $n = m + r$. A n-bit unit containing data and check bits is often referred to as n-bit code word.

Let us suppose, we have two code word say, 10001001 and 10110001, it is possible to determine how many corresponding bits differ. In this case, we determine, how many bits differ, just EX-OR the two code word and count the number of 1 bits in the result.

From the above example,

$$\begin{array}{r} 10001001 \\ 10110001 \\ \hline 00111000 \\ \hline \end{array}$$

The number of bit positions in which two codewords differ is called **Hamming distance**. The significance of this is that if two code words are Hamming distance '*d*' a part, it will require '*d*' single bit errors to convert are in to the other. In this case Hamming distance in 3.

Figure 8.1 shows a data contain '*m*' character bits and '*r*' Hamming bits. Therefore the total number of bits in one data unit, is $m + r$, since the Hamming bits should be able to identity which bit is in error, '*r*' Hamming bits must be able to indicate at least $m + r + 1$ different codes.

One code indicates that no errors have accumed and the remaining $m + r$ codes indicates the bit position where an error has occured. Therefore, $m + r$ bit position must be identified with '*r*' bits. Since '*r*' bits can produce $2r$ different codes, $2r$ must be equal to greater than $m + r + 1$.

$$2r \geq m + r + 1$$

where m = Number of bits in each data character.

r = Number of redundancy or Hamming bits.

For example, to correct a single bit error in an ASCII character, the error correction code must be able to determine which of the '*r*' bits has changed. In this case, we have to distinguish between eight different states; No error, error in position '1' error in position '2' and so on, up to error in position 7.

To do so, we requires enough redundency code to show eight different states. It seems that a 3-bit redundency code should be enough because 3-bits can show eight different states *i.e.* 000 to 111 and can therefore indicate the location of eight different posibilities.

Seven bits of data (ASCII code) and 3-bits of redundancy will become 10-bits. Three bits (redundancy bits) however can identity only 8-possibility. Additional bits are require to cover all posible error locations.

If the total number of bits in a transmittable unit is $m + r$, than '*r*' must be able to indicate at least $m + r + 1$ different states.

So $m + r + 1$ states must be discoverable by r bits; and r bits can indicate $2z$ different states.

If the value of $m = 7$, than the smallest r value that can saitisfy is 4.

$$2^r = 7 + 4 + 1$$

The following Table 8.2, shows some possible m values and the corresponding r values.

TABLE 8.2

Number of data bits (m)	Number of redundancy bits (r)	Total number of bits ($m + r$)
1	2	3
2	3	5
3	3	6
4	3	7
5	4	9
6	4	10
7	4	11

The bits of the code word are numbered consecutively starting with bit 1 at the left end, bit 2 to its immediate right, and so on. The bits which are power of 2 (such as 1, 2, 4, 8, 16, 32 etc.) are Hamming bits and the rest (such as 3, 5, 7, 9, etc.) are filled up with the m data bits, as shown in Fig. 8.12.

11	10	9	8	7	6	5	4	3	2	1
d_1	d_2	d_3	r_8	d_4	d_5	d_6	r_4	d_7	r_2	r_1
			2^3				2^2		2^1	2^0

Fig. 8.12.

d = Data bits,

r = Redundancy bits (2^0, 2^1, 2^2, 2^3...)

In the Hamming code, each r bit is the parity bits for one combination of data bits and that can be determined by the Table 8.3.

We can easily get the position of data bits to be considered for calculating the value of redundancy bit (r). For example, let us get the positions of data combination for determining the value r_1 bit. In table 8.3, check the 1's at LSB (r_1), wherever the value of LSB bit is 1, note down the corresponding decimal number, such as 1, 3, 5, 7, 9 and 11. These decimal numbers are denoting the data combination positions to be considerable to determine the value of r_1 bit. Similarly, get the data combination positions for r_2, r_4 and r_8.

Following bits positions to be considered to determine the parity bits (even) values.

$$r_1 \rightarrow 1, 3, 5, 7, 9, 11$$
$$r_2 \rightarrow 2, 3, 6, 7, 10, 11$$
$$r_4 \rightarrow 4, 5, 6, 7$$
$$r_8 \rightarrow 8, 9, 10, 11$$

TABLE 8.3.

MSB			LSB	Decimal Number
0	0	0	0	0
0	0	0	1	1
0	0	1	0	2
0	0	1	1	3
0	0	0	0	4
	:	:	:	:
1	1	1	1	15

Steps Involved for Calculating the Values of Redundancies Bits

(*i*) The first step is to place each bit of the original character in its appropriate position. For example, we have ASCII code for upper case letter '*A*';

1000001

11	10	9	8	7	6	5	4	3	2	1
1	0	0	r_8	0	0	0	r_4	1	r_2	r_1

Hamming bits r_1, r_2, r_4 and r_8.

(*ii*) We calculate the even parity bits for the various bit combinations. Determine the value of r_1 by considering the following bit positions

1	3	5	7	9	11
r_1	1	0	0	0	1

r_1 ⇓ 0

The value of r_1 will be '0' if we calculate the even parity bit for the above combination.

$$r_1 = 0$$

(*iii*) Replace the r_1 with the parity bit '0'.

(*iv*) Same way, determine the value of r_2 by considerity the following bit positions:

2	3	6	7	10	11
r_2	1	0	0	0	1

r_2 ⇓ 0

The value of r_2 will be '0' if we calculate the value of even parity bit for the above combination.

$$r_2 = 0$$

(*v*) Replace the r_2 with the parity bits '0'.

(*vi*) The value of r_4 will be calculated by considering the following bit positions

4	5	6	7
r_4	0	0	0

r_4 ⇓ 0

$$r_4 = 0$$

(*vii*) Replace the value of r_4 with the '0' bit.

(*viii*) Again calculate the value of r_8 by considering the following bit positions.

8	9	10	11
r_8	0	0	1

r_8 ⇓ 1

The value of r_8 will be '1' bit.

$$r_8 = 1$$

(*ix*) Replace the r_8 value by '1'.

(*x*) Finally, we get the following redundancy bits, $r = 0$, $r_2 = 0$, $r_4 = 0$, $r_8 = 1$

11	10	9	8	7	6	5	4	3	2	1
1	0	0	1	0	0	0	0	1	0	0

Data $\rightarrow$ 1000001

Codeword $\rightarrow$ 10010000100 (To be transmitted)

Steps Involved for Error Detection and Corrections

Let us consider that the 9th bit position value is corrupted when it reached to the receiver. The value of 9th position of the bit was '0' bit it changed to '1' during the transmission. The code word received is

bit values changed from '0' to '1'

1 0 [1] 1 0 0 0 0 1 0 0

(*i*) Recalculate the values of 4 parity bits, using the same sets of bits used by the sender plus the relavent parity *r* bit for each set.

For r_1

1	3	5	7	9	11
0	1	0	0	1	1

Since the value of 9th position is changed '1' that's why the parity value is also changed from '0' to 1, so here

$$r_1 = 1$$

For r_2

2	3	6	7	10	11
0	1	0	0	0	1

The value of r_2 in same as earlier *i.e.* '0'.

$$r_2 = 0$$

For r_4

4	5	6	7
0	0	0	0

$$r_4 = 0$$

For r_8

8	9	10	11
1	1	0	1

Again, find parity bit value is coming wrong because 9th position bit is changed from 0 to 1.

So, $r_8 = 1$

(*ii*) Then assembles the new parity values in to a binary number in order of *r* position *i.e.* r_8, r_4 r_2, r_1. In case there in no error then redundancy bits will be (0000). The method is used only for single bit errors.

In our examples,

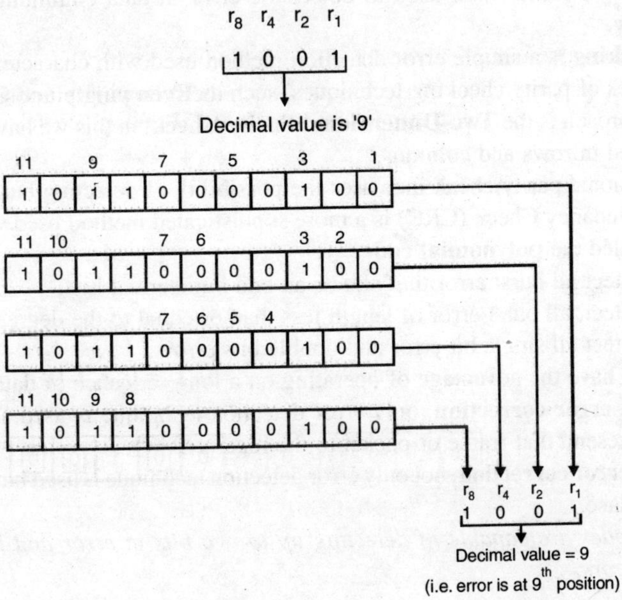

In the example, this step gives us the binary number of 1001 (9 in decimal), which is the precise location of the bit in error.

(iii) Once the bit is identified, it is very simple to correct it just reversing its value. So, by reversing the 9th bit position from '1' to '0'. 9th positions bit value changed from 1 to 0.

Finally,

9th position bit value changed from '1' to '0'

1 0 [0] 1 0 0 0 0 1 0 0

8.7. PROBABILITY OF ERROR AND BIT ERROR RATE

They are often used interchangeably although have slight different meanings.

Probability of error : Pe of error is a theoretical expectation of the bit error rate for a given system.

BIT ERROR RATE (BER): BER is an empirical record of a system's actual bit error performance.

For example, if a system has a Pe of 10^{-5}, this means that mathematically you can expect one bit error in every 100000 bits transmitted.

If a system has a BER of 10^{-5} , that means in past performance there was one bit error for every 100000 bits transmitted.

A bit error is measured and then compared with the expected probability of error to evaluate system's performance.

Summary

- No physical medium can guarantee error-free transmission of data. Errors occur because of interference from the environment surrounding the transmission medium.
- Reliable systems must have a mechanism for detecting and correcting such errors.
- In case of **single-bit error** only one bit of the data is corrupted and rest all the bits are received correctly.
- In case of **Burst error**, multiples bits are corrupted and changed.
- **Error detection** is the first stage of error correction.
- There are some mechanism used to detect the error in data communication based a concept of **redundancy**.
- **Parity checking** is a simple error detection method used with character oriented protocols. There are two types of parity checking techniques, such as **Even parity** and **Odd parity**.
- Another approach is the **Two-Dimensional Parity Check**; in this we have many data units and they are organized in rows and columns.
- Two dimensional parity check increases the possibility of detecting burst errors.
- Cyclic Redundancy Check (CRC) is a more sophisticated method used with bit-oriented protocols. It is also called the **polynomial code.**
- CRC can detect all burst error that affects an odd numbers of bits.
- CRC can detect, all burst error of length less than or equal to the degree of the polynomial.
- CRC can detect all single bit error and double bits errors.
- **Checksums** have the advantage of operating on a long sequence of data.
- In **feedback error correction**, only error detection technique is used and whenever errors occur, sender will resend that frame or packet to the receiver.
- In **forward error correction**, not only error detection technique is used but automatic error correction also takes place.
- *Hamming codes are capable of detecting up to two bits in error and have the ability to correct single bit errors.*

- A mathematician named Richard W. Hamming developed the Hamming code at working at Bell Telephone Laboratories.
- Hamming bits are inserted into a character at random locations. Altogether, the combination of the data bits and hamming bit is known as **hamming code**.

Review Questions

1. Explain briefly different methods of error detection.
2. Discuss CRC code generation.
3. What are the different methods of error correction?
4. What are interleaved codes? How are they good for communication?
5. What do you understand by transmission error and explain the causes of that?
6. What are the types of errors and explain burst error?
7. What do you understand by error correction? Explain the two main approaches of error correction.
8. What are the various types of error detection techniques? Explain parity check with the help of examples. Write the limitations of this method.
9. Explain two-dimensional parity check method and discuss the case when this technique fails.
10. Explain CRC method with the help of example.
11. Explain checksum method with the help of example.
12. What do you understand by forward error correction and explain the method used in this approach.
13. Explain Hamming code method with the help of example.
14. How can a parity bit detect the damaged data unit?
15. Discuss the parity check and the types of errors it can and cannot detect.
16. Explain what are transmission errors? How they can be reduced? Suggest measures.
17. List various transmission errors. Discuss feedback error control approach.
18. Explain error detection and parity check.
19. Write technical notes on hamming codes and their importance.
20. How the error can be detected? Explain any one method for the same.
21. Write the algorithm for computing checksum in CRC and illustrate with example.
22. How does a single bit error differ from a burst error?
23. Discuss the concept of redundancy in error detection.
24. How can a parity bit detect a damaged data unit?
25. What is the difference between even parity and odd parity?
26. Discuss VRC and the types of errors it can and cannot detect.
27. What does the CRC generator append to the data unit?
28. How is VRC related to LRC?
29. Discuss LRC and the types of errors it can and cannot detect?
30. What is the relationship between the size of the CRC remainder and the divisor?
31. How does the CRC checker know that the received data unit is undamaged?
32. What are the conditions for the polynomial used by CRC generator?
33. How is CRC superior to LRC?
34. What is the error detection method used by upper layer protocols?
35. What kind of arithmetic is used to add segments in the checksum generator and checksum checker?
36. List the steps used in creating a checksum.
37. How does the checksum checker know that the received data unit is undamaged?

38. What kind of error is undetectable by the checksum?
39. What is the formula to calculate the number of redundancy bits required to correct a bit error in a given number of data bits?
40. What is the purpose of hamming code?
41. Assuming even parity, find the parity bit for each of the following data units:
1001011
1110111
42. A receiver receives the bit pattern 01101011. If the system is using even parity VRC, is the pattern in error?
43. Find the LRC for the following block of data:
10011001 01101111
44. Given a 10-bit sequence 1010011110 and a divisor of 1011, find CRC. Check your answer.
45. Given a remainder of 111, a data unit of 10110011, and a divisor of 1001, is there an error in the data unit?
46. Find the checksum for the following bit sequence. Assume a 16-bit segment size:
1001001110010011
1001100001001101
47. Find the complement of 1110010001110011.
48. Add 11100011 and 00011100 in one's complement. Interpret the result.
49. For each data unit of the following sizes, find the minimum number of redundancy bits needed to correct one single-bit error:
(*i*) 12
(*ii*) 16
(*iii*) 24
(*iv*) 64
50. Construct the hamming code for the bit sequence 10011101.
51. Calculate the VRC and LRC for the following bit pattern using even parity:
0011101 1100111 1111111 0000000
52. A sender sends 01110001; the receiver receives 01000001. If only VRC is used, can the receiver detect the error?
53. If a divisor is 101101, how many bits long are the CRC?
54. Find the binary equivalent of $x^8 + x^3 + x + 1$.
55. Find the polynomial equivalent of 100001110001.
56. A receiver receives the code 11001100111. When it uses the hamming encoding algorithm, the result is 0101. Which bit is in error? What is the correct code?
57. The code 11110101101 was received. Using hamming code what is the original code sent?
58. In a single-bit error correction, a code of three bits can be in one of the four states: no error, first bit in error, second bit in error, third bit in error. How many of these bits should be redundant to correct this code? How many bits can be the actual data?
59. Write short notes on following:
(*i*) Error correction and detection
(*ii*) Parity check
(*iii*) Frame check sequence
(*iv*) ARQ
(*v*) Checksum
(*vi*) Hamming codes
(*vii*) Cyclic redundancy check
(*viii*) Interleaved codes

□□□

CHAPTER 9

SWITCHING

INSIDE THIS CHAPTER

9.1. INTRODUCTION

After completing this chapter we should be able to understand the nature of network services and use network primitives to describe network service scenarios.

1. Describe how circuit switching works and appreciate its strengths and weaknesses.
2. Describe how packet switching works and distinguishes between the virtual circuits and datagram methods and their packet formats.
3. Understand the three basic routing algorithms (flooding, static routing, and dynamic routing) and their characteristics.

9.2. SWITCHING

We know that a network is set of connected devices; many multiple devices are to be connected to make one-to-one communication. Either we connect each pair of devices in Mesh topology or between central devices in Star topology, to make point to point connection. These methods are impractical and wasteful for a very large network. A better solution is switching. A switched network consists of a series of interlinked nodes, referred as switches.

Switching is the generic method for establishing a path for point-to-point communication in a network. It involves the nodes in the network utilizing their direct communication lines to other nodes so that a path is established in a piecewise fashion. Each node has the capability to 'switch' to a neighboring node that is, a node to which it is directly connected to further stretch the path until it is completed.

Figure 9.1 shows that A and B are end devices and they are connected through the switches or nodes 1 to 7. End devices that wishes to communicate are referred as stations and they may be computer, telephone or any other communicating devices.

In this figure devices are not connected to each other directly but are connected through switches which are used to only route the signal in idle mode. Like in mobile phone, we are always connected to the network but in idle mode. But as we call or receive, the network operator provides a dedicated path to connect to desired user. As soon as we disconnect the call, that path is given to some other user but still you are connected with network by routing signal and this mode is referred as **idle mode**.

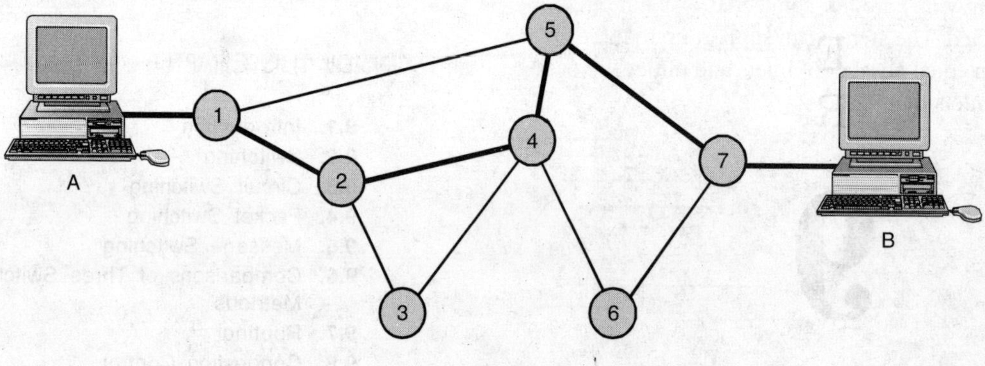

Fig. 9.1. A switched network.

Switches are devices capable of making temporary link or connection between two or more devices linked to the switch but not directly to each other. They may be hardware and/or software devices. Switching nodes are not concerned with the content of the data but their purpose is to provide a switching facility that will transfer the data from node to node until they reach their destination.

DO YOU KNOW►

***Switching** is the generic method for establishing a path for point-to-point communication in a network.*

One of the most important functions of the network layer is to employ the switching capability of the nodes in order to route messages across the network.

There are traditionally three basic methods of switching, namely :

1. Circuit switching
2. Packet switching.
3. Message switching

These are separately discussed below.

9.3. CIRCUIT SWITCHING

In **circuit switching**, two communicating stations are connected by a *dedicated* communication path which consists of intermediate nodes in the network and the links that connect these nodes. *In circuit switching, an electrical dedicated path is established between the source and the destination before any data transfer take place*. The electrical path may be realized by physical wires or radio or satellite links.

What is significant about circuit switching is that the communication path remains intact for the duration of the connection, engaging the nodes and the links involved in the path for that period. However, these nodes and links are typically capable of supporting many channels, so only a portion of their capacity is taken away by the circuit.

Circuit switching takes place at physical layer. Data transmission using a PSTN connection is a typical example of a circuit switched data transfer.

Figure 9.2, shows a simple circuit switch which consists of a 3×3 matrix, capable of connecting any of its inlets (*a*, *b*, and *c*) to any of its outlets (*d*, *e*, and *f*).

Each cross point appears as a circle. A hollow circle means that the cross point is *off i.e.,* the two crossing wires are not connected. A solid circle means that the cross point is *on* i.e., the crossing wires are connected. The switch can support up to three simultaneous but independent connections. Although we have used an equal number of inlets and outlets here, in general, this need not be the case. Switches may also have more inlets than outlets,

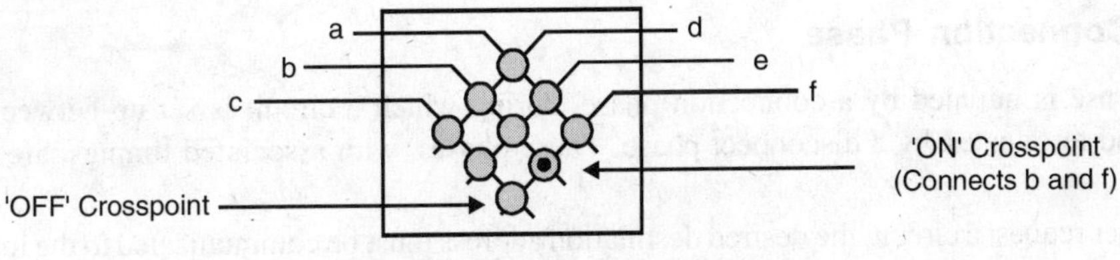

Fig. 9.2. A simple circuit switch.

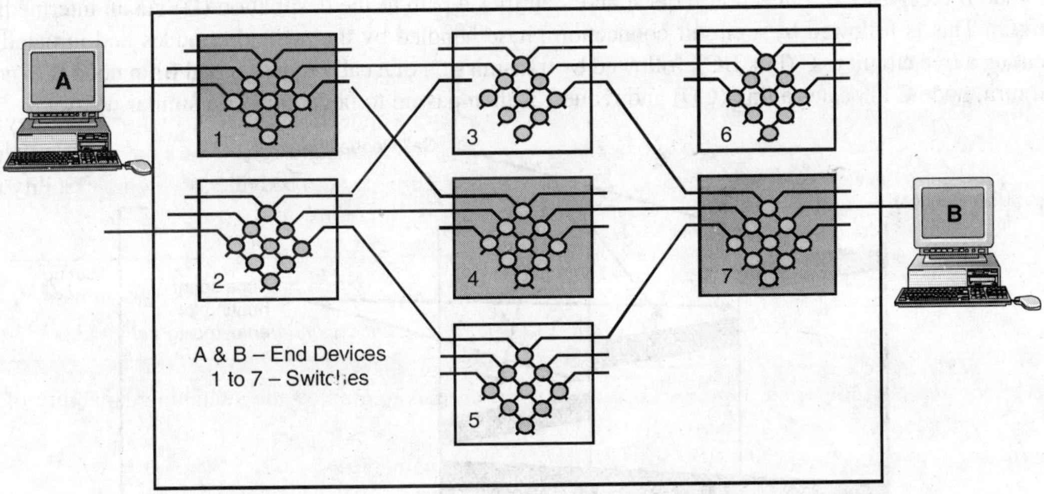

Fig. 9.3. A simple circuit-switched network.

or more outlets than inlets. Each link is normally divided in to n-channels by using FDM (Frequency Division Multiplexing) and TDM (Time Division Multiplexing).

Figure 9.3, shows a simple circuit-switched network built using the switch shown in Fig. 9.2. When the two hosts shown in the figure initiate a connection, the network determines a path through the intermediate switches and establishes a circuit which is maintained for the duration of the connection. When the hosts disconnect, the network releases the circuit.

A circuit switched network has three phases which are as follows :

(*i*) Setup phase
(*ii*) Data transfer phase
(*iii*) Tear down phase

Before starting communication, the station must make provision for the resources to use during the communication, such as follows:

(*i*) Channels Bandwidth (*ii*) Switches buffers
(*iii*) Switch processing time (*iv*) Switches input/outputs.

The operation of the circuit switched network is explained with the help of an example.

Consider communication between two points *A* and *D* in a network. The connection between *A* and *D* is provided using (shared) links between two other pieces of equipment, *B* and *C*.

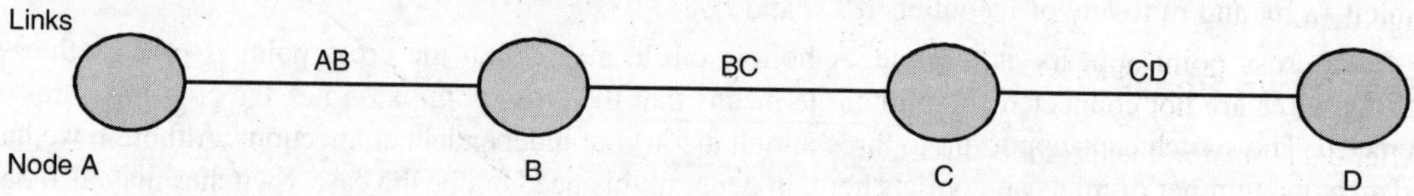

Fig. 9.4. A Connection between Two systems A and D formed from 3 Links.

(a) Setup/Connection Phase

Network use is initiated by a connection phase, during which a circuit is set up between source and destination and terminated by a disconnect phase. These phases, with associated timings, are illustrated in the Fig. 9.5.

After a user request a circuit, the desired destination address must be communicated to the local switching node (B). In a telephone network, this is achieved by dialing a number.

Node B receives the connection request and identifies a path to the destination (D) via an intermediate node (C). This is followed by a circuit connection phase handled by the switching nodes and initiated by allocating a free circuit to C (link BC), followed by transmission of a call request signal from node B to node C. In turn, node C allocates a link (CD) and request is then passed to node D after a similar delay.

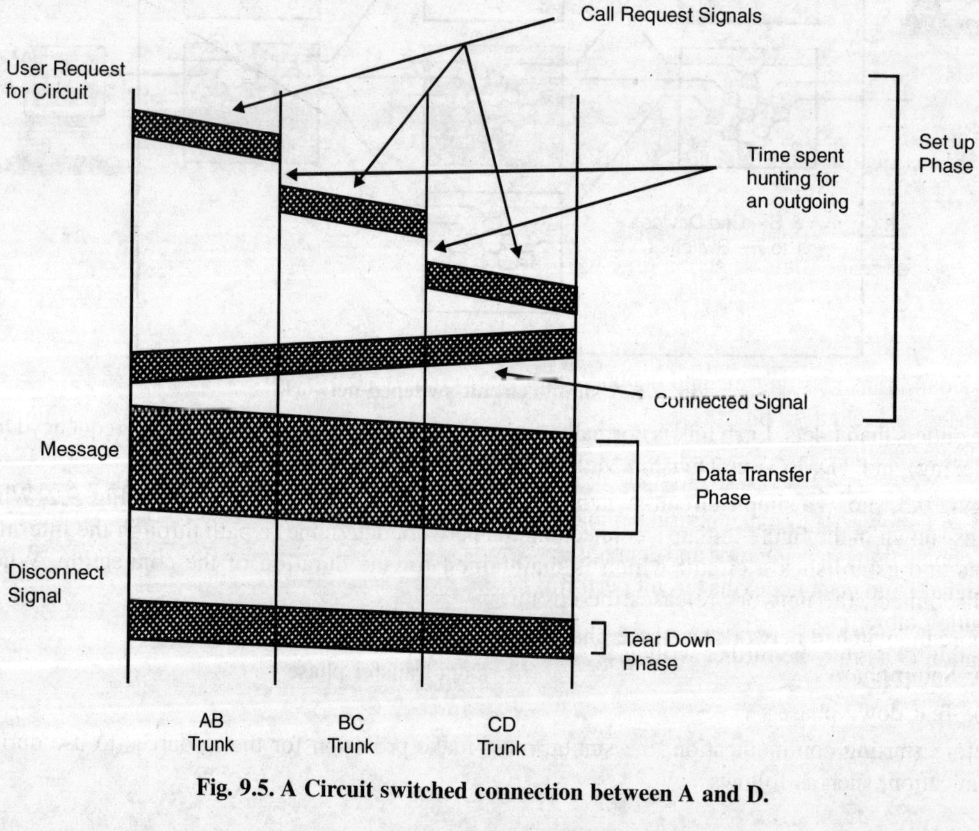

Fig. 9.5. A Circuit switched connection between A and D.

(b) Data Transfer Phase

The circuit is then established and may be used. While it is available for use, resources and capacity on the links between the equipment are dedicated to the use of the circuit.

After completion of the connection, a signal confirming circuit establishment is returned. This flows directly to node A with no search delays since the circuit has been established. Transfer of data in the message then begins.

(c) Tear Down Phase

After data transfer, the circuit is disconnected. A simple disconnect phase is included after the end of the data transmission.

9.3.1. Propagation Time in Circuit Switched Network

Propagation time is dependent upon the medium of transmission and the electrical distance between sources to the destination. Free space propagation speed is 3×10^8 m/s as speed of electromagnetic wave and through coaxial cable its speed is 200 m/μs.

Figure 9.6 shows the variations of delay components in circuit switching.

$$T_{CS} = T_e + T_t + T_r \qquad ...(9.1)$$

where, T_{CS} = Total circuit switching delay

T_e = Connection establishment time

T_t = Data transmission time

T_r = Connection release time Data trans-mission time

$$T_e = (N-1)\,T_m \qquad ...(9.2)$$

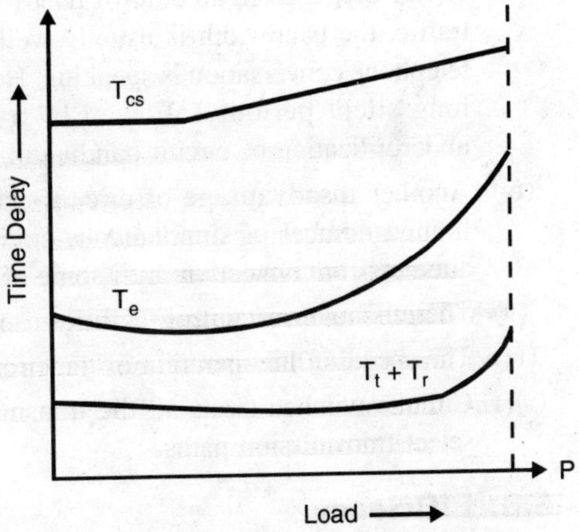

Fig. 9.6. Variation of delay components in circuit switching.

T_e is a set up time and depends upon the number of switches.

where, N = Number of switching nodes

T_m = Average route selection time for each node

$$T_t = \left(\frac{M}{R}\right) \qquad ...(9.3)$$

T_t is a data transmission time and depends upon the data rate and size of message.

M = Message length in bits

R = Data rate in bps

$$T_r = (N\,T_n) \qquad ...(9.4)$$

where, T_n = Time needs to make housekeeping entries by each node.

T_r is connection releasing time and depends upon number of switching nodes. On receiving disconnection request signal from each node, the circuit performs certain housekeeping operations, such as making entries in the routing table.

Equation (9.1) may be further written as

$$T_{CS} = (N-1)\,T_m + \left(\frac{M}{R}\right) + (N\,T_n) \qquad ...(9.5)$$

9.3.2. Advantages/Disadvantages

(*i*) Circuit switching relies on dedicated equipment especially built for the purpose, and is the dominant form of switching in telephone networks. Its main **advantage** lies in its predictable behavior: because it uses a dedicated circuit, it can offer a constant throughput with no noticeable delay in transfer of data. This property is important in telephone networks, where even a short delay in voice traffic can have disruptive effects.

(*ii*) Delays for setting up a circuit connection can be high, especially if ordinary telephone equipment is used. Call setup time with conventional equipment is typically on the order of 5 to 25s after completion of dialing. New fast circuit switching techniques can in theory reduce delays to approximately 140 ms. Trade-offs between circuit switching and other types of switching depend on switching times.

(*iii*) Circuit switching main **disadvantage** is its inflexibility in dealing with computer oriented data. A circuit uses a fixed amount of bandwidth, regardless of whether it is used or not. In case of voice traffic, the bandwidth is usually well used because most of the time one of the two parties in a telephone conversation is speaking. However, computers behave differently; they tend to go through long silent periods followed by a sudden burst of data transfer. This leads to significant underutilization of circuit bandwidth.

(*iv*) Another **disadvantage** of circuit switching is that the network is only capable of supporting a limited number of simultaneous circuits. When this limit is reached, the network blocks further attempts for connection until some of the existing circuits are released.

(*v*) There is no error control facilities used in circuit switched network.

(*vi*) The speed of the operation of the circuit is limited by slowest link in circuit.

(*vii*) Circuit switched treats all the transmission equally. There is problem to prioritize messages and select transmission paths.

9.3.3. Efficiency

The circuit switched networks are not as efficient as the other two types of networks because the resources are allocated during the entire duration of connection. Now these resources are not available to other users. Normally, in telephone network, users terminate the communication when not in use whereas in computer networks, a computer can be connected to another computer even if there is no activity for a long period.

DO YOU KNOW ►

*In **circuit switching**, an electrical dedicated path is established between the source and the destination before any data transfer take place.*

9.3.4. Circuit Switching in Telephone Networks

Early telephone switches were **electromechanical**, many of which are still in use. They are so named because they use electromechanical relays to perform switching functions. The relays respond to the dial pulses from subscriber phones, and hence activate connections. Various switching systems are shown in Fig. 9.7.

Existing electromechanical switches are of two types: step-by-step and crossbar. A **step-by-step switch** (also called Strowger after its inventor) uses step relays capable of assuming ten separate levels, to represent a decimal digit. For example, each of the two switches in **Figure 9.7** would consist of two banks of relays, arranged in a tree-like fashion. Figure 9.8 illustrates how an incoming call to a subscriber whose number is 16 is connected.

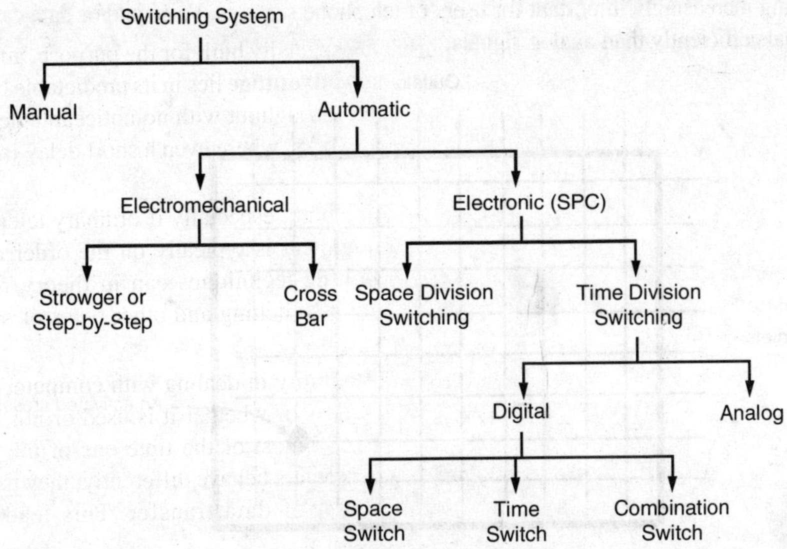

Fig. 9.7. Various Switching Systems

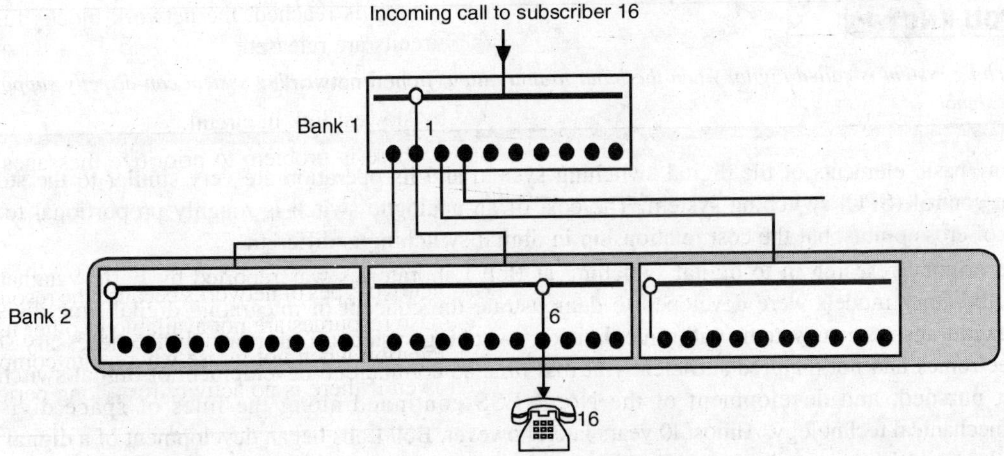

Fig. 9.8. A Step-by-Step Switch.

A **crossbar switch** consists of a matrix of crosspoints between rows of inlets and columns of outlets. Activating the relay of an inlet and the relay of an outlet causes the crosspoint at which they overlap to be activated and therefore result in a physical connection, refer Fig. 9.9.

Modern switches are microprocessor controlled and use digital switching technology. They are generally referred to as **Stored Program Control** (SPC) switches. Unlike electromechanical switches which are inherently hardwired and therefore inflexible, SPC switches achieve significant flexibility by using programmable logic. Because there are no moving parts involved, connections can be established orders of magnitude faster. Furthermore, the functions of the switch can be reconfigured through software (even remotely). Software control has facilitated the introduction of many new service features which were previously beyond the scope of electromechanical switches.

Various facilities of digital switching and transmission are the reason why the analog switching is slowly getting replaced by digital switching. The incorporation of digital switching and transmission technique into telecommunications altered the whole telecommunication industries setup. The reliability of digital switching system is becoming increasingly important for users of telephone services. Voice and/or data can be represented using digital signals efficiently than analog signals.

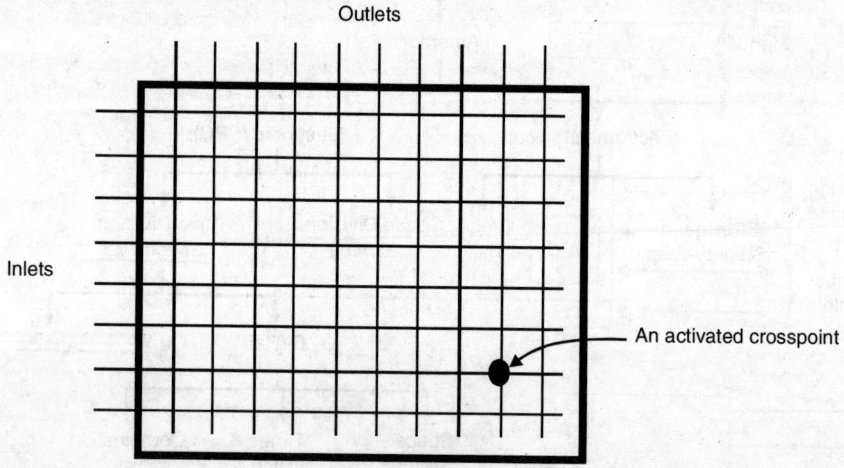

Fig. 9.9. A Crossbar Switch.

> **DO YOU KNOW▸**
>
> *A switching system is called digital when the input to and output from the switching system can directly support digital signal.*

Many basic elements of the digital switching system and its operation are very similar to the stored program control (SPC) switching system. The cost of an analogue switch is roughly proportional to the number of cross points, but the cost relationship in digital switching is different.

The original research in to digital switching at Bell Laboratories was reported by Earle Vaughan in 1959. Laboratory models were developed to demonstrate the concept of integrating digital time division multiplexed transmission systems with time division switching systems. Unfortunately the necessary solid state electronics had not matured sufficiently at that time so commercial development of digital switching was not pursued, and development of the No. 1 ESS continued along the lines of space division Electromechanical technology. Almost l0 years later, however, Bell Labs began development of a digital toll switch, the No. 4 ESS.

The function of the digital switching network is to connect pairs of channels. So that information arriving at the switching centre in a particular channel on one PCM multiplex system can be passed to some other channel on an outgoing PCM multiplex systems.

To achieve this switching, two processes referred to as time switching and space switching is used. The principles of these two switching process are described in this chapter.

The early version of electronic switching system is the stored program control (SPC). The SPC systems have temporary memory for storing transient call information and to carry programming information. The SPC performs line control, trunk control, ancillary control, maintenance control etc. The instructions required for performing these operations are resided in a single processor. For reliability or high availability, the processor may be duplicated. Thus SPC uses centralized software and hardware architectures.

A modern digital switching system employs a number of processors not uses distributed software and hardware architectures. The digital switching system also referred as Electronic Switching System–III generation is purely electronic in operation, the switching process is by time division/digital transmission, the type of control is stored program common control and the network uses pulse code modulation. Figure 9.4 shows the evaluation of digital switching system.

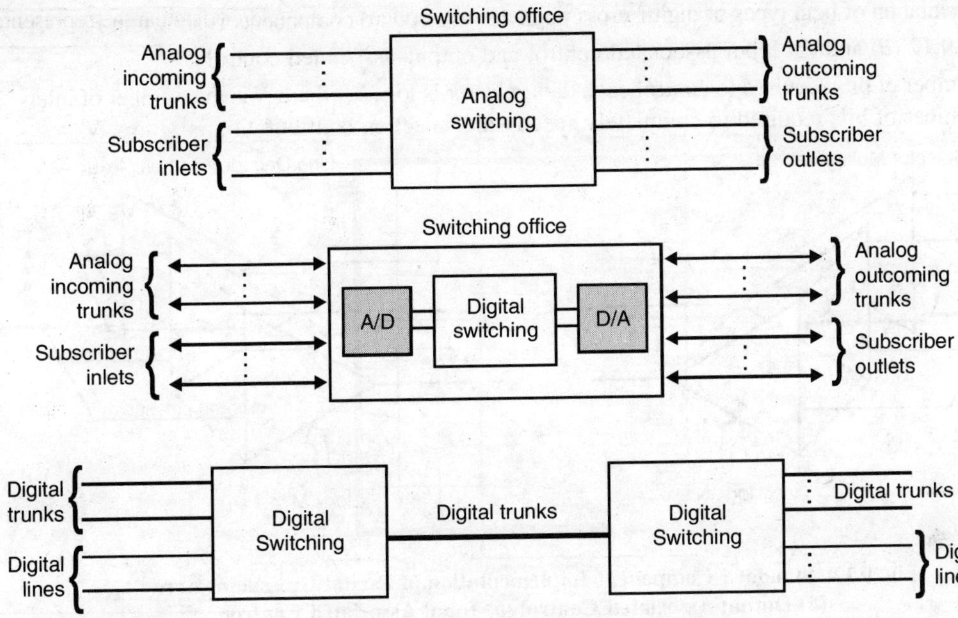

Fig. 9.10. Evolution of Digital Switch.

9.3.4.1. Switching Matrix Control

When an available path through a common control switching network is determined, the control element of the switch transfers the necessary Information to the network to select the appropriate cross points. The control may be associated with the output lines and therefore specify which inputs are to be connected to the associated outputs or the control information may be associated with each input and subsequently specify to which outputs the respective inputs are to be connected. Cross point selection within a matrix is accomplished in one of two ways. The first approach is **output – associated control** while the second is called **input – associated control.** These are shown in Fig. 9.11.

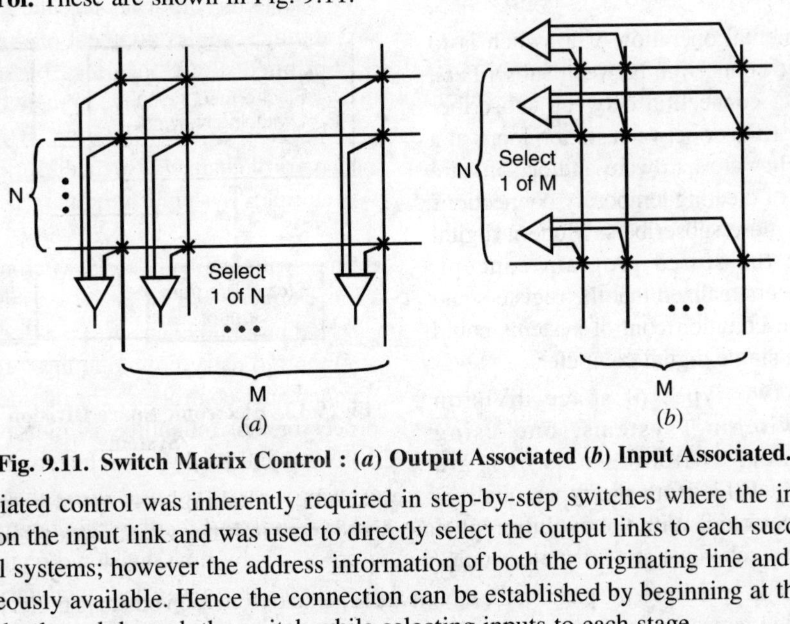

Fig. 9.11. Switch Matrix Control : (*a*) Output Associated (*b*) Input Associated.

Input associated control was inherently required in step-by-step switches where the information (dial pulses) arrived on the input link and was used to directly select the output links to each successive stage. In common control systems; however the address information of both the originating line and the terminating line is simultaneously available. Hence the connection can be established by beginning at the desired outlet and proceeding backward through the switch while selecting inputs to each stage.

Implementation of both types of digital cross point arrays standard components is shown in Fig. 9.12. Figure 9.12 (*a*) and (*b*) Input associated control and output–associated control

The number of bits required to control each data selector is $\log_2 N$, where N is the number of inlets. Thus the total number of bits required to completely specify a connection configuration is $M \log_2 N$.

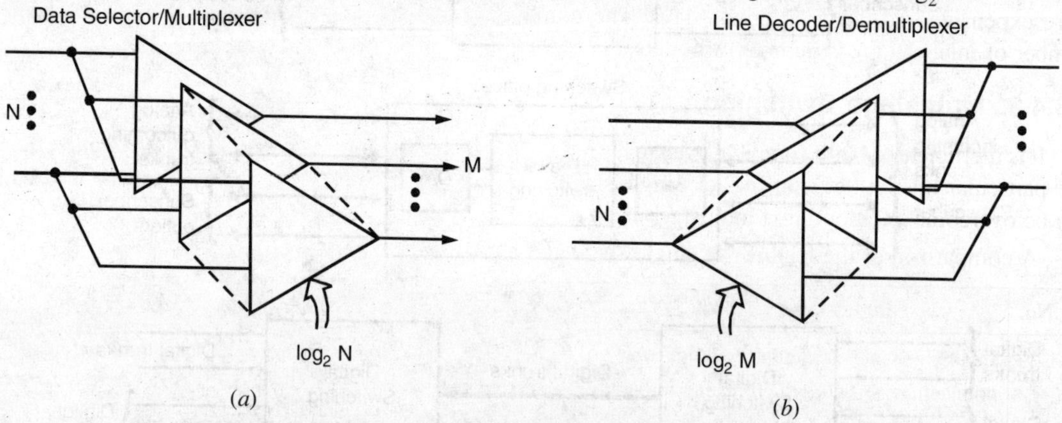

Fig. 9.12. Standard Component Implementation of Digital Crosspoint Array : (*a*) Output Associated Control (*b*) Input Associated Control.

Input-associated control can be implemented using conventional line decoders/demultiplexer. The output are command using a "wired-or" logic function. Hence the total number of bits required to specify a connection configuration in this case is $N \log_2 M$.

A significant drawback of input-associated control arises from the need to disable unused inputs to prevent cross connects when another input selects the same outputs. For this reason and for generally greater speeds of operation, digital switching networks typically use output-associated control. The total amount of information needed to specify a connection configuration with input-associated control is less than that with output control if the number of inputs N is much smaller than the number of outputs M.

9.3.4.2. Space Division Switching

The fundamental operation of a switch is to setup and release connection between subscribers. It involves direct connection between subscriber loops at an end office or between station loops at a PBX. The switches are hardware and/or software devices capable of creating temporary connections between two or more subscribers. Modern digital computers use the stored program concept. Switching engineers realized that the registers and translators of the common control systems could be replaced by a single digital computer.

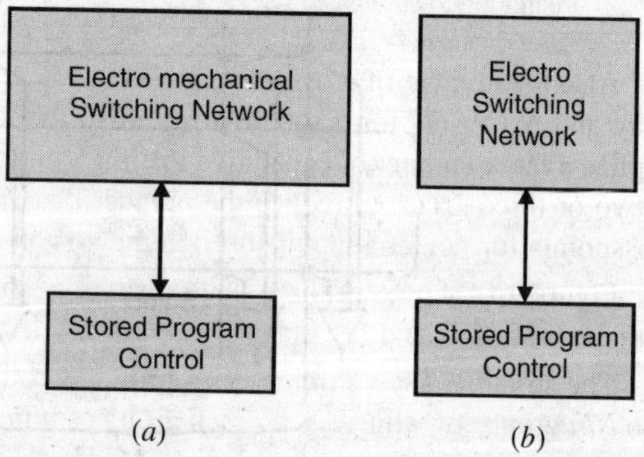

Fig. 9.13. Electronic Space Division Switching Systems.

There are two types of space division electronics switching systems, one using electromechanical switching network, are illustrated in Fig. 9.13. Both the types qualify as electronic switching systems although only one of them is fully electronic. With the evolution of time division switching, which is done in the electronic domain, modern exchanges are fully electronic.

In space division switching, the paths in the circuit are separated from each other spatially. It was originally designed for analog networks, but is used currently in both digital and analog switching. A crosspoint switch is referred to as a space division switch because it moves a bit stream from one circuit/bus to another. For large group of outlets, considerable savings in total crosspoints can be achieved if each inlet can access only a limited number of outlets. Such situation is called limited availability.

By overlapping the available outlet groups for various inlet groups, a technique called "grading" as established. Rectangular crosspoint array is an example of grading. For longer trunk groups, large crosspoints were expensive and not used now-a-days. The numbers of crosspoints required are $M \times N$, where M is number of inlets and N is number of outlets.

9.3.4.3. Multistage Switching

It is inefficient to build complete exchanges in single stages. Single stage can only be used to interconnect one particular inlet outlet pair. In fact, single stage networks suffer from a number of disadvantages which can be overcome by adopting a multistage network configuration.

A comparison of the features of single stage and multistage networks is illustrated below:

S. No.	Single stage	Multistage
1.	In single stage, a single crosspoint per connection is used, which results in better quality link.	Use of multiple crosspoints may degrade the quality of connection.
2.	Inlet to outlet connection is through a single crosspoint.	Inlet to outlet connection is through multiple crosspoint.
3.	A specific crosspoint can be used for only one inlet/outlet pair connection.	Same crosspoint may be used to provide connection between a more numbers of inlet/outlet pairs.
4.	Number of crosspoints is prohibitive.	Number of crosspoints is reduced significantly.
5.	In this crosspoints are used inefficiently.	In this crosspoints are used efficiently.
6.	A large number of crosspoints in each inlet/outlet leads to capacitive loading.	There is no capacitive loading problem.
7.	The network is nonblocking in character.	The network is blocking in character.
8.	Time required establishing a connection is less.	Time required establishing a connection is more.
9.	If a crosspoint fails, associated connection cannot be established.	Alternative crosspoints and paths are available.

Also the number of crosspoints grows as the square of the inputs for grading, $N(N-1)/2$ for a triangular array and $N(N-1)$ for a square array. Also the large number of crosspoints on each inlet and outlet line implies a large amount of capacitive loading on the message paths. Therefore, it is usual to build exchanges in two or three stages to reduce the number of crosspoints and to provide alternative paths. The sharing of crosspoints for potential paths through the switch is accomplished by multiple stage switching.

Figure 9.14 shows the three stage switching structure to accommodate 128 input and 128 output terminals with 16 first stage and last stage.

The structure shown in Fig. 9.14 provides path for N inlets and N outlets. The N input lines are divided into N/n groups of n lines each. Each group of n inputs is accommodated by an n-input, k output matrix. The output matrices are identical to the input matrices except they are reversed. The intermediate stages are k in number and N/n inputs and N/n outputs. The interstage connections are often called junctures. Each of the k paths utilizes a separate center stage array. An arbitrary input can find k alternate output. Thus multistage structure provides alternate paths. Also the switching link is connected to a limited number of crosspoints. This enables the minimized capacitive loading.

The total number of crosspoints N_X for three stages is

$$N_X = 2Nk + k\left(\frac{N}{n}\right)^2 \quad ...(9.6)$$

where,
N = Number of inlets-outlets
n = size of each inlet-outlet group
k = number of second stage.
$2Nk$ = number of cross points in 1st and 2nd stage
$(N/n)^2$ = number of cross points in each array of second stage
$k(N/n)^2$ = number of cross points in second stage

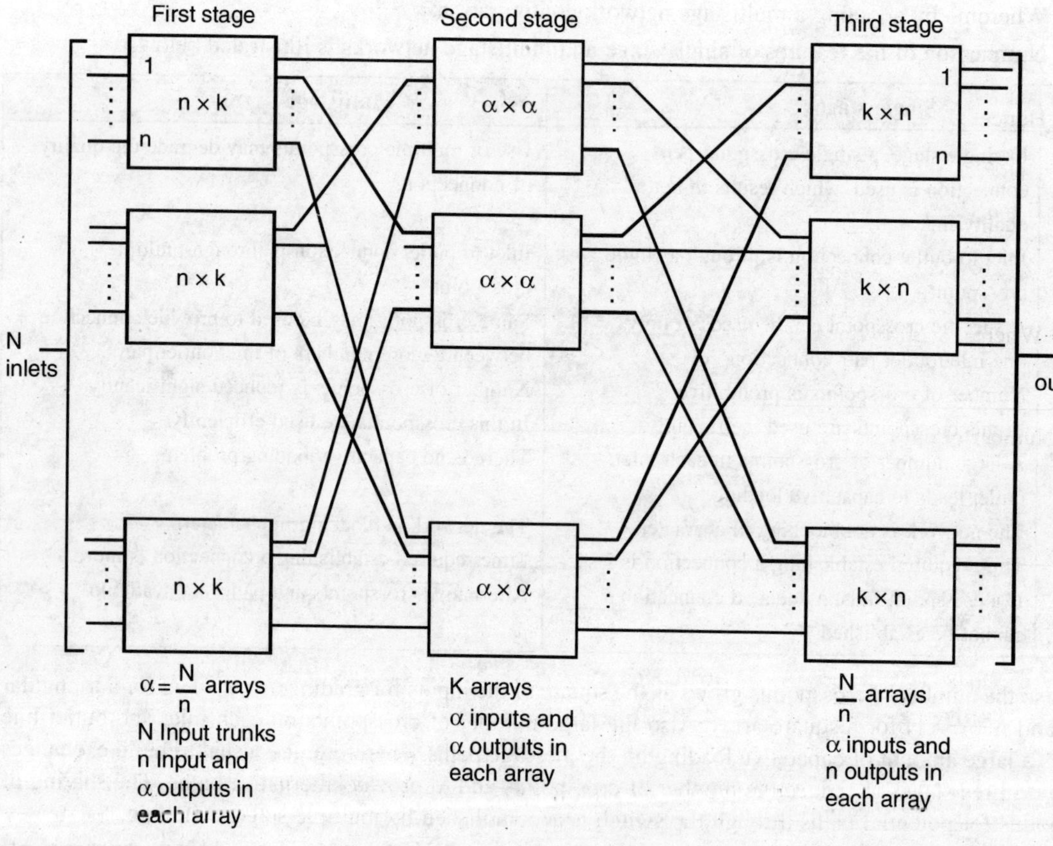

Fig. 9.14. Three Stage Switching Structure.

Also,

$$N_X = k\left\{2N + \left(\frac{N}{n}\right)^2\right\} \quad ...(9.7)$$

Three stage switching matrix require that $k > 2n - 1$ to generate no blocking.

$$k = 2n - 1 \quad ...(9.8)$$

Substituting Eq. (9.8) in Eq. (9.6), we have

$$N_X = 2N(2n-1) + (2n-1)\left(\frac{N}{n}\right)^2 \quad ...(9.9)$$

For large N, $n = \sqrt{\dfrac{N}{2}}$, substitute $n = \sqrt{\dfrac{N}{2}}$ in Eq. (9.9), we get

$$N_X = 2N\left(\frac{2\sqrt{N}}{2} - 1\right) + \left(\frac{2\sqrt{N}}{2} - 1\right)\left(\frac{N}{\frac{\sqrt{N}}{2}}\right)^2$$

$$= 2N(\sqrt{2N} - 1) + 2N(\sqrt{2N} - 1)$$

$$N_{X(\min)} = 4N\left(\sqrt{2N} - 1\right) \quad ...(9.10)$$

$$N_X \approx 4\sqrt{2}N^{3/2} \quad ...(9.11)$$

Where N = total number of inlets-outlets.

Number of cross points for a single stage switching matrix to connect N inlets to N outlets is $N_x(ss) = N^2$.

Hence,

$$\lambda = \frac{N_X \text{ (minimum 3 stages)}}{N_X\,(ss)}$$

$$\lambda = \frac{4\sqrt{2}N^{3/2}}{N^2}$$

$$\lambda = \frac{4\sqrt{2}}{\sqrt{N}} \quad ...(9.12)$$

Where $N_X(ss)$ = Number of cross points in single stage.

TABLE 9.1

Number of lines	Number of cross points		$\lambda = \dfrac{4\sqrt{2}}{\sqrt{N}}$
	Three stage switch $N_X = 4N(\sqrt{2N} - 1)$	Single Stage N^2	
128	7680	16384	0.5
512	63486	262144	0.25
2048	516.096	4.2×10^6	0.125
8192	4.2×10^6	6.7×10^7	0.065
32,768	3.3×10^7	0.1×10^8	0.0313
131,072	2.6×10^8	1.7×10^{30}	0.0156

With the increase in the value of N, the savings in crosspoints becomes more pronounced. Whenever very large number of lines to be accommodated, switching structures with more stages even up to eight stages are used. The probability that all k links are busy, is given by

$$B = [1 - (1 - P)^2]k \quad ...(9.13)$$

Where,

$$P = \frac{np}{k} \quad ...(9.14)$$

Example 9.1. A 3-stage switching structure is to accomodate N = 128 inputs and 128 output terminals. For 32 first stage and 32 last stage, determine the number of crosspoints for non-blocking.

Solution. The number of matrices at first and last stages is given by

$$\alpha = \frac{N}{n}, \text{ Hence } n = \frac{N}{\alpha} = \frac{128}{32} = 4$$

To avoid blocking, $k = 2n - 1 = 2 \times 4 - 1 = 7$

Number of cross points is calculated by

$$N_x = k\left[2N + \left(\frac{N}{n}\right)^2\right] = 7\left[2 \times 128 + \left(\frac{128}{4}\right)^2\right]$$

$$N_x = 8960 \text{ cross points.}$$

Example 9.2. If the number of cross points in the example 1, is to be reduced by the factor of 3 with non-blocking, what is the probability that a call will be blocked ? Suppose the utilization probability $P = 25\%$.

Solution. Total number of cross points = 7680

and reduced by factor of 3 = $\frac{7680}{3} = 2560$

for the cross points 2560, the number of k matrices is calculated from,

$$N_x = k\left(2N + \left(\frac{N}{n}\right)\right)^2$$

$$k = \frac{N_x}{\left[2N + \left(\frac{N}{n}\right)^2\right]} = \frac{2560}{\left[2 \times 128 + \left(\frac{128}{4}\right)\right]^2} = 2$$

$$P = \frac{nP}{k} = \frac{4 \times 0.25}{2} = 0.5$$

The probability that k links are busy is

$$B = [1 - (1 - P)^2]^k = [1 - (1 - 0.5)^2]^2 = 0.5625\%$$

9.3.4.4. Blocking Probability Evaluation Techniques

All the switching systems are designed to provide a certain maximum probability of blocking for the busiest hour of the day. It is one of the aspects of the grade of service of the telephone company. There are varieties of techniques to evaluate the blocking probability of a switching matrix. Depends on the accuracy, required availability, geographical area, priority, complexity and applicability of different network structures, the techniques are varying. Here, two techniques are described.

1. Lee graphs
2. Jacobaeus method.

Lee Graphics

It was proposed by C.Y. Lee. It is a most versatile and straight forward approach of calculating probabilities with the use of probability graphs.

C.Y. Lee's approach of determine the blocking probabilities of various switching system is based on the use of utilization percentage or loadings of individual links.

Let p be the probability that a link is busy.

The probability that a link is idle is denoted by

$$q = 1 - p.$$

When any one of n parallel links can be used to complete a connection, the blocking probability B is the probability that all links are busy is given by

$$B = p^n \quad ...(9.15)$$

When a series of n links are all needed to complete a connection,

$$B = 1 - q^n \quad ...(9.16)$$

For a probability graph of three stage network, shown in Fig. 9.15, the probability of blocking is given by

$$B = (1 - q'^2)^k \quad ...(9.17)$$

where q' = probability that an interstage link is idle = $1 - p'$ ...(9.18)

p' = probability that any particular intersatge link is busy

k = number of centre stage arrays.

If p is known, the probability that an interstage link is busy is given by

$$p' = \frac{p}{\beta} \quad ...(9.19)$$

where

$$\beta = \frac{k}{n} \quad ...(9.20)$$

β is the factor by which the percentage of interstage links that are busy is reduced.

Substituting in Eq. (9.13), we get

$$q' = 1 - \frac{P}{\beta} \quad ...(9.21)$$

Substituting Eq. (9.19) in (9.18), we get complete expression for the blocking probability of a three stage switch in terms p as

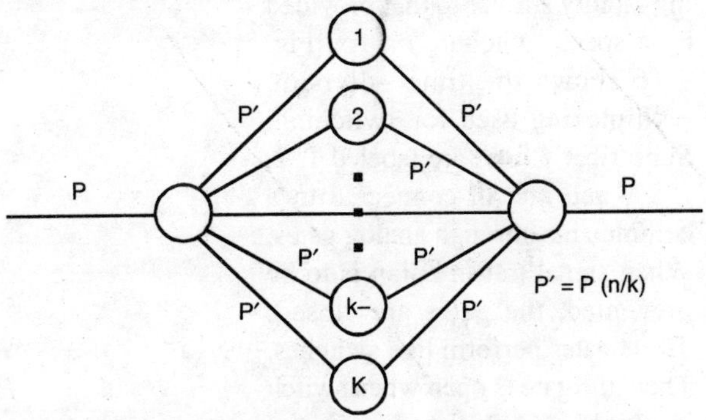

Fig. 9.15. Probability graph of 3-stage network.

$$B = \left[1 - \left[1 - \frac{p}{\beta}\right]^2\right]^k \quad ...(9.22)$$

With inlets of 10% busy, the switch size of N with $n = 8$, $h = 5$, $\beta = 0.625$ requires 2560 crosspoints.

The merits of this method are :

(*i*) It provide accurate results

(*ii*) Its formulas are directly relate to the network structures

(*iii*) It provides insight of the network and thus provides ideas to change the structure for high performance.

Jacobaeus

It was presented in 1950 by C. Jacobaeus. It is more accurate than Lee graph method. The Lee's graph approach is not much accurate because the probability graphs entail several simplifying assumptions. The important one which gives erroneous values of blocking is the assumption that the individual probabilities are independent. In fact the probabilities not independent and highly dependent when significant amounts of expansion are not present. According to C. Jacobaeus the blocking probability of a three stage switch is

$$B = \frac{(n!)^2 \cdot P^x (2 - P)^{2n-x}}{k!(2n-k)!} \quad ...(9.23)$$

where, n = number of inlets (outlets) per first (third) stage array

k = number of second stage array

p = inlet utilization.

More accurate techniques can be used for systems with high concentrations and high blocking. As the high blocking probabilities not having much practical value, those techniques are not considered here.

9.3.4.5. The Concept of Time-Division Multiplexing Used for Switching

Time division multiplexing allows the transmission of many independent signals over a single channel. Time-division multiplexing can be used to proving a needed switching facility similar to that provided by a space switching matrix. Fig. 9.16 shows the time –division multiplexing used for switching. Subscriber's lines are labeled 1, 2, …, *N* and are all connected to a common bus through analog gates. When signal transmission is to be prevented, the gates are closed. These gates perform like switches. Then, the gate is open when switch is closed and the gate is closed when the switch is open. The gates are controlled (*i.e.*, to close and open the switches) by the gate control lines.

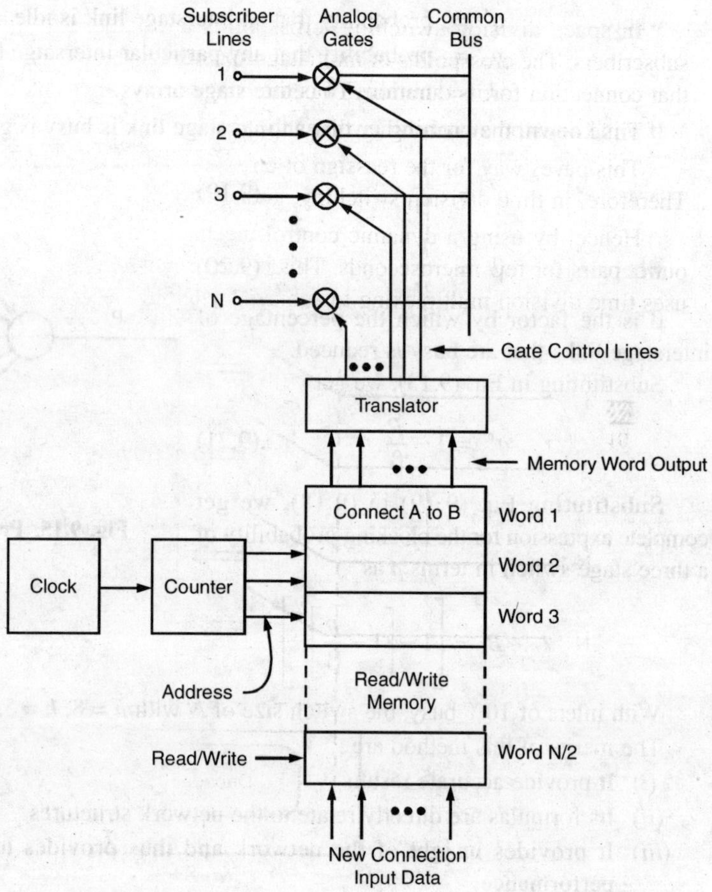

Fig. 9.16. Time-Division Multiplexing Used for Switching

When it is desired to allow subscriber *A* to transmit a signal to subscriber *B* the subscriber *B* must be permitted to sample the signal from subscriber A at the Nyquist rate or more frequently. In the voice telephony, a sampling rate of 8000 Hz (that is, every 125 μs) is sufficient. To allow the sampling, two gates must be opened for signal transmission, one in subscriber *A*'s line and one in the line of subscriber *B*. The time duration (τ) in which this connection can be permitted to persist is, of course, determined by the subscribers who share the bus. If the total number of subscribers is *N*, then there can be at most *N*/2 simultaneous connections.

In Fig. 9.16, the capacity of read/write memory is *N*/2 words. Each word is a coded instruction indicating which analog gates are to be open for transmission. Hence, word 1 holds the instruction that subscriber *A* is to be attached to subscriber *B*. when memory reads out the word, this word is given to a translator which decodes the instruction and sets its output gate control lines so that only gates *A* and *B* are open for transmission. The address specifying the word which is to be read is provided by a counter of modulo *N*/2. The counter is advanced through its states by a clock which runs at a rate to insure that it provides each address from 1 to *N*/2 every 125 μs. in the memory; each word holds an instruction specifying which gates are to be open. When some subscribers lines are idle there will be memory word locations specifying that no gate is to be open. When a call is initiated or terminated, it becomes essential to change the corresponding word. Accordingly the memory is provided with facility not only to be read but also with facility to permit new words to be written into it. This "new connection" data input is shown in Fig. 9.16. Depending on the past history of call connections, the instruction "connect *A* to *B*" can appear in any of the memory locations, there is no reservation of memory locations for particular subscriber pairs.

9.3.4.6. Time Division Switching

In space division switching, crosspoints are used to establish a specific connection between two subscribers. The crosspoints of multistage space switches assigned to a particular connection is dedicated to that connection for its duration. Thus the crosspoints cannot be shared.

Time division switching involves the sharing of crosspoints for shorter periods of time.

This paves way for the reassign of crosspoints and its associated circuits for other needed connections. Therefore, in time division switching, greater savings in crosspoints can be achieved.

Hence, by using a dynamic control mechanism, a switching element can be assigned to many inlet-outlet pairs for few microseconds. This is the principle of time division switching. Time division switching uses time division multiplexing to achieve switching.

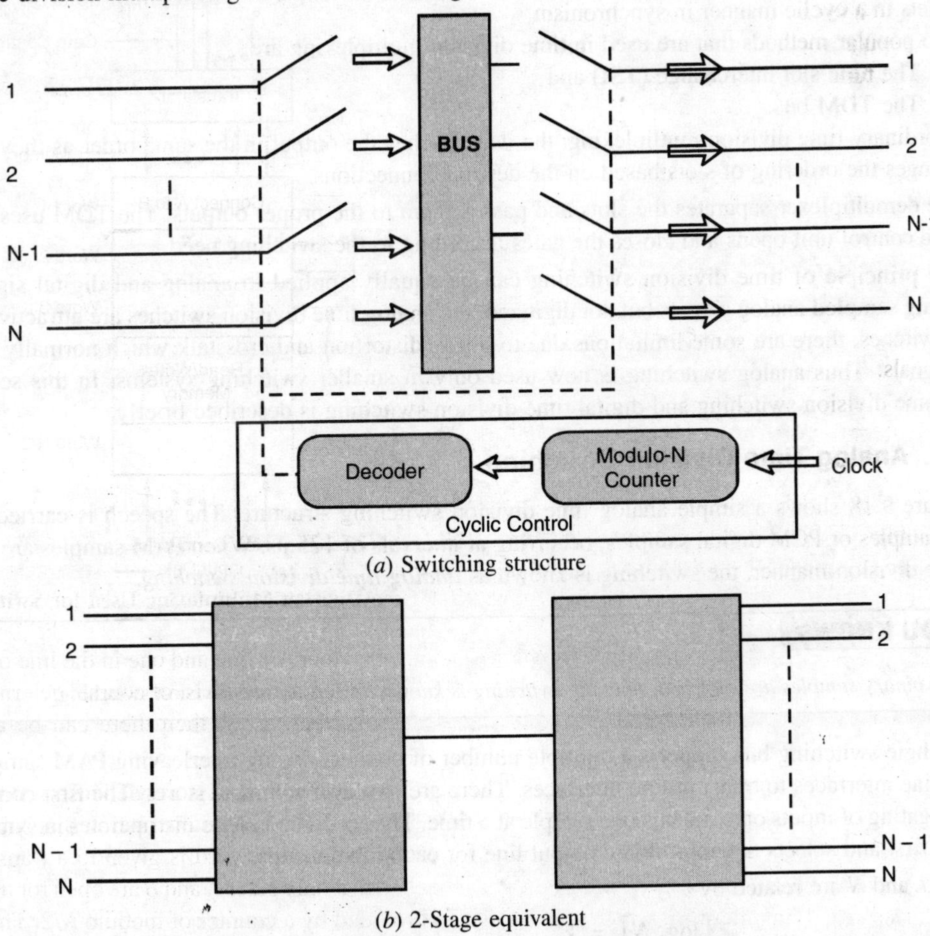

(*a*) Switching structure

(*b*) 2-Stage equivalent

Fig. 9.17. A Simple PAM Time Division Switching.

A simple $N \times N$ time division switch is shown in Fig. 9.17. The switch can be represented in an equivalent form as a two stage network with $N \times 1$ and $1 \times N$ switching matrices for the first and second stages respectively as shown in Fig. 9.17. The network has one link interconnecting the two stages. Each inlet/outlet is a single voice circuit corresponding to a subscriber line. The speech is carried as Pulse Amplitude Modulated (PAM) analog samples or Pulse code Modulated (PCM) digital samples, occurring at 125 μs intervals. When PAM samples are switched in a time division manner, the switching is known as *analog time division switching.*

If PCM binary samples are switched, then the switching is known as *digital time division switching*.

In Fig. 9.17 (*a*), the interconnecting link is shown as a bus to which a chosen inlet/outlet pair can be connected by a suitable control mechanism and the voice sample transferred from the inlet to the outlet. When PAM samples are switched, the bus is analog in nature and if the binary samples are switched, a digital bus is used. The numbers of simultaneous conversions, *Sc* that can be supported on a network of this kind is inversely proportional to the time required to select and connect an inlet-outlet pair and transfer the voice sample from the inlet to the outlet and is given by as

$$Sc = \frac{125}{t_s} \quad ...(9.24)$$

where t_s is the time in microseconds to set up a connection and transfer the sample value.

The selection of inlet/outlet is controlled dynamically. The simplest form of control is to select the inlets and outlets in a cyclic manner in synchronism.

Two popular methods that are used in time division multiplexing are

(*a*) The time slot interchange (TSI) and

(*b*) The TDM bus.

In ordinary time division mutliplexing, the data reaches the output in the same order as they sent. But TSI changes the ordering of slots based on the desired connections.

The demultiplexer separates the slots and passes them to the proper outputs. The TDM uses a control unit. The control unit opens and closes the gates according to the switching need.

The principle of time division switching can be equally applied to analog and digital signals. For interfacing sampled analog signals but not digitized, the analog time division switches are attractive. But for larger switches, there are some limitations due to noise, distortion and crosstalk which normally occurs in PAM signals. Thus analog switching is now used only in smaller switching systems. In this section, the analog time division switching and digital time division switching is described briefly.

9.3.4.7. Analog Time Division Switching

Figure 9.18 shows a simple analog time division switching structure. The speech is carried as PAM analog samples or PCM digital samples, occurring at intervals of 125 μs. When PAM samples are switched in a time division manner, the switching is known as *analog time division switching.*

DO YOU KNOW

If PCM binary samples are switched, then the switching is known as digital time division switching.

A single switching bus supports a multiple number of connections by interleaving PAM samples from receive line interfaces to transmit line interfaces. There are two cyclic control stores. The first control store controls gating of inputs onto the bus one sample at a time. The second control store operates in synchronism with the first and selects the appropriate output line for each input sample.

The *k* and *N* are related by

$$\lceil \log_2 N \rceil = k \quad ...(9.25)$$

where N = number of inlets/outlets

k = decoder size.

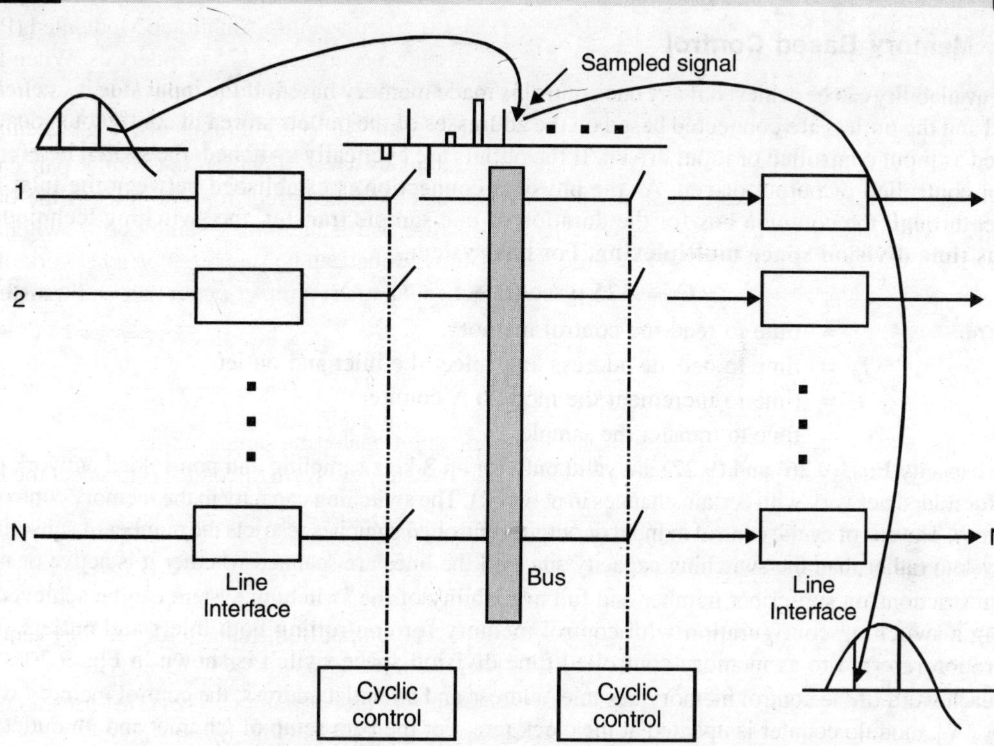

Fig. 9.18. Analog Time Division Switching Structure.

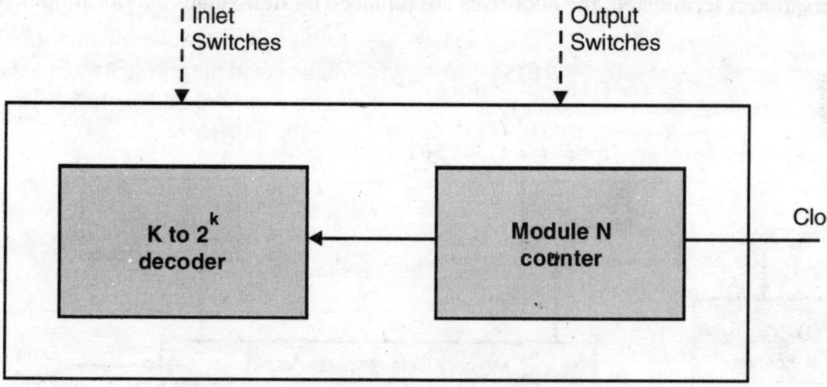

Fig. 9.19. Cyclic Control.

Where, the symbol ⌈⌉ is a ceiling function that gives the lowest integer. It means k may be assumed lowest integer or more than that.

This kind of switching is non-blocking but lack of full availability as it is not possible to connect inlet to any outlet. The switching capacity or number of channel supported by cyclic controlled system is

$$C = \frac{125\ \mu\ \text{sec}}{t_s} \qquad ...(9.26)$$

The numerator 125 μ sec indicates the time taken to scan inlet and outlet and the denominator t_s is the time in micro sec to set-up connection.

9.3.4.8. Memory Based Control

Full availability can be achieved if any one control is made memory based. If the input side is cyclically switched and the outlets are connected based on the addresses of the outlets stored in contiguous location is referred as input controlled or input driven. If the outlets are cyclically switched, the switch is referred as output controlled or output driven. As the physical connection is established between the inlet and the outlet through the common bus for the duration of one sample transfer, the switching technique is known as **time division space multiplexing.** For this system,

$$C = 125\ \mu\text{ sec}/(t_i + t_m + t_d + t_t) \qquad ...(9.27)$$

where t_m = time to read the control memory.

t_d = time to decode address and select the inlet and outlet.

t_i = time to increment the modulo N counter.

t_t = time to transfer the sample.

The capacity Eqs. (9.26) and (9.27) are valid only for an 8 kHz sampling and non-folded network (can be used for folded network with certain changes in network). The switching capacity in the memory controlled is equal to N. The use of cyclic control in input or output controlled switches restricts the number of subscribers on the system rather than the switching capacity since all the lines are scanned whether it is active or not.

No restrictions on subscriber number and full availability of the switching system can be achieved by designing a switching configuration with control memory for controlling both inlets and outlets. This configuration referred to as memory controlled time division space switch is shown in Fig. 9.20.

As each word of the control memory has inlet address and an outlet address, the control memory width is $2\lceil \log_2 N \rceil$. modulo counter is updated at the clock rate. For the path setup of kth inlet and jth outlet, the addresses are entered in control memory and path is made. Then the location is marked busy.

When conversation is terminated, the addresses are replaced by null values and location is marked free. Hence

$$C = \frac{125}{t_s}\ \mu\text{ sec},$$

where $t_s = t_i + t_m + t_d + t_t$

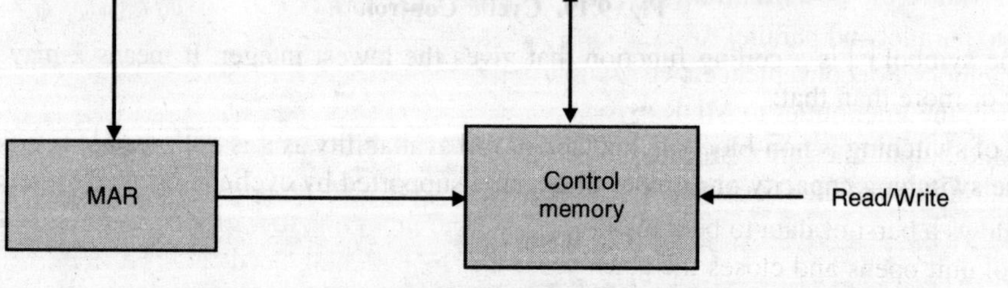

Fig. 9.20. Memory Control for Both Inlet and Outlet.

The switching matrix described above is referred to as time multiplexed switching as the switch in this configuration is replicated once for each time slot.

9.3.4.9. Digital Time Division Switching

The analog time division switching is useful for both analog and digital signals. The digital time division multiplexed signals usually requires switching between time slots as well as between physical lines. The switching between time slots is usually referred as time switching.

Similar to analog time division switching the switching structure can be organized expect the use of memory block in place of the bus. This adds the serial to parallel and parallel to serial bit conversion circuitry's as the input to the memory block should be in parallel form. The time division switch can be controlled in any of the following three ways.

Basic Operation

The basic requirement of time division switching is that the transfer of information arriving at in a time slot of one input link to other time slot of any one of output link. A complete set of pulses, arriving at each active input line is referred to as a frame. The frame rate is equal to the sample rate of each line.

A time switch operates by writing data into and reading data out of a single memory. In the process the information in selected *time slots is interchanged* as shown in Fig. 9.21 and referred as *Time slot interchange (TSI)*.

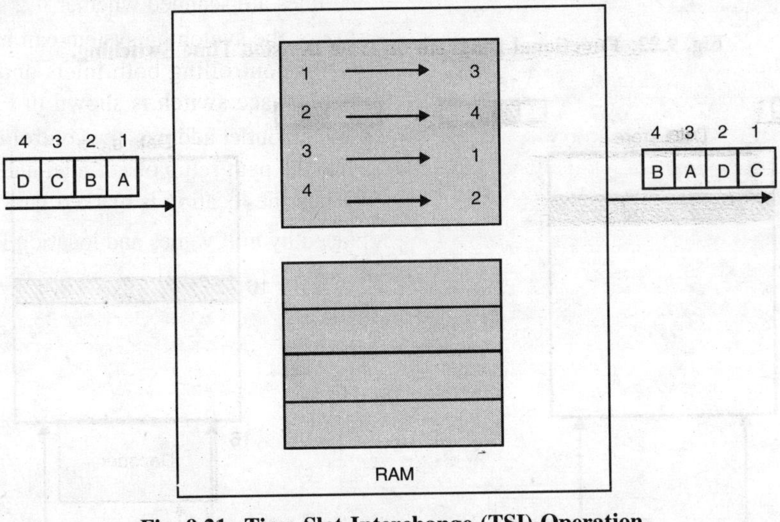

Fig. 9.21. Time Slot Interchange (TSI) Operation.

In TSI operation, inputs are sequentially controlled and outputs are selectively controlled.

The RAM have several memory locations, each size is the same as of single time slot.

Figure 9.22 shows the general arrangement of the time division time switching.

The serial to parallel and parallel to serial converter are used to write the data into the memory and read the data out of memory. For convenience, two MDR are shown, but MDR is a single register. Gating mechanism is used to connect the inlet/outlet to MDR.

The input and output lines are connected to a high speed bus through input and output gates. Each input gate is closed during one of the four time slots. During the same time slot, only one output gate closed. This pair of gates allows a burst of data to be transferred from one input line to a specific output line through the bus. The control unit opens and closes the gates according to switching need.

The time division time switch may be controlled by sequential write/random read or random write/ sequential read. Fig. 9.23 depicts both modes of operation and indicates how the memories are accessed to translate information from time slot 2 to time slot 16. Both methods use a cyclic control.

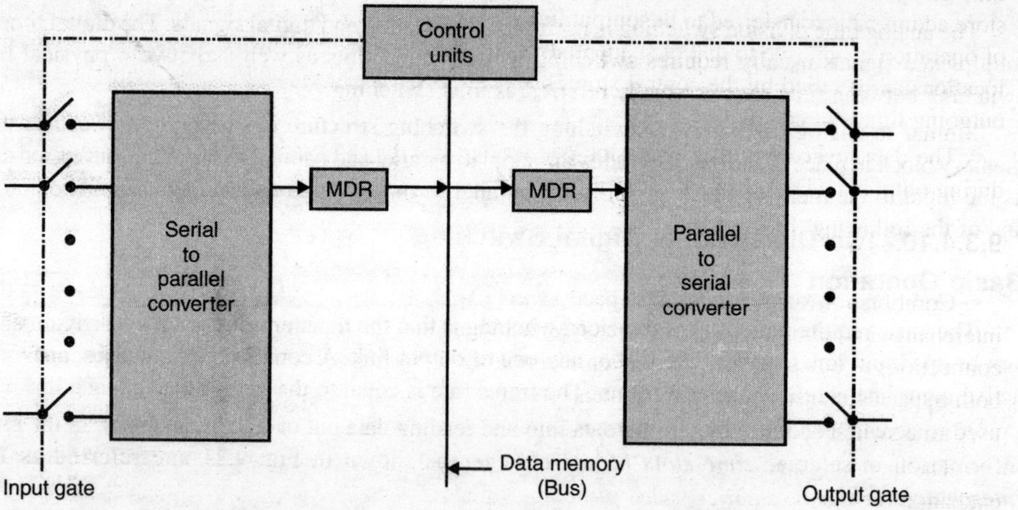

Fig. 9.22. Functional Diagram of Time Division Time Switching.

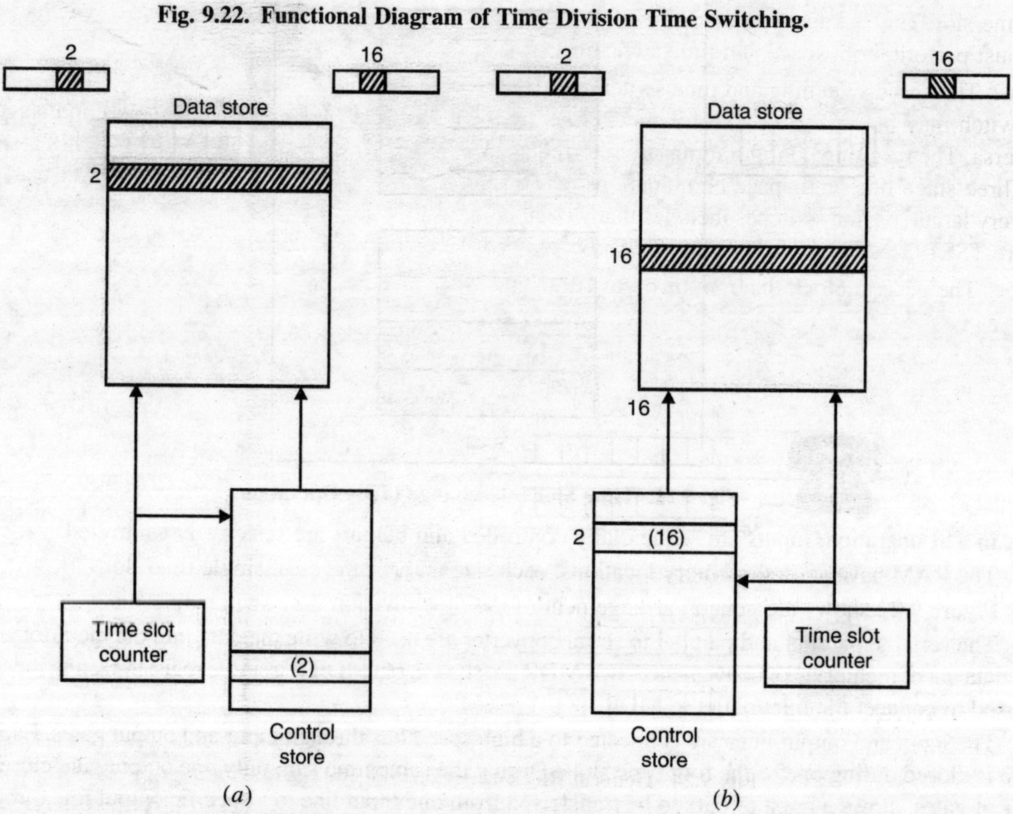

Fig. 9.23. (*a*) Sequential Write/Random Read and (*b*) Random/Sequential Read.

Figure 9.23 (*a*) implies that specific memory locations are dedicated to respective channels of the incoming TDM link. Data are stored in sequential locations in memory by incrementing modulo N counter with every time slot. Thus incoming data during time slot 2 is stored in the second location within the memory. On output, information retrieved from the control store specifies which address is to be accessed for that particular time slot. Thus sixteenth word of control store contains the number 2, implying that the contents of data store address 2 is transferred to the output link during outgoing slot 16. Random write/sequential read mode of operation is opposite to that of sequential write/ random read. Incoming data are written into the memory locations as specified by the control store, but outgoing data are retrieved sequentially under control of an outgoing time slot counter.

The data received during time slot 2 is written directly into data store address 16 and it is retrieved during outgoing TDM channel number 16.

9.3.4.10. Two Dimensional Digital Switching

Combination of the time and space switches leads to a configuration that achieved both time slot interchange and sample switching across trunks. These structures also permit a large number of simultaneous connections to be supported for a given technology. Large digital switches require switching operations in both a space dimension and a time dimension. There are a large variety of network configurations that can be used to accomplish these requirements.

The incoming and outgoing PCM highways are spatially separate. So the connection of one line of local exchange obviously requires space switching to connect to the channel of outgoing highways. Thus the switching network must be able to receive PCM samples from one time slot and retransmit them in a different time-slot. This is known as time slot interchange, or simply as time switching. Thus the switching network must perform both space and time switching.

The space switching and time switching may be accomplished in many ways. A two stage combination switch may be organized with time switch as first stage and the space switch as the second stage or vice versa. The resulting configurations are referred as time space (TS) or space time (ST) switches respectively. Three stage time and space combinations of TST and STS configurations are more popular and flexible. Very large division switches includes many combinations of time and space switches. Typical configurations are TSST, TSSSST, and TSTSTSTS. These switches support 40000 lines or more economically.

The general block diagram involving time and space switching is shown in Fig. 9.24.

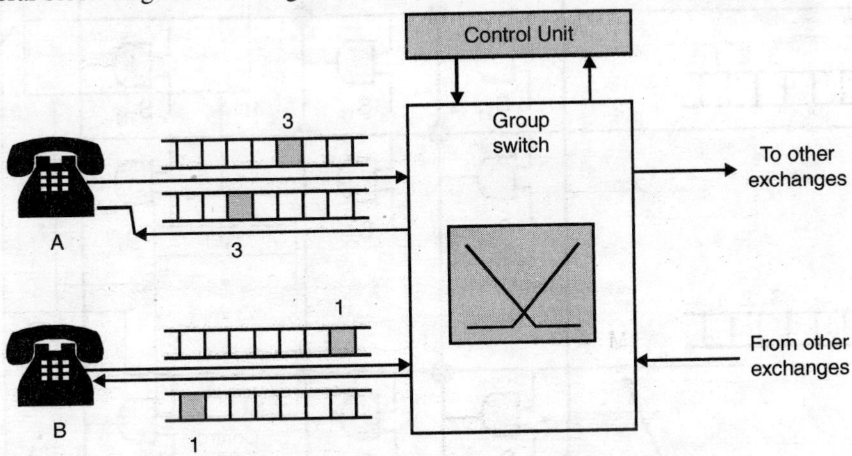

Fig. 9.24. General Block Diagram of Combined Switching.

The main task of the switching part is to interconnect an incoming time slot and an outgoing time slot. The unit responsible for this function is group switch. There are two types of building block in the digital group switch. They are time switch and space switch. In Fig. 9.24, the subscriber makes a local call to B. The control unit has assigned time slot 3 to the call on its way into the group switch, and time slot 1 on its way out of the group switch (to B). This is maintained during the entire call. Similarly B to A also carried out.

The fundamental design and structure of the two switches viz. time switch and space switch are described in the following sections.

9.3.4.11. Space and Time Switches

Space Switch

Figure 9.25 shows a typical space switch. It uses a space array to provide switching generally the space switch consists of a matrix of $M \times N$ switching points where M is number of inlets and N is number of outlets. A connection between an inlet and an outlet is made by the simple logic gates (AND gates). As logic gates are unidirectional, two paths through switching matrix must be established to accommodate a two way conversation. The logic gate array can serve for concentration, expansion or distribution depending on M is larger, smaller or equal to N. Figure 9.25 shows only one voice direction. However, the corresponding components are available for the opposite direction too.

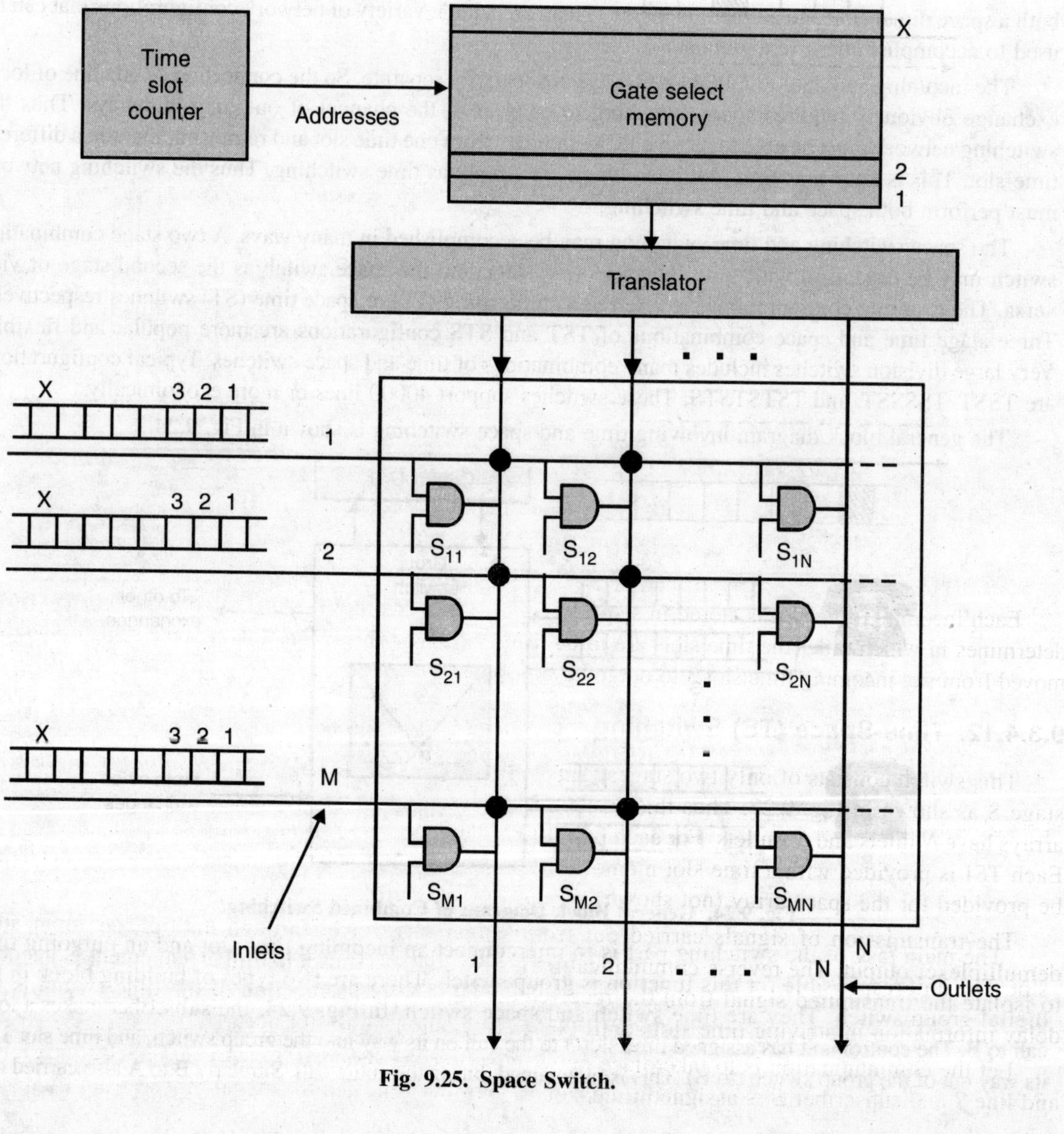

Fig. 9.25. Space Switch.

A number of M, of X slot multiplexers, provide the inputs and the outlets are connected to N, X slot demultiplexers. The gate select memory has X locations. The word containing information about which cross point is to be enabled is decoded by the translator. During each internal time slot, one cross point is activated. In the shift to the next interval time slot, the control memory is incremented by one step, and a new crosspoint pattern is formed in the matrix.

Time Switch

The time-slot interchange (TSI) system is referred to as time switching (T-switching). Figure 9.26 shows the block diagram of time switch.

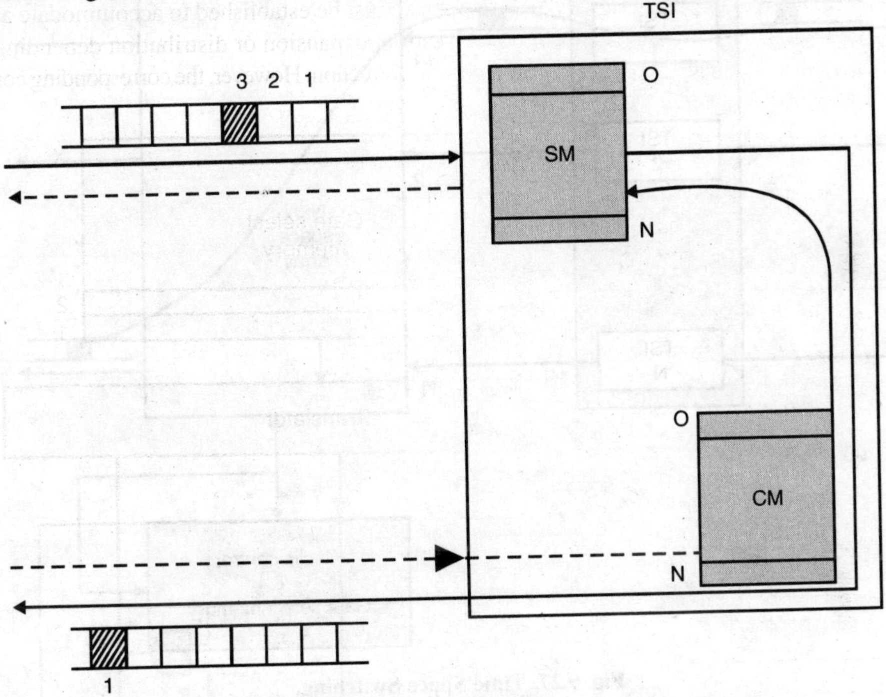

Fig. 9.26. Time Switch.

Each incoming time slot is stored in sequence in a speech memory (SM). The control memory (CM) determines in which order the time slots are to be read from SM. This means that a voice sample may be moved from say incoming time slot 3 to outgoing time slot 1.

9.3.4.12. Time-Space (TS) Switching

This switch consists of only two stages. This structure contains a time stage T followed by a space stage S as shown in Fig. 9.27. Thus this structure is referred to as time-space (TS) switch. The space arrays have N inlets and N outlets. For each inlet line, a time slot interchanger with T slots is introduced. Each TSI is provided with a time slot memories (not shown). Similarly a gate select memory needs to be provided for the space array (not shown).

The transmission of signals carried out from sender to receiver through multiplexer input and demultiplexer output. The reverse communication is also similar. Thus a hybrid arrangement is needed to isolate the transmitted signal from the received signal. The basic function of the time switch is to delay information in arriving time slots until the desired output time slot occurs.

Let the communication is to take place between subscriber A and B. Let A is assigned time slot 2 and line 7 and subscriber B is assigned time slot 16 and line 11. Then the signal moved from time slot

2 to time slot 16 by the time-slot exchanger and is transferred from line 7 to line 11 in the space array. Similarly, the signal originated by B is moved from slot 16 to slot 2 through line 11 to 8.

The cyclic control and gate select memory contains the information needed to specify the space stage configuration for each individual time slot of a frame. The time stage has to provide decays ranging from one time slot to a full frame. During each outgoing time slot, control information is accessed that specifies interstage link number to output link. During other time slots, the space switch is completely reconfigured to support other connections.

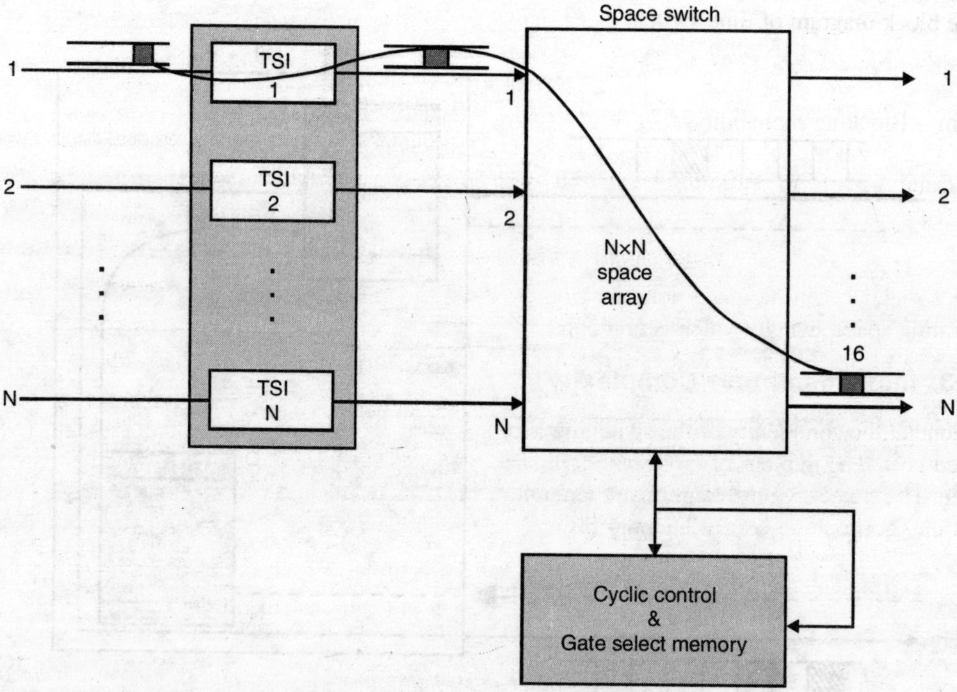

Fig. 9.27. Time Space Switching.

Let each time slot interchanger have T slots. If the space array is an $N \times N$, then the simultaneous connections possible is NT. If $T = 128$ and $N = 16$, 2048 connections can be supported. This structure is not free of blocking. The control store is a parallel end around shift resister. If space array is at the inlet side and time switch is at the output side, the structure is referred as **space time (ST) switching**. Both TS and ST arrangements are equally effective.

TS system is used in DMS 100 digital switching system developed in Canada (1979). It handles 61000 trunks and accommodates 39000 trunks.

Blocking Probability

The blocking probability of TS switching is calculated as follows.

$$\text{The probability that a subscriber } A \text{ is active} = \frac{\rho}{T} \quad \text{...(9.28)}$$

where ρ = fraction of time that a particular link is busy measured in Erlangs

T = number of time slots in a frame.

The probability that any other subscriber is active on the same link

$$= (T-1).\ \frac{\rho}{T} \quad \text{...(9.29)}$$

The probability that a particular called subscriber is chosen by *A*

$$= \frac{1}{NT} \times \frac{1}{T} \quad \text{...(9.30)}$$

where N = Number of inlets (or outlets) for $N \times N$ space array.

NT = Simultaneous connection ...(9.31)

The probability that the same time slot on a different outlet is chosen by the other subscribers on the same inlet

$$= \frac{(T-1)(N-1)\rho}{T(NT-1)} \quad \text{...(9.32)}$$

From Blocking probability, $B = \left\{\frac{\rho}{(T.NT)}\right\}\left\{\frac{(T-1)(N-1)}{T(NT-1)}\right\}$

As T and $N >> 1$, $NT >> 1$

$$B = \frac{P}{NT^3} \quad \text{...(9.33)}$$

The *TS* switch can be made non-blocking by using an expanding time switch (T to T2 slots) and a concentrating space switch (which is complex).

9.3.4.13. Implementation Complexity

In general the complexity of the switching is represented in terms of number of cross points (*N*) and its associated cost. The number of cross points in space stage can be easily calculated which is based on the array size. The time stage uses significant amount of memory which adds the cost of the whole system. To take this into account the cost of memory bit is assumed one hundredth of the cost of cross point. Thus,

$$\text{Implementation complexity} = \frac{N_X + N_B}{100} \quad \text{...(9.34)}$$

where N_X = Number of space stage cross points

N_B = Number of bits of memory.

The N_B not only includes the time stage memory arrays, but also the control memory (store) of the time stage and space stage. Thus,

$$N_B = N_{BX} + N_{BT} \quad \text{...(9.35)}$$

where N_{BX} = Number of memory bits for the space stage control store

= $N \times$ (Number of control words) (number of bits per control word)

N_{BT} = Number of memory bits in the time stage equal to sum of time slot interchange and the control store bits.

= $N \times$ number of channels $\times$ number of bits per channel

+ $N \times$ number of control words $\times$ number of bits per control world.

Example 9.3. If $N = 60$, $NBX = 13, 440$ and $NBT = 24{,}960$ for a typical *TS* switch, calculate the implementation complexity.

Solution.

$$IC = \frac{N_X + (N_{BX} + N_{BT})}{100}$$

$$IC = \frac{60 \times 60 + (13440 + 24960)}{100} = 3600 + 384$$

$$IC = 3984 \text{ equivalent cross points.}$$

As the number of cross points in space array is equal to 3600, the total cost is dominated by the space stage.

9.3.4.14. STS and TST Switching

The TS structure is of blocking nature. Let *A* and *B* are the subscribers using different time slot on the same line want to connect to two subscribers *C* and *D* using same time slot on different lines. *A* and *B* can be moved to the same time slot but during that time slot, the inlet line can be connected to *C*'s line or *D*'s line but not both. This is the significant limitation of the structure. Moreover, time stage switching is generally less expensive than space stage switching as digital memory is much cheaper than digital cross points (AND gates).

The multiple stages overcome the limitations of the individual switches and cost savings can also be achieved. TST, STS, TSST, TSSSST and TSTSTSTSTSTSTS are the switching system configurations used in digital switching system. However, the TST structure is the most common.

STS Switching

In STS switching, the time stage is sandwiched between two space arrays. The digital switching system ITS 4/5 of USA (1976) uses the STS switching configuration.

It handles 3000 trunks and accommodates 1500 Erlangs of traffic. Figure 9.24 shows the space-time-space (S-T-S) switching network for *M* incoming and outgoing PCM highways.

Establishing a path through an STS switch requires finding a time switch array with an available unit's access during the incoming time slot and an available read access during the desired outgoing time slot. The input side space stage as well as the output side space stage is free to utilize any free time switch modules. In the diagram shown in Fig. 9.28, the time slot 2 is connected to the TSM 2 where the time slot allotted is 16 and passed to the $(M-1)^{th}$ line of output space array. Thus the path is provided. This structure is of non-blocking nature.

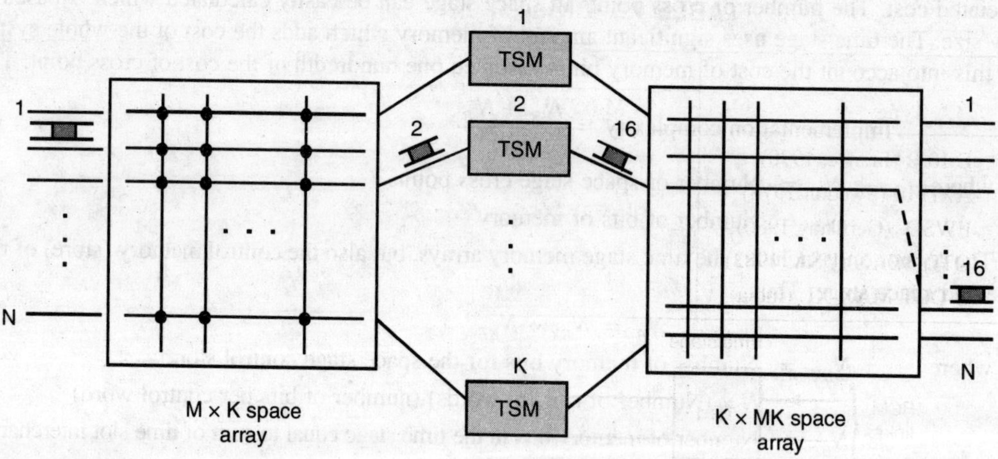

Fig. 9.28. STS Switching Structure.

Blocking Probability

The STS switch is identical to the probability graph of three stage space switches (Fig. 9.12). Similar to that, the blocking probability of an STS switch is

$$B = \left[1 - \left(1 - \frac{p}{\beta}\right)^2\right]^K$$

where p = probability that a link is busy

β = the factor by which the percentage of links that are busy is reduced. ($\beta < 1$)

K = number of center stage TSM.

Implementation Capacity (IC)

While calculating IC, the total number of two space stage cross points, total number of two space stage control bits, number of time stage memory bits and number of time stage control bits are to be considered. Thus,

$$IC = 2\,KN + \frac{2KC\log_2 N + KC(8) + KC\log_2 C}{100} \quad ...(9.36)$$

where K = The minimum number of centre stage TSM to provide desired grade of service, calculated from C = number of channel.

TST Switching

In TST switching the space stage is sandwiched between two time stage switches. Of all the multistage switching, TST is a popular one. Popular digital switching systems using TST are tabulated in Table 9.2.

Some important features of TST switches are :

(*i*) **Low blocking probability :** An incoming channel time slot may be connected to an outgoing channel time slot using any possible space array time slot. Thus there are many alternative paths between two subscribers. This concept reduces the blocking probability of a three stage combination switch.

(*ii*) **Stage independency :** The space stage operates in a time-divided fashion, independently of the external TDM links. The number of space stage time slots L does not coincide with the number of external TDM time slots T.

TABLE 9.2. Digital switching systems using TST.

Type	Characteristics	
	Max. no. of trunks	Traffic (Erlangs)
E 10 B (France, 1970)	3600	1600
AXE 10 (Swedon 1978)	65000	30000
EWSD (Germany 1980)	60000	30000
GTD-SEAX (USA, 1982)	49000	36000
C-DOT MAX-XL (India)	40000	47000

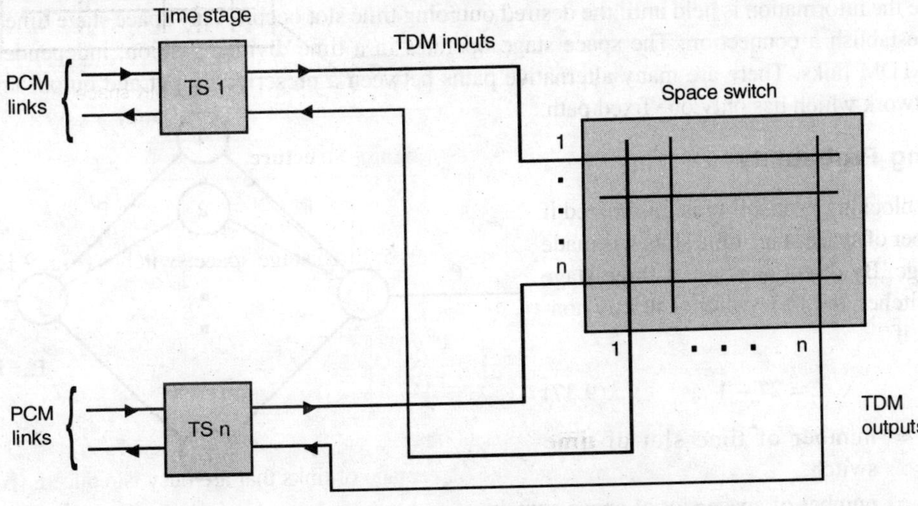

Fig. 9.29. Principle of TST Switching.

(*iii*) **Implementation advantage :** The factors to be considered for switching design and implementation are traffic loads, modularity, testability, expandability and simple control requirements. For large switches with heavy traffic loads, the TST have good implementation advantage.

(*iv*) **More cost effective :** If the input channel loading is high, the time expansion of TST and space expansion of STS are required. Time expansion of TST can be achieved at less cost than space expansion of STS.

In comparison with STS, the TST have certain limitations. For small switches, the STS architectures are less complex to implement than TST. The control requirements of STS are simpler than TST.

The principle of operation of TST switching is shown in Fig. 9.29. In figure, two flows of time slots, one for each direction are connected together.

The functional block diagram which explains the transfer of signals from inlet to outlet is shown in Fig. 9.30. The information arriving at the incoming link of TDM channel is delayed in the· inlet times stage until an appropriate path through the space stage is available. Then the information is transferred through the space stage to the appropriate outlet time stage.

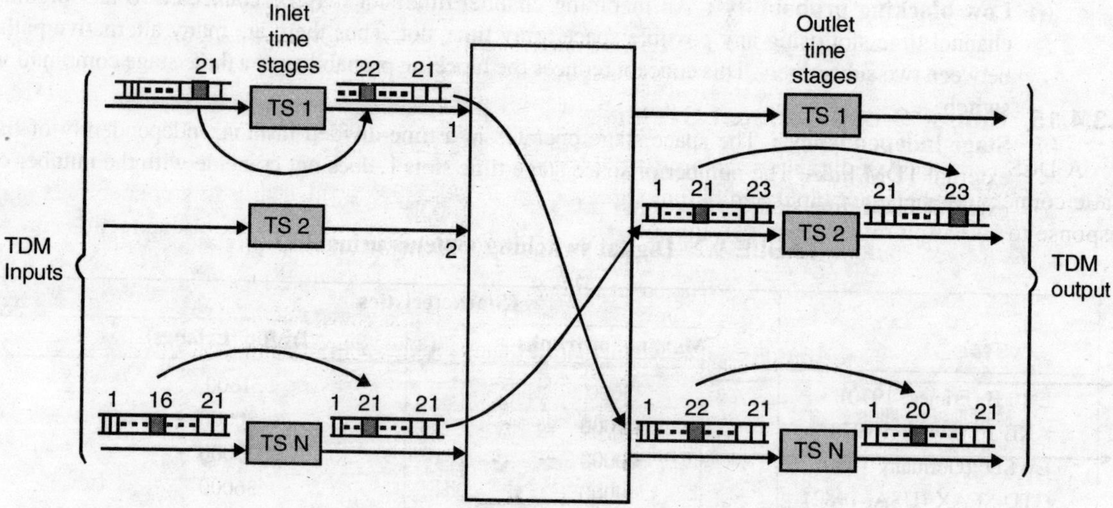

Fig. 9.30. TST Switching Structure.

Here the information is held until the desired outgoing time slot occurs. Any space stage time slot can be used to establish a connection. The space stage operates in a time divided fashion, independently of the external TDM links. There are many alternative paths between a prescribed input and output unlike a two stage network which has only one fixed path.

Blocking Probability

The blocking probability is minimized if the number of space stage time slots L is made to be large. By direct analogy of three stage space switches, the TST switch is strictly non-blocking if

$$L = 2T - 1 \quad \text{...} \qquad \text{...(9.37)}$$

where T = number of time slot of time switch.

L = number of space slot of space switch.

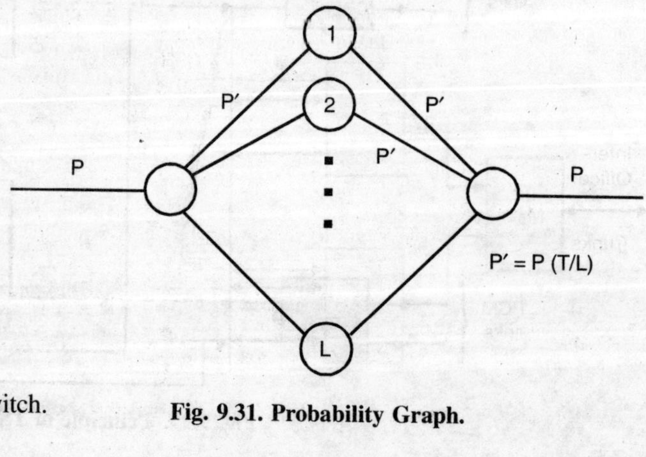

Fig. 9.31. Probability Graph.

The probability graph of TST switch with non-blocking stage is shown in Fig. 9.31.

The general expression of blocking probability for a TST switch with non-blocking individual stage is

For 3-stage

$$B = \left[1-\left(1-\frac{pT}{L}\right)^n\right]^L$$

$$B = \left[1-\left(1-p\left(\frac{T}{2}\right)\right)^2\right]^2 \quad ...(9.38)$$

Implementation Complexity

The implementation complexity (*IC*) of a TST switch can be derived as

$$IC = N^2 + \frac{NL\log_2 N + 2NT(8) + 2NL\log_2 T}{100} \quad ...(9.39)$$

9.3.4.15. Digital Cross-Connect Systems

A DCS is basically a digital switching matrix with an operations interface for setting up relatively static connections between input and output signals or channels. Instead of establishing connections in response to signaling information pertaining to call-by-call connection requests D CS, connections are

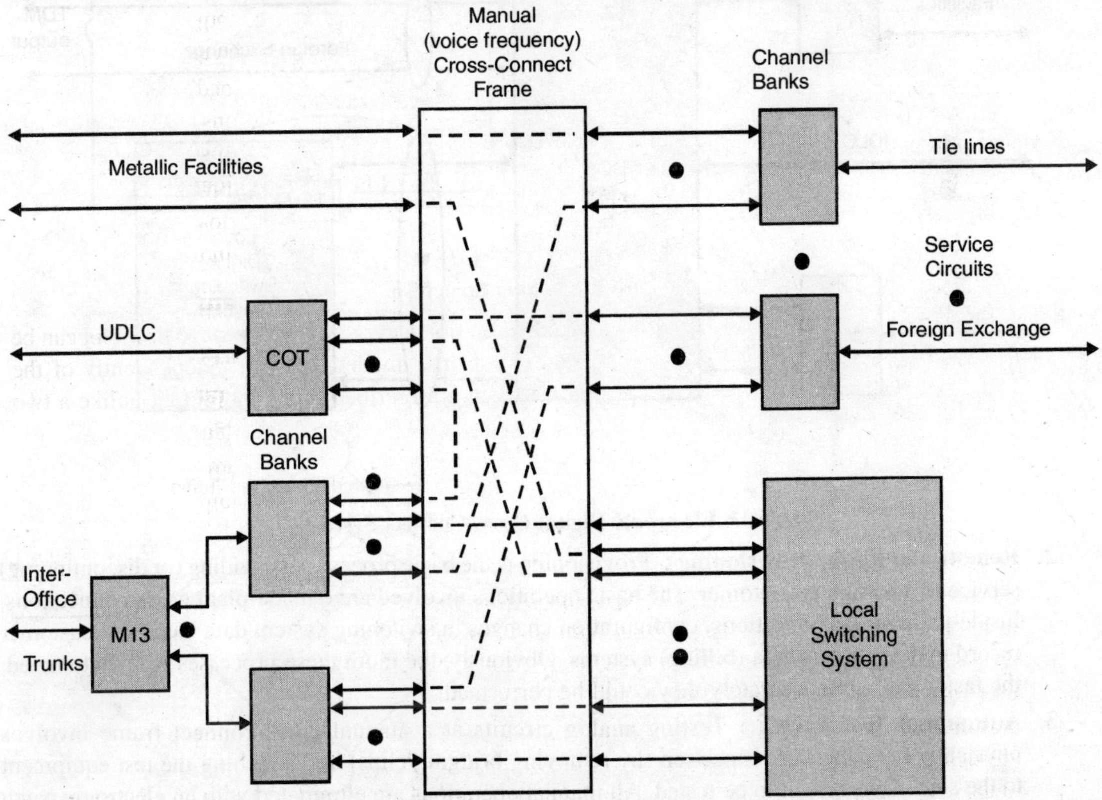

Fig. 9.32. Manual Cross Connect System.

established in response to network configuration needs. The most basic function of a DCS is to act as an electronic patch panel in lieu of a manual cross-connect facility. Manual cross-connect frames were typically installed in switching offices as demarcation points between transmission facilities and switching machines and in wire centers as demarcation points between feeder cables and distribution facilities. In both cases, the major purpose of the cross-connect frame was to allow rearrangement of transmission circuits and access for testing the circuits in both directions.

Figure 9.32 shows a typical manual cross-connect environment while Figure 9.33 shows the same function implemented with a DCS. Manual cross connects are implemented with bridging clips and punch-down wires. Cross-connections within the DCS are established by entering commands at the management console (either locally or remotely).

As indicated in the two figures, a major feature of a DCS is its elimination of back-to-back multiplexing functions when cross-connecting individual channels within TDM transmission links. Additional advantages of a DCS system with respect to manual cross-connect systems are :

1. **Automatic Record Keeping :** Because the cross connects are under processor control, circuit connection reports are readily available through the management interface. In contrast, records in manual systems were inherently error prone and often out of date.

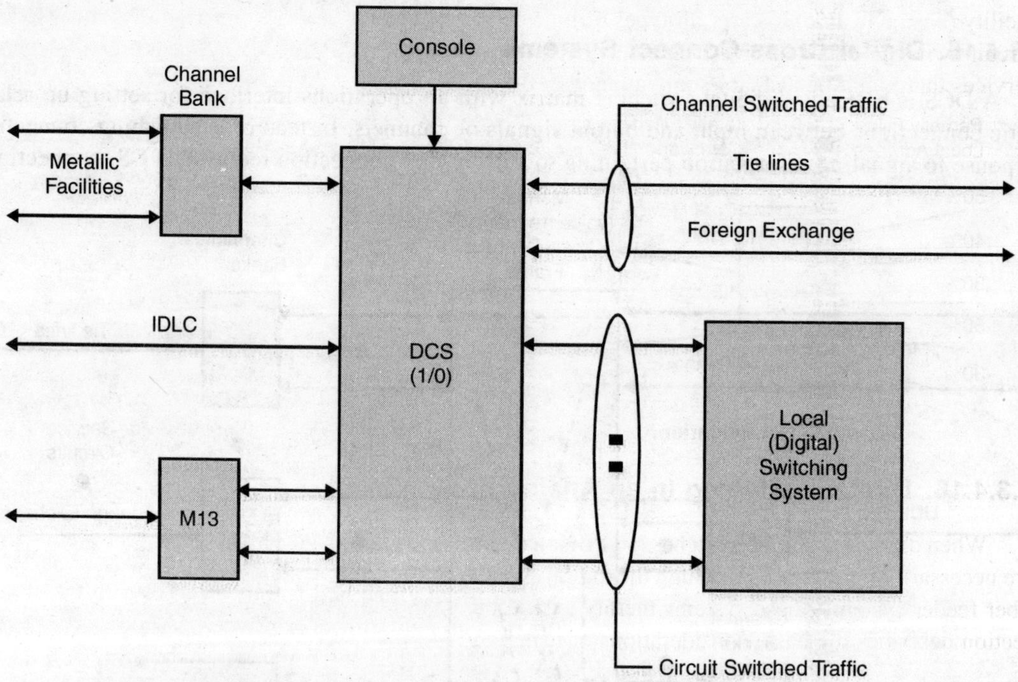

Fig. 9.33. Electronic Digital Cross Connect System.

2. **Remote and Rapid Provisioning :** Provisioning is the basic process of providing (or discontinuing) service to a particular customer. The basic operations involved are outside-plant cross-connections, inside-plant cross connections, configuration changes in switching system data base, and customer record updates in business (billing) systems. Obviously, the more these processes were automated, the faster and more accurately they could be performed.

3. **Automated Test Access :** Testing analog circuits at a manual cross-connect frame involves physically breaking the connection (by removing bridging clips) and attaching the test equipment to the side of the circuit to be tested. All manual operations are eliminated with an electronic patch panel by entering commands at the management console to connect the desired test tones and (DSP) measurement channels to the circuit under test.

Figure 9.33 depicts two types of network traffic: circuit-switched traffic and channel switched traffic. Circuit-switched traffic represents locally switched traffic (DSO circuits typically) and channel switched-traffic refers to leased line equivalents of digital channels. Channel-switched traffic might terminate at another public network switching office as in a foreign exchange (FX) circuit or at a distant customer premise as a tie line of a private network. In the latter case, more than one DSO channel might be concatenated together to form a single higher rate channel referred to as a fractional TI circuit or M × 64-kbps channel. Channel-switched services typically account for over one-half of the transmission bandwidth between U.S. metropolitan offices. The process of separating channel-switched services from circuit-switched services is often referred to as "grooming." Figure 9.28 also shows that universal digital loop carrier (UDLC) becomes integrated digital loop carrier (IDLC) in a digital environment (Figure 9.33).

Consolidation and Segregation

Two basically different DCS grooming functions are depicted in Figures 9.34 and 9.35: consolidation and segregation. When multiple-access lines carrying traffic destined to a common distant node are partially filled, the per-channel costs of transport to the remote node can be reduced by consolidating the traffic. Conversely, when different types of traffic originate at a single location, it is desirable to allow a single facility (*e.g.*, a T1 line) to carry all types of traffic and segregated at the D CS. Examples of such traffic are circuit-switched channels, DDS channels tie lines, multidrop data circuits, F X circuits, or other special services that might be available only at another office.

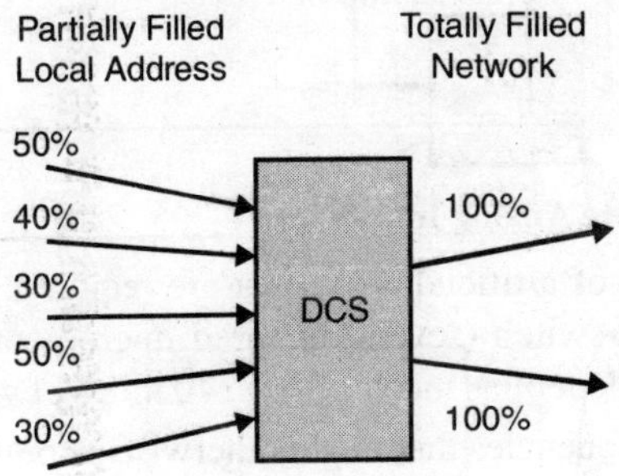

Fig. 9.34. Consolidation.

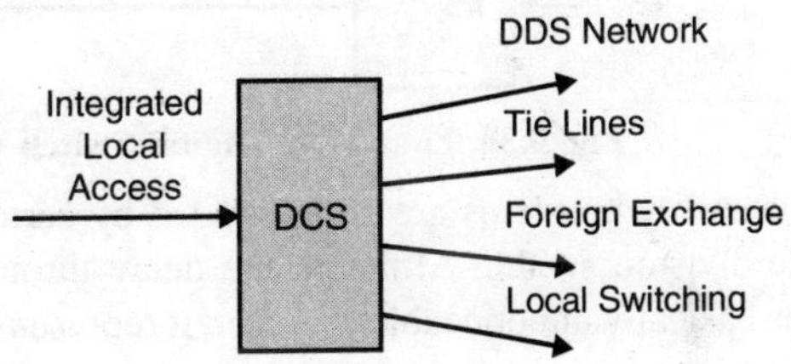

Fig. 9.35. Distribution.

9.3.4.16. Digital Switching in an Analog Environment

When digital end office switches (so PBXs) are installed in an analog environment, the analog interfaces are necessarily unchanged. Although the digital switch may interface with digital subscriber carrier or digital fiber feeder systems, these systems merely extend the analog interface point closer to the subscriber. This section describes the basic considerations of using digital switching in such an analog environment.

Zero-Loss Switching

As already mentioned, a well-designed digital transmission and switching system adds no appreciable degradation to the end-to-end quality of digitized voice. In particular, the derived analog signal coming out of a decoder can be adjusted to the same level as that presented to the encoder at the far end. Curiously, zero-loss transmission presents some significant problems when digital switching is used in the analog transmission environment of a class 5 central office.

Analog end office switches are two-wire switches designed to interconnect bidirectional two-wire customer loops. The voice digitization process, however, inherently requires separation of the go and return signal paths involved in a connection. Thus, a class 5 digital end office or digital PBx must be a four-wire switch. When inserted into a two-wire analog environment, hybrids are required to separate the two directions of transmission. As shown in Figure 9.36, hybrids at each end of internal digital sub network produce a four-wire circuit with the potential for echoes and singing. (Amplifiers are shown in conjunction with the encoders to offset forward path loss inherent in the hybrids.) Instability arises as a result of impedance mismatches at the hybrids causing unwanted coupling from the receive portion to the transmit portion of the four-wire connection. Impedance mismatches occur because of the variability in the lengths and wire sizes of the subscriber loops. In particular, loaded and unloaded wire pairs have markedly different impedance characteristics.

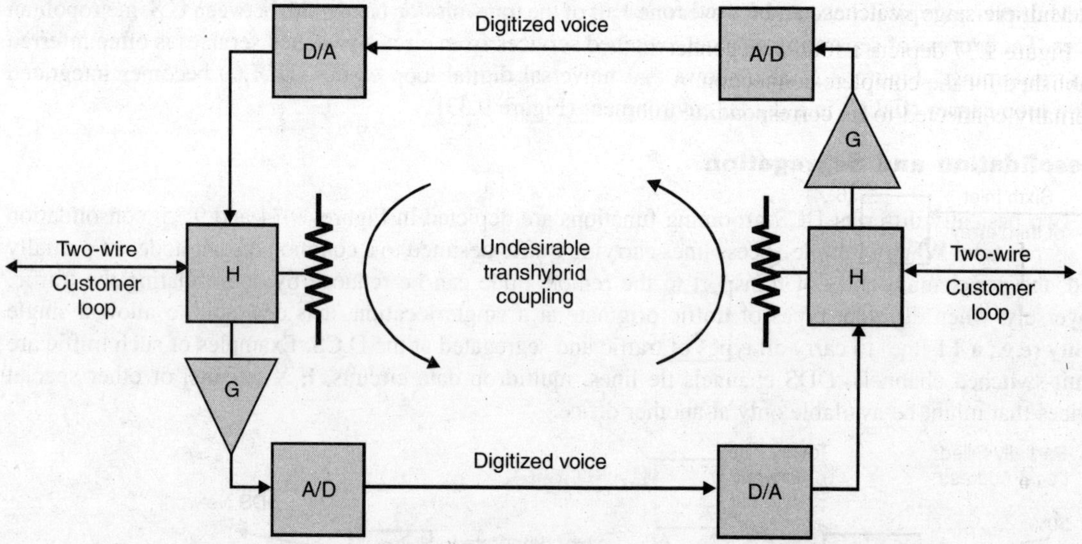

Fig. 9.36. Four-Wire Digital Switch with Two-Wire Analog Interfaces.

The instability problems are compounded by certain amounts of artificial delay that are required in a digital time division switch. Although the delay through a digital switch (several hundred microseconds, typically) is basically unnoticeable to a user, it represents the equivalent of as much as 30 to 40 miles of wire.

This increase delay has the effect of lowering the oscillation frequencies that might otherwise be outside the voice band and effectively be removed by encoder/decoder filters.

Hybrids were traditionally used at the interface to the toll network where the transmission medium is basically four-wire. In these instances, the instability of the four-wire circuit was controlled by designing a prescribed amount of net attenuation (by way of net loss) in to the transmission path of shorter toll network circuits. On the longer circuits, the echoes and singing are eliminated by echo suppressors or echo cancellers. Class 5 digital switches could prevent instability in the same manner: by designing a certain amount of loss into the encoding/decoding process. While the necessary amount of signal attenuation (approximately 2 3 dB in each path) could be tolerated on local connections, the added loss to some toll connections would be unacceptably large.

From the foregoing discussion it can be seen that the use of selectable attenuation is one solution to the instability problems. The necessary loss is inserted in to the talking path for local connections but not used on long-distance calls, which already have loss designed in to them. A second solution involves matching the impedance at the hybrids more closely. Before the advent of DSP nearly complete elimination of the unwanted coupling was prohibitively expensive. However, adequate isolation of the transmission paths could be accomplished with just two different matching networks: one for loaded loops and one for unloaded loops. Impedance matching has been simplified by the use of advanced interface electronics (DSP) that contain trainable and automatic impedance matching circuitry. The impedance matching circuits essentially represent short-delay echo cancellers.

Notice that an all-digital network (with four-wire telephones) avoids instability problems because there are no two-wire analog lines. Voice is digitized at the telephone and uses a separate path from the receive signal all the way to the destination telephone. Thus an end-to-end four-wire circuit completely eliminates echoes and singing, allowing all connections to operate on a zero loss basis.

9.3.4.17. Folded Four-Wire Switches

Multiple-stage switches can be used for either two- or four-wire switching operations.

Figure 9.37 depicts a four-wire connection through a four-stage switch. Notice that two paths must be established for the complete connection. A two-wire connection requires only one path since each outlet is externally connected to its corresponding inlet.

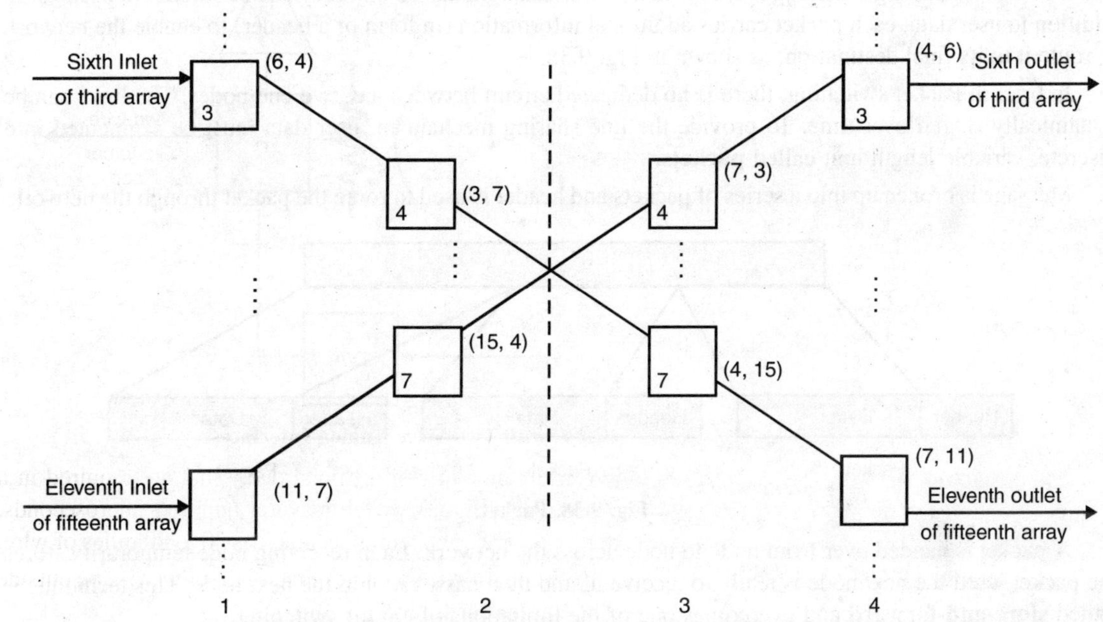

Fig. 9.37. Four Wire Connection Through Four Stage Switch.

The two paths shown in Figure 9.37 demonstrate particularly useful relationship; one path is a mirror image of the other path. If the switch diagram is folded about the vertical center line, the paths coincide. Hence this method of setting up connection is sometimes referred to as a folded operation. When all connections in the switch are set up with a folded relationship, it has several advantages which are as follows:

1. First of all, only one path finding operation is needed since the reverse path is automatically available as a mirror image of the forward path. In essence, every crosspoint on one side is paired with another crosspoint in a corresponding array on the opposite side of the switch. Whenever one crosspoint of a pair is used in a connection, the other crosspoint in the pair is also used.
2. A second advantage of the folded four-wire operation results because the amount of information specifying the status of the switch can be cut in half. Only the status of each pair of crosspoints or associated junction is needed to find an available path through the switch.
3. A third benefit of the folded structure occurs because the blocking probability is one-half of the probability of finding two paths independently. It might seem that pairing the crosspoints in the described manner would restrict the paths available for a particular connection. On the contrary, the crosspoint pairings merely guarantee that a reverse path is automatically available for any selected path in the forward direction.

9.4. PACKET SWITCHING

As we have already discussed about the circuit switched network, in circuit switching, a dedicated channel, called *circuit*, is set up between two stations for the duration of the transmission. The circuit must be established to make a complete the end-to-end connection before any data can be transferred. After some period of data transfer, the connection is terminated.

Packet switching was designed to minimize the shortcomings of circuit switching in dealing with data communication. Unlike circuit switching where communication is continuous along a dedicated circuit, in packet switching, communication is discrete in form of packets. Messages are split at the source host reassembled at the destination host. Each packet is of a limited size and can hold up to certain number of octets of user data. Larger messages are broken into smaller chunks so that they can be fitted into packets. In addition to user data, each packet carries additional information (in form of a header) to enable the network to route it to its final destination, as shown in Fig. 9.38.

In brief in Packet switching, there is no dedicated circuit between any two end nodes. The links can be dynamically shared over time. To provide the line sharing mechanism, user data must be segmented into discrete, variable length unit called **packets.**

Message is broken up into a series of packets and header is used to route the packet through the network.

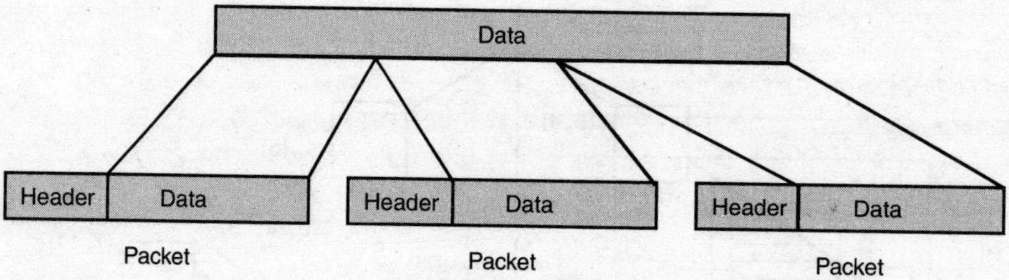

Fig. 9.38. Packet.

A packet is handed over from node to node across the network. Each receiving node temporarily stores the packet, until the next node is ready to receive it, and then passes it onto the next node. This technique is called **store-and-forward** and overcomes one of the limitations of circuit switching.

DO YOU KNOW►

> *In **packet switching**, communication is discrete in form of packets and there is no dedicated circuit between any two end nodes.*

A packet-switched network has a much higher capacity for accepting further connections. Additional connections are usually not blocked but simply slow down existing connections, because they increase the overall number of packets in the network and hence increase the delivery time of each packet.

Figure 9.39 shows a simple packet switch with six I/O channels (*a* through *f*).

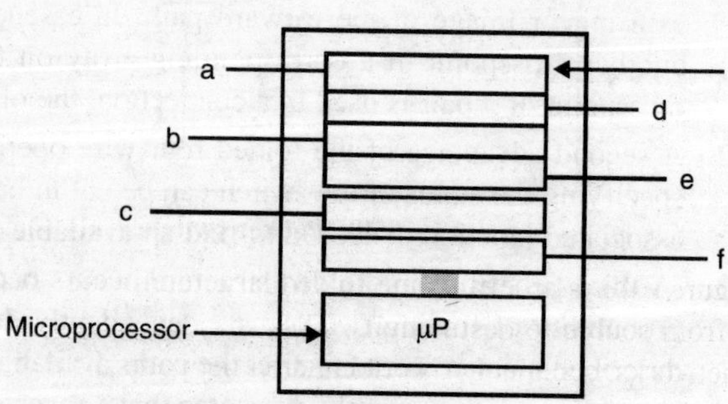

Fig. 9.39. A simple packet switch with six I/O channels.

Each channel has an associated buffer which it uses to store packets in transit. The operation of the switch is controlled by a microprocessor. A packet received on any of the channels can be passed onto any of the other channels by the microprocessor moving it to the corresponding buffer.

Packet switching uses two approaches :

1. **The Virtual Circuit** approach (often used in data link layer)
2. **The Datagram** approach (often used in network layer)

9.4.1. The Virtual Circuit

The virtual circuit method (also known as **connection-oriented**) is closer to circuit switching. Here a complete route is worked out prior to sending data packets.

The route is established by sending a connection request packet along the route to the intended destination. This packet informs the intermediate nodes about the connection and the established route so that they will know how to route subsequent packets. The result is a circuit somewhat similar to those in circuit switching, except that it uses packets as its basic unit of communication. Hence it is called a virtual circuit.

A virtual circuit network is normally implemented in the data link layer, while a circuit switched network is implemented in the physical layer and a datagram network in the network layer.

DO YOU KNOW▸

A virtual circuit network is normally implemented in the data link layer, while a circuit switched network is implemented in the physical layer and a datagram network in the network layer.

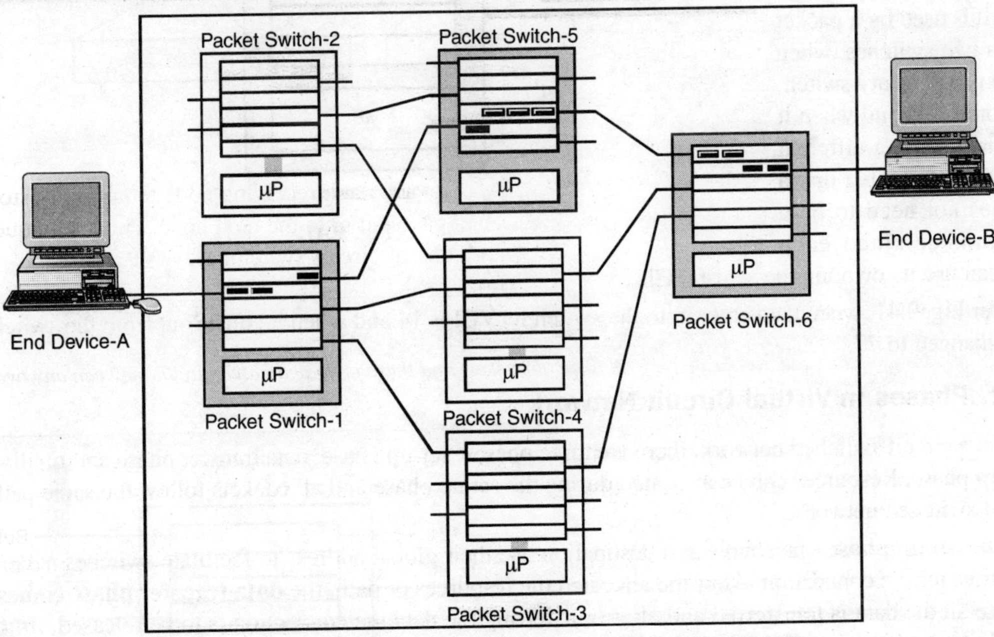

Fig. 9.40. Packet switched network using virtual circuit.

Figure 9.40 is an example of a virtual circuit wide area network. The network has switches that allow traffic from source to destinations. They may be a computer, router bridges, or any other device that links other networks, such as LANs (local area networks), to the switched WAN.

All communications across a data network are carries out over previously established virtual circuits. These can be set up either on demand by user or permanently by network management. With on-demand connections, the user device sends a request for a *switched virtual connection (SVC)* to be set up to a central control unit known as the *Signaling Control Point (SCP).*

The SCP then determines the availability of both the required destination and the transmission bandwidth appropriate to call across the network.

With *Permanent Virtual Circuits* (PVC), the route is already determined when demand is made. The PVC is a logical predefined path or link through a carrier network.

To understand the operation of virtual circuit network, we should discuss some important terms used in this mechanism.

9.4.1.1. Addressing

There are two types of addressing involves in virtual circuit networks, namely:

(*i*) Global addressing

(*ii*) Local addressing (Virtual circuit identifier)

Global addressing means an address that can be unique in the scope of network, a source or destination required to have global address. However, we will see that global addressing in virtual circuit networks is used only to create a virtual circuit identifier.

Local addressing or Virtual circuit identifier (VCI) means the identifier that is actually used for data transfer from switch to switch. It is used by a packet between two switches, when a packet is arrived at a switch; it has some VCI and when it leaves then it has a different VCI. It is to be noted that a VCI does not need to be a large number since each switch can use its own unique set of VCIs.

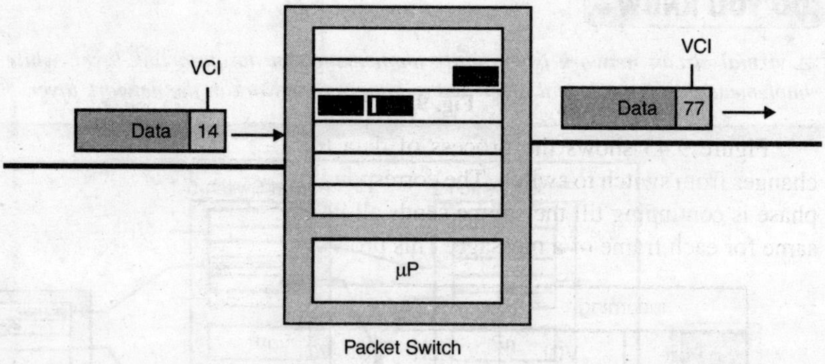

Fig. 9.41.

Refer Fig. 9.41, when data enter in to the switch its VCI is 14 and when it comes out from the switch its VCI is changed to 77.

9.4.1.2. Phases in Virtual Circuit Network

As in a circuit switched network, there are three phases; set-up phase, data transfer phase and finally the tear down phase. Resources can be allocated during the set up phase and all packets follow the same path as in circuit switched network.

In **the set up phase,** the source and destination use their global address to facilitate switches make the table entries for the connection. After the allocated the resources or path, the **data transfer phase** comes. In this phase all the data is transferred and after transferring the data the connection is to be released. In **tear down phase**, the source and destination informs the switches to delete the corresponding entry.

9.4.1.3. Data Transfer Phase

In virtual circuit network, when data has to be transferred from source to the destination, all switches required to have a table entry for this virtual circuit. The table has four columns as shown in Fig. 9.42. The switch holds four pieces of information for each virtual circuit that is already set up. All switches must have a table with entries for all virtual circuits.

Let us take one example, shown in Fig. 9.41, a frame arriving at port-1 with a VCI of 24, when the frame arrives, the switch looks in its table to find port-1, and VCI-24. When it is located, the switch knows that VCI to be changed to 32 and send out the frame from port-2. Another data enter from the same port with a VCI 17 and moves out from port-3 with the changed VCI 55.

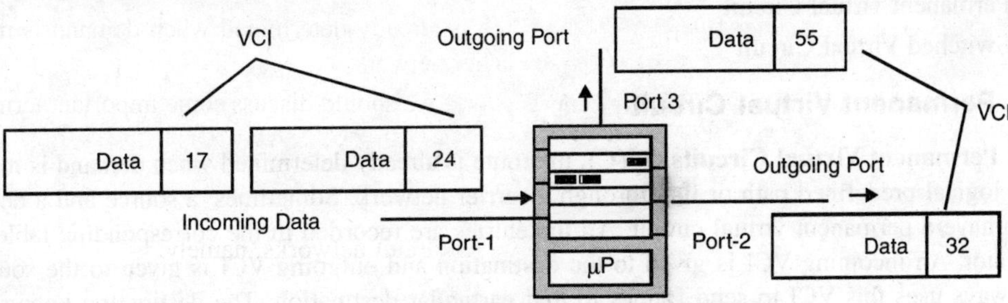

Incoming		Outgoing	
Port	VCI	Port	VCI
1	24	2	32
1	17	3	55

Fig. 9.42. Switch and Table in virtual circuit.

Figure 9.43 shows the process of data transfer from source, A, to destination, B, and how its VCI changes from switch to switch. The corresponding table's entries are shown for each switch. The data transfer phase is continuing till the source sends all its frames to the destination. The activities at the switch are the same for each frame of a message. This process creates a virtual circuit between the source and destination.

Incoming		Outgoing	
Port-	VCI	Port-	VCI
1	24	2	32

Incoming		Outgoing	
Port-	VCI	Port-	VCI
2	55	3	17

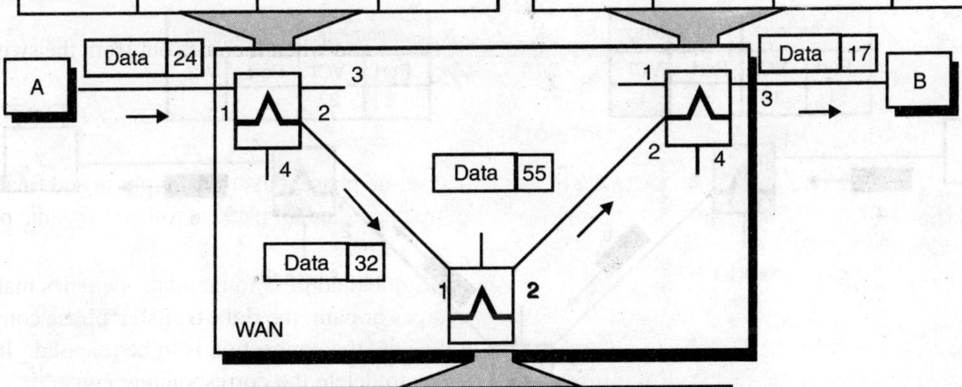

Incoming		Outgoing	
Port-	VCI	Port-	VCI
1	32	2	55

Fig. 9.43. Source to destination data transfer in virtual circuit.

9.4.2. Types of Virtual Circuit

During the set up phase a switch create an entry for the virtual circuit. Now there are two approaches to create the virtual circuits.

1. Permanent virtual Circuit
2. Switched Virtual Circuit

9.4.2.1. Permanent Virtual Circuit

With **Permanent Virtual Circuits (PVC),** the route is already determined when demand is made. The PVC is a logical predefined path or link through a carrier network. Sometimes, a source and a destination choose to have a permanent virtual circuit. All the entries are recorded in the corresponding tables by the administrator. An incoming VCI is given to the destination and outgoing VCI is given to the source. The source always uses this VCI to send frames to that particular destination. The destination knows that the frame is coming from that particular source if the frame carries corresponding incoming VCI. Permanent virtual circuit is used in a leased telephone line. In case of leased telephone line, it is not required to dial the number, user can just pick up the receiver and talk to another user.

9.4.2.2. Switched Virtual Circuit

There are few disadvantages of PVC network. Such as

(*i*) It is costly because both the users pay for the connection all the time even when it is not in use.

(*ii*) The other major limitation is that the connection is created from one source to the one single destination. If it is require to have many destinations than it needs PVC for each connection.

Another approach is the **switched virtual circuits** where the circuit creates a temporary connection that exists only when data are being transferred between source and destination. It requires set up phase or connection phase.

Let us suppose X needs to create a virtual circuit to Y. We need to take two steps, the set up request and the acknowledgement.

The Set-up Request

A set-up request frame is sent from the source, X, to the destination, Y, shown in the Fig. 9.44.

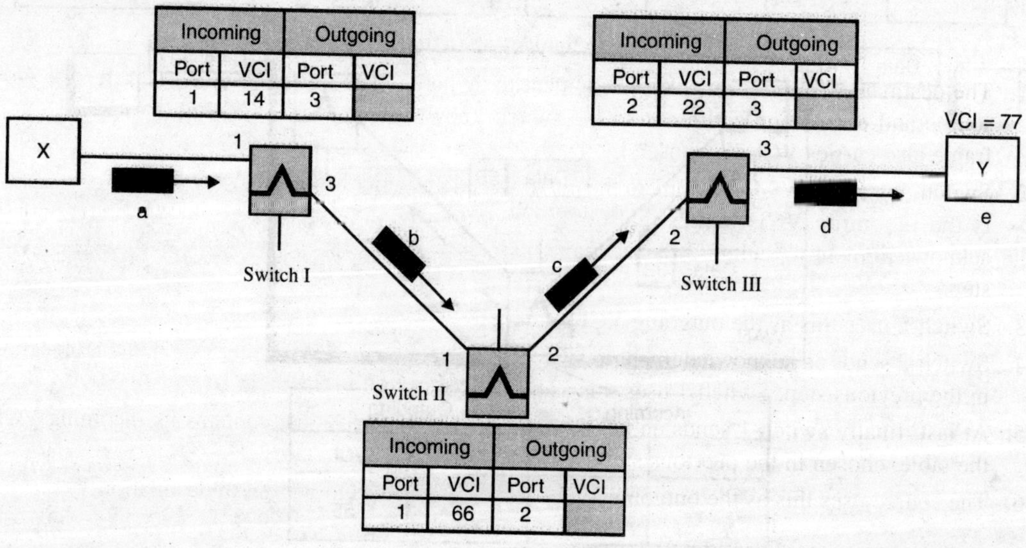

Fig. 9.44. A set-up request in virtual circuit network.

1. Source X sends a setup frame to switch-1.
2. Switch-1 receives the setup request frame. This is known that a frame moving from X to Y goes out through port-3.
3. The switch creates an entry in its table for this virtual circuit. But it is only capable to fill only three columns, one column has been left blank.
4. The switch-1 assigns the incoming port-1 and chooses an available incoming VCI-14 and the outgoing port-3. This switch does not yet know the outgoing VCI, which will be found during the acknowledgment step. Now the switch forwards the frame through port-3 to switch-2.
5. Switch-2 gets the request frame. Now in switch-2, all three columns of the table are filled; in this case, incoming port-1, incoming VCI-66, and outgoing port-2.

Acknowledgement

The acknowledgment frame filled the entries which are unfilled during setup phase, refer Fig. 9.45.

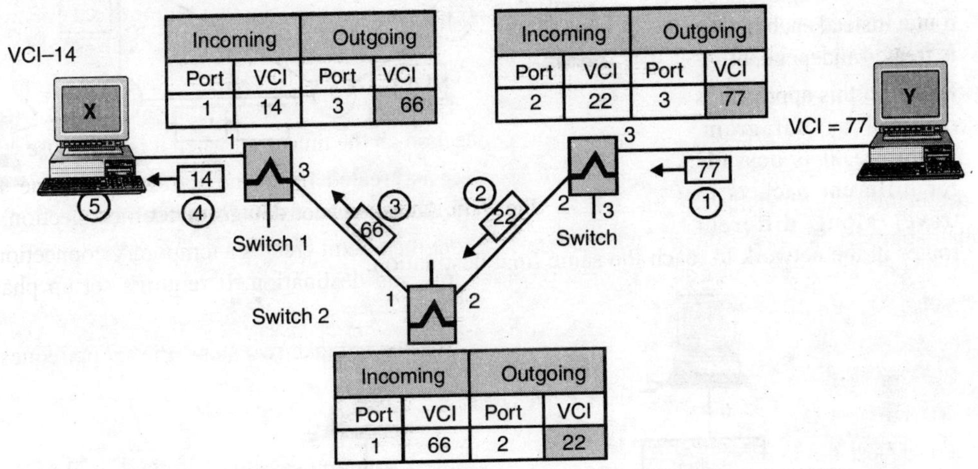

Fig. 9.45. The acknowledgment in virtual circuit network.

1. The destination Y sends an acknowledgment to switch-3. The acknowledgment carries the global source and destination addresses so the switch knows which entry in the table to be filled. The frame also carries VCI-77, chosen by the destination as the incoming VCI for frames from X.
2. Switch-3 uses this VCI to complete the outgoing VCI column for this entry. It is to be noted that 77 is the incoming VCI for destination Y, but outgoing VCI for switch-3. Switch-3 sends an acknowledgment to switch-2 that contains its incoming VCI in the table, chosen in the previous step.
3. Switch-2 uses this as the outgoing VCI in the table.
4. Switch-2 sends an acknowledgment to switch -1 that contains its incoming VCI in the table, chosen in the previous step. Switch-1 uses this as the outgoing VCI in the table.
5. At last, finally switch-1 sends an acknowledgment to source X that contains its incoming VCI in the table, chosen in the previous step.
6. The source uses this as the outgoing VCI for the data frames to be sent to destination Y.

Teardowns Phase

When all the frames from the source X are send to the destination Y, sends a special frame referred as teardown request. Destination Y responds with a teardown confirmation frame. All switches erase the corresponding entry from their tables.

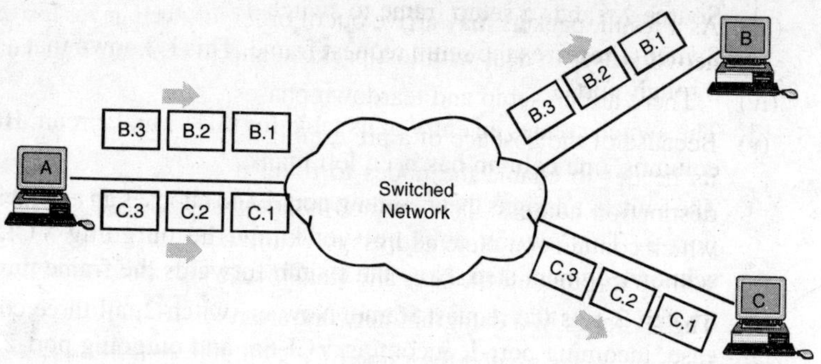

9.4.3. The Datagram

(*i*) The **datagram** method (also known as **connectionless**) does not rely on a pre established route; instead each packet is treated independently.

(*ii*) Packet in this approach is referred as datagram. Therefore, it is possible for different packets to travel along different routes in the network to reach the same final destination.

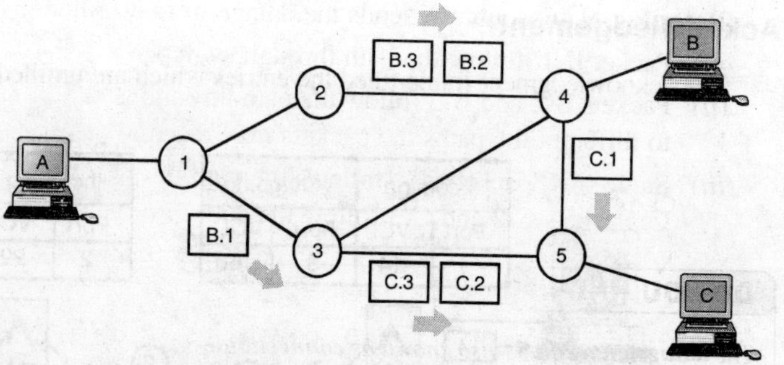

Fig. 9.46. An example of datagram network.

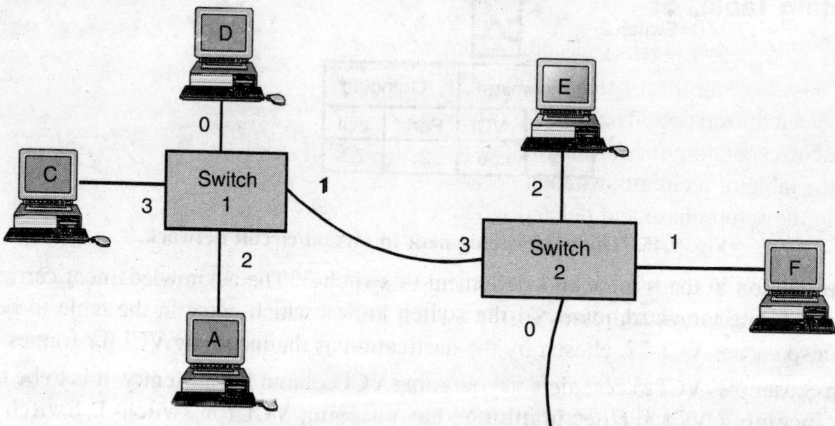

Fig. 9.47. An example of Datagram network.

(*iii*) As a result, packets may arrive out of order, or even never arrive (due to node failure). It is up to the network user to deal with lost packets, and to rearrange packets to their original order.

(*iv*) There are no setup and teardown phases.

(*v*) Because of the absence of a pre established circuit, each packet must carry enough information in its header to enable the nodes to route it correctly. That is, every packet contains the complete destination address. This method is analogies to mail (inland letter/ post card / registered letter) which contain complete address to whom it sends. As we drop the mail to the post office, the letter will go through all the way and finally reaches to the addressee. No matter which path it follows.

(*vi*) The switches in a datagram network are referred as **routers.**

(*vii*) Each packet is treated the same by a switch regard less of its source and destination.

Refer Fig. 9.46, source A sends the datagram to destinations B and C.

(*i*) Packet B-1, follow the path through switches 1, 3 and 4 and reaches to destination B.

(*ii*) Packets B-2 and B-3 follow the path through switches 1, 2 and 4; reaches to the destination B. Due to different the paths the packets may reach to destination B, in out of order.

(*iii*) Same way, packets, C-1follows the path through switches 1, 2, 4 and 5; reaches to the destination C. Packets C-2 and C-3 follow the path through switches 1, 3 and 5.

DO YOU KNOW▶

*The **datagram** method (also known as **connectionless**) does not rely on a pre established route; instead each packet is treated independently.*

9.4.3.1. Routing Table

TABLE 9.3. Forwarding table for switch-2

Destination Address	Output Port
A	3
B	0
C	3
D	3
E	2
F	1
G	0
H	0

In this type of switched network, each switch (router) has a routing table which is based on the destination address. The routing tables are dynamic and updated time to time. The **routing tables** record the destination address and the corresponding forwarding output ports. This table is quite different from the table of a circuit switched network in which each entry is created during the setup phase and deleted when the teardown phase is over.

Destination address in the header of a packet in a datagram network remains the same during the entire journey from source to destination of the packet.

Consider the example network shown in Fig. 9.47, in which hosts have addresses, *A, B, C, D, E, F, G* and *H.* To decide how to forward a packet, a switch consults a forwarding table or may be refer as routing table, an example of which is given in Table 9.3. This table shows the forwarding information that switch- 2 need to forward datagram's in the example. It is quite easy to make such table when you a complete map of simple network as shown in the example; we could imagine a network operator configuring the table statically. It is difficult to create the forwarding tables in large, complex networks with dynamically changing topologies and multiple paths between destinations. We use routing, it is a process that takes place in the background so that, when a data packet turns up, we will have the right information in the forwarding table to be able to forward or switch, the packet.

9.4.3.2. Advantages of the Datagram

1. The advantage of the datagram approach is that because there is no circuit, congestion and faulty nodes can be avoided by choosing a different route.

2. Also, connections can be established more quickly because of reduced overheads. This makes datagram's better suited than virtual circuits for brief connections. For example, database transactions in banking systems are of this nature, where each transaction involves only a few packets.

9.4.3.3. Advantages of the Virtual Circuit

1. The advantage of the virtual circuit approach is that because no separate routing is required for each packet, they are likely to reach their destination more quickly; this leads to improved throughput.
2. Packets always arrive in order.
3. Virtual circuits are better suited to long connections that involve the transfer of large amounts of data.

9.4.3.4. Advantages of Packet Switching

There are few important benefits from packet switching.

1. The first and most important benefit is that since packets are short, the communication links between the nodes are only allocated to transferring a single message for a short period of time while transmitting each packet.
2. Longer messages require a series of packets to be sent, but do not require the link to be dedicated between the transmission of each packet. The implication is that packets belonging to other messages may be sent between the packets of the message being sent. This provides a much fairer sharing of the resources of each of the links.
3. Another benefit of packet switching is known as pipelining.
4. The simultaneous use of communications link represents a gain in efficiency; the total delay for transmission across a packet network may be considerably less than for message switching, despite the inclusion of a header in each packet rather than in each message.

9.5. MESSAGE SWITCHING

There is an alternative switching technique known as message switching. There is no pre-established physical path between the source and the destination. In message switching type of switched networks uses

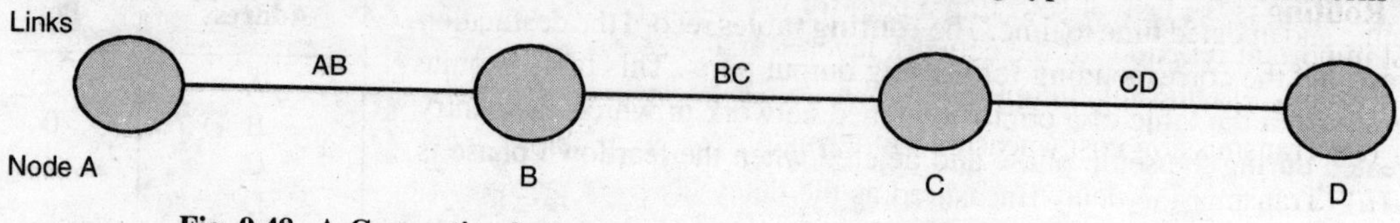

Fig. 9.48. A Connection between two systems A and D formed from 3 Links.

a technique which is known as **store-and-forward.** Whenever the sender has blocks of data to be transmitted, it is stored in the first switching office (or router); each block is received in it's entirely inspected for errors and then forwarded to the next switching office. The network uses this technique is referred as store and forward network.

Sometimes there is no need for a circuit to be established all the way from the source to the destination. Consider a connection between the users (A and D) in the Fig. 9.48. The connection is represented by a series of links (AB, BC, and CD).

When a message (telex or email) is sent from A to D, it first passes over a local connection (AB). It is then passed at some later time to C (via link BC), and from there to

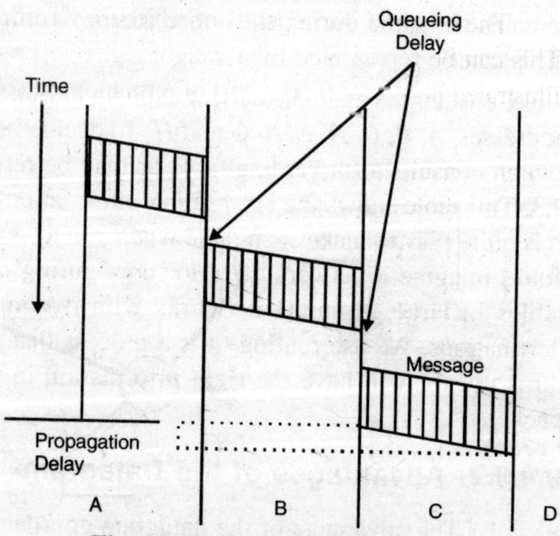

Fig. 9.49. Message switching.

the destination (via link CD). At each message switch, the received message is stored, and a connection is subsequently made to deliver the message to the neighboring message switch. The Fig. 9.49 illustrates message switching. The transmission of only one message is illustrated for simplicity. A complete message is sent from node A to node B when the link interconnecting them becomes available. Since the message may be competing with other messages for access to facilities, a queuing delay may be incurred while waiting for the link to become available.

The use of message switching to communicate between A and D (see Figure 9.49).

The message is stored at B until the next link becomes available, with another queuing delay before it can be forwarded. It repeats until reaches its destination. The Circuit setup delays are replaced by queuing delays. Considerable extra delay may result from storage at individual nodes.

This type of network is not so popular nowadays because there is lot of delay at each switch.

9.6. COMPARISONS OF THREE SWITCHING METHODS

Message Switching	Circuit Switching	Packet Switching
(*i*) Long queuing delays and message can take a long time to transmit (*ii*) A long message may block system for other messages	(*i*) Long call setup times (*ii*) The setup times may be negligible compared to data length (*iii*) Inefficient channel utilisation for burst traffic	(*i*) Break message into packets. (*ii*) Can interpolate packets from other nodes, therefore cannot block system (*iii*) Efficiently handles asymmetric traffic.

9.7. ROUTING

Routing is the task of selecting a path for the transport of packets across the network, and is one of the most important functions of the network layer. Routing is generally viewed as an optimization problem with the objective of choosing an optimal path according to certain criteria :

(*i*) Transmission cost (measured in terms of tied up network resources).

(*ii*) Transmission delay (measured as the delay involved in delivering each packet).

(*iii*) Throughput (measured as the total number of packets delivered per unit of time).

The overall cost depends on all these three, and an optimal route is one that minimizes the overall cost. This can be represented by a weighted network, where an abstract cost figure is associated with each link, as illustrated in Fig. 9.50. The cost of a route is the sum of the cost of its links.

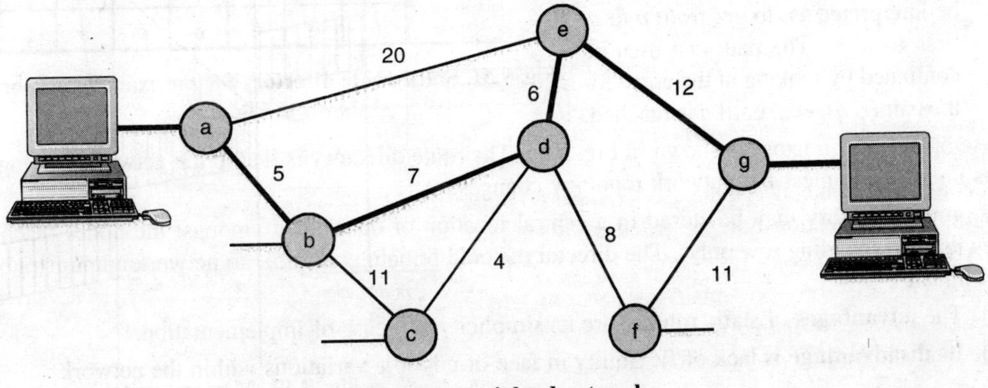

Fig. 9.50. A weighted network.

There are three classes of routing algorithms:

(*i*) Flooding

(*ii*) Static Routing

(*iii*) Dynamic Routing

Of these, only dynamic routing makes any serious optimization attempts.

9.7.1. Flooding

(*i*) In **flooding**, every possible path between the source and the destination station are exercised.

(*ii*) Each node, upon receiving a packet, forwards copies of it to all its neighboring nodes (except the one from which it received the packet).

(*iii*) Flooding is a highly robust technique since it offers the best chance of at least one packet copy reaching the destination.

(*iv*) Its major disadvantage, however, is that it quickly congests the network.

(*v*) To avoid packets being indefinitely copied, each packet is assigned a limited lifetime which when expired will cause it to be destroyed.

(*vi*) Because of its limitations, use of flooding is confined to specialized applications that require very high levels of robustness (*e.g.*, military networks).

(*vii*) Flooding is only suited to the datagram approach.

DO YOU KNOW►

*In **flooding**, every possible path between the source and the destination station are exercised.*

9.7.2. Static Routing

(*i*) In **static routing**, a fixed routing directory is used to guide the selection of a route which remains unchanged for the duration of the connection.

(*ii*) The directory consists of a table which for each node pair (*p*, *q*) in the network suggests a partial path by nominating the first intermediate node, *r*, along the path. This should be interpreted as 'to get from *p* to *q*, first go to *r*'. The path can then be continued by looking at the entry for the pair (*r*, *q*), etc., until *q* is reached.

		To						
		a	**b**	**c**	**d**	**e**	**f**	**g**
From	a	–	b	b	b	b	b	b
	b	a	–	c	d	d	d	d
	c	b	b	–	d	d	d	d
	d	b	b	c	–	e	f	e
	e	d	d	d	d	–	d	g
	f	d	d	d	d	d	–	g
	g	e	e	e	e	e	f	–

Fig. 9.51. Static route directory for the example network shown in Fig. 5.18.

A sample route directory is shown in Fig. 9.51. The route directory is tied to the network topology and remains unchanged unless the network topology changes.

The route directory may be stored in a central location or distributed amongst the nodes (each node requires its corresponding row only). The directory should remain accessible to network administrators for manual updates.

(*i*) The **advantages** of static routing are its simplicity and ease of implementation.

(*ii*) Its **disadvantage** is lack of flexibility in face of possible variations within the network.

(*iii*) Static routing can be used with both the datagram and the virtual circuit approach.

(*iv*) Although static routing takes some form of cost into account, these cost measures are fixed and predetermined.

(*v*) Variations in traffic load have no impact on route selection, which makes this method unsuitable for large networks with significant traffic fluctuations. For small networks with relatively constant traffic, however, static routing can be very effective.

DO YOU KNOW▶

*In **static routing**, a fixed routing directory is used to guide the selection of a route which remains unchanged for the duration of the connection.*

9.7.3. Dynamic Routing

(*i*) **Dynamic routing** attempts to overcome the limitations of static routing by taking network variations into account when selecting a route.

(*ii*) Each node maintains a route directory which describes the cost associated with reaching any other node via a neighboring node. A sample directory is shown in Figure 9.52.

	To						
	a	b	c	d	e	f	g
From d	b	b	c	–	e	f	e
Cost	7	7	4	0	6	8	6

Fig. 9.52. Dynamic route directory for node-d in Fig. 5.19.

(*iii*) The nodes periodically calculate estimates for the costs of the links to their neighboring nodes according to the statistical data that they have collected (queue lengths, packet delays, traffic load, etc.), and share these estimates with other nodes.

(*iv*) This enables the nodes to update their route directories in an objective manner so that they reflect the current state of the network.

(*v*) To select a route between two stations, dynamic routing employs a graph optimization algorithm to find an optimal (or close to optimal) path.

(*vi*) The advantage of dynamic routing is its potential to improve performance and reduce congestion by choosing more optimal routes.

(*vii*) Its disadvantage is its inherent complexity. Nevertheless, dynamic routing algorithms enjoy widespread use because they have proved to be generally more effective than other algorithms.

(*viii*) Like static routing, dynamic routing can be used with both the datagram and the virtual circuit approach.

DO YOU KNOW▶

***Dynamic routing** is the attempts to overcome the limitations of static routing by taking network variations into account when selecting a route.*

9.8. CONGESTION CONTROL

A network has a certain carrying capacity, denoted by the maximum number of packets that it can hold at any point in time. When this limit is approached, considerable delays are experienced in packet delivery, and the network is said to be **congested**. Congestion can occur in all types of networks. Uncontrolled congestion can lead to outright network failure.

At the node level, congestion manifests itself as packet buffers that have approached their full capacity. This happens because the node is receiving more packets than it is passing on to the next nodes, which in turn are presumably unable to receive more packets due to their buffers being full. When this situation occurs, the chances are that it will progressively get worse.

The best way to deal with congestion is to avoid it. This is facilitated by putting in place measures that prevent buffer overflow.

These measures include the following:

1. **Reducing the load on a node by disposing packets :** For example, a node that receives a packet for which it has almost no buffer space may destroy it immediately.
2. **Reducing the traffic destined for a heavily-utilized link :** Nodes can monitor the traffic on their outgoing links and ask the source host to reduce the transmission rate when they feel that a link is approaching its capacity. The request can be put to the source host using a special packet.
3. **Imposing a limit on the total number of packets in the network :** This approach requires some means of keeping a count of the packets in the network.

Furthermore, the nodes will have to communicate to ensure that the count is kept up-to-date. Although, this approach ensures that the network cannot be overloaded with too many packets, it does not prevent an individual node from being overloaded.

9.9. POLLING

Master-slave protocols are used in situations where a primary station (such as a host) controls one or more secondary stations (such as peripheral devices).

Communication between the master and its slaves is governed by a general technique called **polling**. Polling is typically used in a multidrop line configuration where, for example, a set of terminals are connected to a host via a shared communication line.

In its simplest form, polling works as follows. The master can send data in form of a message to any of the slaves. The message contains an address which uniquely identifies the slave for which it is intended. To receive data from the slaves, the master 'polls' each slave in turn by sending a message to it and asking if it has any data to send. The slave responds by either sending data or by sending a rejection message to indicate it has nothing to send.

9.10. HYBRID SWITCH

A hybrid switch integrates the features of store – and – forward (reliable frame transmission) and cut-through (low latency) designs. A hybrid switch can be configured on a per port basis to change automatically from cut-through switching to store-and-forward switching if error rates exceed a user-defined threshold. When error rates fall below this threshold, the switch reverts the cut-through switching. An additional capability is a "runt-free" mode in which the switch discards frames smaller than the mandated 64 bytes minimum size for Ethernet. This ensures the filtering of collisions fragments while maintain the low-latency characteristics of cut-through switching.

Switch Performance

Switches provide high throughput with low or fixed latency. Used in the context, latency is the amount of delay a network introduced when data frames pass through it. Thus, latency is amount of time a frame spends inside a network device such as a bridge or switch. Switch latency is usually measured from the instant the first bit of a frame enters the device to the time this bit leaves the outbound (i.e., destination) port. Depending on a switch's architecture, its latency is usually less than 100 μs. in compression, bridge latency of 400 μs is common, and routers have latency as high as 1500 μs (i.e.,

1.5 ms). Low latency is too high, and then time-sensitive network protocols such as SNA or IPX can time out. Latency also is an issue for time-sensitive application such as full-motion video.

Store-and-forward switches provide both traffic isolation and error immunity to any destination port. The trade-off for error immunity, through, is high latency, which can be in milliseconds rather than microseconds. Since store-and-forward switches wait until an entire frame has arrived before forwarding it, latency, which is dependent on frame size, can range from 61 μs to 1200 μs over a 10-Mbps channel. This could be a serious concern for some applications.

Cut-through switches have extremely low latency, usually on the order of 20 to 40 μs. Furthermore, since cut-through switches begin forwarding frames the moment the destination address is known latency is independent of frame size. Thus, a maximum-size Ethernet/802.3 frame (1518 bytes) has the same latency as a maximum-size Ethernet/802.3 frame (64 bytes). This is not the case with store-and-forward switches, where latency varies directly with frame size. Considering that the total round-trip bit propagation speed of a standard 10-Mbps Ethernet/802.3 network is approximately 51.2 microseconds, the speed of a cut-through switch approximates cable or wire speed, which is 14880 packets per second using 64-bytes packets. The trade-off, through, is there is no opportunity for the switch to check the integrity of a frame before it is forward to its destination port. Thus, if corrupt data messages are contained in the field that follow the destination and source address fields of a data frame, the frame is still forwarded to the destination port. Errors on a specific port are propagated to other ports on the switch or throughout the network. This reduces bandwidth and can delay forwarding error-free frames.

Summary

- **Switching** is the generic method for establishing a path for point-to-point communication in a network.
- **Switches** are devices capable of making temporary link or connection between two or more devices linked to the switch but not directly to each other.
- One of the most important functions of the network layer is to employ the switching capability of the nodes in order to route messages across the network.
- In **circuit switching**, two communicating stations are connected by a *dedicated* communication path which consists of intermediate nodes in the network and the links that connect these nodes.
- *Circuit switching takes place at physical layer.* Data transmission using a PSTN connection is a typical example of a circuit switched data transfer.
- A circuit switched network has three phases which are as follows :
  - (*i*) Setup phase
  - (*ii*) Data transfer phase
  - (*iii*) Tear down phase
- Circuit switching main **disadvantage** is its inflexibility in dealing with computer oriented data.
- **Packet switching** was designed to minimize the shortcomings of circuit switching in dealing with data communication.
- In packet switching, communication is discrete in form of packets.
- In Packet switching, there is no dedicated circuit between any two end nodes.
- A packet is handed over from node to node across the network. Each receiving node temporarily stores the packet, until the next node is ready to receive it, and then passes it onto the next node. This technique is called **store-and-forward.**
- Packet switching uses two approaches :
  1. The Virtual Circuit approach (often used in data link layer)
  2. The Datagram approach (often used in network layer)

- A virtual circuit network is normally implemented in the data link layer, while a circuit switched network is implemented in the physical layer and a datagram network in the network layer.
- With *Permanent Virtual Circuits* (PVC), the route is already determined when demand is made. The PVC is a logical predefined path or link through a carrier network.
- There are two types of addressing involves in virtual circuit networks, namely:
  (*i*) Global addressing
  (*ii*) Local addressing (Virtual circuit identifier)
- **Global addressing** means an address that can be unique in the scope of network,
- **Virtual circuit identifier (VCI)** means the identifier that is actually used for data transfer from switch to switch.
- Now there are two approaches to create the virtual circuits.
  1. Permanent virtual Circuit
  2. Switched Virtual Circuit
- The PVC is a logical predefined path or link through a carrier network.
- The **switched virtual circuits** where the circuit creates a temporary connection that exists only when data are being transferred between source and destination. It requires set up phase or connection phase.
- The **datagram** method (also known as **connectionless**) does not rely on a pre established route; instead each packet is treated independently.
- The switches in a datagram network are referred as **routers.**
- The routing tables are dynamic and updated time to time. The **routing tables** record the destination address and the corresponding forwarding output ports.
- The advantage of the datagram approach is that because there is no circuit, congestion and faulty nodes can be avoided by choosing a different route.
- The advantage of the virtual circuit approach is that because no separate routing is required for each packet, they are likely to reach their destination more quickly; this leads to improved throughput.
- In message switching type of switched networks uses a technique which is known as **store-and-forward.**
- **Routing** is the task of selecting a path for the transport of packets across the network, and is one of the most important functions of the network layer.
- There are three classes of routing algorithms:
  (*i*) Flooding
  (*ii*) Static Routing
  (*iii*) Dynamic Routing
- In **flooding**, every possible path between the source and the destination station are exercised.
- In **static routing**, a fixed routing directory is used to guide the selection of a route which remains unchanged for the duration of the connection.
- Static routing can be used with both the datagram and the virtual circuit approach.
- **Dynamic routing** attempts to overcome the limitations of static routing by taking network variations into account when selecting a route.
- ISDN is the integrating of a wide range of services in to a single multipurpose network.
- ISDN or integrated services digital network is an international standard for end to end digital transmission of voice, data and signaling.
- In North America and Europe, ISDN has a bandwidth of 80 kHz. In Japan, ISDN has a bandwidth of 320 kHz.
- **B channels** are used for carrying user data (digitized voice or computer generated data) at 64 kbps.

- **D channels** are primarily used for common channel signaling purposes.
- **H channels** are used in a high-speed trunk capacity.
- The **basic rate access** package (also called 2B+D) is primarily intended for residential subscribers.
- The **primary rate access** (PRA) package is aimed at business users with greater bandwidth requirements.
- ISDN provides three types of services :
  1. Bearer services 2. Teleservices 3. Supplementary services
- In 1988, the CCITT released a document that described a new set of **broadband ISDN (B-ISDN)** services.
- Master-slave protocols are used in situations where a primary station (such as a host) controls one or more secondary stations (such as peripheral devices).
- Communication between the master and its slaves is governed by a general technique called **polling**.
- **Hybrid switching** is the combination of circuit and packet switching in order to provide the more advantages over these switching.
- According to the International Softswitch Consortium (ISC). **Softswitch** is defined as a software band entity that provides the call control functionality in the IP network.
  The term "softswitch" is almost as generic as call agent, but seems to be winning out over that term because it is more descriptive of the way these systems generally are designed as software running on a standard computing platform.
- An X.25 network transfers data via packet switching. With this method, information is taken from many different users and combined into discrete data packets.
- A *virtual circuit* is a logical connection created to ensure reliable communication between two network devices. A virtual circuit denotes the existence of a logical, bidirectional path from one DTE device to another across an X.25 network.
- *Switched virtual circuits (SVCs)* are temporary connections used for sporadic data transfers.
- *Permanent virtual circuits (PVCs)* are permanently established connections used for frequent and consistent data transfers.
- Early telephone switches were **electromechanical**, many of which are still in use.
- Existing electromechanical switches are of two types: step-by-step and crossbar.
- A **crossbar switch** consists of a matrix of crosspoints between rows of inlets and columns of outlets. Activating the relay of an inlet and the relay of an outlet causes the crosspoint at which they overlap to be activated and therefore result in a physical connection.
- A switching system is called digital when the input to and output from the switching system can directly support digital signal.
- The function of the digital switching network is to connect pairs of channels.
- To achieve this switching, two processes are referred to as Time switching and Space switching.
- The fundamental operation of a switch is to setup and release connection between subscribers.
- It is inefficient to build complete exchanges in single stages. Single stage can only be used to interconnect one particular inlet outlet pair.
- In fact, single stage networks suffer from a number of disadvantages which can be overcome by adopting a multistage network configuration.
- All the switching systems are designed to provide a certain maximum probability of blocking for the busiest hour of the day. It is one of the aspects of the grade of service of the telephone company.
- Two techniques are
  1. Lee graphs 2. Jacobaeus method.

- **Lee graphics :** It was proposed by C.Y. Lee. It is a most versatile and straight forward approach of calculating probabilities with the use of probability graphs.
- **Jacobaeus.** It was presented in 1950 by C. Jacobaeus.
- **Time division switching** involves the sharing of crosspoints for shorter periods of time.
- PAM analog samples or PCM digital samples, occurring at intervals of 125 μs.
- When PAM samples are switched in a time division manner, the switching is known as *analog time division switching.*
- If PCM binary samples are switched, then the switching is known as *digital time division switching*.
- Combination of the time and space switches leads to a configuration that achieved both time slot interchange and sample switching across trunks.
- A DCS is basically a digital switching matrix with an operations interface for setting up relatively static connections between input and output signals or channels.
- When multiple-access lines carrying traffic destined to a common distant node are partially filled, the per-channel costs of transport to the remote node can be reduced by consolidating the traffic.
- Conversely, when different types of traffic originate at a single location, it is desirable to allow a single facility (*e.g.*, a T1 line) to carry all types of traffic and segregated at the DCS. Examples of such traffic are circuit-switched channels, DDS channels tie lines, multidrop data circuits, F X circuits, or other special services that might be available only at another office.
- When digital end office switches (so PBXs) are installed in an analog environment, the analog interfaces are necessarily unchanged.
- Multiple-stage switches can be used for either two- or four-wire switching operations.

Review Questions

1. What do you understand by terminal handling and polling?
2. Explain the principle of working of hybrid switching technique.
3. What do you understand by switching? What are the various techniques used for switching.
4. Explain and write the comparison of virtual circuit subnet and datagram subnet.
5. What is the essential difference between Message switching and Packet switching?
6. Explain the circuit switching network and write the advantages and disadvantages of this type of network.
7. Explain the packet switching and compare it with circuit switching.
8. What do you understand by store-and-forward technique in switching?
9. Why is circuit switching inefficient for the transmission of non-voice data?
10. What is fundamental difference between circuit switching and packet switching?
11. A message is broken up into three pieces. Discuss transmission of the packets using datagram approach to packet switching?
12. A message is broken up into three pieces. Discuss transmission of the packets using a permanent virtual circuit?
13. A message is broken up into three pieces. Discuss transmission of the packets using a switched virtual circuit?
14. Why has message switching been replaced by other switching methods?
15. Which is more efficient, circuit switching or virtual circuit switching? Why?
16. What are the two types of switching used in circuit switching?
17. Give the comparison between circuit switching, packet and message switching.
18. Write shorts on following:
    (*i*) Phases used in circuit switching (*ii*) Global and local address

(iii) Permanent and switched virtual circuits
(iv) VCI
(v) Datagram
(vi) Congestion control
(vii) Routing
(viii) Dynamic routing and static routing
(ix) Flooding
(x) Router
(xi) ISDN
(xii) Broad ISDN
(xiii) Polling
(xiv) X.25 Devices
(xv) The X.21bis Protocol
(xvi) Softswitch architecture

19. Describe the evaluation of digital switching.
20. Explain the process of digitization with neat diagram.
21. List the advantages and disadvantages of digital transmission.
22. Explain a three stage switching (general) with neat diagram.
23. What is probability graph?
24. Describe various blocking probability evaluation techniques.
25. Explain the principle of time division switching.
26. Distinguish analog time division switching and digital time division switching.
27. Write short notes on combinational switching.
28. With neat diagrams explain time switch and space switch.
29. Explain the TS switch with neat diagram.
30. What is internal complexity?
31. What are the features of TST?
32. List the practical system which uses TST, STS and TS.
33. Explain the operations of four wire switching in detail.
34. Explain TS1 in detail and give its advantages.
35. Discuss briefly the following switching :
(i) Space switching, (ii) Time spacing time switching
36. Write a short note on the blocking probabilities.
37. Explain the operation of four wire switching in details.
38. Calculate the maximum access time that can be allowed for the data and control memories in a TSI switch with a single input and single output trunk multiplexing 2500 channels. Also estimate the cost of switch and compose if with that a single stage space division switch. [**Ans.** 25 ns, 5000 units, 625×10^4 units, 1250]
39. Calculate the number of frames that can be supported on a time multiplexed space switches, given that
(i) 42 channels are multiplexed in each stream
(ii) control memory access time is 200 ns
(iii) Bus switching and transfer time is 50 ns per transfer.

[**Hint :** $N = \frac{125 \times 10^{-6}}{Mts}$; **Ans.** $N = 12$]

40. Write short notes on :
(i) Digital Switching function
(ii) Space division switching
(iii) Multistage switching
(iv) Non-blocking switches
(v) Blocking probabilities
(vi) Four wire switches
(vii) Switch matrix control
(viii) Time division switching—analog and digital
(ix) Two dimensional switching
(x) Multistage time and space switching
(xi) STS and TST switching
(xii) Digital cross connect systems

□□□

GLOSSARY OF TELECOMMUNICATION TERMS

LIST OF ABBREVIATIONS FOR TELECOMMUNICATIONS TERMS

AAL : ATM Adaptation Layer
ADPCM : Adaptive Differential Pulse Code Modulation
ADSL : Asymmetric Digital Subscriber Line
AIN : Advanced Intelligent Network
ALI : Automatic Location Information
AMA : Automatic Message Accounting
ANI : Automatic Number Identification
ANSI : American National Standards Institute
API : Applications Programming Interface
ATM : Asychronous Transfer Mode
BHCA : Busy Hour Call Attempts
BHCC : Busy Hour Call Completions
B-ISDN : Broadband Integrated Services Digital Network
B-ISUP : Broadband ISDN User's Part
BLV : Busy Line Verification
BNS : Billed Number Screening
BRI : Basic Rate Interface
CAC : Carrier Access Code
CCS : Centi Call Seconds
CCV : Calling Card Validation
CDR : Call Detail Record
CIC : Circuit Identification Code
CLASS : Custom Local Area Signaling
CLEC : Competitive Local Exchange Carrier
CO : Central Office
CPE : Customer Provided/Premise Equipment
CPN : Called Party Number
CTI : Computer Telephony Intergration
DLC : Digital Loop Carrier System
DN : Directory Number

DSL : Digital Subscriber Line
DSLAM : Digital Subscriber Line Access Multiplexer
DSP : Digital Signal Processor
DTMF : Dual Tone Multi-Frequency
ESS : Electronic Switching System
ETSI : European Telecommunications Standards Institute
GAP : Generic Address Parameter
GT : Global Title
GTT : Global Title Translations
HFC : Hybrid Fiber Coax
IAD : Integrated Access Device
IAM : Initial Address Message
ICP : Integrated Communications Provider
ILEC : Incumbent Local Exchange Carrier
IMT : Inter-Machine Trunk
IN : Intelligent Network
IP : Internet Protocol
ISDN : Integrated Services Digital Network
ISP : Internet Service Provider
ISTP : Internet Signaling Transport Protocol
ISUP : ISDN User Part
IXC : Inter-Exchange Carriers
LAN : Local Area Network
LATA : Local Access Transport Area
LCR : Least Cost Routing
LEC : Local Exchange Carrier
LERG : Local Exchange Routing Guide
LES : Loop Emulation Service
LIDB : Line Identification Database
LNP : Local Number Portability
LRN : Local Routing Number
MF : Multi Frequency
MFJ : Modified Final Judgement
MGCP : Media Gateway Control Protocol
MMF : Multi-Mode Fiber
MTP : Message Transfer Part
NANP : North American Numbering Plan
NEBS : Network Equipment Building Standards
NFAS : Non-Facility Associated Signalling
N-ISDN : Narrowband Intergrated Services Digital Network
NP : Number Portability
OAM : Operations, Administration and Maintenance

OLNS : Originating Line Number Screening
OSI : Open Systems Interconnection Model
OSS : Operations Support System
PBX : Private Branch Exchange
PCM : Pulse Code Modulation
PODP : Public Office Dialing Plan
POI : Point of Interface
POP : Point of Presence
POTS : Plain Old Telephone Service
PRI : Primary Rate Interface
PSTN : Public Switched Telephone Network
PVC : Permanent Virtual Circuit
RBOC : Regional Bell Operating Company
SAC : Service Access Code
SAM : Service Access Multiplexer
SCCP : Signaling Connection Control Part
SCE : Service Creation Environment
SCP : Service Control Point
SDH : Synchronous Digital Hierarchy
SIBB : Service Independent Building Block
SIP : Session Initiation Protocol
SLEE : Service Logic Execution Environment
SMF : Single Mode Fiber
SMS : Service Management System
SONET : Synchronous Optical Network
SS7 : Signaling System 7
SSP : Service Switching Point
STM : Synchronous Transport Mode
STP : Signal Transfer Point
SVC : Switched Virtual Circuit
TCAP : Transaction Capability Application Part
TDM : Time Division Multiplexing
UM : Unified Messaging
VC : Virtual Circuit
VoATM : Voice over ATM
VoDSL : Voice over DSL
VoIP : Voice over IP
VoP : Voice over Packet
4ESS : Class 4 toll switch made by Lucent
5ESS : End Office switch made by Lucent

GLOSSARY

A

AAL : ATM Adaptation Layer. Protocol used on top of ATM to support high-level service requirements, converting non-ATM bit streams into ATM cells.

AAL2 : ATM Adaptation Layer 2. Used for carrying both CBR traffic and VBR traffic simultaneously, usually in support of voice over ATM.

AAL5 : ATM Adaptation Layer 5. AAL functions in support of VBR, delay-tolerant connection-oriented traffic requiring minimal sequencing or error detection support.

Access Tandem : A tandem switch that is used to interconnect between carriers for equal access. Typically, this used to interconnect ILECs with IXCs, but now also includes CLECs.

Adaptive Differential Pulse Code Modulation (ADPCM) : A ITU-TS standard technique for voice encoding and compression. It allows analog traffic to be carried within a 32 kbps digital channel.

Advanced Intelligent Network (AIN) : An evolving, service-independent architecture that allows a carrier to quickly and economically create and modify telecommunication services for its customers.

A-Link : In the SS7 world, an A-link is a signalling link that connects a ST? to a SSP or SCP. Alinks operate at a transmission speed of 56 kbps.

American National Standards Institute (ANSI) : A non-government organization which develops and distributes standards for transmission codes, protocols and high-level languages for suggested use in the U.S.

Applications Programming Interface (API) : Software that an application program uses to request and carry out lower level services performed by the computersj or a telphone systems operating system. For Windows, the API also helps applications manage windows, menus, icons and other GUI elements. In short, an API is a "hook' into software. An API is a set of standard software interrupts, calls and data formats that applications programs use to initiatc contact with network services, mainframe communications programs, telephone equipment or program-to-program communications. For example, applications use APIs to call services that transport data across a network. Standardization of APIs at various layers of a communications protocol stack provides a uniform way to write applications. NetBIOS is an early example of a network API. Applications use APIs to call services that transport data across a network.

Asymmetric Digital Subscriber Line (ADSL) : Technology using digital filtering to remove noise from twisted-pair copper lines, enabling broadband transmission. There are several varieties of ADSL using varying hardware, modulation software and compression techniques. ADSL-2 can deliver up to four VCR-quality video signals but has limited upstream response.

ADSL can only work over distances of less than 12,000 feet, a requirement 60% of U.S. homes meet.

Asynchronous Transfer Mode (ATM) : An international standard for high-speed broadband packet-switched networks operating at broadband digital transmission speeds. The technology is based on fixed-length, 53-byte cells. ATM includes protocols that specify how diverse kinds of traffic are transformed into standardized packets whose transport can be managed uniformly within the network.

Automatic Location Identifier (ALI) : A feature of E911 systems that provides information such as name, phone number, address, nearest cross street, to agents answering E911 calls.

Automatic Message Accounting (AMA) : The network functionality that measures, collects, formats, and outputs subscriber network-usage data to upstream billing and other operating systems.

Automatic Number Identification (ANI) : More colloquially called Caller ID. A service provided by local exchange carriers in which the telephone number of a caller is sent to the called-party's telephone between the first and second ring. This is one of several CLASS services, all of which require SS7 interoffice signalling.

B

Backbone : Part of a network used to connect smaller segments of networks together.

Bandwidth : The relative range of frequencies that can be passed by a transmission medium. Greater bandwidths mean a higher information carrying capacity of the transmission circuit. Usually measured in Hertz, bandwidth is assessed as the number of bits that can be transferred per second.

Basic Rate Interface (BRI) : The ISDN interface standard for single-line ISDN service. This standard provides for two message-bearing 64 kbps B channels for speech and data, plus a 16 kbps D channel for network signalling and data.

B channel : Message-bearing 64 kbps digital channel used for digital transmission of high speed data and video.

Billed Number Screening (BNS) : When consumers decide who can and who cannot charge a call to their phone based on an agreement with their local telephone company to screen calls.

Bit Rate : The number of bits transmitted over a telephone line per second.

B-Links, D-Links and B/D Links : Links interconnecting two mated pairs of STPs are referred to as B-links, D-links or B/D links.

Broadband : A term used to describe a channel with more bandwidth than a standards voice grade channel. Broadband channels are used to carry multiple high-speed voice and data transmissions on a common communications path.

Broadband Integrated Services Digital Network (B-ISDN) : An evolving standard for the second generation of integrated services digital networks. Broadband ISDN services employ packet switching to integrate voice and data services over a high-speed, packet-based infrastructure.

Broadband ISDN User's Part (B-ISUP) : An SS7 protocol defining the signalling messages to control connections and services.

Bursty : Data transmitted in short, uneven bursts with relatively long, silent intervals between.

Busy Hour : An uninterrupted 60-minute period during which the average volume of telecommunications traffic is at its maximum.

Busy Hour Call Attempts (BHCA) : A measure of dynamic traffic calls that can be attempted in an average Busy Hour.

Busy Hour Call Completion (BHCC) : A measure of dynamic traffic calls that can be completed in an average Busy Hour.

Busy Line Verification (BLV) : A feature that allows an attendant to verify the busy or idle state of telephone lines and to break into the conversation.

C

Call Detail Record (CDR) : A billing feature of a telephone system which allows the system to collect and record information on outgoing and incoming phone calls such as who made/received them, where they went, what time, how long, etc.

Called Party Number (CPN) : When a call is set up over an ISDN network, SS7 send an initial address message which contains the as part of the ISUP protocol

Calling Card Validation (CCV) : Process of verifying that a calling card is valid and then processing a call to be billed to that account.

Carrier Access Code (CAC) : The sequence that an end user dials in order to access the carrier's switch service. The codes are composed of 7 digits in the form of 101xxxx, where xxxx is the CAC.

Centi Call Seconds (CCS) : In telecommunications traffic engineering terminology, CCS represents (centi call seconds) one hundred call seconds or one hundred seconds of telephone conversation. One hour of telephone traffic is equal to 36 CCS ($60 \times 60 = 3600$ divided by $100 = 36$) which is equal to one erlang. CCS are used in network optimization. CCS can also mean Cumulative call seconds which is measure of trunk occupancy.

Central Office (CO) : The facility of a telecommunications common carrier where subscribers' lines are joined to switching equipment for connecting other subscribers to each other, locally and long distance.

Centrex : A type of phone service offered by local exchange carriers that provide PBX like functions to a group of users without the need for a PBX. Despite having individual phones connected to the central office, users are able to dial each other by extensions, transfer calls, etc.

Circuit Identification Code (CIC) : An SS7 term used to identify a particular circuit within a trunk group.

Circuit Switch : A switching system that establishes a dedicated physical connection between end points, in a network, for the duration of the communication session.

Class 4 Office : A switching center for toll calls. A class 4 office switches toll traffic originating at class 5 offices to other class 4 offices, or offices of a higher class. In addition, a class 4 office relays toll traffic from class 4 toll offices to the class 5 end office serving the destination address.

Class 5 Office : The lowest level in a hierarchy of central offices. Class 5 offices serve as the network entry point for user access lines and are a switching center for local calls.

C-Links : Links that interconnect mated STPs. They are used to enhance the reliability of the signalling network in cases where one or multiple links are unavailable.

Competitive Local Exchange Carrier (CLEC) : These are new local carriers, typically formed after the US Telecommunications Act of 1996, to compete with the incumbent RBOCs.

Compression : Reducing the size of the data, image, voice or video file sent over a telephone line, lessening the bandwidth needed to transmit the file.

Computer Telephony Integration (CTI) : The combining of data with voice systems in order to enhance telephone service. Examples include the delivery of Caller ID information via a PC, and the ability to access mail via the PC.

Custom Local Area Signaling Services (CLASS) : A grouping of optional features to basic local exchange telephone service or enhanced telephony services that utilize the signalling system 7 (SS7) channel to carry data about a call. CLASS provides subscribers with the ability to screen and selectively reject, forward, trace, and redial incoming call Caller fl) is one example.

Customer Premise Equipment (CPE) : Equipment which resides on the customer's premise, such as a PBX or IAD.

D

Default Routing : The ability of the switch to continue the call based on the dialed number when the SCP cannot be accessed due to abnormal circumstances.

Digital Signal Processor (DSP) : A digital microprocessor that calculates digitized signals that were originally analog (e.g. voice) and then sends the results on. DSPs are used in telecommunications for tasks such as echo cancellation, call progress monitoring, voice processing and for compression.

Directory Number (DN) : A unique complement of digits associated with the name of a subscriber in a telephone directory. Your phone number.

Digital Loop Carrier (DLC) : An access provisioning system by which a telephone switch is able to remotely deploy telephony interfaces to customers. Typically, the connection between the switch and DLC is via a digital or fibre connection, and user interfaces such as POTS lines are deployed out of the DLC to customers.

Digital Subsciber Line (DSL) : DSL is the technology that is employed between a customer location and the carrier's network that enables more bandwidth to be provided by using as much of the existing network infrastructure as possible.

Digital Subscriber Line Access Multiplexer (DSLAM) : A network device at a telephone company central office that receives signals from multiple customer DSL connections and puts the signals on a high-speed backbone line using multiplexing techniques.

DMS 100 : Class 5 end office switch made by Nortel.

DMS 250 : Access tandem switch made by Nortel.

DMS 500 : Class 5 end office switch made by Nortel.

Donor Exchange : The switch the Directory Number was initially ported from.

DS-0 : Digital signal level zero. A single digital 64 kbps, pulse code modulated, transmission channel which represents the starting point for a digital multiplexing hierarchy.

DS-1 : Digital signal level one. A 1.544 Mbps digital signal comprised of 24 multiplexed 64 kbps DS-0 digital channels.

DS-3 : Digital signal level three. A 44.6 Mbps digital signal comprised of 28 multiplexed DS-1 signals that is carried over a T-3 facility.

Dual Tone Multi-Frequency (DTMF) : A term used to describe push button or touchtone dialing.

E

E-1 : The European equivalent of a T-1, however an E-1 line can support 32 64kbps channels, versus 24 64kbps of a T-1.

E-3 : The European standard for T-3.

E911 : Enhanced emergency reporting service. 911 service becomes enhanced 911 when there is a minimum of two special features added to it : Automatic Number Identification (ANI) and Automatic Location Information (ALI).

Echo cancellation : A technique that allows for the isolation and filtering of unwanted signals caused by echoes from the main transmitted signal.

E-links : Links that provide backup connectivity to the SS7 network in the event that the 'home' STPs are unreachable using the A-links.

End Office : The location where carriers place telecom equipment closest to the customer. Typically, this is where customers are provisioned. Class 5 switches are typically located here.

Equal Access : A condition where the local exchange access service offered by a carrier is made available in equal kind, quality, and price to all long distance companies.

Electronic Switching System (ESS) : One type of AT&T/Lucent's family of stored, programcontrolled central office switches, including the 4ESS and 5ESS switches.

European Telecommunications Standards Institute (ETSI) : The European counterpart of ANSI, tasked with paving the way for telecommunications integration in the Eurpoean community as part of the single European market program. ETSI's main goal is the unrestricted communication between all the member states by provisioning European standards.

EWSD : Switching system made by Siemens.

Exchange : Another term for switch.

F

F-links : A link used to connect two SS7 signalling points.

Facilities Based Service Provider : Identifies owners of switches and infrastructure -a reseller of switched services is not a facility based service provider.

Feature Group : In switched access tariffs, a Feature Group denotes a specific and uniform type and quality of local exchange access available to inter-exchange carriers and other types of telecommunications companies.

Feature Group A : A line-side switched access connection for originating and terminating traffic. Customers of a long distance company that use Feature Group A for originating access must dial a seven-digit local number to reach an inter-exchange carrier, and then use a touch tone phone or tone dialer to dial an identification code plus the phone number they want to reach. In areas where equal access is not available, Feature Group A is provided at discounted rates.

Feature Group B : A trunk-side switched access connection for originating and terminating traffic. Transmission quality is superior to that of Feature Group A. Customers of a long distance company using Feature Group B for originating access must dial "950" followed by a "1" or a "0" and the three-digit carrier identification code of their chosen company. Mostly superseded by Feature Group D.

Feature Group C : A trunk-side switched access connection that directly links local phone company end offices with the long distance network of AT&T. Only AT&T has Feature Group C access connections, which offer the highest transmission quality and a complete array of access features for originating and terminating long distance traffic.

Feature Group D : The equal access connection; a trunk-side switched access connection equal in quality, features, and price to the Feature Group C connection of AT&T. Customers in exchanges where Feature Group D is available can pre-subscribe to any one long distance company. They can reach their chosen company by dialing "1" plus the phone number they want to reach. Customers in an equal access area can use other long distance companies by dialing "1" and "0" and the five-digit carrier identification code of the carrier they want.

Frame Relay : A packet-switched method of data communications provided by telecommunications carriers and Internet service providers. Frame relay can provide guaranteed bandwidth at no additional charge if the lines are open during periods of low traffic.

Fully Associated Link (F-links) : A link which is used to connect two SS7 signalling points when there is a high community of interest between them and it is economical to link them.

G

Global Title (GT) : An address such as customer-dialed digits that does not contain information that would allow the SS7 network to route it.

Global Title Translations (GTT) : The process of translating a GT from dialed digits to a point code address and application address by means of a STP.

GR-303 : A set of technical specifications from Bellcore that define the interface between the DLC and a Class 5 switch.

G-Recommendations : A series of standards defined by the ITU-T covering transmission facilities. Including, but not limited to :

H

H.323 : A framework of protocols for inter-working voice, video and data across an IP network.

Hot Swappable : The ability of a component to be added to or removed from a piece of equipment without powering down the device, providing maximum uptime.

Hunt Group : A series of telephone lines organized so that if the first line is busy the next line is hunted and so on until a free line is found.

Hybrid Fiber Coax (HFC) : An access infrastructure typically used by cable companies and multiple system operators (MSOs) to deliver services to their customers. Most of the access infrastructure is via fiber, with the last portion (street to home) being coaxial cable.

I

In-Band Signaling : A method of controlling information in a telecommunications network by using tones or other signals carried within the same band or channel as the information being carried. For example, in a telephone call, tones can be used to control the transmission, receipt and disconnection of the call.

Initial Address Message (IAM) : A SS7 signalling message that contains the address and routing information required to establish a point-to-point telephone connection.

Integrated Access Device (LAD) : An access device located on the customer premises that can handle both voice and data services.

Integrated Communications Provider (ICP) : A company that provides bundled communications services, including voice and data services.

Interexchange Carrier (IXC) : Long distance companies that sell toll-free 800, international and outgoing telephone service on an interstate basis.

Incumbent Local Exchange Carrier (ILEC) : This is the incumbent local phone company which owns most of the local loops and facilities in a serving area; frequently an RBOC.

Integrated Services Digital Network (ISDN) : A switched network providing end-to-end digital connectivity for simultaneous transmission of voice and/or data over multiple multiplexed communications channels and employing transmission and out-of-band signalling protocols that conform to internationally-defined standards.

Intelligent Network (IN) : A telecommunications network architecture in which processing capabilities for call control and related functions are distributed among specialized network nodes rather than concentrated in a switching system. The SS7 network forms part of the IN infrastructure.

Inter-Machine Trunk (IMT) : These are switch-to-switch trunks that are used to carry calls between carriers. Signaling is not performed in-band on these trunks; instead, they are coordinated via the overlayed SS7 network.

Inter-Exchange Carrler (IXC) : A carrier that is allowed to carry traffic from one LATA to another, typically long distance inter-state traffic, but can also include intea-state toll traffic.

Intermediate Exchange : A tandem switch.

Internet Protocol (IP) : A network layer (Layer 3) standard for data transmission that performs the addressing function and contains some control information to allow packets to be routed through networks.

Internet Signaling Transport Protocol (ISTP) : A standard used to provide signalling interconnection between packet networks and the PSTN.

Internet Service Provider (ISP) : An ISP connects end-users to the Internet via telephone lines.

Intea-LATA : Telecommunications services that originate and terminate in the same Local Access and Transport Area.

ISDN User Part (ISUP) : The portion of SS7 that handles call control for ISDN-type calls.

ITU-T : The Telecommunication Standardization Sector of the International Telecommuni-cations Union.

L

Local Area Network (LAN) : A geographically localized network located on an individual organization's premise. A LAN enables computer devices to communicate with each other as well as share and have access to peripherals such as printers, fax services, modem services and centralized databases.

Local Access Transport Area (LATA) : A geographical area within which a divested RBOC is permitted to offer regional toii and access services.

Local Exchange : Geographic area determined by the appropriate state regulatory authority in which calls generally are transmitted without toll charges to the calling or called party. Several local exchanges may exist within a LATA.

Least Cost Routing (LCR) : A telephone system feature that automatically chooses the lowest cost phone line to the destination. The "lowest cost" is deteremined by algorithms, equations and decision trees programmed into a PBX.

Line Identification Database (LIDB) : Database developed by RBOCs and all local telephone companies that contain all the valid telephone and calling card numbers in their regions, and have the necessary information to perform billing validation.

Line Served by Switch : Any Directory Number that is connected to the switch or subtends the switch. The DN may be a physical subscriber port or a virtual DN.

Local Number Portability (LNP) : The ability of telephone subscribers to maintain their phone numbers when they change local telephone companies. Dependent on SS7 in order to implement.

Local Exchange Carrier (LEC) : Any company authorized by the state public utility commission to sell local telephone service.

Local Exchange Routing Guide (LERG) : A Bellcore document which lists all North American Class 5 central offices or end offices and describes their relationship to Class 4 tandem offices. It is often used by carriers for network design.

Local Loop : The local loop is the telephone line that runs from the local telephone company's end office switch to the end user's premise.

Location Routing Number (LRN) : A 10 digit number used to uniquely identify a switch that has ported numbers.

Loop Emulation Service (LES) : An ATM Forum specification designed to emulate a customer's local loop using ATM by extending class 5 services capabilities, along with highspeed data services, to customers.

M

Media Gateway : Communications switch equipment operating at the edge of multi-service packet networks.

Modified Final Judgment (MFJ) : The 1984 US Department of Justice ruling that resulted in the divestiture of the Bell Operating Companies from AT&T.

Media Gateway Control Protocol (MGCP) : An IETF draft standard for a protocol that allows Voice Gateways to control external call control elements. MGCP assumes a call control architecture where the call control 'intelligence' is outside the gateways and handled by external call control elements.

Message Transfer Part (MTP) : Level 1 through 3 protocols of the SS7 protocol stack. It provides functions for the basic routing of signalling messages between signalling points.

Multi-Frequency (MF) : Inband, analog trunk signalling.

Multi-Mode Fiber (MMF) : Fiber in which the ultrapure glass that forms the core transmission medium is between 50 and 62.5 microns. This fiber has less carrying capacity than single-mode fiber.

Multiplexing : A process that concentrates traffic by combining a large number of lower-speed transmission lines into one high-speed line by splitting the total available bandwidth of the highspeed line into narrower bands (frequency division), or by allotting a common channel to several different transmitting devices, one at a time in sequence (time division).

MUX : A multiplexing device.

N

Narrowband Integrated Services Digital Network (N-ISDN) : Standards-based voice and data network that operate over today's TDM-based switches and provides 144K and 1.544 Mbps interfaces.

Network Equipment Building Standards (NEBS) : NEB S defines a set of performance, quality, environmental and safety requirements for carrier class telecommunications equipment. NEBS compliance is usually required by telecommunications service providers for equipment installed in their switching offices. Level 3 represents the highest ranking.

Next Generation (Next Gen) : Used to describe emerging technologies.

Non-Facility Associated Signalling (NFAS) : Out of band signalling which allows signalling to be completed without using up excessive bandwidth.

North American Numbering Plan (NANP) : The numbering architecture in which every station in an NANP area is identified by a unique 10-digit address consisting of a three-digit area code, a three-digit central office code, and a four-digit subscriber number.

Number Portability (NP) : The ability for end-users to retain their telephone number when they change service providers, location or their service.

O

OC-1 : Optical Carrier One. Bit rate of 51.84 Mbps and capacity of 28 DS- Is.

OC-3 : Optical Carrier Three. Bit rate of 155.52 Mbps and capacity of 84 DS-Is.

OC-12 : Optical Carrier Twelve. Bit rate of 622.08 Mbps and capacity of 336 DS-Is.

Open Standard : A computer or communications standard whose technical specifications are readily available to equipment manufacturers and other parties that want to incorporate the standard into their products or systems.

Open Systems Interconnection Model (OSI) : An international set of rules for computer networking that creates open standards to allow a computer on any network to share information with any other computer on that network or a connected network.

Operations, Adminstration, and Maintenance (OAM) : A group of network management functions that provide information and specifics to manage a system or network such as performance information, network fault indications, and data and diagnosis functions.

Operations Support System (OSS) : Methods and procedures that support the daily operation of a carrier's infrastructure, including order processing, equipment assignment, etc.

P

Packet Switching : The technique by which a stream of data is broken into standardized "packets," each of which contains address, sequence, control, size, and error checking information, in addition to the user data. Packet switches operate on this added information to move the packets to their destination in the proper sequence and again present them in the correct continuous stream.

Permanent Virtual Circuit (PVC) : This is a connection between end points that is defined in advance and requires little if any setup time, and end points are normally defined for the carrier, by the customer, in advance.

Point of Interface (POI) : The point in a network at which carriers interface with one another.

Point Code : A network address used within the SS7 network to represent a Service Switching Point (SSP) for routing purposes. Typically, each SSP switch within an SS7 network has at least one point code.

Point of Presence (POP) : A long distance company's switch that is connected to the local telephone company's central office. The POP is the point at which telephone and data calls are handed off between local telephone companies and long distance telephone companies.

Port : An interface location that provides a point of access for peripheral equipment, such as central office lines.

Plain Old Telephone Service (POTS) : The traditional telephone service for the transmission of speech across the telephone network.

Primary Inter-LATA Carrier Code (PIC) : This code is associated with the customer profile of every phone subscriber, and is used to route to the customer's pre-selected long distance carrier.

Primary Rate Interface (PRI) : TThis is the narrowband ISDN interface standard for high-speed ISDN service. Within the US, this provides 23 channels of data and/or voice traffic.

Private Branch Exchange (PBX) : Equipment used to switch telephone calls within a business or closed environment and also for that environment to outside lines.

Public Switched Telephone Network (PSTN) : The current narrowband-based telephone network that was designed for voice traffic.

Pulse-Code Modulation (PCM) : An analog to digital conversion technique. It is used to convert voice for transmission over digital facilities. It is also used to convert voice analog data to digital data for transmission in a multiplexed voice and data stream over T1 or other digital circuit.

Q

Q.2931 : An ITU standard for basic call control/connection across an ATM network.

R

Reciprocal Billing : A process by which the carrier who locally terminates a call to a customer gets compensated by the carrier who delivered the call for termination.

Redundancy : Having back-up systems available to provide uninterrupted continuous service in the case of a failure in the main system.

Regional Bell Operating Company (RBOC) : Regional companies formed after the divestiture of AT&T in 1984. At the time of the divestiture, there were seven companies, but now that number has been reduced to four. In today's competitive environment, they are typically referred to as Incumbent Local Exchange Carriers (ILECs).

S

Scripting Language : Simplest form of computer programming using nearly plain English commands. A high-level programming language which uses a language that is recognizable as something like natural language. JAVAScript, developed jointly by SUN and Netscape for writing Web applets, is an example of a scripting language, as is Hypertalk and Supertalk.

Service Access Code (SAC) : These are the special codes that replace the area code and are used for special network services. (Examples include 500, 700, 800, 888, 877 and 900.)

Service Access Multiplexer (SAM) : Generic name for a central office located multiplexer that aggregates multiple customers via lower speed line to a higher speed trunk connection.

Service Control Point (SCP) : A remote computer database within the SS7 network that receives queries from SSPs in order to process applications such as 800 and LNP number lookups and calling card verification.

Service Creation Environment (SCE) : Supports the creation, management, and execution of various test cases that allow a user to verify and maintain IN services.

Service Independent Building Block (SIBB) : Software modules that can be combined together to allow new computer telephony services to be developed.

Service Logic Execution Environment (SLEE) : A functional group residing in an signal control point or adjunct that contains the service logic and control, information management automatic message accounting and operations functional entity.

Service Management System (SMS) : Allows provision and updating of information on subscribers and services in near real time for billing and administrative purposes.

Service Switching Point (SSP) : Within the SS7 network, an SSP is a telephone central office switch that inter-works with the network.

Session Inititation Protocol (SIP) : A protocol for transporting call setup, routing, authentication and other feature messages to endpoints within the IP domain, whether those messages originate from outside the IP cloud over PSTN resources or within the cloud.

Signalling Connection Control Part (SCCP) : Part of the ITU-T signalling protocol and of the SS7 protocol. It provides routing and management functions for the transfer of messages other than call set-up between signalling points. It typically supports the MTP.

Signalling System 7 (SS7) : The SS7 network allows call control and transaction messages from the integrated voice and data network to be transferred on communications paths that are separate from the voice and data connections. It delivers out-of-band signalling that provides fast call setup by means of high-speed, circuit-switched connections and transaction capabilities which deal with remote database interactions. SS7 makes such enhanced telephony features as caller 113, call forwarding, and call waiting widely available. SS7 also plays an integral role in the deployment of ISDN. The SS7 protocol consists of four basic sub-protocols :

(*i*) Message Transfer Part (MTP), which provides functions for basic routing of signalling messages between signalling points;

(*ii*) Signaling Connection Control Part (SCCP), which provides additional routing and management functions for transfer of messages other than call setup between signalling points;

(*iii*) Integrated Services Digital Network User Part (ISUP), which provides for transfer of call setup signalling information between signalling points;

(*iv*) Transaction Capabilities Application Part (TCAP), which provides for transfer of non-circuit related information between signalling points.

Signaling Transfer Point (STP) : An STP is a packet switch within the SS7 network that routes network call information among other circuit switches and between SSPs and SCPs.

SIGTRAN Standards : Standards to define an architecture and related protocols to support transport and interworking of signalling over IP.

Single-Mode Fiber (SMF) : Fiber in which the ultrapure glass that forms the core transmission medium is between 8-10 microns. This fiber makes up the majority of today's long distance network. Only one ray or mode can travel down the strand, making a simpler job of regenerating the signal at points along the span.

Softswitch : As defined by the International Softswitch Consortium, a softswitch (a.k.a. call agent, call server or media gateway controller) is a device that provides, at a minimum : Intelligence that controls connection services for a media gateway, and/or native IP endpoints. It has the ability to select processes that can be applied to a call. It can provide routing for a call within the network based on signalling and customer database information. It has the ability to transfer control of the call to another network element. It provides interfaces to and supports management functions such as provisioning, fault, billing, etc.

Switched Virtual Circuit (SVC) : A temporary connection between end points. Connections last only as long as necessary and are disconnected when the session is complete.

Synchronous Digital Hierarchy (SDH) : Term used by the International Telegraph and Telephone Consultative Committee to refer to SONET.

Synchronous Optical Network (SONET) : The ANSI standard that allows interworking of transmission protocols from multiple vendors. SONET is the ANSI standard for transmitting digital information over optical networks.

Synchronous Transport Mode (STM) : The synchronous transmission capability of a system that is capable of both synchronous and asynchronous capabilities of B-ISDN service.

T

T-1 : A digital transmission link capable of handling 1.544 Mega bits per second.

T-3 : 28 T-1 lines.

Tandem Network : An arrangement of voice switches that enables calls to be routed through two or more switching centers in tandem fashion, such that each end office switch does not need to be directly connected to each other.

Tandem Switch : A voice switch that is designed primarily with trunk interfaces rather than subscriber interfaces.

Telephony-Grade : 99.9994% uptime or the higher reliability standard circuit switches require (translating into 3 minutes of downtime a year), as opposed to the "carrier-class" reliability standard of 99.999% (called "five-nines").

Time Division Multiplexing (TDM) : A multiplexing scheme in which numerous signals are combined for transmission on a single communications line or channel. Each signal is broken up into many segments, each having a very short duration and specific time slots within the channel. The slots are assigned whether or not any signals are available for transmission.

TR-303 : A Bellcore standard for communication between telephony switches and Digital Loop Carrier systems.

Transaction Capabilities Application Part (TCAP) : The portion of the SS7 protocol that is used to make database queries to SCPs. It is used to support services such as 800 and LNP number translation, as well as other functions.

Trunk : The line of communication between switching systems.

U

Unbundling : 1. As part of telecom deregulation, ILECs were required to unbundle their network elements and lease them at wholesale rates to CLECs, in order to be allowed into the long distance market. 2. Removing the service intelligence from the switching function.

Unified Messaging (UM) : Voice, fax, electronic mail, image and video all on one platform available to consumers via a local area network.

V

V5 : A standard approved by ETSI for the interface between the access nework and the carrier switch to provide basic telephony, ISDN and semi-permanet leased lines. The Eurpoean equivalent of GR.303.

Virtual Circuit (VC) : A link between two or more end stations on a packet switched mesh network. It provides a temporary or dedicated connection-oriented session between two end points. The defining characteristic is a predefined path through a network that has many paths.

Voice Over ATM (VoATM) : The process of transmitting voice traffic across an ATM-based packet network.

Voice Over DSL (VoDSL) : The process of transmitting voice traffic across a DSL-based packet network. Using a greater range of frequencies over the existing copper line makes this increase in bandwidth possible.

Voice Over IP (VoIP) : The process of transmitting voice traffic across an IP-based packet network.

Voice Over Packet (VoP) : The process of transmitting voice traffic across any kind of packet network.

❑❑❑

INDEX

D

E

F

❑❑❑

Win Prizes

ATTENTION READERS

Win Prizes

We request for your frank assessment regarding various aspects of the book as given below :

Fundamentals of

Voice and Data Communication

By : K. K. SHARMA (RGTU)

Please Fill up the Sheet in Neat and Clean Writing

(i) What topics of your syllabus, important for your examination point of view, are not covered in this book ?

....

....

....

....

(ii) What are the chapters wherein the treatment of the subject matter is not systematic or organised or updated ?

....

....

....

....

(iii) Have you come across misprints/mistakes/factual inaccuracies in the book ? Please specify the chapters and the page numbers.

....

....

....

....

(iv) What other books on the same subject have you found/heard better than the present book ? Please specify in terms of price and quality.

....

....

....

....

(v) Further suggestions and comments for the improvement of this book.

....

....

....

....

(PS : If need be please attach a separate sheet to write your views)

OTHER DETAILS

(i) Who recommended you this book ? (Please tick (✓) in the box near the option relevant to you.

☐ Teacher ☐ Friends ☐ Bookseller

....

....

....

....

(ii) Name of the recommending teacher, his designation and address:

....

....

....

....

(iii) Name and address of the bookseller you purchased the book from :

....

....

....

....

(iv) Name and address of your institution (Please mention the University or Board as the case may be)

....

....

....

....

(v) Your name and complete address :

....

....

....

....

(vi) Course you are enrolled in :

....

....

....

....

The best assessment will be awarded each month : The award will be in the form of our publications as decided by the Editorial Board.

Please mail the filled up coupon at the vearliest to :

S.K. KATARIA & SONS®

4760-61/23, Ansari Road,

Daryaganj, New Delhi-110002

Phone: +91-11-23243489, 23269324

TeleFax : +91-11-23243489

email : katariabooks@yahoo.com